Stem Cells in Marine Organisms

Baruch Rinkevich · Valeria Matranga
Editors

Stem Cells in Marine Organisms

Springer

Editors
Prof. Dr. Baruch Rinkevich
Israel Oceanographic &
Limnological Research
31 080 Haifa
Israel
buki@ocean.org.il

Dr. Valeria Matranga
Istituto di Biomedicina e
Immunologia
Molecolare "Alberto Monroy"
Consiglio Nazionale delle
Ricerche
Via La Malfa, 153
90146 Palermo
Italy
valeria.matranga@ibim.cnr.it

ISBN 978-90-481-2766-5 e-ISBN 978-90-481-2767-2
DOI 10.1007/978-90-481-2767-2
Springer Dordrecht Heidelberg London New York

Library of Congress Control Number: 2009927004

Cover illustration:
Front Cover: Botryllus schlosseri, a colonial tunicate, with extended blind termini of vasculature in the periphery. At least two disparate stem cell lineages (somatic and germ cell lines) circulate in the blood system, affecting life history parameters. Photo by Guy Paz.

Back Cover: Paracentrotus lividus four-week-old larvae with fully grown rudiments. Sea urchin juveniles will develop from the echinus rudiment which followed the asymmetrical proliferation of left set-aside cells budding from the primitive intestine of the embryo. Photo by Rosa Bonaventura.

Printed on acid-free paper

Springer is part of Springer Science+Business Media (www.springer.com)

Preface

Stem cell biology is a fast developing scientific discipline. Based on solid basic aspects, this emerging cell biology field is associated with highly popular applied features. Growing attention is devoted in recent years toward studies on stem cells biology in model organisms (mainly mammalians) including stem cells differentiation and gene expression in selective vertebrates. These studies, in addition to their inherent interest in the general biological phenomenon of stem cells, are motivated by the rationale that stem cells might serve in fighting against human diseases, or in regenerating damaged organs in humans. While the literature on stem cell from vertebrates is rich and expanding in an exponential rate, investigations on marine organisms' stem cells are very limited and scarce. This is in spite of the results, pointing that marine organisms' stem cells are important in various biological disciplines that involve understanding of mechanisms promoting cell growth and differentiation, in developmental biology aspects such as regeneration and budding processes in marine invertebrates, body maintenance of marine organisms (including those that may live for decades), aging and senescence.

It is unfortunate that the research on marine organisms is lagging behind the studies on vertebrates and some invertebrate model organism (like *Drosophila melanogater* and *Caenorhabditis elegans*), in spite the discoveries that very potent stem cells exist, even in the most primitive multicellular marine organisms (like sponges and cnidarians). Recent studies further revealed similarities between the biological properties of stem cells in marine organisms and the vertebrates, and confirmed the existence of unique properties associated with stem cells from marine organisms that lead to phenomena such as somatic and germ cell parasitism, or whole body regeneration. These results also showed that marine invertebrates' stem cells are characterized by an efficient motility within an organism and between organisms, that they have the ability to parasitize effectively host organisms, in the same species or in other species, with stemness and unlimited replication characterization. Marine invertebrate stem cells have the ability to blend with other cell lineages, blurring the line between distinct somatic and germ cell lineages, co-opting and managing somatic cells to collaborate in parasitic agenda and in many cases (like sponges and cnidarians) they lack of a distinct niche where they reside.

The book is the outcome of discussions raised during and original presentations at the Exploratory Workshop on Stem Cells in Marine Organisms held in Palermo

(Italy), in November 2006 (http://www.ibim.cnr.it/stemcell/index.htm), under the umbrella of the NoE Marine Genomics Europe and hosted by the CNR Institute of Biomedicine and Molecular Immunology "Alberto Monroy". In addition, solicited contributions were brought together to gather recent knowledge in the identification and functional characterization of stem cells in marine organisms and genes expressed in growing and differentiating cells of these organisms. The book holds 14 chapters on marine stem cells, including theoretical chapters, overview essays and specific research outcomes. In addition to the aim in summarizing much of the knowledge on marine invertebrates' stem cells biology, another important aim is to reach a larger scientific community in order to encourage the study on marine invertebrate stem cells, theoretically, practically and through innovative technology. This might lead to biotechnologically relevant discoveries or breakthrough in the mainstream research on stem cell biology.

Last we would like to thank all the contributors to this book as we are confident that the high scientific value of their reviews on past and current findings will serve as a forum of ideas for the exploitation of marine invertebrate stem cells biology and will promote the development of studies in the new exciting field.

In honor of the bicentenary of Charles Darwin's birth (February 12th, 1809), we would like to close this preface by the recommendation to look at the sea as a source of knowledge and life experience. Darwin used to say "It is not the strongest of the species that survive, nor the most intelligent, but the one most responsive to change". Thus, very much can be learnt from marine invertebrates, have been surviving on Earth for so many geological periods, posing major challenges to scientists that try to reveal their most intimate secretes.

Special warm thanks go Francesca Zito for her generous assistance and friendly collaboration in handling all the matters related to the editorial work. The Editors

Haifa, Israel — Baruch Rinkevich
Palermo, Italy — Valeria Matranga

Contents

Contributors

Antoine Aze Université Pierre et Marie Curie-Paris6, CNRS, UMR 7628, Banyuls-sur-mer, BP44, Avenue du Fontaulé, F-66651 France, antoine.aze@obs-banyuls.fr

Bertil Åkesson Zoology, Göteborg University, SE 405 30 Göteborg, Sweden, bertil.akeson@zool.gu.se

Loriano Ballarin Department of Biology, University of Padova, Via U.Bassi 58/B, 35121 Padova, Italy, ballarin@mail.bio.unipd.it

M. Daniela Candia-Carnevali Dipartimento di Biologia, Università degli Studi di Milano, Via Celoria 26, 20133 Milano, Italy, daniela.candia@unimi.it

Amélie Châtel Institut für Physiologische Chemie, Abteilung Angewandte Molekularbiologie, Universität, Duesbergweg 6, D-55099 Mainz, Germany, amelie.chatel@univ-brest.fr

Edwin L. Cooper Laboratory of Comparative Neuroimmunology, Department of Neurobiology, David Geffen School Of Medicine at UCLA, Los Angeles, CA 90095-1763, USA, cooper@mednet.ucla.edu

Márcio Reis Custódio Departamento Fisiologia – IB/USP. Sala 300, Rua do Matão. Trav. 14, n. 321, São Paulo (SP) – CEP 05508-900, Brazil, mcust@usp.br

Giuseppina D'Ancona Lunetta Institute of Hystology and Embryology, Department of Biology, Viale delle Scienze, Edificio 16, University of Palermo, Palermo, Italy, dancona@unipa.it

Yasmine Even The Biomedical Research Centre, University of British Columbia, 2222 Health, Sciences Mall, Vancouver, BC, V6T 1Z3, Canada, yasmine@brc.ubc.ca

Uri Frank School of Natural Sciences and Martin Ryan Marine Science Institute, National University of Ireland, Galway, Ireland, uri.frank@nuigalway.ie

Anne-Marie Genevière Université Pierre et Marie Curie-Paris6, CNRS, UMR 7628, Banyuls-sur-mer, BP44, Avenue du Fontaulé, F-66651 France, anne-marie.geneviere@obs-banyuls.fr

Maria Imschenetzky Department of Biochemistry and Molecular Biology, Universidad de Concepcion, Concepcion, Chile, mimschn@udec.cl

Lucia Manni Department of Biology, University of Padova, Via U.Bassi 58/B, 35121 Padova, Italy, lmanni@bio.unipd.it

Valeria Matranga Consiglio Nazionale delle Ricerche, Istituto di Biomedicina e Immunologia Molecolare "Alberto Monroy", Via La Malfa 153, 90146 Palermo, Italy, matranga@ibim.cnr.it

Isabel M. Müller Institut für Physiologische Chemie, Abteilung Angewandte Molekularbiologie, Universität, Duesbergweg 6, D-55099 Mainz, Germany, imuellerzahn@aol.com

Werner A. Müller Heidelberg Institute of Zoology, Dept. of Mol. Dev. Biology & Physiology University of Heidelberg, Im Neuenheimer Feld 230 69120 Heidelberg, Germany, w.muller@zoo.uni-heidelberg.de

Werner E.G. Müller Institut für Physiologische Chemie, Abteilung Angewandte Molekularbiologie, Universität, Duesbergweg 6, D-55099 Mainz, Germany, wmueller@mail.uni-mainz.de

Clara Nervi Department of Histology & Medical Embryology, Università "Sapienza" Rome, Italy, clara.nervi@uniroma1.it

Matthias Obst Kristineberg Marine Research Station, Marine Ecology, Göteborg University, Kristineberg 566, SE 450 34 Fiskebäckskil, Sweden, matthias.obst@zool.gu.se

Enzo Ottaviani Department of Animal Biology, University of Modena and Reggio Emilia, Via Campi 213/D, 41100 Modena, Italy, ottaviani.enzo@unimore.it

Günter Plickert Institute of Zoology, University of Cologne, Weyertal 119, 50931 Cologne, Germany, g.plickert@uni-koeln.de

Baruch Rinkevich Israel Oceanographic & Limnological Research, National Institute of Oceanography, Tel Shikmona, Haifa 31080, Israel, buki@ocean.org.il

Yuval Rinkevich Institute for Stem Cells Research and Regenerative Medicine, Stanford University School of Medicine, Stanford, CA 94305 USA, ryuval@stanford.edu

Amanda L. Rychel Department of Biology, Center for Developmental Biology, and Institute for Stem Cell and Regenerative Medicine, University of Washington, Seattle, WA 98195, USA; Friday Harbor Laboratories, University of Washington, Friday Harbor, WA 98250-9299, USA; Kewalo Marine Laboratory, University of Hawaii, Honolulu, HI 96813, USA, arychel@u.washington.edu

Michel Salzet Centre de Biologie Cellulaire, Laboratoire de Phylogénie, moléculaire des Annélides EA DRED 1027, Groupe de Neuroendocrinologie des

Hirudinées, Université des Sciences et Technologies de Lille, 59655 Villeneuve d'Ascq Cédex, France, salzet@pop.univ-lille1.fr

Heinz C. Schröder Institut für Physiologische Chemie, Abteilung Angewandte Molekularbiologie, Universität, Duesbergweg 6, D-55099 Mainz, Germany, hschroed@uni-mainz.de

Helen Nilsson Sköld Department of Marine Ecology - Kristineberg, Göteborg University, Kristineberg 566, SE 450 34 Fiskebäckskil, Sweden, helen.skold@marecol.gu.se

Mattias Sköld Swedish Board of Fisheries, Inst. Marine Research, SE 453 21 Lysekil, Sweden, mattias.skold@fiskeriverket.se

George B. Stefano Neuroscience Research Institute, State University of New York, College at Old Westbury, Old Westbury, NY 11568 USA, gstefano@optonline.net

Billie J. Swalla Department of Biology, Center for Developmental Biology, and Institute for Stem Cell and Regenerative Medicine, University of Washington, Seattle, WA 98195, USA; Friday Harbor Laboratories, University of Washington, Friday Harbor, WA 98250-9299, USA; Kewalo Marine Laboratory, University of Hawaii, Honolulu, HI 96813, USA, bjswalla@u.washington.edu

Michael C. Thorndyke The Royal Swedish Academy of Sciences, Sven Lovén Centre for Marine Sciences, Kristineberg 566, SE-450 34, Fiskebäckskil, Sweden, Mike.Thorndyke@marecol.gu.se

Luigi Vitelli Department of Hematology, Oncology and Molecular Medicine, Istituto Superiore di Sanità, Viale Regina Elena 299, 00161 Roma, Italy, luigi.vitelli@iss.it

Ayelet Voskoboynik Institute of Stem Cell Biology and Regenerative Medicine, Stanford University School of Medicine, Stanford, CA 94305, USA; Department of Developmental Biology, Stanford University Hopkins Marine Station, Pacific Grove, CA 93950 USA, ayeletv@stanford.edu

Irving L. Weissman Institute of Stem Cell Biology and Regenerative Medicine, Stanford University School of Medicine, Stanford, CA 94305, USA; Department of Developmental Biology, Stanford University Hopkins Marine Station, Pacific Grove, CA 93950 USA, irv@stanford.edu

Matthias Wiens Institut für Physiologische Chemie, Abteilung Angewandte Molekularbiologie, Universität, Duesbergweg 6, D-55099 Mainz, Germany, wiens@uni-mainz.de

Carla Zilberberg Departamento Biologia Celular e Genética, Universidade do Estado do Rio de Janeiro, Rua São Francisco Xavier 524 PHLC, sala 205, Rio de Janeiro, RJ, Cep: 20550-013, Brazil, carlazilber@yahoo.com.br

Francesca Zito Consiglio Nazionale delle Ricerche, Istituto di Biomedicina e Immunologia Molecolare "Alberto Monroy", Via La Malfa 153, 90146 Palermo, Italy, zito@ibim.cnr.it

Chapter 1
Stem Cells: Autonomy Interactors that Emerge as Causal Agents and Legitimate Units of Selection

Baruch Rinkevich

Abstract Can stem cells be defined as legitimate units of selection (UOS)? Here I review traits of hallmark stem cells (many of which are manifested in marine organisms), which collectively reveal that some types of stem cells behave as real UOS entities. The two existing categories of stem cells maintain the soma and the germ line, have infallible high capacity for self-renewal, produce diverse lines of differentiated progenies and exist *ad infinitum* (within the organism's life span). Because of their primitive, undifferentiated state, stem cells are pliable to adopt totipotentiality, but that is not enough to be considered as genuine UOS agents. The discussion on highly transmitted cancer, germ and somatic stem cells, the autonomous traits of somatic stem cells and their stemness, stem cells consortia in biological chimerism, and parasitism vs. cooperation on the level of stem cells, point to a group of characters that denote some types of stem cells as UOS. Prime features of these stem cells include: (a) efficient motility within an organism and between organisms (conspecifics/different taxa; vertical/horizontal transmission); (b) ability to parasitize effectively host organisms, in the same species or in other species; (c) stemness and unlimited replication; (d) ability to blend with other cell lineages, blurring the line between distinct somatic and germ cell lineages; (e) ability to designate a variety of different soma as hosts; (f) ability to co-opt and manage somatic cells to collaborate in parasitic agenda; (g) lack of a distinct niche where they reside; and (h) manifestation of the power of holism. Events in which stem cells acquire the traits of legitimate UOS as real Darwinian individuals, are widespread in multicellular organisms, even when lacking structural cohesion; a phenomenon that confounded accepted dogma for UOS entity. It is therefore concluded that some stem cell lines foster the traits of UOS, as efficiently as whole organisms or genes.

Keywords Cancer · Chimerism · Germ line · Soma · Stem cell · Stemness · Transmission · Unit of selection

---

B. Rinkevich (✉)
Israel Oceanographic & Limnological Research, National Institute of Oceanography, Tel-Shikmona, Haifa 31080, Israel
e-mail: buki@ocean.org.il

B. Rinkevich, V. Matranga (eds.), *Stem Cells in Marine Organisms*,
DOI 10.1007/978-90-481-2767-2_1, © Springer Science+Business Media B.V. 2009

1.1 Introduction: Can we Consider Stem Cells as Legitimate Units of Selection?

Stem cell biology is a fast developing discipline, representing both applied and basic scientific aspects of generation and perpetuation of cellular diversity. Results confirmed the concept that multicellular organisms depend on stem cells for survival and maintenance. Those cells have infallibly high capacity for self-renewal and produce diverse lines of differentiated progenies during early phases of ontogeny or throughout adult lifespan (Hall and Watt 1989; Fuchs and Segre 2000; Weissman 2000; Rando 2006). Furthermore, stem cells are spatiotemporally localized entities, existing *ad infinitum* (within the organism's life span) with robust maintenance of their internal organization. Because of their primitive, undifferentiated state, stem cells are pliable to adopt totipotentiality or pluropotentiality. However, stem cells are not only part of the modular units of the multicellular organization, responsible for the development of tissue and organ systems, but are also considered as independent units of evolution and as legitimate units of selection (UOS; Rinkevich 2000; Weissman 2000). This term is not easy to grasp. Whereas according to Webster's Dictionary (New International Dictionary, 2nd ed.) a unit is defined as "a distinct part or member analyzable in an aggregate or whole", the basic term of the "unit of selection" is polemic, continuously subjected to scientific and philosophical dispute (Lewontin 1970; Wilson and Sober 1994; Gould and Lloyd 1999). Different types of UOS have been identified as possessing the requisite properties to be "units" of selection. Furthermore, clarifying the UOS concept turned into a major issue because much of the ambiguity, the confusion, and the controversy engendered when discussing this tenet seemed to be emerging from failure to identify the biological entities upon which natural selection operates (Rinkevich 2000).

When should stem cells be defined as UOS? For further evaluation of the biological entities that are the objects of natural selection, Rinkevich (2000) defined three requisite criteria for UOS properties (holism, minimalism, and functionalism) that revealed, in addition to the self-evident case of the "individual", at least, six distinct types of UOSs. These UOSs can differ from each other substantially in their characteristics, in their morphology, or in their structural organization. Some UOSs even blend into the somatic background of conspecifics or other organisms. One of these types that clearly evince a kind of holism, with properties of independent functionalism, is the "stem cells".

Two types of stem cells should be considered. (1) the cells of the germline that are selected for their ability to act in the somatic environment, independent of the general perpetuation trait of cellular diversity, and for their ability to deliver their genetic components through sexual reproduction to proceeding generations of stem cell lineages; (2) the somatic stem cell lineages. In many organisms, the boundaries between germ- and somatic- stem cells are blurred. This is manifested in situations when the germ-line is sequestered from somatic cells either late in ontogeny or not completed at all during the lifespan of an organism (epigenetic and somatic embryogenesis modes of development; Buss 1982, 1987; Blackstone and Jasker 2003). One example is plants that throughout their life propagate their germ-line from cells of the meristematic region at the tips of branches, together with somatic cells. Another example is *Hydra*, whose totipotent cell lineage can undergo enormous number of cell divisions before sequestering the germ line (Buss 1987).

Considering the biological organization of some stem cell lines as justifiable UOS candidates, calls for further exploration of the roles of stem cells as modules of selection. Not only are stem cells of an individual organism subjected to natural selection forces (see below for contracting outcome in within-chimera interactions), but their level of cell lineage with its special traits, such as boundless cellularity, lack of structural cohesion, harmonious and coordinated functional integration, and efficient transmission (see below), render them as most interesting UOS types, central to the concept of Darwinian natural selection. Below is a concise review of striking stem cell traits of which many are related to marine organisms, collectively shedding more light on the nature of some types of stem cells as real UOS. This encourages the use of comparative epistemology for sclarifying the complex biological notion of stem cells as a legitimate UOS.

1.2 Cancer Stem Cells – Acquiring the Stemness of UOS

While it is accepted that cancer develops according to the rules of Darwinian selection (Michor et al. 2003; Breivik 2005; Crespi and Summers 2005; however, because metastasis is usually lethal, it appears to many investigators as a mere end point of successive cellular changes), very little has been accomplished in research on their UOS properties and stemness of cancer cells. In contrast, many resources have been invested in selective traits of cancer cells, their proliferative advantage over neighboring cells, their ability to form "self-organizing" foci, and the required plasticity to increase cellular fitness in terms of malignancy. Considering the above characteristics, Hanahan and Weinberg (2000), although not directly interpreting stem cells properties, have suggested that a cancer cell must become independent from external growth stimuli, be resistant to inhibitory signals, should acquire limitless proliferative potential that leads to rapidly growing clones, and become capable of tissue invasion and metastasis. These cancer cell properties share many characteristics with normal stem cells. Furthermore, recent studies (reviewed in Lobo et al. 2007) have uncovered the presence of cancer stem cells that have the exclusive ability to regenerate tumors, a phenomenon validated in a growing list of tumors. As in normal tissues that maintain homeostasis by stem cells, tumors would eventually degenerate without being sustained by cancer stem cells. Cancer stem cells are progenitor cells capable of self-renewing cancer stem cells and non self-renewing cells, progenies that are more differentiated and destined to reduced proliferation.

There are, however, at least two examples for ecologically and evolutionary successful stem cell tumors that have acquired the ultimate autonomy and traits of UOS. The first is Canine Transmissible Venereal Sarcoma (CTVS), a worldwide common sexually transmitted neoplasm of the external genitalia of dogs (in some regions it is the most common dog tumor), transmitted from one dog to another, usually during coitus (reviewed in Das and Das 2000). This pathogenic lineage of cancerous cells has been known as infectious since 1876, when a veterinarian successfully transplanted the tumor from one dog to another by rubbing the excised tumor onto the scarified genital mucosa of a susceptible dog. This was followed by the 17-year experiment of Karlson and Mann (1952) that succeeded in passing on the same line of tumors through 40 generations of dogs. It has been suggested that all current CTVS are direct descendents from the tumor cells of the original animal and that the tumor cell line originated 200–2,500 years ago in a wolf, coyote, or an old Asian dog breed (Das and Das 2000; Murgia et al. 2006), possibly the world's longest living cloned mammalian cells. The tumor had probably been initiated by the insertion of a LINE (Long Iinterspersed Nuclear Element) retrotransposable element upstream of c-MYC oncogene (Katzir et al. 1985). It is the continuity of a single cell lineage that distinguishes CTVS from other transmissible cancers (originating by viruses, such as in human cervical cancer). Once on a new host, the cells reproduce to form tumor-like structures, usually around the genitals, but occasionally on the skin, and in and around the mouth. The cell lineage can be transmitted to another host, e.g., any breed of dogs or foxes, and then, even without medical treatment, tumors usually regress after 1–3 months. This "naturally occurring allograft" or a "natural chimera" can be regarded as a highly degenerated mammal (Leroi et al. 2003) or as a highly domesticated stem cell parasitic form.

Another type of transmissible tumor is thought to be endangering the Tasmanian devil (*Sarcophlus harrissi*), a carnivorous marsupial whose numbers have plummeted by at least a third since the mid-1990s due to the spread of unsightly and often fatal facial cancer. This facial cancer, which is transmitted from one individual to another by biting (Hawkins et al. 2006; Murgia et al. 2006; Lachish et al. 2007; Siddle et al. 2007), was identified nearly a decade ago and called the Devil Facial Tumor Disease (DFTD). This is a highly aggressive disease (no evidence of recovery or natural immunity documented as yet; Hawkins et al. 2006), which will eventually be domesticated (as the dog CTVS), unless it first eradicates the vulnerable Tasmanian devil population. As in CTVS, DFTD is also transferred between conspecifics as tissue allograft, since tumor cells from different Tasmanian devils contain identical unique chromosomal rearrangements (Pearse and Swift 2006; Siddle et al. 2007). It was further suggested (Siddle et al. 2007) that the DFTD, which is transmitted due to depleted HMC diversity in the Tasmanian devil, results in an infection that does not invoke an immune response, and T lymphocytes of infected devils do not infiltrate the tumors.

1.3 The Trait of Transmission – Autonomous Somatic Stem Cells

Can any differentiated cell, which exists as an integral component of the organism's soma, acquire stemness properties and be regarded as a legitimate UOS? The discussion on cancer stem cells and the two examples of transmittable cancers (CTVS in dogs, DFTD in the Tasmanian devil) revealed that somatic cells could adopt the feature of continuous propagation, forming specific cell lines that undergo countless mitosis and host-to-host multiple transfers. Evidently, these successful cellular parasitic forms (termed here as somatic cell parasites) have become independent from their soma of origin, and outlived their original genotypes.

However, they are not alone. HeLa cells, to which I refer as an experimental counterpart of CTVS and DFTD, are reminiscent of many characteristics of somatic cell parasites. In October 1951, the body of Hennrietta Lacks, a young mother of five, was laid to rest after succumbing to cervical cancer. However, she has been "survived" by a small biopsy of this cancer that has been growing ever since in hundreds of laboratories worldwide (named HeLa in her memory; reviewed in Finkel et al. 2007). The HeLa cell line lacks only the acquired feature of host-to-host transfer to become a definite UOS.

Corroboration for the evolutionary importance of transmission comes also from occult tumors in donor organs, following iatrogenic transplantation (Kauffman et al. 2002; Mackie et al. 2003). There are, possibly, other routes for host-to-host tumor transfers in humans and mammals. One example is represented by a vertical transplacental transmission of cancers. In contrast, the fetal tumors have not been reported to metastasize to the mother (reviewed in Tolar and Neglia 2003). Also the horizontal (intraplacental) transmission of leukemia between sets of identical twins, their engraftment and survival has been described (Tolar and Neglia 2003). In other mammals, a transmissible reticulum cell sarcoma reported in a colony of Syrian hamsters (Banfield et al. 1965, 1966). In this case, studies revealed the possibility of contingency through mosquito bites (1–2% of mosquito), with unusually consistent karyotype shown by the tumor and tumor cells that remained viable in the mosquito mid-gut (the main site of digestive enzymes secretion) up to eight hours after their ingestion (Banfield et al. 1965, 1966). As in regular tumors (Hanahan and Weinberg 2000), transmissible tumors should conscript and subvert host cells to collaborate in their UOS agenda, but before that they should acquire stemness. Theoretically, any differentiated somatic cell that enters a bona fide de-differentiation process, allows the lineage-reversed cell to regain and revert stemness (Chi et al. 2007) prior to turning into a genuine autonomous UOS, by acquiring the trait of efficient transmission.

A different mode of somatic stem cell transmission, as autonomous UOS, is the evolutionary route for vegetative propagation of certain invertebrate taxa. In the bdelloid rotifers, as well as in the cosmopolitan marine clam *Lasaea* (O'Foighil and Smith 1995; Judson and Normark 1996), asexual lineage propagations, each of which started from a totipotent uni-stem cell lineage, have probably persisted for millions of years, manifesting vigor of transmission and mobility. It would, therefore, appear that many somatic single cell entities are equipped with traits of evolutionary functions (not always expressed under the control of multicellularity configurations) and with the traits of stemness transmission and mobility.

1.4 The Trait of Transmission – Autonomous Germ and Somatic Stem Cells

In contrast to somatic stem cells that express various transmission pathways, germ stem cells that reveal the UOS autonomy are transmitted exclusively after fusion between conspecifics, the formation of chimerism or genetically non-homogeneous entities. The literature reveals that chimeras of different forms are widely documented in nature (Gill 1977; Benirschke 1981; Buss 1982), distributed in at least nine phyla of protists, animals and plants (Buss 1982; Santelices et al. 1999; Poudyal et al. 2007), as well as in a variety of vertebrates and mammals, including humans (reviewed in Gill 1977; Tippett 1984; Maloney et al. 1999; Rinkevich 2001; Nelson 2002; Reed 2003). Above vertebrate and invertebrate chimeras were found to carry significant costs, such as reduced growth rates, morphological resorption and necroses in invertebrates (Frank et al. 1997; Barki et al. 2002), autoimmune diseases, freemartins and other abnormal syndromes in vertebrates (Rinkevich 2001). All these outcomes are probably related to the activities of autonomous stem cells, a phenomenon that has been elucidated in a few cases.

The most studied invertebrate model system is chimerism in the colonial tunicate, *Botryllus schlosseri*. This species, as other botryllid ascidians, possesses a unique type of allorecognition. Pairs of colonies that meet each other extend peripheral termini of blood vessels and either fuse to form a vascular parabiont (cytomictical chimeras; Rinkevich and Weissman 1987a), or develop cytotoxic lesions in the contact zones (points of rejections, PORs; reviewed in Teneda et al. 1985; Weissman et al. 1990; Rinkevich 2002a,2004b, 2004a, 2005b). It takes up to two weeks from initial fusion until blood cells are completely merged. The formation of a chimera is only the start of a long biological journey. Following fusion, a second allorecognition phenomenon begins, which leads to resorption (morphological elimination) of one partner in the chimera and then, the third allorecognition phenomenon is manifested as somatic/germ cell parasitism (Pancer et al. 1995; Stoner and Weissman 1996; Magor et al. 1999; Stoner et al. 1999; Rinkevich and Yankelevich 2004; Simon-Blecher et al. 2004; Laird et al. 2005). This phenomenon takes place even in forced chimeras, established between allogeneic non-compatible partners and xenografts (Rinkevich et al. 1998; Simon-Blecher et al. 2004). Results demonstrated that, in many cases, regardless of the directionality of morphological resorption (allorecognition hierarchy), the blood, the soma and the germ cells of the "winner" partner in the resorption phenomenon are chimeric, pointing to cell lineage parasitism. In addition, results revealed cases where the whole mass of gonads, as well as the soma, were derived from the genotype of the resorbee. The germ cell parasitism events are reproducible, show hierarchical patterns, and are sexually inherited, as demonstrated by breeding experiments (Stoner et al. 1999; Rinkevich and Yankelevich 2004). Somatic cell parasitism, on the other hand, does not reveal the trait of sexual inheritance through a pedigree (Stoner et al. 1999). Therefore, gametic and somatic competitive routes, although reproducible, appear to be unlinked (Stoner et al. 1999; Magor et al. 1999; Rinkevich and Yankelevich 2004).

The phenomenon of intraspecifc somatic- and, mainly, germ- cell parasitism in colonial tunicates (Pancer et al. 1995; Stoner et al. 1999; Laird et al. 2005) may challenge, theoretically, the concept of Darwinian selection (Rinkevich 2002b). In this system, autonomous cells of parasitic germ line hitchhike with positively selected genotypes, passing through successive generations without being visible to natural selection forces. These cells, which fulfill the four criteria cited by Pancer et al. (1995) for germ cell parasitism in *Botryllus* (transitive/non transitive hierarchies, disproportionate share of gametic output, frequencies increase, somatic embryogenesis type of development) reveal therefore a clear agenda for successful transmission. Hitchhiking onto the soma of positively selected genotypes provides the parasitic forms with an advantage for establishing new progenies and may lead to the development of super parasitic stem cells. Eventually, this may turn into a Pyrric victory, as the inferior partners in "germ line war" (Rinkevich 2002a,b) develop several specific characters to combat super parasitic stem cells (reviewed in Rinkevich 2002a,b, 2004a, 2005b). This urochordate biological system combines natural tissue transplantation phenomena with stem cell biology in complex ecological mechanisms that involve various evolutionary concepts and autonomy UOS players.

"Stem cell wars" are also recorded in vertebrate systems (Rinkevich 2004b), manifested by several types of naturally occurring transplantation events. These include phenomena associated with bone-marrow transplantation in dizygotic twins, fetal implantation, including human "whole body chimerism" (spontaneous early fusion of dizygotic twins into one body), germ cell chimerism and proliferations of the fetal cells in the maternal blood system decades after parturition (reviewed in Benirschke and Kaufman 1990; Rinkevich 1998, 2004b).

The feto-maternal chimerism is one of the prevailing examples. From implantation in the mother's womb to parturition, mammalian embryos develop as semiallogeneic grafts. The placenta, the interface between the circulatory systems of the mother and the embryo is an incomplete anatomical separation, permeable to cellular exchanges. In human pregnancies, fetal cells cross this feto-maternal barrier bi-directionally, and probably in the vast majority of cases, they proliferate, establishing natural chimerism based on long-term existence of exchanged stem cells. Studies (Rinkevich 1998, 2001; Evans et al. 1999; Maloney et al. 1999; Bianchi 2000; Khosrotehrani et al. 2000) documented the persistence and proliferation of fetal cells in maternal blood circulation for several decades postpartum and revealed that non-shared maternal specific DNA is found in healthy subjects for, at least, five decades after birth. A variety of autoimmune diseases such as scleroderma, systemic lopus erythematosus, polychondritis, inflammatory myopathy, systemic sclerosis, polymorphic eruptions of pregnancy, pre-eclampsia, primary biliary cirrhosis, Sjögren's syndrome, polymyositis, etc., were found to be associated and linked to this human microchimerism (Evans et al. 1999; Bianchi 2000; Leung et al. 2001; Lambert and Nelson 2003; Adams and Nelson 2004), although a causal link has yet to be proved. Considering the other vectorial direction, maternal cells may also pass into the fetus, where they can live on into adulthood, proliferate, triggering or perpetuating autoimmune diseases in offspring (Stevens 2007). Clearly, the roles of fetal or maternal stem cells in autoimmune diseases or other biological phenomena (such as somatic stem cell parasitism in human) are not fully understood. However, recent studies have also revealed the opposite biological situation, the likelihood that maternal cells may cooperate intimately with host soma (similar to somatic stem cells cooperation in botryllid ascidians where a fine-tuning of the chimera's genetic components is achieved in response to changes in environmental conditions; Rinkevich and Yankelevich 2004). One example is Nelson et al. (2007) study, confirming that maternal stem cells are helping to regenerate damaged tissues in host progenies. By investigating the potential existence of maternal cells in children with type 1 diabetes, an immune-mediated disorder, they confirmed the existence of a small number of female islet beta cells that produced insulin, in male pancreatic tissue (procured from autopsies).

Human cytomictical (blood sharing, sensu Rinkevich 2001) chimeras, also called "blood chimeras" and "twin chimeras" (Tippett 1984; Benirschke and Kaufman 1990) are of more evolutionary importance. These chimeras appear following anastomoses of blood vessels between genetically non-identical twins in uterine life. By engrafting stem cells and by virtue of acquired tolerance, long-lasting allogeneic natural transplantation takes place decades after morphological resorption of one of the dizygotic twins or after their vasculature is separated. In contrast to other mammals, human chimeric twins do not show sexual abnormalities and therefore they are usually detected when blood group tests have peculiar results that necessitate full genetic investigations. The literature (Gill 1977; Tippett 1983; Mosebach et al. 2006; and literature therein), however, indicates that within human natural chimeras, cells of different origins may interact through life in a genotypic pre-determined hierarchy, in some cases characterized by swift changes within periods of months. For example, in one chimera of dizygotic twins, when first tested, the twin brother had 86% of his own blood cell type, which was reduced to 63% after 15 years. In his twin sister, only 1% of her brother's cell type was recorded constantly over the whole period. Above examples of stem cells cannot be regarded as legitimate UOS because they carry only a limited trait of transmission. However, this biological situation is most relevant to the UOS concept as it may develop towards germ cell parasitism. Indeed, one of the most intriguing situations is the possibility that totipotent stem cells exchanged through blood may enter into a germ cell parasitism scenario.

There are many human examples in which the phenotypic spectrum of 46,XX/46,XY chimeras is recorded in patients. These chimeras are commonly called whole body chimeras or dispermic chimeras (Gill 1977; Tippett 1983) and are the outcome of the fertilization of two maternal nuclei (or the ovum and the polar body) by two sperm cells, followed by the amalgamation of two different zygotes in a single embryo, a self-contained twin. Morphologically, it ranges from normal male or female genitalia to different degrees of ambiguous genitalia (a more detailed account of human chimerism is outlined in Rinkevich 2001).

Germ cell parasitism is considered when one genotype in the chimera gains evolutionary benefits (successfully developing the next generation of offspring) at the expense of the other (Pancer et al. 1995; Stoner et al. 1999; Laird et al. 2005), by seeding its primordial germ cells in the gonads of a conspecific partner. Although such a scenario may be deemed imaginary in the vertebrates, it is not. About three decades ago, Ford et al. (1975) revealed in the mouse system, the possible functionality of XY oocytes by producing a male offspring from an aggregated chimera, a study that was pursued by Evans et al. (1977) documentation for the transformation of an XY germ cell into an oocyte. Such a situation was documented in cattle (Ohno et al. 1962) and in primates, where male marmosets spermatocysts originated from germ cells with an XX sex chromosome constitution (Hampton 1973).

Humans display very interesting cases of effective germ cell chimerism with clear characteristics of true UOS entities and cases where human soma was genetically different from the gonads (Mayr et al. 1979; Tippett 1983; Ramsay et al. 1988; Talerman et al. 1990; Strain et al. 1995; Bjornson et al. 1999). It could possibly indicate a wider distributed phenomenon, because in humans, natural germ cell chimeras are almost never detected unless the offspring have abnormalities such as male/female or hermaphrodite characteristics or skin discoloration. However, even with the limited genetic tools available three decades ago, Mayr et al. (1979) revealed a case of a human female chimera that was detected as such, after finding that none of her four children matched genetically her genome and none of the children's 21 unique genetic markers was detected in this woman. This study puts forward the conclusion that this female was the outcome of a dispermic chimera or a chimera of two embryos. Following an extreme elimination process, this woman subject possessed two populations of allogeneic cells, one in the soma and the second in her gonads, having the germ line of only one original partner. Talerman et al. (1990) reported a phenotypic female, true hermaphrodite with bilateral ovotestis, and a successful pregnancy. Although not evaluating the possibilities for germ cells parasitism following chimerism, Tippett (1983) presented several cases of male dispermic chimeras that had children. Another situation is the status of congenital tetragametic chimerism formed through the fertilization of two ova by two sperm, followed by the fusion of the zygotes and the development of an organism with intermingled cell lines. Tetragametic chimerism is usually identified, unexpectedly, after at least partial histocompatibility testing of family members. In a more recent case (Yu et al. 2002), a fertile XX/XX female with no evident chimerism in the peripheral blood, was found to be the non-biologic mother of two of her three children, a clear case of mixed germ cell parasitism. In another recent case of twinning, Souter et al. (2007) reported a case of 46,XX/46,XY twins that were chimeric and shared a single genetic contribution from their mother but two genetic contributions from their father. Twin A was determined to be a true hermaphrodite and twin B, apparently a normal male. This situation was evaluated by using more than 6,000 DNA markers on skin fibroblast samples from the twins and on peripheral blood samples from both parents. It is therefore evident that primitive stem cells, in the right environment, may differentiate into cell lineage lines never recorded in sites of origin (Johnson et al. 2005), and probably a variety of healthy, normal germ cell chimeras of different ontogenic routes were not disclosed.

1.5 Cooperation Between Cell Lineages – United we Stand

Could different conspecific stem cells expressing the autonomous UOS agenda develop a multi-clonality super-entity favorable to the state of genetically homogenous entities? The positive answer points to the model case of Rhizocephalan barnacles, a cosmopolitan group of parasites infesting exclusively a range of other crustaceans. Rhizocephalan barnacles are taxonomically placed within the primitive crustacean class Cirripedia, as the free-living barnacles, only because they all share a unique larval type, called cyprid. It is difficult to group the adults with any crustaceans or arthropods because they lack appendages, segmentation, alimentary canal and almost all organs characteristic to crustaceans, including respiratory and excretory organs (Høeg 1992). Adults show an extraordinarily simplified morphology with an external (on the surface of the host body) sac containing an ovary and two male receptacles, and ramified internal nutrient-absorbing network of rootlets, from which numerous projections extend into the haemolymph spaces of the host. The female cyprid settles on the host integument, metamorphoses into an inoculating parasitic entity, forming a hollow cuticle-enforced syringe that eventually penetrates into the host and injects the parasitic material (a few totipotent cells or groups of undifferentiated cells, called vermigon, each surrounded by a thin cuticle with worm-like movement capabilities; Glenner and Høeg 1995) into the host haemocoel.

Inoculation by the parasitic form of the female cyprid proceeds through the development of the internal roots (interna) and the formation of the "externa" structures with female gonads. Male receptacles that develop spermatogenesis are formed by two male cyprids after settling on juvenile externae. In some species of the genus *Thompsonia*, a multitude of dimorphic, ova- or sperm-producing externae (male and female) are created by asexual propagation (Jespersen and Lützen 1992). The entire parasitic form is therefore a multi-clonality super-entity, a product of aggregation of at least three different genotypes in which males are reduced to spermatogenic tissues, grown by the female in specific male receptacles. Moreover, feeding the male cells is performed by the agency of adjacent female cells, so that spermatogenesis in male receptacles is synchronized with oogenesis. In fact, this whole multi-clonality super-entity (including the externa, the internal stolons, germ cells, male spermatogenic cells, and female oogenic cells) is a "germ cell entity", because all parts express *vasa*-related genes (Shukalyuk et al. 2007), members of the DEAD-box protein family, exclusively expressed in the germ cells of metazoans. This type of aggregated stem cell lineages, morphologically structured as a parasitic rhizocephalan barnacle, is therefore a cooperation of self-renewing pools of totipotent stem cells with UOS characteristics providing the cell with sources for both sexual and asexual reproduction. The free-swimming larva (also carrying pools of stem cells) is the dispersion arm of the entity, revealing the trait of autonomous stem cells transmission.

The acrothoracican barnacle *Trypetesa lampas* represents another case of a unified UOS shared by conspecific genotypes, only one serves as a female. In this species, dwarf males (weighing just about 1/500 of the female; up to 15 males per female) are attached permanently to the exterior mantle surface of the female, fertilizing her through the mantle cavity opening. Males compete with one another for attachment sites on a female (Gotelli and Spivey 1992). This resembles male-male competition in certain Rhizocephala where the first arriving male sheds a cuticle armed with spines that permanently blocks the female receptacle and prevents other males from entering and fertilizing the female (Høeg 1991). Immature females excavate permanent burrows in gastropod shells held by hermit crabs and reach sexual maturity within 6–9 months. Male larvae enter the female burrow aperture, select the settlement sites and each gets attached by a knob of cement and exoskeleton secreted onto the mid-anterior margin of the ovarian disc. After molting, males reach sexual maturity within a few days of settlement. The male body consists of only a single testis, vas deferens, and extensible penis (summarized in Gotelli and Spivey 1992).

The germ-line consortia in the aforementioned examples are probably generated from pressures imposed on the organism's life history portrait to ensure availability of a mate for breeding, when needed. The parasitic barnacle solution is the formation of an intermingled consortium (a multi-partners chimera) of stem cells; altogether in the form of a parasitic UOS on other taxa. The two following examples simplify a different solution pathway, formation of a parasitic male (reduced down to stem cells in extreme cases) on host conspecific female.

The first example is the anglerfishes, a group of deep-sea ceratioid fishes (distributed throughout the world's oceans below a daytime depth of 300m, the most species-rich vertebrate taxon within the bathypelagic zone) whose males are reduced to tiny blobs of testes, attached to the female host as parasitic entities. Using highly developed olfactory organs and large eyes, males can locate females by vision, but homing is performed primarily by smell, towards a female-emitted, species-specific pheromone. While normal jaw and teeth are lost during metamorphosis, the post-metamorphosed dwarf male is equipped with a set of pincer-like denticles at the anterior tips of the jaws for grasping and holding fast to the gigantic female (reviewed in Pietsch 2005). In many ceratioid anglerfishes, attachment of the male on the female (the literature reveals the common existence of multiple attachments of males to a single female; Pietsch 2005) is followed by fusion of tissues (the dermis of the male is completely fused with that of the female) and, apparently, by connecting the circulatory systems. The parasitic males derive nourishment from female hosts, as no alimentary canal exists in males. Blood-borne sex hormones that are continuously secreted by the female fish control and secure the synchronic maturation of sperm and eggs, and males remain alive and reproductively functional as long as the female lives, participating in repeated spawning events. The host female becomes, therefore, a kind of self-fertilizing hermaphrodite, a UOS that simultaneously contains several genotypes, only one in the form of a functional female.

The second example is the bottom-dwelling detritus-eating echiuran worm, *Bonellia viridis*. The female of this species exceeds 1m in length (including extended proboscis), whereas the male is dwarfed to just a few millimeters. Only the female has developed blood vascular system and gut, including mouth and anus (Schuchert and Rieger 1990). In this species, undifferentiated larvae metamorphose into males when exposed to females (usually settling on the proboscis of the adult female, surviving as semi-parasitic sperm producing entities. The larvae can also settle on the body trunk of the adult and differentiate into males), but become females in female-free seawater. Males live more than eight months (Baltzer 1931) within a specialized chamber in the female body close to the female oviduct. Since the male gut fails to develop beyond the larval stage (is degenerated and serves as a storage organ, since it was found to contain only an oily liquid; Schuchert and Rieger 1990), as do the sensory eyespots, it is plausible to assume that feeding by the male takes place in the female cavity. Therefore, larvae settling on female *Bonellia* can be regarded as primitive germ cell vehicles (sensu Rinkevich 2000), autonomous UOS that carry the germ line in the form of "flexible to environmental conditions" germ stem cells. In this case, sex is determined by specific environment scenario (the presence or absence of females).

1.6 Stem Cells in Multichimeras

Many sedentary marine organisms possess the trait of gregarious settlements of kin larvae (Grosberg and Quinn 1986; Rinkevich and Weissman 1987b; Barki et al. 2002). When allorecognition dictates chimerism, following tissue-to-tissue contacts of allogeneic compatible partners, multichimeras (chimeras consisting of more than two partners) can evolve. Such scenarios were documented in cnidarians (Duerden 1902; Barki et al. 2002; Poudyal et al. 2007) and in botryllid ascidians (Rinkevich 1996; Rinkevich and Shapira 1999; Paz and Rinkevich 2002). It was suggested (Rinkevich 2002b, 2004a, 2005a) that increasing the numbers of conspecifics in such an entity is also associated with augmentation of contradictory cell lineage interactions in a way that eventually increases the stability of the whole entity. This argument, not supported by soft corals multichimerism (Barki et al. 2002), was confirmed in other cases like stony corals (unpubl.) and multi chimeras of botryllid ascidians (Rinkevich 1996; Rinkevich and Shapira 1999), where more equilibrated entities formed. Multichimeras form larger colonies than single entities but with measurable cost (such as reduced body size per genotype and morphologically resorbed genotypes). However, it was suggested that potentially losing genotypes (defined as such in bi-chimeras scenario) might survive in multichimera entities as stem cells in refugia and successfully propagate to succeeding generations. Therefore, natural selection acts on the multichimeras at the group level (Rinkevich and Shapira 1999) where stem cells within these consortia serve as the UOS vehicles. Thus, the multichimera fitness is actually dependant on the joint fitness of the mixed stem cell lineages. It is unfortunate, however, that the contribution of fitness of each stem cell lineage in multichimerism has not yet been evaluated. Yet, it was suggested (Rinkevich 2002b, 2004a, 2005a) that the domesticated manifestation of stem cell wars in multichimeras is one of the routes to reduce the possibilities of super parasitic stem cells (like the dog CTVS and the Tasmanian devil DFTD) from developing.

It is, however, important to note that in either bi-chimerism or in multi-chimerism scenarios, parasitic germ cell lines in the form of autonomous UOS entities are supposed to hitchhike along with positively selected allogeneic soma, passing throughout successive generations, seemingly without obeying the rules of natural selection as they are not seen by environmental parameters (Rinkevich 2002a,b, 2004a, 2005a).

1.7 Synthesis

Richard Lewontin's (1970) seminal publication on the UOS concept initiated the conceptual and theoretical debates that led to the spectrum of perspectives on the different types of biological units that can be regarded as real UOS (i.e., Rinkevich 2000). The goal of the present contribution is to illuminate the roles of stem cells, primarily in marine invertebrates but not exclusively, as causal agents and legitimate UOS entities. The comparative epistemology of autonomous stem cell in various organisms may leave the reader contemplating the evolution potency of these cells at levels distinct from individual organisms, of which most UOS are referred (Gould and Lloyd 1999; Rinkevich 2000). This manuscript points repeatedly to a group of characters that mark some types of stem cells as USO: (a) the efficient motility of these stem cells within the organism and between organisms (conspecifics or different taxa); (b) the ability to parasitize effectively host organisms of the same species or other species, vertical and horizontal transmission; (c) the unlimited replication through repeated mitotic cycles; (d) the ability to mix with other cell lineages, blurring the differences between distinct somatic and germ cell lineages that commonly destine specific traits (e.g., whole organism's expression of *vasa* protein; Shukalyuk et al. 2006); (e) the utilization of a variety of different soma as hosts; and (f) indiscrimination between somatic or germ cell parasitism. Events in which stem cells acquire the traits of legitimate UOS are widespread in multicellular organisms and should therefore be considered accordingly.

It is not surprising that many of the characters mapped for legitimate UOS share traits with cancers. In a recent study, Crespi and Summers (2005) have shed light on the contrasts between the evolution of individuals within a population and cancer cells within an individual, revealing many of the characters of legitimate UOS. Indeed, many of the hallmarks of cancers (Hanahan and Weinberg 2000) are also listed as stem cells agenda. For example, successful tumor cells, as parasitic stem cells in chimeras of botryllid ascidians, acquire the ability to co-opt and induce somatic cells to release various signals important to the maintenance of the parasitic cells. In both cases, normal cells collaborate actively in parasitic agenda. One of the striking examples for such collaboration (Rinkevich and Yankelevich 2004) is the fine-tuning of somatic genetic components within the botryllid chimera, in response to changes in environmental conditions. The plastic combination of the chimera soma does not affect the hierarchy at the level of germ stem cells. Therefore, conscripting the soma of all stem cell lineages for the germ stem cell agenda (hitchhiking on positively selected soma) provides the more parasitic form of germ cell lineage with the unavoidable advantage of establishing new progenies.

Another trait shared by all legitimate stem cells that were evaluated above as legitimate UOS, is the lack of any visible or distinct niche(s) where they reside. The idea that stem cells are controlled by a particular microenvironment known as "niche" (although it was revealed in a relatively small number of stem cell types because of the difficulty to identify and manipulate individual stem cells and their niches) is widely accepted (Spradling et al. 2001). Studies on stem cells from diverse systems have shown that stem cells' functions are supported, controlled and regulated largely, by extracellular cues from surrounding tissues moieties. Several characteristics were assigned to the stem cell niche. These characteristics reveal that the stem cell niche is composed of a group of cells in a special tissue location, that the niche functions as a physical anchor for stem cells and that the niche generates extrinsic factors that control stem cell fate and number (Li and Xie 2005; Scadden 2006). However, the contradictory trait revealed in this essay, i.e., the lack of "niche" in stem cells regarded as UOS, is not new. The niche-independent stem cells, were found in some stem cells of adult mammals (Fuchs and Segre 2000), in invertebrates, such as in planarian neoblasts (Sánchez Alvarado and Kang 2005) or in tunicates (Akhmadieva et al. 2007) and are the common character for transmissible diseases like the dog CTVS and the Tasmanian devil's DFTD. The interplay between stem cells and their niche(s) creates the dynamics necessary for sustaining tissues. Since most agenda of stem cells acting as legitimate UOS is not targeted towards maintenance of existing tissues, there is no need to allocate to this type of stem cells a specific stem cell compartment within targeted soma. This also supports other UOS traits for better mobility and characters of faster and efficient engraftment capabilities in foreign tissues, regardless of the existence of any controlling cellular microenvironment and without tethering of these stem cells to their surroundings.

The interesting discovery of a putative germline stem cell population in mouse bone marrow, revealing that bone marrow transplantation may restore oocyte production in wild-type mice sterilized by chemotherapy (Johnson et al. 2005), suggests that even in mammals, stem cells in vascular circulation behave like legitimate UOS, by generating oocytes throughout adult life. This outcome resembles the behavior of blood-borne germline stem cells in the urochordate *Botryllus* system (Rinkevich 2002b, 2004a, 2005a). If such a pool of circulating germline stem cells in mammals use the trait of an efficient transmission by crossing the placenta (as documented for hematopoietic stem cells; Bianchi et al. 1996; Bianchi 2000; O'Donoghue et al. 2004), the potential bi-directional parasitism of the maternal germline by the fetus or the fetus germline by mother stem cells (Laird et al. 2005) is emerging. This is specifically true in cases where there is no early differentiation of germline cells, and their segregation from somatic stem cells occurs along the entire ontogeny (somatic embryogenesis). Such a line may easily adopt and facilitate UOS characteristics.

While the classical perception of stem cells and much of the wealth of already gathered information has concentrated on their pliability to adopt different cell fates, this chapter focuses on other traits of those stem cells that are considered genuine

UOS. Resulting from our current meager knowledge on the biology of these stem cells, additional hallmark UOS characters that have not been discussed earlier are probably concealed behind somatic morphological and biochemical characteristics. Not all stem cells acquire the characters of UOS. For those that adopt UOS traits, the perception of a UOS as a group of dispersed stem cells raises a conceptual dilemma: this is a physically/morphologically no coherent UOS, which confounds accepted dogma for the entity of UOS (see also Rinkevich 2000). However, above elucidated common themes provide the background for further research targeted on stem cells (many of them from marine invertebrates) as unique UOS entities, real Darwinian individuals (sensu Gould and Lloyd 1999; Rinkevich 2000). Therefore, some entities at the cellular level of organization foster the traits of UOS, as efficient as whole organisms or genes, manifesting the power of holism, stemness, lack of structural cohesion mobility within and between individual organisms, niche-independency and invasiveness as prime features, experiencing changes in somatic microenvironments during their journeys.

References

Adams KM, Nelson JL (2004) Microchimerism. An investigative frontier in autoimmunity and transplantation. J Am Med Assoc 291:1127–1131

Akhmadieva AV, Shukalyuk AI, Aleksandrova YN, Isaeva VV (2007) Stem Cells in Asexual Reproduction of the Colonial Ascidian *Botryllus tubaratus* (Tunicata: Ascidiacea). Russian J Mar Biol 33:181–186

Baltzer F (1931) Echiurida. In: Kukenthal W, Krumbach T (eds) Handbuch der Zoologie, vol. 2(9). De Gruyter, Berlin and Leipzig, pp 62–168

Banfield WG, Woke PA, Mackay CM (1966) Mosquito transmission of lymphomas. Cancer 19:1333–1336

Banfield WG, Woke PA, Mackay CM, Cooper HL (1965) Mosquito transmission of a reticulum cell sarcoma of hamsters. Science 148:1239–1240

Barki Y, Gateno D, Graur D, Rinkevich B (2002) Soft coral natural chimerism: A window in ontogeny allows the creation of entities, compounding incongruous parts. Mar Ecol Progr Ser 231:91–99

Benirschke K (1981) Hermaphrodites, freemartins, mosaics and chimaeras in animals. In: Austin CR, Edwards RG (eds) Mechanisms of sex differentiation in animals and man. Academic Press, New York, pp 421–463

Benirschke K, Kaufman P (1990) Pathology of the human placenta. Springer-Verlag, New York

Bianchi DW (2000) Fetomaternal cell trafficking: A new cause of disease? Am J Med Genet 91:22–28

Bianchi DW, Zickwolf GK, Weil GJ, Sylvester S, DeMaria MA (1996) Male fetal progenitor cells persist in maternal blood for as long as 27 years postpartum. Proc Natl Acad Sci USA 93: 705–708

Bjornson CRR, Rietze RL, Reynolds BA, Magli MC, Vescovi AL (1999) Turning brain into blood: A hematopoietic fate adopted by adult neural stem cells in vivo. Science 283:534–537

Blackstone NW, Jasker BD (2003) Phylogenetic considerations of clonality, coloniality, and mode of germline development in animals. J Exp Zool (Mol Dev Evol) 297B:35–47

Breivik J (2005) The evolutionary origin of genetic instability in cancer development. Semin Cancer Biol 15:51–60

Buss LW (1982) Somatic cell parasitism and the evolution of somatic tissue compatibility. Proc Natl Acad Sci USA 79:5337–5341

Buss LW (1987) The evolution of individuality. Princeton University Press, Princeton
Chi S, Fu X, Sheng Z (2007) Differentiation: A new approach in stem cell research. BioSci 57: 655–662
Crespi B, Summers K (2005) Evolutionary biology of cancer. Trends Ecol Evol 20:545–552
Das U, Das AK (2000) Review of canine transmissible venereal sarcoma. Veter Res Comm 24: 545–556
Duerden JE (1902) Aggregated colonies in madreporarian corals. Am Nat 36:461–471
Evans EP, Ford CF, Lyon MF (1977) Direct evidence of the capacity of the XY germ cell in the mouse to become an oocyte. Nature 267:430–431
Evans PC, Lambert N, Maloney S, Furst DE, Moore JM, Nelson JL (1999) Long-term fetal microchimerism in peripheral blood mononuclear cell subsets in healthy women and women with scleroderma. Blood 93:2033–2037
Finkel T, Serrano M, Blasco MA (2007) The common biology of cancer and ageing. Nature 448:767–774
Ford CE, Evans EP, Burtenshaw MD, Clegg HM, Tuffrey M, Barnes RD (1975) A functional "sex-reversed" oocyte in the mouse. Proc Royal Soc Lond B 190:187–197
Frank U, Oren U, Loya Y, Rinkevich B (1997) Alloimmune maturation in the coral *Stylophora pistillata* is achieved through three distinctive stages, four months post metamorphosis. Proc Royal Soc Lond B 264:99–104
Fuchs E, Segre JA (2000) Stem cells: A new lease on life. Cell 100:143–155
Gill TJ (1977) Chimerism in humans. Transplant Proc 9:1423–1431
Glenner H, Høeg JT (1995) A new motile, multicellular stage involved in host invasion by parasitic barnacles (Rhizocephala). Nature 377:147–150
Gotelli NJ, Spivey HR (1992) Male parasitism and intrasexual competition in a burrowing barnacle. Oecologia 91:474–480
Gould SJ, Lloyd EA (1999) Individuality and adaptation across levels of selection: How shall we name and generalize the unit of Darwinism? Proc Natl Acad Sci USA 96:11904–11909
Grosberg RK, Quinn JF (1986) The genetic control and consequences of kin recognition by the larvae of a colonial marine invertebrate. Nature 322:456–459
Hall PA, Watt FM (1989) Stem cells: the generation and maintenance of cellular diversity. Development 106:619–633
Hampton SH (1973) Germ cell chimerism in male marmosets. Am J Physic Anthropol 38:265–268
Hanahan D, Weinberg RA (2000) The hallmarks of cancer. Cell 100:57–70
Hawkins CE, Baars C, Hesterman H, Hocking GJ, Jones ME, Lazenby B, Mann D, Moony N, Pemberton D, Pyecroft S, Restani M, Wiersma J (2006) Emerging disease and population decline of an island endemic, the Tasmanian devil *Sarcophlus harrissi*. Biol Conserv 131:307–324
Høeg JT (1992) Rhizocephala. In: Harrison FW, Hurnes AG (eds) Microscopic Anatomy of Invertebrates, vol. 9. Wiley-Liss, New York, pp 313–345
Høeg JT (1991) Functional and evolutionary aspects of the sexual system in the Rhizocephala (Thecostraca: Cirripedia) In: Bauer RT, Martin JW (eds) Crustacean Sexual Biology. Columbia University Press, New York, pp 208–227
Jespersen A, Lützen J (1992) Thompsonia dofleini, a colonial akentrogonid rhizocephalan with dimorphic, ova- or sperm-producing, externae (Crustacea, Cirripedia). Zoomorphol 112: 105–116
Johnson J, Bagley J, Skaznik-Wikiel M, Lee HJ, Adams GB, Niikura Y, Tschudy KS, Tilly JC, Cortes ML, Forkert R, Spitzer T, Iacomini J, Scadden DT, Tilly JL (2005) Oocyte generation in adult Mammalian ovaries by putative germ cells in bone marrow and peripheral blood. Cell 122:303–315
Judson OP, Normark BB (1996) Ancient asexual scandals. Trends Ecol Evol 11:41–46
Karlson AG, Mann FC (1952) The transmissible venereal tumour of dogs: observation of forty generations of experimental transfer. Ann New York Acad Sci 54:1197–1213

Katzir N, Rechavi G, Cohen JB, Unger T, Simoni F, Segal S, Cohen D, Givol D (1985) "Retroposon" insertion into the cellular oncogene c-myc in canine transmissible venereal tumor. Proc Natl Acad Sci USA 82:1054–1058

Kauffman HM, McBride MA, Cherikh WS, Spain PC (2002) Transplant tumor registry: donors with central nervous system tumors. Transplantation 73:579–582

Khosrotehrani K, Johnson KL, Cha DH, Salomon RN, Bianchi DW (2000) Transfer of fetal cells with multilineage potential to maternal tissue. J Am Med Assoc 292:75–80

Lachish S, Jones M, McCallum H (2007) The impact of disease on the survival and population growth rate of the Tasmanian devil. J Animal Ecol 76:926–936

Laird DJ, De Tomaso AD, Weissman IL (2005) Stem cells are units of natural selection in a colonial ascidian. Cell 123:1351–1360

Lambert N, Nelson JL (2003) Microchimerism in autoimmune disease: more questions than answers? Autoimmun Rev 2:133–139

Leroi AM, Koufopanou V, Burt A (2003) Cancer selection. Nat Rev Cancer 3:226–231

Leung TN, Zhang J, Lau TK, Chan LYS, Dennis LYM (2001) Increased maternal plasma fetal DNA concentrations in women who eventually develop preeclampsia. Clinical Chem 47:137–139

Lewontin RC (1970) The units of selection. Ann Rev Ecol Syst 1:1–18

Li L, Xie T (2005) Stem cell niche: structure and function. Ann Rev Cell Dev Biol 21:605–631

Lobo NA, Shimono Y, Qian D, Clarke MF (2007) The biology of cancer stem cells. Ann Rev Cell Dev Biol 23:675–699

MacKie RM, Reid R, Junor B (2003) Fatal melanoma transferred in a donated kidney 16 years after melanoma surgery. New Engl J Med 348:567–568

Magor BG, De Tomaso AW, Rinkevich B, Weissman IL (1999) Allorecognition in colonial tunicates: Protection against predatory cell lineages? Immunol Rev 167:69–79

Maloney S, Smith A, Furst DE, Myerson D, Rupert K, Evans PC, Nelson JL (1999) Microchimerism of maternal origin persists into adult life. J Clin Invest 104:41–47

Mayr WR, Pausch V, Schnedl W (1979) Human chimaera detectable only by investigation of her progeny. Nature 277:210–211

Michor F, Frank SA, May RM, Iwasa Y, Nowak MA (2003) Somatic selection for and against cancer. J Theor Biol 225:377–382

Mosebach M, Parkner A, Jakubiczka S, Wieacker P, Heim MU (2006) Dispermic chimerism identified during blood group determination and HLA typing. Transfusion 46:1978–1981

Murgia C, Pritchard JK, Kim SY, Fassati A Weiss RA (2006) Clonal origin and evolution of a transmissible cancer. Cell 126:477–487

Nelson JL (2002) Microchimerism: incidental byproduct of pregnancy or active participant in human health ? Trends Mol Med 8:109–113

Nelson JL, Gillespie KM, Lambert NC, Stevens AM, Loubiere LS, Rutledge JC, Leisenring WM, Erickson TD, Yan Z, Mullarkey ME, Boespflug ND, Bingley PJ, Gale EAM (2007) Maternal microchimerism in peripheral blood in type 1 diabetes and pancreatic islet β cell microchimerism. Proc Natl Acad Sci USA 104:1637–1642

O'Donoghue K, Chan J, de la Fuente J, Kennea N, Sandison A, Anderson JR, Roberts IA, Fisk NM (2004) Microchimerism in female bone marrow and bone decades after fetal mesenchymal stem-cell trafficking in pregnancy. Lancet 364:179–182

O'Foighil D, Smith MJ (1995) Evolution of asexuality in the cosmopolitan marine clam Lasae. Evolution 49:140–150

Ohno S, Trujillo JM, Stenius C, Christian C, Teplitz RL (1962) Possible germ cell chimeras among newborn dizygotic twin calves (Bos taurus). Cytogen 1:258–265

Pancer Z, Gershon H, Rinkevich B (1995) Coexistence and possible parasitism of somatic and germ cell lines in chimeras of the colonial urochordate *Botryllus schlosseri*. Biol Bull 189: 106–112

Paz G, Rinkevich B (2002) Morphological consequences for multi-partner chimerism in Botrylloides, a colonial urochordate. Dev Comp Immunol 26:615–622

Pearse AM, Swift K (2006) Allograft theory: transmission of devil facial-tumour disease. Nature 439:549

Pietsch TW (2005) Dimorphism, parasitism, and sex revisited: modes of reproduction among deep-sea ceratioid anglerfishes (Teleostei: Lophiiformes). Ichthyological Res 52:207–236

Poudyal M, Rosa S, Powell AE, Moreno M, Dellaporta SL, Buss LW, FG Lakkis (2007) Embryonic chimerism does not induce tolerance in an invertebrate model organism. Proc Natl Acad Sci USA 104:4559–4564

Ramsay M, Bernstein R, Zwane E, Page DC, Jenkins T (1988) XX true hermaphroditism in Southern African blacks: an enigma of primary sexual differentiation. Am J Human Gen 43:4–13

Rando TA (2006) Stem cells, ageing and the quest for immortality. Nature 441:1080–1086

Reed A (2003) Chimerism in myositis. Curr Rheumatol Rep 5:421–424

Rinkevich B (1996) Bi-vs multi-chimerism in colonial urochordates: a hypothesis for links between natural tissue transplantation allogenetics and evolutionary ecology. Exp Clin Immunogen 13:61–69

Rinkevich B (1998) Immunology of human implantation: From the invertebrate point of view. Human Reprod 13:55–459

Rinkevich B (2000) A critical approach to the definition of Darwinian units of selection. Biol Bull 199:231–240

Rinkevich B (2001) Human natural chimerism: An acquired character or a vestige of evolution? Human Immunol 62:651–657

Rinkevich B (2002a) The colonial urochordate *Botryllus schlosseri*: from stem cells and natural tissue transplantation to issues in evolutionary ecology. BioEssays 24:730–740

Rinkevich B (2002b) Germ cell parasitism as an ecological and evolutionary puzzle: Hitchhiking with positively selected genotypes. Oikos 96:25–30

Rinkevich B (2004a) Will two walk together, except they have agreed? J Evol Biol 17:1178–1179

Rinkevich B (2004b) Primitive immune systems: are your ways my ways? Immunol Rev 198:25–35

Rinkevich B (2005a) Natural chimerism in colonial urochordates. J Exp Mar Biol Ecol 322:93–109

Rinkevich B (2005b) Rejection patterns in botryllid ascidian immunity: The first tier of allorecognition. Canad J Zool 83:101–121

Rinkevich B, Shapira M (1999) Multi-partner urochordate chimeras outperform two-partner chimerical entities. Oikos 87:315–320

Rinkevich B, Weissman IL (1987a) A long-term study of fused subclones of a compound ascidian. The resorption phenomenon. J Zool (London) 213:717–733

Rinkevich B, Weissman IL (1987b) The fate of *Botryllus* (Ascidiacea) larvae cosettled with parental colonies: beneficial or deleterious consequences. Biol Bull 173:474–488

Rinkevich B, Weissman IL, De Tomaso AW (1998) Transplantation of Fu/HC incompatible zooids in *Botryllus schlosseri* results in chimerism. Biol Bull 195:98–106

Rinkevich B, Yankelevich I (2004) Environmental split between germ cell parasitism and somatic cell synergism in chimeras of a colonial urochordate. J Exp Biol 207:3531–3536

Sánchez Alvarado A, Kang H (2005) Multicellularity, stem cells, and the neoblasts of the planarian*Schmidtea mediterranea*. Exp Cell Res 306:299–308

Santelices B, Correa JA, Aredo D, Flores V, Hormazábal M, Sanchez P (1999) Convergent biological processes in coalescing rhodophyta. J Phycol 35:1127–1149

Scadden DT (2006) The stem-cell niche as an entity of action. Nature 441:1075–1079

Schuchert P, Rieger RM (1990) Ultrastructural observations on the dwarf male of Bonellia viridis (Echiura). Acta Zoologica (Stockholm) 71:5–16

Shukalyuk AI, Golovnina KA, Baiborodin SI, Gunbin KV, Blinov AG, Isaeva VV (2007) vasa-related genes and their expression in stem cells of colonial parasitic rhizocephalan barnacle

Polyascus polygenea (Arthropoda: Crustacea: Cirripedia: Rhizocephala). Cell Biol Int 31: 97–108

Siddle HV, Kreiss A, Eldridge MD, Noonan E, Clarke CJ, Pyecroft S, Woods GM, Belov K (2007) Transmission of a fatal clonal tumor by biting occurs due to depleted MHC diversity in a threatened carnivorous marsupial. Proc Natl Acad Sci USA 104:16221–16226

Simon-Blecher N, Achituv Y, Rinkevich B (2004) Protochordate concordant xenotransplantation settings reveal outbreaks of donor cells and divergent life span traits. Dev Comp Immunol 28:983–991

Souter VL, Parisi MA, Nyholt DR, Kapur RP, Henders AK, Opheim KE, Gunther DF, Mitchell ME, Glass IA, Montgomery GW (2007) A case of true hermaphroditism reveals an unusual mechanism of twinning. Human Genet 121:179–185

Spradling A, Drummond-Barbosa D, Kai T (2001) Stem cells find their niche. Nature 414:98–104

Stevens AM (2007) Do maternal cells trigger or perpetuate autoimmune diseases in children? Pediatric Rheumatology 5:9 (8 pp). doi:10.1186/1546-0096-5-9

Stoner DS, Rinkevich B, Weissman IL (1999) Heritable germ and somatic cell lineage competitions in chimeric colonial protochordates. Proc Natl Acad Sci USA 96:9148

Stoner DS, Weissman IL (1996) Somatic and germ cell parasitism in a colonial ascidian: possible role for a highly polymorphic allorecognition system. Proc Natl Acad Sci USA 93: 15254–15259

Strain L, Warner JP, Johnston T, Bonthrone DT (1995) A human parthenogenetic chimaera. Nat Genet 11:164–169

Talerman A, Verp MS, Senekjian E, Gilewski T, Vogelzang N (1990) True hermaphrodite with bilateral ovotestes, bilateral gonadoblastomas and dysgerminomas, 46 XX/46, XY karyotype, and a successful pregnancy. Cancer 66:2668–2672

Teneda Y, Saito Y, Watanabe H (1985) Self or nonself discrimination in ascidians. Zool Sci 2: 433–442

Tippett PT (1983) Blood group chimeras. Vox Sanguinis 44:333–359

Tippett P (1984) Human chimeras. In: Le Douarin N, McLaren A (eds) Chimeras in developmental biology. Academic Press, London, pp 165–178

Tolar J Neglia JP (2003) Transplacental and other routes of cancer transmission between individuals. J Pediatric Hematol/Oncol 25:430–434

Weissman IL (2000) Stem Cells: Units of development, units of regeneration, and units in evolution. Cell 100:157–168

Weissman IL, Saito Y, Rinkevich B (1990) Allorecognition histocompatibility in a protochordate species: Is the relationship to MHC semantic or structural? Immunol Rev 113:227–241

Wilson DS, Sober E (1994) Reintroducing group selection to the human behavioral sciences. Behavioral Brain Sci 17:585–654

Yu N, Kruskall MS, Yunis JJ, Knoll JHM, Uhl L, Alosco S, Ohashi M, Clavijo O, Husain Z, Yunis EJ, Yunis JJ, Yunis EJ (2002) Disputed maternity leading to identification of tetragametic chimerism. New Engl J Med 346:1545–1552

Chapter 2
Putative Stem Cell Origins in Solitary Tunicates

Edwin L. Cooper

Abstract Stem cells are primordial cells in complex metazoans, inside a developing embryo (blastocyst) and are understood to be totipotent (capable of differentiating into every cell type). Hematopoietic stem cells (HSC) are characterized by their capacity of self-renewal, multi-lineage differentiation. Because solitary tunicates are protochordates, they are believed to share certain developmental features of vertebrates and as such they are excellent, inexpensive, non-controversial animal models, when searching for stem cell strategies that are not likely to cause ethical, political or moral concerns. In the pharyngeal region and in the body wall there are collections of cells that form nodules capable of responding to antigenic stimulation. Turning to another approach related to differentiation in nodules vs. circulating cells, lymphocyte-like cells (LLCs) proliferate m response to allogeneic stimuli. *In vitro* labeling of proliferative hemocytes (blood cells) revealed significantly greater proliferative activity among individuals immunized with allogeneic tissue in contrast to autogeneically primed and naive animals. Enhanced proliferation was restricted to discrete crypts of dividing cells within the body wall of recipients. Here, increased proliferative activity was specifically associated with LLCs. A discrete circulatory hemocyte population mediates the recognition of histocompatibility antigens in *Styela plicata*. These immunocompetent cells, which bear an overt morphological similarity to vertebrate lymphocytes, specifically infiltrate incompatible tissue transplants prior to obvious allograft rejection. Host lymphocyte-like cells (LLC's) accumulate around and within graft vascular tissue and appear to initiate rejection by means of cytotoxic activity. Yet to be tested are experiments that would show regeneration or transfer of immunological memory by removing these cells, characterizing their appearance with the aim of confirming whether they are terminally differentiated or may be cells with greater potential, i.e. protochordate stem cells.

E.L. Cooper (✉)
Laboratory of Comparative Neuroimmunology, Department of Neurobiology, David Geffen School of Medicine at UCLA, University of California, Los Angeles, CA 90095-1763, USA
e-mail: cooper@mednet.ucla.edu

B. Rinkevich, V. Matranga (eds.), *Stem Cells in Marine Organisms*,
DOI 10.1007/978-90-481-2767-2_2, © Springer Science+Business Media B.V. 2009

Keywords Putative · Stem cells · Origins · Solitary Tunicates · Lymphocyte Like Cells · Totipotent

2.1 A General View of Stem Cells from Embryos and Adults

Stem cells are primordial cells in the human body as well as the bodies of other complex metazoans. The first stem cells occur inside a developing embryo (blastocyst). These embryonic stem cells are understood to be totipotent (capable of differentiating into every cell type). First, they divide and then differentiate to ultimately construct an organism. Many diseases of humans that are currently incurable or unmanageable are associated with degeneration of specific cell types. These include but are not limited to: cancer, infectious diseases, heart disease, diabetes, neurodegenerative diseases, auto immune diseases, and skin disorders. Stem Cell Therapy (SCT) when fully developed will involve the introduction of healthy new stem cells to potentially repair and replace damaged or lost cells. This therapy, often referred to as Regenerative Medicine (RM), provides much promise for the treatment of what were previously regarded as incurable diseases. Understanding the evolutionary origin of stem cells or their precursors is essential for applicable universtal information and potential application. This approach has proven to be of immense value throughout the history of biology and its applications (Leri et al. 2008; Horsley et al. 2008; Wyss et al. 2008).

2.2 A Cursory View Concerned with the Evolution of Regeneration of Lost Parts

From the historical viewpoint stem cells have often been associated with regeneration especially in multicellular animals beginning with flatworms, notably aquatic free-living planaria (Cebrià 2007; Gurley et al. 2008; Rossi et al. 2008; Petersen and Reddion 2008). From this group to more complex vertebrates, we view certain better-known examples of regeneration especially the limbs of urodele amphibians or the salamanders and newts. Amphibians are instructive in this regard since there are drastic differences between the well-examined urodeles and the anurans (frogs and toads) (Kawakami et al. 2006; Endo et al. 2007)Whereas adult salamanders will regenerate almost any structure after amputation, only larvae not adult anurans show limited regeneration of lost limbs. Lacertilian reptiles, i.e., lizards can often regenerate a lost tail following an encounter with a prey – an escape mechanism (Rumping and Jayne 1996, Clause and Capaldi 2006).

It is in mammals where the capacity to regenerate, especially of whole organs, seems lost. Only in the case of bilateral organs such as the kidney will a contra lateral kidney increase in size after removing the other one – a form of compensatory hypertrophy (Lash et al. 2006).Epithelial cells in the integument, gastrointestinal tract and certain components of the reproductive system are regularly replaced. And in this instance the source of the replacing cells are the non-surface cells, those that are deeply embedded and that are stem cells. So popular, the mere mention of stem cells provides a convenient entry into the rather new world of speculation in which the technology and conceptual approaches often clash over regeneration (RM) and stem cells (SCT). The resulting controversy has been largely sparked by our attempts to repair damaged or effete cells in certain human diseases of the aged (Alzheimer's, Parkinson's) and this often engages polemics between varying groups involved in politics, ethics and religion. The resulting struggle often hampers objective scientific inquiry (Shapira 2006; Alikani 2007; Sadeghi 2007).

2.3 Is There a Likely Candidate in Evolution that Could be a Source of Putatively Homologous Stem Cells?

Marine animals figure prominently in our understanding of the evolution of immune responses (Cooper et al. 1992). Tunicate hemocytes (leukocytes) play essential roles in immune-defense mechanisms through active phagocytosis (Rowley 1981, 1982), allogeneic recognition in solitary tunicates, allogeneic discrimination during colony fusion of colonial species (Watanabe and Taneda 1982), and the opposite of incompatible or rejection response as revealed by discharge of their cytoplasmic components (Azumi et al. 1991). Despite a wealth of information on immune strategies, certain functional details and differentiation pathways of individual hemocytes have only been sparsely analyzed. For example, hemocyte classification remains to be established for many species, and precise details concerning regulation of hemopoiesis and hemocyte differentiation require more modern explanations that could be better clarified through the use of specific reagents some of the essential tools that foster advancement. As related to stem cells, both hemopoiesis and differentiation are important aspects for a more realistic comprehension of cellular immune responses as they relate to earlier stem cell origins.

2.4 Tunicate Immune Responses Aid in a Search for Homologous Immune Cells and Their Putative Equivalents and Stem Cells

Although related and still not resolved, Raftos and Cooper did crucial experiments that suggested stem cells in tunicates (Raftos and Cooper 1991). Tunicate pharyngeal cells include lymphocyte-like cells and granular amoebocytes. They are involved in specific allergenic and phagocytic reactions but little is known about their origin, regulation or control. A tunicate interleukin 1 (IL-1)-like fraction has been shown to stimulate cell proliferation *in vitro*. This fraction, designated tunicate IL-1β, has been isolated from tunicate hemolymph by gel filtration and chromatofocusing chromatography. Mitogenic responses to tunicate IL-1β appeared to be dose dependent and could be eliminated rapidly by removing tunicate IL-1β from the culture medium. A second tunicate hemolymph fraction seemed to have no effect on tunicate cell proliferation even though it exhibited IL-1-like activity in a mouse thymocyte proliferation assay. Phytohemagglutin (PHA) did not act synergistically with either fraction, suggesting specificity (Raftos et al. 1991).

Earlier Raftos and colleagues posed interesting questions concerning possible homology of organs, which if confirmed would help decipher stem cell origins (Raftos et al. 1990). They created closer linkages designed to support homology to the vertebrate thymus. Pharyngeal explants and circulatory hemocytes from the tunicate *Styela clava* were cultured in a medium containing tunicate plasma, artificial seawater, RPMI 1640, and antibiotics. Pharyngeal tissue remained viable and proliferated for up to 72 days *in vitro*. Proliferative activity maintained the pool of hemocytes within explants and facilitated the migration of pharyngeal hemocytes from explants into culture supernatants. The diversity of morphologically distinct cell types within the hemocyte pool of pharyngeal cultures indicated that cell division was followed by regulated differentiation. In marked contrast to the behavior of pharyngeal fragments in cultures, suspensions of circulatory hemocytes (some of which were pharyngeal derived) did not survive for prolonged periods *in vitro*. Thus proliferative activities not detected in cultures of circulatory hemocytes suggest that long-term viability must be dependent upon the integrity of whole organs. These results provide a wealth of approaches with respect to differentiation state of hemocytes within nodules or those that are circulating and the efficacy of culture conditions.

Turning to another approach related to differentiation in nodules vs. circulating cells, according to Raftos and Cooper, Lymphocyte-like hemocytes (LLCs) of solitary tunicates proliferate in response to allogeneic stimuli (Raftos and Cooper 1991). *In vitro* labeling of proliferative hemocytes from the solitary species *Styela clava* revealed significantly greater proliferative activity among individuals immunized with allogeneic tissue as opposed to autogeneically primed and naïve animals. Enhanced proliferation was restricted to discrete crypts of dividing cells within the body wall of recipients. Here, increased proliferative activity was specifically associated with LLCs. These data support previous results that implicated LLC activity with immunological memory that is evident in mediating allograft rejection. Hence, it is postulated that adaptive histoincompatibility responses in solitary tunicates depends upon the specific proliferation of immunocompetent cells. Whether these are derived from a particular lineage of stem cells is still an open question and now seems to be related to localization, i.e. nodular vs. circulatory cells.

2.5 Ancient Origins: Early Cells and Their Products

The present failure to detect molecular similarities in humoral and cellular defensive responses has led certain comparative immunologists to suggest that tunicates are incapable of vertebrate type adaptive immunological reactivity (Klein 1989). Models restricting immunological recognition to vertebrates ignore many studies demonstrating that invertebrates are capable of adaptive immune reactions essentially similar to those controlled by the Ig-related recognition molecules of vertebrates. In particular, analyses of tissue transplantation, as opposed to anti-pathogenic defense mechanisms (Monniot 1990), have shown that many invertebrates, including tunicates, possess a precise capacity to recognize and reject foreign tissue.

2.6 Histoincompatibility is Mediated by Lymphocyte-Like Cells

Invertebrate graft rejection systems often exhibit immunological memory and specificity, definitive characteristics of vertebrate type immune reactivity, (Raftos 1990a,b, 1991). Tissue transplantation studies have identified a sensitive histocompatibility system in the solitary tunicate, *Styela plicata*. Grafts taken from the outer tunic are actively rejected with concomitant alloimmune memory and significant immunological specificity. This response obeys the transplantation "laws" of mammalian-like histocompatibility systems suggesting the existence of specific anti-allogeneic receptors. Population genetic analysis also suggests that such receptors are directed against discrete histocompatibility antigens encoded by a single gene locus incorporating at least six independent alleles. Immunization studies have confirmed that these histocompatibility antigens have a cellular restriction and are not localized in the circulating plasma.

A discrete circulatory hemocyte population mediates the recognition of histocompatibility antigens in *Styela plicata*. These immunocompetent cells, which bear an overt morphological similarity to vertebrate lymphocytes, specifically infiltrate incompatible tissue transplants prior to obvious allograft rejection. Host lymphocyte-like cells (LLC's) accumulate around and within graft vascular tissue and appear to initiate rejection by means of cytotoxic activity (Raftos and Cooper 1991). Similar to the kinetics of graft rejection, LLC's respond specifically to second-set immunization that results in enhanced proliferation and migratory activity. This central role of LLC's in allograft rejection has been confirmed by the adoptive transfer of alloimmune memory using presensitized fractions specifically enriched in LLC's (Raftos 1996).

2.7 Long Term Viability in a Marine Model that Suggests Sites of Hemopoiesis and Stem Cells

Going still further with respect to nodular vs. circulating cell origins, according to Sawada and colleagues, in order to acquire more in-depth evidence for the site(s) of hemopoiesis in marine animals, they continued previous approaches and successfully cultured pharyngeal explants from *Styela clava* over a period of 82 days (Sawada et al. 1994). Transmission electron microscopy of resident hemocytes within explants revealed changes in hemocyte composition. Hemoblast-like cells suggestive of stem cells increased shortly after beginning cultivation (8.1% at Day 0 increasing to a maximum of 28.7% at Day 7). Autoradiography using ^{3}H-thymidine incorporation confirmed that hemocyte proliferation in pharyngeal explants still continued after 37 days culture.

During culture, the migration of many free cells into the medium resulted in sparse, residential hemocytes in the pharyngeal explants. Hemocyte migration increased by up to 4.3×10^5 cells/explant (max.) at days 17–24, but finally decreased to 4.9×10^4 cells/explant at 75×82 days. Vital neutral-red staining revealed that many emerging cells were not hemoctyes such as those found in normal hemolymph. These successes suggested that continued development of *in vitro* approaches would strengthen analyses of immune-defense responses as revealed in tunicates that, as protochordates, are the immediate invertebrate ancestors of vertebrates. Clearly this is an important conclusion with implications to stem cell origins and possible homology.

2.8 Changes in Hemocyte Composition of Explants as Revealed by TEM

Sawada and colleagues (Sawada et al. 1994) examined hemocyte population inside explants by TEM up to Day 56. Eosinophilic hemocytes were most abundant throughout the culture period. But differentiated gametocytes, such as eosinophilic and basophilic granulocytes, gradually decreased in number. In contrast, the proportion of hemoblasts, i.e. stem cells, increased after a short time following initiation of cultures (from 8.5% of all hemocytes at the beginning of culture up to 22.0% by the third day). Hyaline cells decreased, but phagocytic cells increased so that the sum of hyaline cells and phagocytic cells did not change significantly. After 40 days of culture, degenerating cells and others that could not be identified because of the irregularity of their morphology had increased significantly.

2.9 Hemopoietic Nodules Inside Explants

In addition to changes of hemocyte composition inside explants, morphological changes of hemocyte "nodules" became apparent after 30–40 days of culture. In the fresh pharynx (at the beginning of culture), hemoblasts (suggestive of stem cells) were present in the middle of hemocyte nodules, and were surrounded by many differentiated hemocytes such as granulocytes and hyaline cells with their typical cytoplasmic contents. Differentiated hemocytes comprising nodules began to decrease after 30–40 days of culture, paralleling increased numbers of solitary or small aggregates of hemoblasts not surrounded by differentiated hemocytes. After 37 days of culture, pharyngeal explants in another group were incubated in ^{3}H-thymidine, and proliferative cells within them examined by light microscopic autoradiography. Radioactivity was detected in only a few cells, indicating those cells, considered as proliferating that, specifically incorporated 3H-thymidine. Hemocytes accounted for 69.1% and epithelial cells for 30.9% of the radiopositive cells. Basophilic hemocytes constituted 91.6% and eosinophilic hemocytes 8.4% of the radiopositive hemocytes. The percentage of hemocytes that were radioactive was 1.6%. Identification of precise hemocyte types for all radiopositive hemocytes was not possible for basophilic cells, but >90% of them were small with a narrow cytoplasm.

2.10 Implications for Analyzing Hemopoietic Cells in Culture

To avoid contamination, culture conditions require further improvement, and the addition of cytokines or hemopoietic growth factors may be necessary to maintain normal hemopoiesis within explants for longer periods. Herein lies the relevance of understanding the precise role of cytokine- like factors (Beck et al. 1989; Raftos et al. 1991). In summary, hemocyte proliferation in tunicate pharyngeal explants for 5 weeks *in vitro* represents a major technical breakthrough. Without improved cell and molecular techniques, analysis of stem cells in relation to invertebrate immune systems will remain a classical morphofunctional analysis despite the need o probe deeper. Only then will we be able to extend our understanding of stem evolution in relation to competent immune cells. Of course seeking answers in tunicates seems to be of immediate relevance as ancestors of vertebrates (Cooper et al. 1992).

2.11 Tunicate LLC and Thymocytes Possess Lyt-1-2/3

Lymphocyte-like cells have been described in tunicates (Peddie and Smith 1995). They possess badges of identification, i.e. cell surface markers that identify them specifically. An Lyt-1 homolog in tunicate (protochordate) hemocytes has been demonstrated by using a panel of monoclonal antibodies specific to murine Lyt-l allotypic and framework determinants has been used (Negm et a1. 1991a,b). In immunoprecipitation experiments, antigenic activities are associated with a major 67 kD component on tunicate hemocytes and *C57BL6* mouse thymocytes. At the polypeptide level, putative tunicate and mouse Lyt-l molecules seem to be structurally similar and each may contain at least one intra-chain disulfide bridge. Binding assays and quantitative absorption experiments established the expression of Lyt-1 cross reacting determinants on a distinct population of tunicate hemocytes, known as hemoblasts and lymphocytes that suggest an early phylogenetic emergence of an Lyt-1 homolog.

Several lines of evidence suggest that the murine Lyt-2/3 molecular complex and its homologous Leu-2/T8 and MRC-OX8 antigens in humans and rats constitute together a set of conserved molecules that plays an important role in mammalian T cell functions. At a more fundamental level, genes encoding Lyt-2 and Leu-2/T8 are closely linked to the IgV locus in each species (Gottlieb 1974; Sukhatme et al. 1985a), and both express significant sequence homology to each other and to members of the Ig superfamily (Williams and Gagnon 1982; Mansour and Cooper 1984;

Williams 1984; Littman et al. 1985; Mansour et al. 1985; Nakauchi et al. 1985; Sukhatme et al. 1985b; Zamoyska et al. 1985; Kurosawa and Hashimoto 1996; Negm et a1. 1992). After finding Lyt-l, another study was initiated to identify a tunicate homolog to be murine Lyt-2/3 molecular complex, by using a monoclonal antibody specific to Lyt-2 framework determinants (mAb 53-6.7) (Negm et al. 1992). This antibody labeled a distinct population of tunicate hemocytes, as determined in indirect immunofluorescence and FACS analysis, and immunoprecipitated disulfide-bonded subunits from hemocytes equivalent to the 38 KD (α), 34 KD (δ) and 30 KD (β) subunits of murine Lyt-2/3 molecules. Observations suggest the phylogenetic emergence of a Lyt-213 homolog in tunicates. Along with previous work on Thy-1 (Mansour and Cooper 1984; Mansour et a1. 1985) and Lyt-1 glycoproteins (Negm et al. 1991a,b), it has been proposed that these antigens may display a high degree of conservation (Cooper and Parrinello 2001).

2.12 Perspectives on Regeneration and Stem Cells Nearly 50 Years Ago

According to Goss, the possible relationships between stem cells, differentiating cells, and mitosis have been reduced to two relatively simple models that are not fully explained (Goss 1964; Mitsiadis et al. 2007; Lin 2008). In view of the necessity of not depleting the reservoir of stem cells, it has often been proposed that primary cell divisions might be unequal in that one daughter cell would remain as an undifferentiated stem cell capable of undergoing more primary divisions, whereas the other would undergo differentiation (which might or might not be accompanied by mitoses). An alternative possibility would involve the differentiation of both daughter cells resulting from the replenishment of the sacrificed parent cells. Conceivably, this could be achieved by differentiation of new stem cells from an exogenous source or by occasional dedifferentiation of some of their descendants. Each of the above models is probably an oversimplification of the actual kinetics of proliferation. It is quite possible that stem cells may occasionally give rise by division to daughter cells with different fates, but they may at other times produce division products both of which might differentiate or both of which might remain stem cells. The important view is that on the average the rates of mitotic activity and cellular differentiation should remain in balance, and if the latter takes place at a constant rate, then growth must be governed by factors affecting cell division.

2.13 More on Regeneration, Whole Organisms in Relation to Stem Cells

Goss went further to suggest if there were no regeneration there could be no life (Goss 1969). If everything regenerated there would be no death. All organisms exist between these two extremes. Other things being equal, they tend toward the latter end of the spectrum, never quite achieving immortality because this would be incompatible with reproduction. It is by reproducing that organisms regenerate themselves and their species. At this level of organization everything regenerates as it occurs likewise at the molecular level. In between, some components regenerate whereas others do not. This is particularly true of whole fractions of the body and of such appendages as may produce from it. There are many kinds of regeneration. In the broadest sense regeneration is synonymous with reproduction, especially vegetative regeneration – in which case the prefix becomes redundant. Yet nothing new can be made without generating additional units of structure. To make a cell, new organelles must be produced: this in turn requires the synthesis of molecules.

All organisms, of course, can synthesize molecules. Without this constant biochemical turnover, cells could never adapt to the changing physiological demands of their existence. All living beings can also make new cells, some better than others. They can also heal wounds, renew tissues, indulge in compensatory hypertrophy, and reproduce. So important are these phenomena that they have undergone little or no decline in the course of evolution. What declined is the capacity to regenerate substantial parts of the body, or appendages. To achieve this requires more than just making more molecules and cells. Goss concluded, to be sure, these units must be multiplied, but they must also be put together the right way. Cells have to be assembled into tissues, and tissues into organs, because what is produced will be of no use unless it is organized into something that works.

2.14 A Cautionary Note on Possible Fate of Stem Cells

Finally a firm admonition seems appropriate and that concerns the somewhat sure view that stem cells will be welcomed after transplantation and eventual engraftment. These are microenvironments and they are the locales where certain deficits develop causing debilitating diseases in humans. Clearly on the road to normal development without any deficits cells destined to become certain structures usually do so due to being located in a totally new embryonic organism, in this case humans. This situation may not exist *in toto* in "terminally differentiated humans" and there is no guarantee that those highly plastic stem cells whether from embryos or autogeneically (self) from the same individual will upon transplantation to a diseased site be welcomed (O'Donoghue and Chan 2006; Cheung et al. 2006; Grünert et al. 2008; Yin et al. 2008). That microenvironment may still house the toxic or unfriendly milieu, a possible source of the original demise, could conceivably repel even autogeneic or self stem cells. How to make it work (successful engraftment, redifferentiation *qua* regeneration) for restoration and perhaps rejuvenated physiologic activity is the challenge. In fact knowing more about stem cells and regeneration in invertebrates and non-mammalian vertebrates will prove instructive. After all is this not in a sense a recapitulation of the history of biology, biomedicine and ultimate translation into the broader sphere of application to humans (Raftos 1991; Cooper and Nisbet-Brown 1993).

References

Alikani M (2007) The debate surrounding human embryonic stem cell research in the USA. Reprod Biomed Online 2:7–11

Azumi K, Yokozawa H, Ishii S (1991) Lipopolysaccharide induces release of a metallo-protease from hemocytes of the ascidian, Halocymthia roretzi. Dev Comp Immunol 15:1–7

Beck G, Vasta G, Marchalonis JJ, Habicht GS (1989) Characterization of interleukin-1 activity in tunicates. Comp Biochem Physiol 92B:93–98

Cebrià F (2007) Regenerating the central nervous system: how easy for planarians! Dev Genes Evol l217:733–48

Cheung AM, Kwong YL, Liang R, Leung AY (2006) Stem cell model of hematopoiesis. Curr Stem Cell Res Theories 1:305–315

Clause AR, Capaldi EA (2006) Caudal autotomy and regeneration in lizards. J Exp Zool 305: 965–973

Cooper EL, Rinkevich B, Uhlenbruck G, Valembois P (1992) Invertebrate immunity: another viewpoint. Scand J Immunol 35:247–266

Cooper EL, Nisbet-Brown E (1993) Developmental immunology. Oxford University Press, USA

Cooper EL, Parrinello N (2001) Immodefense. In: Sawada H, Yokosawa H, Lambert CC (eds) Tunicates: Cells and molecules, the biology of ascidians. Springer-Verlag, Tokio

Endo T, Yoshino J, Kado K, Tochinai S (2007) Brain regeneration in anuran amphibians. Dev Growth Diff 49:121–129

Goss RJ (1964) Adaptive growth. Basic Book, London

Goss RJ (1969) Principles of regeneration. Basic Book, London

Gottlieb PD (1974) Genetic correlation of a mouse light chain variable region marker with a thymocyte surface antigen. J Exp Med 140:1432–1437

Gurley KA, Rink JC, Alvarado AS (2008) β-catenin defines head versus tail identity during planerian regeneration and homeostasis. Science 319:323–327

Grünert M, Nurcombe V, Cool SM (2008) Stem cell fate decisions: the role of heparan sulfate in the control of autocrine and paracrine signals. Curr Stem Cell Res Theories 3:1–8

Horsley V, Aliprantis AO, Polak L, Glimcher LH (2008) Fuchs E.NFATc1 balances quiescence and proliferation of skin stem cells. Cell 2:299–310

Kawakami Y, Rodriguez Esteban C, Raya M, Kawakami H, Martí M, Dubova I, Izpisúa B (2006) Wnt/beta-catenin signaling regulates vertebrate limb regeneration. JC Genes Dev 20: 3232–3237

Klein J (1989) Are invertebrates capable of anticipatory immune responses? Scand J Immunol 29:499–505

Kurosawa Y, Hashimoto K (1996) The immunoglobulin superfamily; Where do invertebrates fit in? Adv Comp Environ Physiol 23:151–184

Lash LH, Putt DA, Zalups RK (2006) Influence of compensatory renal growth on susceptibility of primary cultures of renal cells to chemically induced injury. Toxicol Sci 94:417–427

Leri A, Kajstura J, Anversa P, Frishman WH (2008) Myocardial regeneration and stem cell repair. Curr Problems Cardiol 33:91–153

Lin H (2008) Cell biology of stem cells: an enigma of asymmetry and self-renewal. J Cell Biol 180:257–260

Littman DR, Thomas Y, Maddon PJ, Chess L, Axel R (1985) The isolation and sequence of the gene encoding T8: A molecule defining functional classes of T lymphocytes. Cell 40: 237–246

Mansour MH, Cooper EL (1984) Serological and partial molecular characterization of a Thy-l homolog in tunicates. Europ J Immunol 14:1031–1039

Mansour MH, DeLange R, Cooper EL (1985) Isolation, purification and amino acid composition of the tunicate hemocyte Thy-1 homolog. J Biol Chem 260:2681–2686

Mitsiadis TA, Barrandon O, Rochat A, Barrandon Y, De Bari C (2007) Stem cell niches in mammals. Exp Cell Res 313:3377–3385

Monniot C (1990) Diseases urochordata. In: Kinne O (ed) Diseases of marine animals III. Biologische Anstalt Helgoland, Hamburg
Nakauchi H, Nolan GP, Hsu C, Huang HS, Kavathas P, Herzenberg LA (1985) Molecular cloning of Lyt-2, a membrane glycoprotein marking a subset of mouse T lymphocytes: Molecular homology to its human counterpart, Leu 2/T8 and to immunoglobulin variable regions. Proc Natl Acad Sci 82:5126–5130
Negm HI, Mansour MH, Cooper EL (199la) Identification and structural characterization of Lyt-l glycoproteins from tunicate hemocytes and mouse thymocytes. Comp Biochem Physiol 99B:741–749
Negm HI, Mansour MH, Cooper EL (I991b) Serological characterization and partial purification of a Lyt-l homolog in tunicate hemocytes. Biol Cell 72:249–257
Negm HI, Mansour MH, Cooper EL (1992) Identification and structural characterization of a Lyt-2/3 homolog in tunicates. Comp Biochem Physiol 101B:55–67
O'Donoghue K, Chan J (2006) Human fetal mesenchymal stem cells. Curr Stem Cell Res Theories 1:371–386
Petersen CP, Reddien PW (2007) Smed-betacatenin-1 is required for anteroposterior blastema polarity in planarian regeneration. Science 319:327–330
Peddie CM, Smith VJ (1995) 'Lymphocyte-like' cells in ascidians: precursors for vertabrate lymphocytes? Fish Shellfish Immunol 5:613–629
Raftos DA (1990a) Cellular restriction of histocompatibility responses the solitary urochordate, Styela plicata. Dev Comp Immunol 15:241–249
Raftos DA (1990b) Morphology of integumentary allograft rejection in the solitary urochordate, Styela plicata. Cell Tissue Res 261:389–396
Raftos DA (1991) Development of primitive recognition systems in invertebrates. In: Cooper EL, Nisbet-Brown E (eds) Developmental immunobiology. Oxford University Press, New York
Raftos DA (1996) Histocompatibility reactions in invertebrates. In: Cooper EL (ed) Invertebrate Immune Responses: Cells and Molecular Products. Advances in Comparative & Environmental Physiology, vol. 23. Springer-Verlag Berlin Heidelberg
Raftos DA, Cooper EL (1991) Proliferation of lymphocyte-like cells from the solitary tunicate, Styela clava, in response to allogeneic stimuli. J Exp Zool 260:391–400
Raftos DA, Cooper EL, Habicht GS, Beck G (1991) Invertebrate cytokines: tunicate cell proliferation stimulated by an interleukin 1-like molecule. Proc Natl Acad Sci 88:9518–9522
Raftos DA, Stillman DL, Cooper EL (1990) In vitro culture of tissue from the tunicate Styela clava. In Vitro Cell Dev Biol 26:962–970
Rossi L, Salvetti A, Batistoni R, Deri P, Gremigni V (2008) Molecular and cellular basis of regeneration and tissue repair: Planarians, a tale of stem cells. Cell Mol Sci 65:16–23
Rowley AF (1981) The blood cells of the sea squirt, Ciona intestinalis: morphology, differential counts and in vitro phacocytic activity. J Invert Pathol 37:91–200
Rowley AF (1982) Ultrastructural and cytochemical studies on the blood cells of the sea squirt, Ciona intestinalis. I. Stem cells and amoebocytes. Cell Tissue Res 223:403–414
Rumping JM, Jayne BC (1996) Muscle activity in autotomized tails of a lizard (Gekko gecko): a naturally occurring spinal preparation. J Comp Physiol 179:525–538
Sadeghi M (2007) Islamic perspectives on human cloning. Hum Reprod Gen Ethics 13:32–40
Sawada T, Zhang J, Cooper EL (1994) Sustained viability and proliferation of hemocytes from the cultured pharynx of Styela clava. Mar Biol 119:597–603
Shapira A (2006) Law and bioethics in Israel: between liberal ethical values and Jewish religious norms. J Int Bioéthique 17:115–123, 165
Sukhatme VP, Sizer KC, Vollmer AC, Hunkapiller T, Parnes JR (1985a) The T cell differentiation antigen Leu-2/T8 is homologous to immunoglobulin and T cell receptor variable regions. Cell 40:591–597
Sukhatme VP, Vollmer AC, Erikson J, lsobe M, Crose C, Parnes JR (1985b) Gene for the human T cell differentiation antigen Leu-2/T8 is closely linked to the light chain locus on chromosome 2. J Exp Med 161:429–434

Watanabe H, Taneda Y (1982) Self or non-self recognition in compound ascidians. Am Zool 22:775–782

Williams AF (1984) The immunoglobulin superfamily take shape. Nature 308:12–13

Williams AF, Gagnon J (1982) Neuronal cell Tby-l glycoproteins: Homology with immunoglobulin. Science 216:696–703

Wyss BK, Meyers JL, Sinn AL, Cai S, Pollok KE, Goebel WS (2008) A novel competitive repopulation strategy to quantitate engraftment of ex vivo manipulated murine marrow cells in submyeloablated hosts. Exp Hematol 36(4):513–521

Yin L, Fu SL, Shi GY, Li Y, Jin JQ, Ma ZW, Lu PH (2008) Expression and Regulation of Major Histocompatibility Complex on Neural Stem Cells and Their Lineages. Stem Cells Dev 17(1):53–65

Zamoyska R, Vollmer AC, Sizer KC, Liaw CH, Parnes JR (1985) Two Lyt-2 polypeptides arise from a single gene by alternative splicing patterns of mRNA. Cell 43:153–163

Chapter 3
Cnidarian Interstitial Cells: The Dawn of Stem Cell Research

Uri Frank, Günter Plickert, and Werner A. Müller

Abstract The first stem cells described in the biological literature were those of hydroid cnidarians; their detection by Weismann in 1883 gave rise to his germ line and "germ plasm" theory (with "germ plasm" meaning what is called genome today). Somatic cells preserving properties of eggs (called Stammzellen, i.e. stem cells, by him) were considered by him to be the cellular source of regeneration and reproduction. Weismann's studies have been the foundation of modern cnidarian stem cell research. In the latter, hydroid stem cells have been referred to as interstitial cells (shortly i-cells), and have mostly been studied in two cnidarian genera: the freshwater polyp Hydra and the colonial marine hydroid Hydractinia. In these animals, i-cells constitute a complex system of multipotent (in Hydra) or totipotent (in Hydractinia) stem cells and their derivatives. I-cells' potencies have been investigated by specific elimination of stem cells and reintroduction of i-cells from donors. The complement of stem cells confers potential immortality to the genetic individual. Cnidarians' cells in general have an unmatched capability of re- and transdifferentiation. Isolated, fully differentiated striated muscle cells of hydroid medusae may resume features of multipotent stem cells and give rise to almost all cell types including germ cells. Reverse development of adult stages back into juveniles is a further manifestation of cnidarian developmental plasticity. Typical i-cells have not been described in other cnidarian groups. In these taxa the source of new nematocytes nerve and germ cells may be differentiated cells that preserve plasticity. Following a historical perspective we review recent advances in hydroid i-cell research, and discuss the potential of invertebrate stem cell work.

Keywords *Cnidaria* · *Hydra* · *Hydractinia* · *Podocoryne* · Germ cells · Germ plasm theory · Nematocytes · Nerve cells · Reverse development · Redifferentiation · Transdifferentiation

---

U. Frank (✉)
School of Natural Sciences and Martin Ryan Marine Science Institute
National University of Ireland, Galway, Ireland
e-mail: uri.frank@nuigalway.ie

B. Rinkevich, V. Matranga (eds.), *Stem Cells in Marine Organisms*,
DOI 10.1007/978-90-481-2767-2_3, © Springer Science+Business Media B.V. 2009

3.1 Introduction

Hydroid cnidarians were the first animals in which stem cells have been described. These cells, which today are referred to as interstitial cells, or shortly, i-cells, constitute a pool of migratory, undifferentiated cells, capable of self-renewal as well as of differentiation into specialized cell types. I-cells are lodged in interstitial spaces of epithelial cells (mostly ectoderm), that form their niche, and hence their name.

Surprisingly, the presence of i-cells, and stem cells altogether, in other cnidarians has been unclear (see Section 3.8). Even within the Hydrozoa most i-cell research has been conducted on only two genera, the solitary freshwater polyp, *Hydra*, and the colonial marine hydroid, *Hydractinia*. These studies (see refs below) revealed the complexity, developmental potencies and some of the molecular mechanisms that control i-cell fate. Although still in its infancy, i-cell research at the molecular level provides exciting perspectives for both basic as well as applied biomedical research. First evidences (e.g. Teo et al. 2006; Khalturin et al. 2007 suggest that many components of the genetic networks controlling i-cell fate are shared with vertebrate stem cell systems. With the rapidly growing cnidarian developmental biology community, and the newly established resources, such as full genome sequences, transgenic and knockdown technologies now available, it is anticipated that the coming years will bring great advances to the field.

In this chapter we review studies done on i-cells in some hydroid species. After a historical view, we discuss different aspects of i-cell biology. In particular we concentrate on morphology, staining properties, developmental potencies, gene expression, and the molecular mechanisms controlling their fate. Finally, we address the perspectives of i-cell research in basic biology and biomedicine.

3.2 Historical View

It was the physician and zoologist August Weismann (1834–1914) from the University of Freiburg (Germany) who first detected stem cells in the animal kingdom, and these were the interstitial stem cells in hydroids (Weismann 1883). The term "Stammzellen" (stem cells) in the sense of committed precursors with capacity to multiply appears in paragraphs dealing with the origin of germ cells in *Hydractinia echinata* (Weismann 1883, p. 82), and *Coryne pusilla* (Weismann 1883, p. 238). In *Coryne* "stem cells of the Glockenkern (the jellyfish's entocodon tissue) would migrate out into the endoderm, and would do this singly" (translations by WAM). These putative stem cells were thought to include the germ cell precursors. In a paragraph related to *Corydendrium* the term "interstitielle Zellen" (interstitial cells) refers to the precursors of nematocytes (Weismann 1883, p. 38).

Weismann's (1883) extensive investigations on the origin of germ cells in many colonial hydroids aimed at detecting the first identifiable "Stammzellen der Keimzellen" (stem cells of germ cells, p. 239) for which he coined the new term "Ur-Keimzellen" or "Urkeimzellen" (*primordial germ cells*; p. 230). He stated "histologically differentiated cells" never transform into germ cells but "only cells

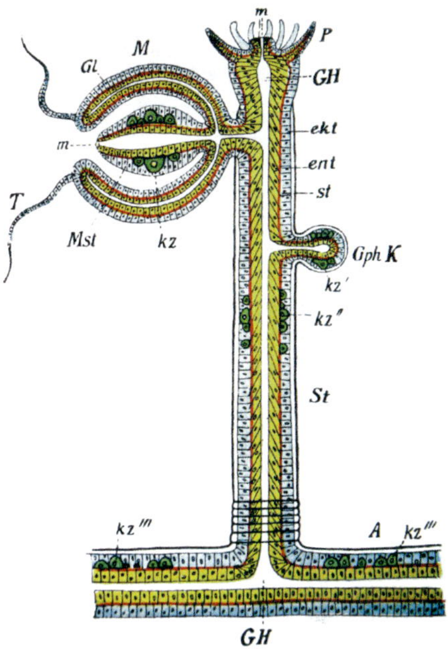

Fig. 3.1 Origin of germ cell precursors in a typified hydrozoan according to Weismann 1883, 1904. In terms of phylogeny, the origin of germ cell precursors (kz) is shifted from the manubrium (Mst) of the medusa (M) into the gonophore bud (Gph K) and further into the stalk (St) of the polyp and eventually into the stolon (A). GH gastral cavity. The drawing shown is reproduced from Weismann 1904, p. 94. In retrospect, one correction should be made: The precursors in the stolons are not strictly germ line-restricted but multipotent

having preserved *embryonic characteristics*". He described their presence in the blastostyles (gonozoids, sexual polyps/hydranths) and in the stolonal compartment of marine, colonial species such as *Coryne pulsilla*, *Eudendrium racemosum*, *Hydractinia echinata* and *Podocoryne carnea* (and many others). Weismann (1904) described the location of germ cell precursors at the base of epithelial cells contacting the basal lamina (mesoglea), their migration into budding medusae, and their final residence in the manubrium (feeding organ) of the medusae in *Eudendrium* and *Podocoryne*, or in the phylogenetic derivatives of medusae, the gonophores in *Hydractinia* (Fig. 3.1).

Weismann concluded that "there are distinct generations of cells, determined in advance, that undergo transformation into germ cells". For Weismann this observation was the incentive to infer the existence of a *particular germ line* separated from mortal somatic cell lineages in animals in general and to propose his influential but disputed "Keimplasmatheorie" (germ plasm theory, Weismann 1885, 1892a,b).

It may appear trivial that in multicellular organisms a cell line exists leading from the fertilized egg through series of cell divisions to new primordial germ cells. The question is when in ontogeny a particular cell line separates from cell lines leading to terminally differentiated somatic cell types such as nerve cells or muscle cells. Corroborating Weismann's expositions, recent reviews (Extavour and Akam 2003) and textbooks tell of two different modes of germ line separation: In some animals, represented by the classical model organisms *Drosophila*, *Caenorhabditis* and *Xenopus*, primordial germ cells are specified very early in embryogenesis by

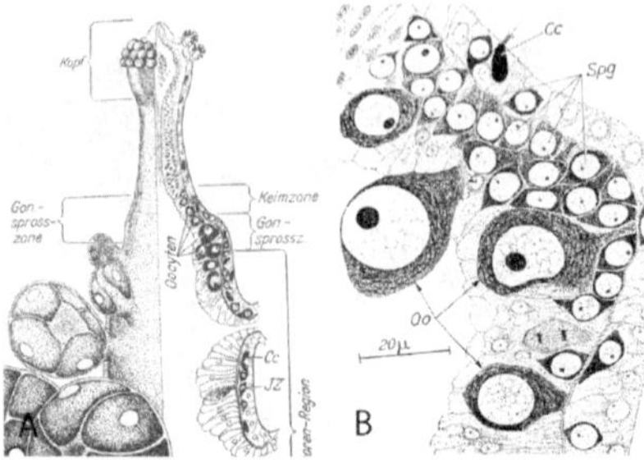

Fig. 3.2 Primordial germ cells in the sexual polyp. (**A**) Blastostyle (gonozoid, sexual polyp) in a colony of *Hydractinia echinata*. Keimzone = germ zone. Gon.-spross-zone = gonophore budding zone. Gonophores are the ball-shaped containers of the germ cells (here of eggs). In terms of phylogeny, they are reduced, sessile medusae. In light-induced spawning the wall of gonophores opens at a prespecified spot to release the mature eggs. Detail from a drawing by Müller 1964. (**B**) Intersexual germ zone of a blastostyle of *Hydractinia echinata*. The multiplying spermatogonia (Spg) displace the oocytes (Oo); therefore, oocytes do not mature and the polyp will funtionally be male. Cc = cnidocyte (nematocyte). From Müller 1964

maternal cytoplasmatic determinants stored in the egg and allocated during cleavage to one or few daughter cells. These cells are definitively programmed to become primordial germ cells and set aside for future use in the adult animal. In other animals, represented by the annelid *Platynereis* (Rebscher et al. 2007), by a parasitic arthropod barnacle (Shukalyuk et al. 2006), and the mouse embryo, specification of separate germ lines occurs later in the development (Extavour and Akam 2003), probably by inductive signals emitted by the surrounding tissues. It is ironic that the animals in which the existence of germ line-restricted stem cells was first proposed (i.e. hydroids) probably do not have an early committed germ line, but rather, germ cells differentiate continuously from multi- or totipotent stem cells that can also give rise to somatic cells.

Weismann hypothesized that in the evolution of the hydroids the time point of separation moved gradually back from the gonad of the medusa into to the polyp and, in colonial forms such as *Hydractinia*, into the stolons (as exposed below in Section 2.3, the origin of the multipotent or totipotent interstitial stem cells is even more shifted back into the gastrula). Weismann also described the transitory migration of the germ cell precursors from the ectoderm into the endoderm in the germ zone of the hydranth (polyp) that produces medusae or gonophores, and their eventual emigration from the endoderm to their final location (Figs. 3.1 and 3.2). As Weismann already stated, in other animals a separate germ line can be traced back to a few cells in the embryo. It was Weismann who coined the term "Polzellen" (pole cells) to designate the primordial germ cells in the early embryo of certain dipterans (later also found in *Drosophila*).

What determines cells to become germ cell precursors? In the model organisms mentioned before, i.e. *Drosophila*, *Caenorhabditis* and *Xenopus*, the determinants are called "germ plasm". This consists of non-vesicular aggregates of various types of RNAs including mitochondrial RNA in the cytoplasm, and/or of RNA bound to accumulations of fibrous material near the nucleus, known as nuage (French for cloud). The germ plasm also embodies transcripts of *vasa*. The term "Keimplasma" (germ plasm) was coined by Weismann (1885, 1892a,b, 1904) but the meaning of his term was very different from its present usage. In his theory, proposed before the dawn of classic genetics, germ plasm means the entirety of the putative "Vererbungssubstanzen" (hereditary substances), consisting of different material units, which he called "Iden". When chromosomes and their behavior in cell division and meiosis were described, Weismann did not hesitate to attribute his "Vererbungssubstanz" to these structures. He equated his Iden with chromosomes. Thus, his germ plasm is identical with the whole set of chromosomes. In his germ plasm theory he proposed that the cells of the germ line always are supplied with the entire set of chromosomes and thus with the whole complement of determinants, now known as genome.

However, in the further elaboration of his theory, Weismann made a wrong prediction. He thought cellular differentiation in the somatic cell lines would result from the discriminate distribution of the different "Iden" (chromosomes) among the various somatic cell precursors. "Biophores" would then export different information from the chromosomes of the nucleus into the cytoplasm. Asexual reproduction by budding and regeneration would be possible by cells, which inherit more chromosomes from the egg cells. This reserve supply of chromosomes would be silent until activated by external influences such as wounding. Although the idea of cells equipped with varying numbers of chromosomes soon turned out to be invalid (except for certain nematodes such as *Parascaris* and a few other invertebrates), the idea of multipotent stem cells preserving features of embryonic cells was born. From now on, stem cells with properties not much different from those of fertilized eggs were considered to be the source for regeneration and asexual reproduction.

Weismann himself applied this hypothesis to regeneration in hydroid polyps (and other animals). Stem cells distributed among the entire body would harbor a large complement of hereditary determinants enabling renewal and restoration. But only the germ line would retain the whole complement (Weismann 1904).

The origin of germ cells from interstitial stem cells was inferred by Weismann from the interpretation of histological preparations only and was described in more detail by a contemporary study (Bunting 1894). Experimentally, the origin of germ cells from i-cells was first shown by Hauenschild (1956, 1957) who transplanted i-cells into asexual clones of the medusa *Eleutheria dichotoma*. The recipients were enabled to produce germ cells. This pioneering study was followed by transplantation studies in *Hydractinia echinata*, a species of reference for Weismann (Müller 1964, 1967; Müller et al. 2004a). In the genus *Hydra* subpopulations of interstitial cells were shown to be stem cells with potencies limited to the germ line by introducing few stem cells in stem cell-depleted recipients (Littlefield and Bode 1986; Bosch and David 1987; Littlefield 1991; Nishimiya-Fujisawa and Sugiyama 1993). The experimental setup for eliminating i-cells is described below.

3.3 Hydroid Interstitial Stem Cells: Morphology, Cellular Properties, and Gene Expression

3.3.1 Staining and Morphology

In hydroids, interstitial stem cells have traditionally been identified by their morphology and staining properties. The i-cells of *Hydractinia echinata* (see Fig. 3.3

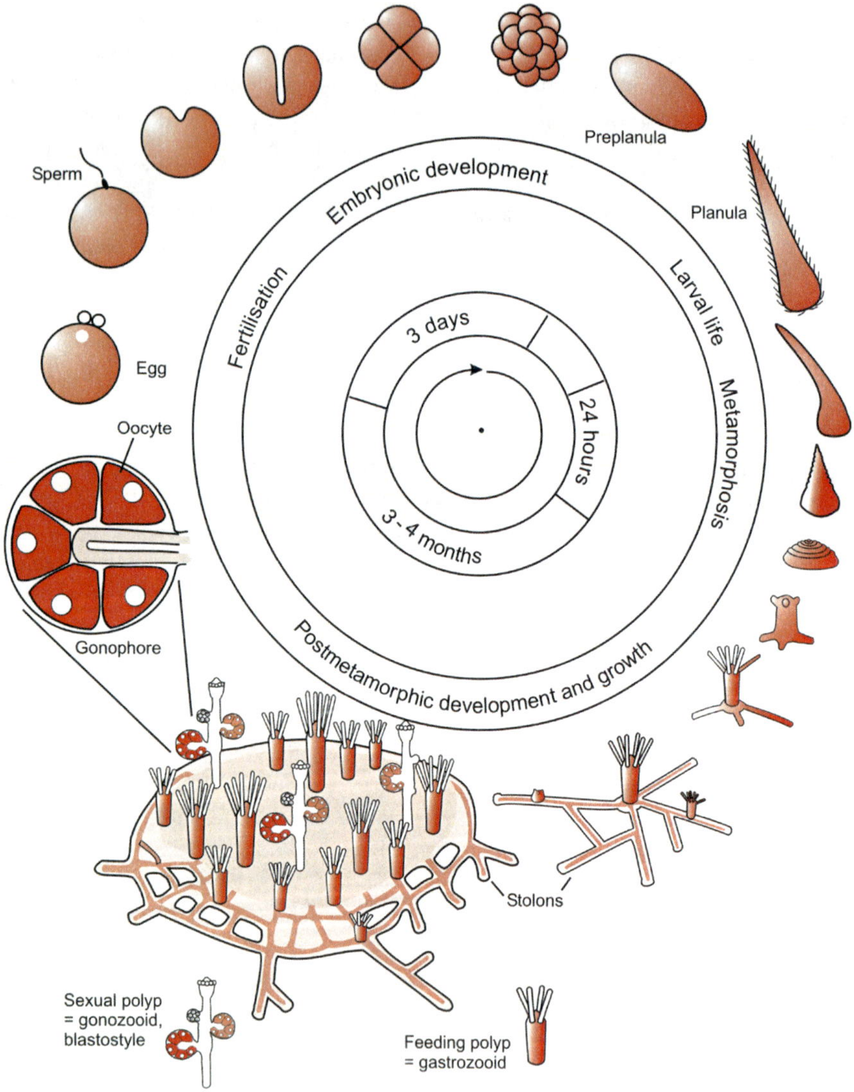

Fig. 3.3 Life cycle of *Hydractinia echinata*. From Müller: Entwicklungsbiologie, 4th ed., Springer Verlag Heidelberg New York, 2006

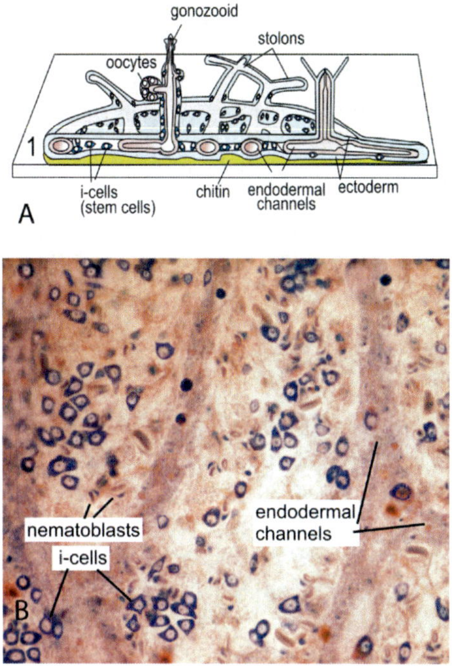

Fig. 3.4 Morphology and location of i-cells in *Hydractinia*. (**A**) Structure of the stolon plate of a mature colony with i-cells between two ectodermal epithelial layers. The upper ectodermal epithelium is translucent and allows detection of i-cells in whole mount preparations. (**B**) Detail of a stolon plate stained with blue basic dyes (May Grünwald + Giemsa). Source: Müller et al. (2004a)

for life cycle) are shown in Fig. 3.4. I-cells are small, rounded or spindle-shaped, and approximately 7–10 μm in diameter. Their nuclei are large, taking a significant proportion of the cells' volumes, and the chromatin is less dense than in differentiated somatic cells. By contrast, their cytoplasm is very dense and filled with ribosomes (Lentz 1965, 1966; Weis et al. 1985; Martin and Archer 1986, 1997). This high content of nucleic acids confers stainability with basic dyes. I-cells can be stained with toluidine blue or methylene blue and particularly set off from the surrounding tissue after a consecutive staining first with eosine-methylene blue solution after May-Grünwald followed by an azur-eosine-methylene blue staining after Giemsa (Müller 1964). Labeling the cytoplasm with acridine orange allows observing migrating cells for short periods of time. Over long periods and distances migrating cells can be traced by labeling cycling i-cells with BrdU and detection with anti-BrdU antibodies (Plickert and Kroiher 1988; Plickert et al. 1988; Kroiher et al. 1990; Müller et al. 2004a). Transgenic interstitial stem cells expressing GFP can be traced in living animals (Miljkovic et al. 2002; Wittlieb et al. 2006; Kalthurin et al. 2007; Plickert, unpubl).

3.3.2 Cell Cycle

Labeling with ^{3}H-thymidine enables measurement of the duration of the cell cycle and its phases. This has been carried out in *Hydra* (David and Campbell 1972; Campbell and David 1974; David and Gierer 1974; Campbell 1976; David and

Murphy 1977; David and MacWilliams 1978; David and Plotnick 1980; Herrmann and Berking 1987; Holstein and David 1990a,b; David et al. 1991; Holstein et al. 1991; also reviewed by David et al. 1987; Bode 1996; Bosch 2007a,b). I-cells are characterized by a short G1, an S-phase of 12 h, and a variable G2. The majority of the interstitial stem cells divide with an average cell cycle time of 24 h (Campbell and David 1974). In descendants, commitment to a particular differentiation pathway occurs at the S/G2 boundary. The pathways leading to terminally differentiated nematocytes and nerve cells take 3–5 days. In this time mitotically active epithelial cells undergo one round of division on average.

3.3.3 Ontogeny

In *Hydractinia* i-cells first appear in the inner cell mass of the gastrula (Fig. 3.5). Previous reports on other hydroids identified i-cells only later, in the endoderm of planula larvae (van de Vyver 1964, 1967; Weis et al. 1985; Martin and Archer 1986, 1997). The discrepancy probably arose from the densely packaging of cells in the gastrula's inner cell mass, rendering the identification of single cells in histological preparations difficult. New analyses with specific i-cell markers clearly show

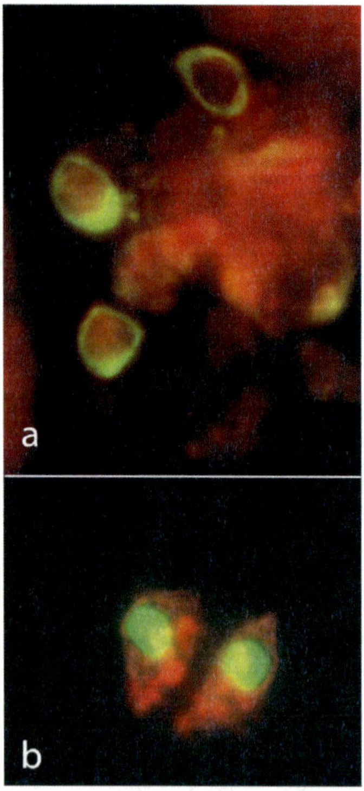

Fig. 3.5 I-cells isolated from a gastrula/preplanula. (**A**) stained with anti-cytoplasmic ß-catenin (**B**) stained with anti-nuclear ß-catenin. Source: Müller unpubl

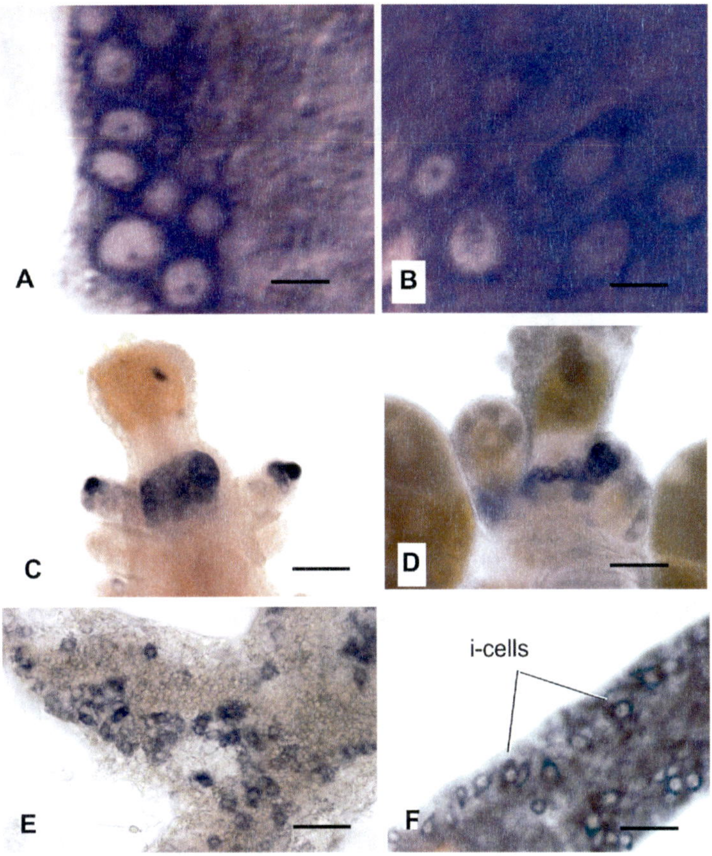

Fig. 3.6 *In situ* hybridization of genes specifically expressed in i-cells. (**A**) An *Oct4*-like POU domain gene, expressed in i-cells in stolons; (**B**) *PL10*, expressed in i-cells in stolons (**C**) *GalTSP*, a galectin expressed in oocyte precursors; (**D**) *HeGal3*, expressed in oocyte precursors; (**E**) *HyGal4*, a galectin, expressed in i-cells in stolons; (**F**) *frizzled*, the Wnt receptor, expressed in i-cells in stolons. Scale bars approx. 10 μm in (**A**) and (**B**), 60 μm in (**C**) and (**D**), 100 μm in (**E**), and 40 μm in (**F**). Source: Mali, Millane, Frank unpubl

these cells as early as the gastrula (Figs. 3.5 and 3.6). During larval development some of these cells cross the mesoglea and emigrate into the ectoderm. Here they give rise to larval-type nematocytes that, in *Hydractinia*, are used to anchor the larvae onto shells of hermit crabs (Müller and Leitz 2002), and to larval neurons. The latter enable phototactic behavior. A subset of neurosecretory cells confers competence to metamorphose (Schmich et al. 1998; Müller and Leitz 2002; Plickert et al. 2003). Stimulated by external cues the larva will undergo metamorphosis to a primary polyp, the founder of a new colony (Fig. 3.3).

During metamorphosis most i-cells migrate into the ectoderm of the developing stolons. In post-metamorphic development i-cells are predominantly found in the stolonal compartment, enclosed between two ectodermal epithelia (Fig. 3.4). These epithelia provide a meshwork of interstitial spaces through which the cells can migrate to populate new parts of a growing colony. Committed descendants of these stem cells, mainly precursors of nematocytes, nerve cells, germ cells and gland cells, emigrate from the stolonal tissue into newly emerging polyps. The fate of the precursors is sometimes morphologically indicated as follows: nematoblasts by their developing capsules, oogonia by their location in the endodermal germ zone of sexual polyps and their increasing size. Further evidence was provided by gene expression studies and by experimental inferences (see below). The i-cell stem cell niche, at least in *Hydractinia*, is thought to be at the interstices of stolonal ectodermal epithelia. Polyps, at least in older colonies, may only contain committed cells. These assumptions are partly based on the observation that most i-cells are indeed found in stolons, but they are also supported by some experimental data: While able to regenerate heads, isolated *Hydractinia* polyps mostly do not regenerate stolons (Müller et al. 1986), whereas stolons bud new polyps regularly. There is an exception to this rule in very young primary polyps (Müller et al. 1986). This could be interpreted as a lack of totipotent stem cells in polyps (young polyps may still contain stem cells).

Sex in *Hydractinia* is genetically determined (Hauenschild 1954). The occurrence of intersexual polyps harboring both oogonia and spermatogonia, but being functional males results from a particular genetic heritage (Hauenschild 1954) or from the fusion of histocompatible male and female sibling colonies of different sex (Müller 1964). Recent studies suggest that intersexuality may be more widespread than previously thought (unpublished, see below). The evolution of histoincompatibility in *Hydractinia* (and in colonial invertebrates in general) has been interpreted as a mean to avoid parasitism of an individual colony by allogeneic germ cell precursors immigrating from neighboring colonies (Buss 1982, 1983).

3.3.4 Gene Expression in Stem Cells and Derivatives

Many i-cells in *Hydra* and *Hydractinia* express the gene *vasa* (Mochizuki et al. 2001; Rebscher et al. 2008). In *Hydractinia* oocytes the protein is found in a perinuclear cloud (nuage). In mature oocytes the mRNA disappears but the protein remains present. In fertilized eggs Vasa is redistributed and located at the polar-body-forming pole where also transcripts of the Wnt pathway (*wnt* and *tcf*) are concentrated (Plickert et al. 2006). During cleavage Vasa protein is equally allocated to the first four blastomeres; no early specification and segregation of a germ line is evident, in line with the situation in other clonal invertebrates (Buss 1999). New *vasa* transcripts and Vasa protein appear in the gastrula and planula, in parallel to the appearance of new i-cells. In primary polyps and colonies derived from them, *vasa* is expressed by i-cells in stolons and gonozoids. This suggests that in *Hydractinia*, *vasa* expression characterizes multipotent or totipotent stem cells with the capability to give rise to somatic and germ cells in mature colonies. This is different from animals that sequester the germ line already during early embryonic development. In the latter, *vasa* is only expressed in cells restricted to the germ line. In *Hydractinia*, germ cell restriction of stem cell descendants presumably occurs continuously in the stolon plate of mature colonies or during the migration of i-cells into the gonozoids.

In eggs of the hydrozoan jellyfish *Podocoryne carnea* transcripts of the germ line and stem cell specific gene *Piwi* (called *Cniwi* in cnidarians) were identified (Seipel et al. 2004a). The expression of this gene marks stem cells in many other animals as well. In *Podocoryne*, *Cniwi* is upregulated during transdifferentiation of somatic cells into multipotent stem cells. Another interesting gene, expressed in *Hydractinia* oocytes, is *GalTSP*. This gene, which encodes a galectin with a number of TSP domains, is expressed in female germ cell precursors (Mali and Frank unpubl). Interestingly, in an *in situ* hybridization analysis of over 16 male colonies, a few cells expressing the gene were found in all colonies (Fig. 3.6). A related gene in *H. symbiolongicarpus*, a sibling species to *H. echinata*, seems to function in immunity and has a very different expression pattern (Schwarz et al. 2007).

In addition to germ line-specific genes, oocytes of *Hydractinia* contain mRNAs of genes of the canonical Wnt-signaling system (Müller et al. 2007; Plickert et al. 2006; Teo et al. 2006). These genes are also expressed in i-cells in mature animals. In an ongoing study, we (unpubl) have identified many i-cell specific genes. Of particular interest are orthologues of genes known to be stem cell specific in higher animals such as a POU domain gene resembling *oct4*, a *PL10*-like gene (unpubl), and more. Studies done on *Hydra* and *Hydractinia* identified many genes that mark early nematocytes and nerve cells (see below). The function of these genes is still unclear; however, they are very useful in mapping different, morphologically indistinguishable, sub-populations of i-cell in polyps and colonies.

Cells entering the pathway to neuronal differentiation express proneural genes with sequence similarity to genes of the *achaete-scute* complex (*ac-sc*) family of bHLH transcription factors in *Hydra* and *Podocoryne* (Grens et al. 1995; Seipel et al. 2004b). Nematocyte precursors, which are considered to represent a particular type of sensory neurons, also express *ac-sc* (Hayakawa et al. 2004). Many proneuronal genes are under the upstream control of the homeobox gene *cnox-2* (Miljkovic-Licina et al. 2007). The actual number of neuronal sub-populations is unknown. Of particular interest are genes that mark only a small fraction of neurons. One example is the *CTRN* gene (Cnidarian Tachylectin-Related gene in Neurons), expressed in only few neurons around the *Hydractinia* polyp mouth (Mali et al. 2006) and in their immediate i-cell precursors. It also marks early polyp buds on stolons. In contrast to the above genes, marking only a small neuronal fraction, *FMR1*, the *Hydractinia* homolgue of the fragile X mental retardation gene, is a broad neuronal marker (Guduric-Fuchs et al. 2004).

Specific marker genes for nematoblasts are genes that are activated when cells just enter this pathway of differentiation such as a homologue of *zic/odd-paired* gene (Lindgens et al. 2004) a *Dickkopf-3*-related gene (Fedders et al. 2004), or genes used to produce the capsule such as *spinalin* (Hellstern et al. 2006).

3.4 Developmental Potencies of i-cells

Like stem cells in other organisms and many cancer cells, cycling i-cells of hydroids are particular susceptible to agents that interfere with DNA replication and nucleic acid metabolism. This offers an opportunity to study the developmental potency of i-cells by selectively depleting animals of their i-cells and repopulate them with defined donor cells.

3.4.1 Studies on Hydra

Like in other hydroids, the complement of stem cells in *Hydra* confers potential immortality to the whole individual (genotype). All aged cells are replaced by newly produced cells. The contribution of the various stem cells to this process has been inferred from experimental interferences.

I-cells are lodged in the interstices mainly of the ectodermal epithelium, fewer numbers are found in interstices of the gastrodermis. In *Hydra* all i-cells can be eliminated by irradiation with x-rays (Strelin 1928; Brien and Reniers-Decoen 1950), or by treatment with nitrogen mustard (Diehl and Burnett 1964; Diehl 1973) or colchicine (Campbell 1976, 1979). The treated animals lose the ability to regenerate, apparently in support of the traditional notion that i-cells being the source of *Hydra*'s enormous capacity to regenerate (Strelin 1928; Brien and Reniers-Decoen 1950; Diehl and Burnett 1964; Diehl 1973). The evidence these early authors presented was considered invalid by later studies where harsh agents were replaced by the milder acting chemical hydroxy urea (e.g. Sacks and Davis 1979). With time, the treated animals become impoverished of nematocytes and nerve cells – which are derivatives of i-cells. Hence, treated hydras lose the capacity to catch prey and engulf food. Surprisingly these epithelial hydras survive if force-fed, continue to bud new polyps and restore the removed head by regenerating a new one (albeit with inferior quality; Campbell 1976, 1979; Marcum and Campbell 1978; Sacks and Davis 1979; Wanek et al. 1980; Bode et al. 1987; Bosch and David 1987; Holstein and David 1990a). Sf-1, a mutant strain of *Hydra magnipapillata* that loses its i-cells following a heat shock, displays similar deficiencies but retained regeneration capacity (Terada et al. 1988). Transplanting i-cells from wild type, untreated donors results in reappearance of all lost cell types.

Another experimental set up made use of the ability to dissociate hydras into single viable cell suspensions and the preparation of pellet aggregates by centrifugation (Bode et al. 1973). Aggregates prepared from i-cell-depleted animals were inoculated with few cells from healthy donors (David and Murphy 1977; Bosch and David 1986, 1987, 1990). The i-cells "cloned" in this way gave rise to various cell types or to uni-cultures, e.g. developing a single type of nematocytes or gametes, paralleling the clone-forming units in the mammalian hematopoietic system (see also David and Gierer 1974; Bode et al. 1976, 1987, 1990; Bode and Smith 1977; Bode and David 1978; David and MacWilliams 1978; Berking 1979, 1980; Sugiyama and

Fujisawa 1979; Fujisawa and Sugiyama 1980; Bosch and David 1984, 1987; David et al. 1987, 1991; Terada et al. 1988; Fujisawa 1989; Holstein and David 1990a,b; Bosch et al. 1991; Holstein et al. 1991; Zeretzke and Berking 2002). In a recent publication GFP-labeled donor cells were used to trace their migration (Khalturin et al. 2007).

Above results and older studies showing that cellular transitions from gastrodermis to ectodermis or back (reported by Haynes and Burnett 1963; Burnett et al. 1966, 1973; Davis 1970, 1973), were not reproduced unambiguously (Macklin 1968), have led to the conclusion (reviewed by Bode 1996; Bosch 2004, 2007a,b) that three separate compartments of cell lineages exist in the genus Hydra (Fig. 3.7): (1) ectodermal epithelial cells, (2) endodermal epithelial cells, and (3) the interstitial cell lineages. Both epithelial cell lines contain functional epithelio-muscular cells capable of proliferation and self-renewal (sometimes called "epithelial stem cells") and this enables epithelial hydras to regenerate and bud new polyps. No evidence for "true" epithelial stem cells exists, to the best of our knowledge, in the *Hydra* literature. By definition, stem cells are undifferentiated whereas *Hydra*'s "epithelial stem cells" are functional, differentiated cells displaying characteristics of epithelio-muscular cells and/or digestive cells. In well-fed animals the gastric region supplies supernumerary cells that are used to produce new animals by budding.

The third *Hydra* cell lineage comprises the interstitial cells. These cells are "true" stem cells, being undifferentiated. The majority of these cells are lodged in the ectodermal interstices, predominantly in the gastric region. A few are found in the head region; they are never found in the tentacles and peduncle. A type of large i-cells is considered to represent the stem cells proper (while the smaller ones are considered to be committed descendants; reviewed by Bode 1996). The interstitial stem cells are multipotent and give rise to four types of nematocytes, to sensory cells and ganglionic nerve cells, to endodermal gland cells, and to primordial germ cells (Fig. 3.7). The descendants migrate to their destinations in the terminal regions of the polyp, the hypostome with the tentacles (i.e., the head) or the basal disk (foot), and into buds. All hydra cells are displaced and move toward the oral and aboral pole, sloughing off at the tip of tentacles or at the basal disk.

Two features in the production of stem cell descendants are noteworthy: (1) Cells committed to become nematoblasts that undergo up to five rounds of division. The resulting two, four, eight, 16 or 32 nematoblasts adhere to each other by cytoplasmatic bridges, forming nests. Connection by cytoplasmic bridges enables synchronous differentiation. All nematoblasts of a nest form the same type of capsule. The type is determined at the beginning of the pathway (Shimizu and Bode 1995). Only shortly before the nematocytes reach full maturation do the cells separate and quickly migrate to their final sites in the tentacles (Kalthurin et al. 2007, and references therein). (2) Oocytes that engulf sister cells to increase their own size, a feature known from *Hydra* and some marine hydroids (*Cladocoryne*, *Pennarla*, *Tubularia*; Weismann 1883). In *Hydra*, oocytes are transitorily multinuclear, but only the original nucleus of the oocyte is thought to survive (Alexandrova et al. 2005).

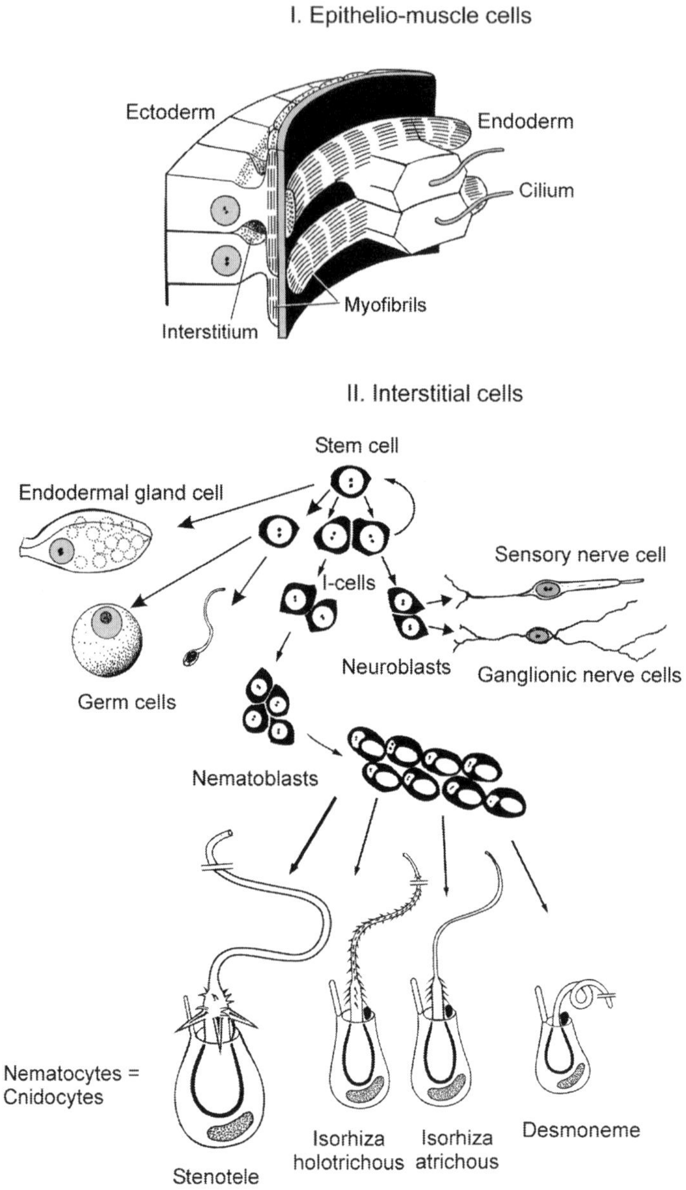

Fig. 3.7 Fixture of cells in a hydra. There are three compartments with self-renewal capabilities: (1) the ectodermal, (2) the endodermal epithelial cells, and (3) the interstitial cell lineages. Note: each nest of nematoblasts gives rise to only one type of nematocytes. After Müller WA, Developmental Biology, Springer New York 1996

I-cell descendants with potencies restricted to germ cell pathways were identified in several members of the genus *Hydra* (Bosch and David 1986; Nishimiya-Fujisawa and Sugiyama 1993). Two i-cell subpopulations, that were isolated from the phenotypic male strain *nem-1* of *Hydra magnipapillata*, were examined for their roles in sex determination. One i-cell subpopulation was restricted to the sperm differentiation pathway, and the other subpopulation restricted to the egg differentiation pathway. Hydras containing only sperm- or egg-restricted stem cells, but no other interstitial stem cell types were maintained by force-feeding for two years. Sex reversals occurred three times during this period. These observations suggest that strain *nem-1* (male) contains both sperm and egg-restricted i-cells. Differentiation of eggs, however, is normally suppressed, and only sperm are produced by the sperm-restricted stem cells. Evidence is presented which suggests that similar "phenotypic males", which normally only produce sperm but contain the stem cell types capable of differentiating into eggs, occur widely in *Hydra magnipapillata*.

Particular features of *Hydra*, probably shared with polyps of other hydroid polyps, are the plasticity of the differentiated state of the epithelial and nerve cells. All cells in *Hydra* are continuously moving from the middle of the body column upward or downward, to the extremities. While migrating, cells have to adopt the phenotypic features of the new region they enter. For example, the last epithelial cell in the tentacle-generating zone is still dividing while its neighbor across the border in the tentacle does no longer divide (Holstein et al. 1991). While crossing this border the cells abruptly express different molecular markers. In the lower body column ectodermal cells that cross the border to the peduncle undergo a conversion into gland cells.

Sensory and ganglionic nerve cells are displaced along the body column together with the epithelial cells. It has been observed that nerve cells in different locations express different neuropeptides. This spatial repertoire appears to be accomplished partly through the replacement of nerve cells by immigrating neuroblasts, and partly by transdifferentiation (Koizumi et al. 1988; Koizumi and Bode 1991).

3.4.2 Studies on Hydractinia

This colonial marine species offers many advantages compared to other cnidarians (Frank et al. 2001). Sexual reproduction, embryonic development from eggs to planula larvae and their metamorphosis into primary polyps, and asexual cloning of colonies can routinely be performed in the laboratory. Mutant strains are available (Müller 2002) and were used as stem cell donors for i-cell-depleted wild-type colonies. I-cells were eliminated using alkylating agents such as mitomycin C (Müller 1966, 1967). Like in *Hydra*, mild treatments (i.e. low mitomycin C concentration) yield i-cells, nematocytes and germ cells free colonies, that are able to bud new (epithelial) polyps. To unravel the complete potency of migratory stem cells, a harsh treatment (i.e. high mitomycin C concentration) was used to deprive also epithelial cells of their capability of self-renewal, and in the long term (about two to three weeks) causing their death (Müller 1964; Müller et al. 2004a). The dying colonies were used as scaffolding for transplanted i-cells from untreated, BrdU-labeled mutant colonies. The donor colonies differed from the recipient also by sex. The transplanted i-cells migrated into all areas of the dying colonies and differentiated into all cell types of the animal as indicated by the mutant phenotype of the regenerated colonies, BrdU-containing nuclei of epithelial cells, showing that they originated from donor i-cells. The recipient also underwent sex reversal following i-cell transplantation showing again the origin of regeneration (Müller et al. 2004a). The sequence of cell replacement is shown in Fig. 3.8. It is concluded that in

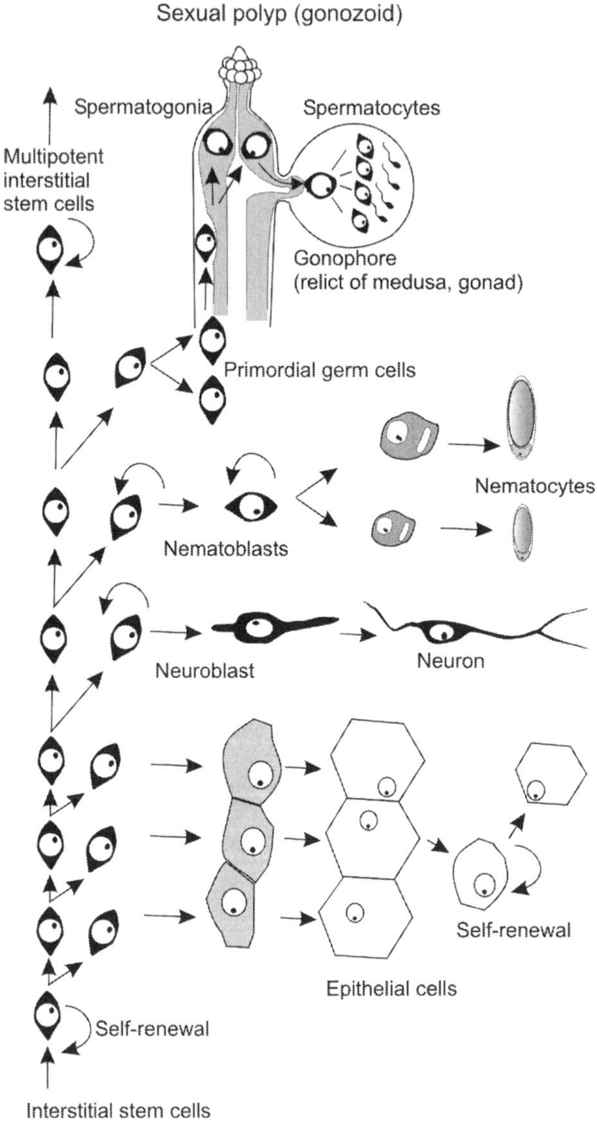

Fig. 3.8 Sequence of cell replacent by immigrating i-cells introduced into stem cell-depleted colonies of *Hydractinia*. First epithelial cells are replaced, finally new germ cells are produced. From Müller 1967, modified

Hydractinia i-cells are capable to contribute to all cell types of the animal including epithelia. This does in no way rule out the existence of descendant interstitial cells with more restricted potencies and the immigration of committed descendants with limited abilities to proliferate, namely nematoblasts, neuroblasts and germ cell precursors.

3.5 Signals Controlling Self-Renewal and Differentiation

The signals controlling self-renewal and differentiation in hydroids have been subjected to speculations and are poorly defined. Key words used to indicate such signals are terms like "positional information" and "feed back signals" (Bode and Smith 1977; Bode and David 1978; Fujisawa et al. 1986, 1990; David et al. 1987, 1991; Fujisawa 1989; Bosch et al. 1991; Teragawa and Bode 1991; cited in Khalturin et al. 2007). An inhibitor suppressing the formation of stenoteles, a type of nematocyte in hydra, is interpreted as a feedback-signaling molecule regulating the density of this cell type (Fujisawa 1988). Oligopeptides have been suggested to control differentiation. For example, a role in regulating neuron differentiation has been attributed to the peptide Hym-355 (Takahashi et al. 2000; further references therein and in Bode 1996). However, in many cultured mammalian neuroblasts or embryonic stem cells, withdrawal of serum with its mitogenic factors stimulates neuronal differentiation (Howard et al. 1993; Wang et al. 2004; Zhang et al. 2005; Shigeta et al. 2006). Unspecific blocking of receptors for mitogenic signals is expected to induce similar effects and there is no easy way to discriminate between stimulation of differentiation by positive signals or by unspecific prevention of further proliferation. Remarkably, i-cells of *Hydractinia* express a gene for a serum-response factor and a role for this factor in stem cell decision-making has been postulated (Hoffmann and Kroiher 2001).

The Wnt pathway plays a major role in directing i-cells out of self-renewal. Wnt plays a crucial role also in axis formation and patterning of the *Hydra* polyp (Hobmayer et al. 2000; Broun et al. 2005) and in embryos and polyps of *Hydractinia* (Müller et al. 2004b, 2007; Plickert et al. 2006). In *Hydractinia*, a subtype of interstitial cells, probably cells already committed to their final fate (putative nematoblasts, neuroblasts and oogonia) express *wnt* and the Wnt receptor *frizzled* (Teo et al. 2006; Müller et al. 2007). Experimental interference was done by treating the animals with pulses of paullones, specific inhibitors of GSK-3β. Blockage of GSK-3β is expected to activate Wnt target genes. The treated colonies produced huge numbers of nematocytes and the polyps developed enlarged heads with high numbers of nerve cells (Fig. 3.9). It should be pointed out that this stimulation of nematocyte and nerve cell production occurs only in a small range of optimal concentrations. With respect to nematocyte production, similar results were recently described for hydra (Kalthurin et al. 2007). Evidence has also been presented for the involvement of the Notch signaling system in the control of the stem cell fate (Kaesbauer et al. 2007).

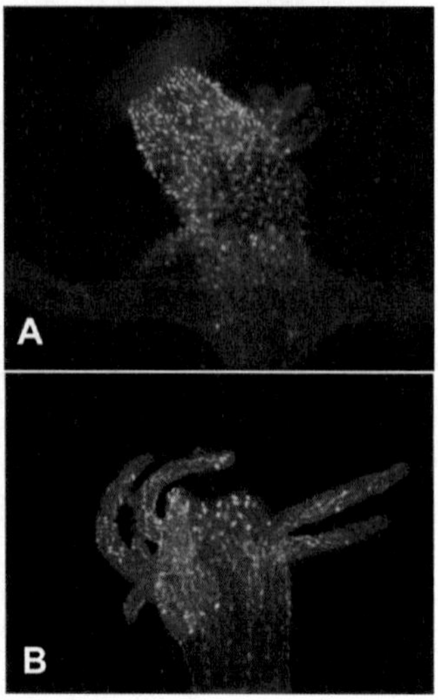

Fig. 3.9 Head of polyps of *Hydractinia* with RFamide-positive nerve cells. (**A**) alsterpaullone treated; (**B**) control animal

3.6 Redifferentiation, Transdifferentiation, and Reverse Development in Hydroids

Isolated mononucleated muscle cells of hydroid medusae may transdifferentiate *in vitro* to new cell types and even form a complex regenerate (Schmid et al. 1982, 1993; Alder and Schmid 1987; Schmid and Plickert 1990; Schmid and Reber-Müller 1995; Yanze et al. 1999). Transdifferentiation events follow a strict pattern, starting with cell type that resembles smooth muscle cells without a preceding DNA replication. This cell type behaves like a stem cell and produces all other cell types of the animal, including new interstitial stem cells and their derivatives, nematocytes, nerve cells and germ cells, and also new epithelial cells. Some preparations developed an inner and an outer layer separated by a basal lamina eventually resulting in a complete manubrium, the feeding organ of a medusa (Fig. 3.10).

Another manifestation of cnidarian developmental plasticity is the ability to "reverse develop" (Schmich et al. 2007, and references therein). Some cnidarians are unique in the animal kingdom because of their unequalled potential to undergo reverse development, such as cases of medusae buds or free swimming medusa that undergo reverse development into the polyp stage (e.g. in *Podocoryne carnea*) or isolated tentacles in *Aurelia* that can convert into metamorphosis-competent planula larvae (Lesh-Laurie et al. 1991). These examples and others of reverse development are stress-induced. However, in several cnidarian species reverse development is part of the normal life cycle: *Stephanoscyphus planulophorus* is a member of the scyphozoan group Coronata. The animals lack germ cells. Instead young medusae, known as ephyrae, transform into planula larvae that metamorphose into polyps (Werner and Hentschel 1983). Likewise, in the scyphozoan *Cassiopea andromeda*

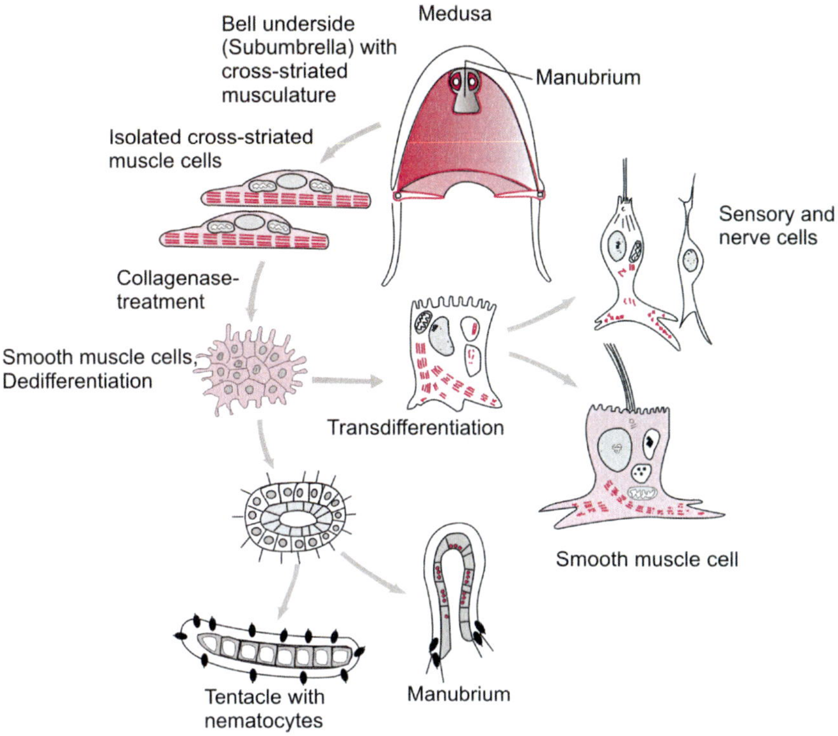

Fig. 3.10 Transdifferentiation pathways of striated muscle cells isolated from medusae. Copies of figures originally drawn by Volker Schmid, redrawn and modified

asexually produced buds look and behave like sexually produced planula larvae. Sexual and asexual borne larvae undergo metamorphosis into polyps, induced by the same external cues (Hofmann et al. 1978; Neumann et al. 1980; Thieme and Hofmann 2003). Reverse development exemplifies an unusual high plasticity of the differentiated state of these organisms.

It is reasonable to hypothesize that in cnidarians, which apparently lack morphologically identifiable interstitial stem cells (e.g. Anthozoa), terminally differentiated cells such as nematocytes, nerve cells and germ cells may arise through transdifferentiation of epithelial or other somatic cells having preserved high plasticity (Aochi and Kato 1980; Minasian and Mariscal 1980). Gene expression studies with marker genes such as *vasa* and *nanos* for the germ line (Extavour et al. 2005; Torras and Gonzalez-Crespo 2005), and other stem cell markers, may help to trace cell lineages in these animals. The unmatched reversibility of the differentiated state may help to identify genes and factors controlling reversal to stem cell characteristics.

Finally, dedifferentiation and transdifferentiation of fully differentiated epithelial cells, postulated in old literature (Davis 1970, 1973; Haynes and Burnett 1963; Lowell and Burnett 1969), is a topic recently again proposed for *Hydra*: During head regeneration an early dedifferentiation of digestive cells into blastema-like cells was described (Galliot et al. 2006).

3.7 Stem Cells in Other Cnidarians

Typical i-cells as described here are only known from Hydrozoa. Old textbooks occasionally speak of interstitial or amoeboid cells in Octocorallia (e.g. Kükenthal 1923) but a role as stem cells has not yet been shown or attributed to them. Drawings in old literature point to epithelial cells as source for nematocytes. The dividing precursors of nerve cells and gametes are not identified so far.

With the emergence of the sea anemone, *Nematostella vectensis*, as a new cnidarian model organism it is to hope that new research will address this issue. Indeed, anti-Vasa antibodies recognize i-cell-like cells in *Nematostella* (Fig. 3.7 in Extavour et al. 2005). *Nematostella* is an anthozoan. The class Anthozoa is considered by many Cnidaria researchers to represent the basal group within the Cnidaria. If stem cells and the mechanisms that control their fates are indeed conserved between hydroids and bilaterians, than it would be likely to find i-cells also in sea anemones.

3.8 Perspectives

Stem cell research is at the cutting edge of current research in biology and biomedicine. This is not only due to the basic interest of developmental and cell biologists, but also, and perhaps mainly, because of the great potential of stem cells in regenerative medicine. Mammalian stem cells can be cultured in defined media *in vitro* for prolonged periods of time, during which they remain in an undifferentiated state. It is also possible to manipulate them, instigating them to differentiate into specialized cell types. Mammalian stem cells can roughly be divided into two main types: embryonic stem cells, and adult (or tissue) stem cells. Embryonic stem cells are derived from the inner cell mass of the blastocyst (the mammalian blastula). They are pluripotent, i.e. able to give rise to any cell type of the embryo proper, but not to all extra-embryonic tissues (they do not contribute to the trophectoderm). Adult stem cells are more restricted in their developmental potencies, generally able to differentiate into a single cell type (= unipotent, e.g. skin stem cell) or several cell types (= multipotent, e.g. the hematopoietic stem cell). The term totipotent in vertebrates refers to the ability to contribute also to all extra-embryonic tissues and this ability is restricted to the fertilized egg and the blastomeres resulting from the first few cleavages. At present these cells are not culturable *in vitro*. In most invertebrates, in contrast, pluripotency equals totipotency due to the lack of extra-embryonic tissues.

Work on invertebrate stem cells could circumvent several problems associated with working on mammalian (especially human) cells. First, work on invertebrate cells is simpler in technology and, hence, cheaper. Hydroid i-cells, for example, can be observed and manipulated *in vivo* in the small, translucent animal without the need for costly cell culture technology (the animals grow in seawater). Furthermore, cell behavior *in vivo* is more likely to reflect "real life" than cells growing in a culture dish. Second, invertebrate stem cell research is interesting from a comparative, evolutionary, perspective (Yanze et al. 1999; Galliot and Schmid 2002; Holstein et al. 2003; Bosch 2004, 2007b). Finally, working on invertebrates minimizes the ethical issues associated with working on vertebrate systems, especially human embryonic stem cells. Clearly, invertebrate stem cell research cannot substitute for work on human cells. It can, however, complement the latter by enabling us to identify universal stem cell genes, both the genes that control differentiation, and those that maintain "stemness" (see Okita et al. 2007; Wernig et al. 2007; Yu et al. 2007). Many of both groups of genes will be shared with mammalian systems, but identifying them in invertebrates would be faster, cheaper, and free of ethical issues.

Acknowledgements We thank Peter Schuchert (Muséum d'histoire naturelle, Geneva) and Klaus Sander (University of Freiburg) for providing us with copies of Weissmann's 1883 work, which had been considered lost for decades. UF is supported by SFI (Science Foundation Ireland), by a Beufort Marine Biodiscovery Grant, and by the Irish Higher Education Authority's PRTLI Cycle 4 Award.

References

Alder H, Schmid V (1987) Cell cycles and in vitro transdifferentiation and regeneration of isolated, striated muscle of jellyfish. Dev Biol 124:358–369

Alexandrova O, Schade M, Bottger A, David CN (2005) Oogenesis in *Hydra*: nurse cells transfer cytoplasm directly to the growing oocyte. Dev Biol 28:91–101

Aochi M, Kato KI (1980) The polyp reconstitution in ectoderm and-or endoderm isolated from *Arelia aurita*. Dev Growth Diff 22:717

Berking S (1979) Control of nerve cell formation from multipotent stem cells in *Hydra*. J Cell Sci 4:193–205

Berking S (1980) Commitment of stem cells to nerve cells and migration of nerve cell precursors in preparatory bud development in *Hydra*. J Embryol Exp Morphol 60:373–387

Bode HR (1996) The interstitial cell lineage of hydra: a stem cell system that arose early in evolution. J Cell Sci 109:1155–1164

Bode HR, Berking S, David CN, Gierer A, Schaller H, Trenkner E (1973) Quantitative analysis of cell types during growth and morphogenesis in *Hydra*. Wilhelm Roux' Arch 171:269–285

Bode HR, David CN (1978) Regulation of a multipotent stem cell, the interstitial cell of hydra. Progr Biophys Molec Biol 33:189–206

Bode HR, Flick KM, Smith GS (1976) Regulation of interstitial cell differentiation in *Hydra attenuata*. I. Homeostatic control of interstitial cell population size. J Cell Sci 20:29–46

Bode HR, Gee LW, Cow MA (1990) Neuron differentiation in hydra involves dividing intermediates. Dev Biol 139:231–334

Bode HR, Heimfield S, Chow MA, Huang LW (1987) Gland cells arise by differentiation from interstitial cells in *Hydra attenuata*. Dev Biol 122:577–585

Bode HR, Smith GS (1977) Regulation of interstitial cell differentiation in *Hydra attenuata*. II. Correlation of the axial position of interstitial cells with nematocyte differentiation. Wilhelm Roux' Arch 181:203–213

Bosch TCG (2004) Control of asymmetric cell divisions: will cnidarians provide an answer? Bioessays 26:929–931

Bosch TCG (2007a) Symmetry breaking in stem cells of the basal metazoan *Hydra*. In: Macieira-Coelho (ed) Asymmetric cell division series, Progr Mol Subcell Biol. Springer, Heidelberg, pp 61–78
Bosch TCG (2007b) Why polyps regenerate and we don't: towards a cellular and molecular framework for Hydra regeneration. Dev Biol 303:421–433
Bosch TCG, David CN (1984) Growth regulation in *Hydra*: Relationship between epithelial cell cycle length and growth rate. Dev Biol 104:161–171
Bosch TCG, David CN (1986) Male and female stem cells and sex reversal in *Hydra*polyps. Proc Natl Acad Sci USA 83:9478–9482
Bosch TCG, David CN (1987) Stem cells of *Hydra magnipapillata*can differentiate somatic and germ cell lines. Dev Biol 121:182–191
Bosch TCG, David CN (1990) Cloned interstitial stem cells grow as contiguous patches in hydra. Dev Biol 138:513–515
Bosch TCG, Rollbühler R, Scheider B, David CN (1991) Role of the cellular environment in interstitial stem cell proliferation in hydra. Dev Genes Evol 200:269–276
Brien P, Reniers-Decoen M (1950) La significance des cellules interstitielles des hydres d'eau douce et le problème de la résèrve embryonaire. Bulletin Biol France Belgique 89:285–325
Broun M, Gee L, Reinhardt B, Bode HR (2005) Formation of the head organizer in hydra involves the canonical wnt pathway. Development 132:2907–2916
Bunting M (1894) The origin of sex cells in *Hydractinia*and *Podocoryne*and the development of *Hydractinia*. J Morphol 9:203–236
Burnett AL, Davis LE, Ruffin FE (1966) A histological and ultrastructural study of germinal differentiation of interstitial cells arising from gland cells in *Hydra viridis*. J Morphol 120:1–8
Burnett AL, Lowell R, Cyrlin M (1973) Regeneration of a complete hydra from a single differentiated somatic cell type. In: Burnett AL (ed) Biology of Hydra. Academic Press, New York, pp 225–270
Buss LW (1982) Somatic cell parasitism and the evolution of somatic tissue compatibility. Proc Natl Acad Sci USA 79:5337–5341
Buss LW (1983) Evolution, development, and the units of selection. Proc Natl Acad Sci USA 80:1387–1391
Buss LW (1999) Slime molds, ascidians, and the utility of evolutionary theory. Proc Natl Acad Sci USA 96:8801–8803
Campbell RD (1976) Elimination of hydra interstitial and nerve cells by means of colchicine. J Cell Sci 21:1–13
Campbell RD (1979) Development of hydra lacking interstitial and nerve cells ("epithelial hydra"). In: Subtelny S, Konigsberg IR (eds) Determinants of Spatial Organization. 37th Symp Soc Dev Biol. Academic Press, New York, pp 267–293
Campbell RD, David CN (1974) Cell cycle kinetics and development of *Hydra attenuata*. II. Interstitial cells. J Cell Sci 16:349–358
David CN, Bosch TCG, Hobmeyer B, Holstein T, Schmidt T (1987) Interstitial stem cells in hydra. In: Loomis WF (ed) Genetic control of development. Alan R Liss, New York, p 389
David CN, Campbell RD (1972) Cell cycle kinetics and development of *Hydra attenuata*. I. Epithelial cells. J Cell Sci 11:557–568
David CN, Fujisawa T, Bosch TC (1991) Interstitial stem cell proliferation in hydra: evidence for strain-specific regulatory signals. Dev Biol 148:501–507
David CN, Gierer A (1974) Cell cycle kinetics and development of *Hydra attenuata*. J Cell Sci 16:359–376
David CN, MacWilliams H (1978) Regulation of the self-renewal probability in hydra stem cell clones. Proc Natl Acad Sci USA 75:886–890
David CN, Murphy S (1977) Characterization of interstitial stem cells in hydra by cloning. Dev Biol 58:372–383
David CN, Plotnick I (1980) Distribution of interstitial cells in *Hydra attenuata*. Dev Biol 76: 175–184

Davis LE (1970) Cell division during dedifferentiation and redifferentiation in regenerating isolated gastrodermis of *Hydra*. Exp Cell Res 60:127–132

Davis LE (1973) Ultrastructural changes during dedifferentiation and redifferentiation in regenerating, isolated gastrodermis. In: Burnett AL (ed) Biology of Hydra. Academic Press, New York, pp 171–219

Diehl FA (1973) The developmental significance of interstitial cells during regeneration and budding. In: Burnett AL (ed) Biology of Hydra. Academic Press, New York, pp 109–141

Diehl FA, Burnett AL (1964) The role of interstitial cells in the maintenance of hydra. I. Specific destruction of interstitial cells in normal, asexual, non-budding animals. J Exp Zool 155: 253–260

Extavour CG, Akam M (2003) Mechanisms of germ cell specification across the metazoans: epigenesis and preformation. Development 130:5869–5884

Extavour CG, Pang K, Matus DQ, Martindale MQ (2005) Vasa and nanos expression patterns in a sea anemone and the evolution of bilaterian germ cell specification mechanisms. Evol Dev 7:201–215

Fedders H, Augustin R, Bosch TC (2004) A Dickkopf-3-related gene is expressed in differentiating nematocytes in the basal metazoan Hydra. Dev Genes Evol 214:72–80

Frank U, Leitz T, Müller WA (2001) The hydroid *Hydractinia*: a versatile, informative cnidarian representative. Bioessays 23:963–971

Fujisawa T (1988) Inhibition of stenotele commitment by an endogenous factor in Hydra. J Cell Sci 91, 361–366

Fujisawa T (1989) Role of interstitial cell migration in generating position-dependent patterns of nerve cell differentiation in Hydra. Dev Biol 133:77–82

Fujisawa T, David CN, Bosch TCG (1990) Transplantation stimulates interstitial cell migration in hydra. Dev Biol 138:509–512

Fujisawa T, Nishima C, Sugiyama T (1986) Nematocyte differentiation in hydra. Curr Top Dev Biol 20:281–290

Fujisawa T, Sugiyama T (1980) Nematocyte differentiation from interstitial cells newly introduced into interstitial cell-deficient hydra. In: Tardent P, Tardent R (eds) Developmental and cellular biology of coelenterates. Elsevier/North-Holland Biomedical Press, Amsterdam, pp 319–324

Galliot B, Miljkovic-Licina M, de Rosa R, Chera S (2006) Hydra, a niche for cell and developmental plasticity. Sem Cell Dev Biol 17:492–502

Galliot B, Schmid V (2002) Cnidarians as a model system for understanding evolution and regeneration. Int J Dev Biol 46:39–48

Grens A, Mason E, Marsh JL, Bode HR (1995) Evolutionary conservation of a cell fate specification gene: the Hydra achaete-scute homolog has proneural activity in Drosophila. Development 121(1):4027–4035

Guduric-Fuchs J, Möhrlen F, Frohme M, Frank U (2004) A fragile X mental retardation-like gene in a cnidarian. Gene 343:231–238

Hauenschild C (1954) Genetische und entwicklungsphysiologische Untersuchungen über Intersexualität und Gewebeverträglichkeit bei dem Hydroidpolypen Hydractinia echinata. Wilhelm Roux' Archiv Entwicklungsmech Org 147:132–138

Hauenschild C (1956) Experimentelle Untersuchungen über die Entstehung asexueller Klone bei der Hydromeduse Eleutheria dichotoma. Zeitschrift Naturforschung 11b:394–402

Hauenschild C (1957) Ergänzende Mitteilung über die asexuellen Medusenklone bei Eleutheria dichotoma. Zeitschrift Naturforschung 12b:412–413

Hayakawa E, Fujisawa C, Fujisawa T (2004) Involvement of *Hydra achaete-scute*gene *CnASH*in the differentiation pathway of sensory neurons in the tentacles. Dev Genes Evol 214:486–492

Haynes J, Burnett AL (1963) Dedifferentiation and redifferentiation in *Hydra viridis*. Science 142:1481–1483

Hellstern S, Stetefeld J, Fauser C, Lustig A, Engel J, Holstein TW, Ozbek S (2006) Structure/function analysis of spinalin, a spine protein of *Hydra*nematocysts. FEBS J 273:3230–3237

Herrmann K, Berking S (1987) The length of S-phase and G2-phase of epithelial cells is regulated during growth and morphgenesis in *Hydra attenuata*. Development 99:33–39

Hobmayer B, Rentzsch F, Kuhn K, Happel CM, Cramer von Laue C, Snyder P, Rothbächer U, Holstein TW (2000) WNT signaling molecules act in axis formation in the diploblastic metazoan *Hydra*. Nature 407:186–189

Hoffmann U, Kroiher M (2001) A possible role for the cnidarian homologue of serum response factor in decision making by undifferentiated cells. Dev Biol 236:304–315

Hofmann DK, Neumann R, Henne K (1978) Strobilation, budding and initiation of scyphistoma morphogenesis in the rhizostome *Cassiopea andromeda*(Cnidaria: Scyphozoa). Marine Biol 47:161–176

Holstein TW, David CN (1990a) Cell cycle length, cell size and proliferation rate in *Hydra*stem cells. Dev Biol 142:392–400

Holstein TW, David CN (1990b) Putative intermediates in the nerve cell differentiation pathway in *Hydra*have properties of multipotent stem cells. Dev Biol 142:401–405

Holstein TW, Hobmayer E, David CN (1991) Pattern of epithelial cell cycling in hydra. Dev Biol 148:602–611

Holstein TW, Hobmayer E, Technau U (2003) Cnidarians: an evolutionary conserved model system for regeneration. Dev Dyn 226:257–267

Howard MK, Burke LC, Mailhos C, Pizzey A, Gilbert CS, Lawson WD, Collins MK, Thomas NS, Latchman DS (1993) Cell cycle arrest of proliferating neuronal cells by serum deprivation can result in either apoptosis or differentiation. J Neurochem 60:1783–91

Kalthurin K, Anton-Erxleben F, Milde S, Plötz C, Wittlieb J, Hemmerich G, Bosch TCG (2007) Transgenic stem cells in *Hydra*reveal an early evolutionary origin for key elements controlling self-renewal and differentiation. Dev Biol 309:32–44

Kaesbauer T, Towb P, Alexandrova O, David CN, Dall'Armi E, Staudigl A, Stiening B, Boettger A (2007) The Notch signaling pathway in the cnidarian *Hydra*. Dev Biol 303:376–390

Khalturin k, Anton-Erxleben F, Mildel S (2007) Transgenic stem cells in *Hydra* reveal an early evolutionary origin for key elements controlling self-renewal and differentiation. Dev Biol 309(1):32–44

Koizumi O, Bode HR (1991) Plasticity of the nervous system of adult hydra. III. Conversion of neurons to expression of a vasopressin-like immunoreactivity depends on axial location. J Neurosci 11:2011–2020

Koizumi O, Heimfeld S, Bode HR (1988) Plasticity of the nervous system of adult hydra. II. Conversion of ganglionic cells of the body column into epidermal sensory cells of the hypostome. Dev Biol 129:358–371

Kroiher M, Plickert G, Müller WA (1990) Pattern of cell proliferation in embryogenesis and planula development of *Hydractinia echinata*predicts the postmetamorphic body pattern. Roux's Archiv Dev Biol 199:156–163

Kükenthal W (1923) Octocorallia. In: Krumbach T (ed) Handbuch der Zoologie I. Walter de Gruyter, Leipzig, pp 695, 776

Lentz TL (1965) The fine structure of differentiating interstitial cells in hydra. Zeitschrift Zellforschung mikroskop Anatomie 67:547–560

Lentz TL (1966) The cell biology of hydra. North-Holland Publ, Amsterdam

Lesh-Laurie GE, Hujer A, Suchy P (1991) Polyp regeneration from isolated tentacles of *Aurelia scyphistomae*: a role for gating mechanisms and cell division. Hydrobiologia 216–217:91–97

Lindgens D, Holstein TW, Technau U (2004) Hyzic, the Hydra homolog of the zic/odd-paired gene, is involved in the early specification of the sensory nematocytes. Development 131:191–201

Littlefield CL (1991) Cell lineages in Hydra: Isolation and characterization of an interstitial stem cell restricted to egg production in *Hydra oligactis*. Dev Biol 143:378–88

Littlefield CL, Bode HR (1986) Germ cells in *Hydra oligactis*males. II Evidence for a subpopulation of interstitial stem cells whose differentiation is limited to sperm production. Dev Biol 116:381–386

Lowell RD, Burnett AL (1969) Regeneration of complete hydra from isolated epidermal transplants. Biol Bull 137:312–320

Macklin M (1968) Reversal of cell layers in hydra: a critical re-appraisal. Biol Bull 134:465–472

Mali B, Soza-Ried J, Frohme M, Frank U (2006) Structural but not functional conservation of an immune molecule: a tachylectin-like gene in *Hydractinia*. Dev Comp Immunol 30: 275–281

Marcum BA, Campbell RD (1978) Developmental roles of epithelial and interstitial cell lineages in hydra: analysis of chimeras. J Cell Sci 32:233–247

Martin VJ, Archer WE (1986) Migration of interstitial cells and their derivatives in a hydrozoan planula. Dev Biol 116:486–496

Martin VJ, Archer WE (1997) Stages of larval development and stem cell population changes during metamorphosis of a hydrozoan planula. Biol Bull 192:41–52

Miljkovic M, Mazet F, Galliot B (2002) Cnidarian and bilaterian promoters can direct GFP expression in transfected hydra. Dev Biol 246:377–90

Miljkovic-Licina M, Chera S, Ghila L, Galliot B (2007) Head regeneration in wild-type hydra requires de novo neurogenesis. Development 134:1191–1201

Minasian LL, Mariscal RN (1980) Tissue-specific differentiation of cnidoblasts in a sea anemone (*Haliplanella luciae*). Am Zool 20:803

Mochizuki K, Nishimiya-Fujisawa C, Fujisawa T (2001) Universal occurrence of the *vasa*-related genes among metazoans and their germline expression in *Hydra*. Dev Genes Evol 211:299–308

Müller WA (1964) Experimentelle Untersuchungen über Stockentwicklung und Sexualchimären bei *Hydractinia echinata*. Roux' Arch. Entwicklungsmechanik 155:181–268

Müller WA (1966) Elimination der I-Zellen und Hemmung der Planulametamorphose durch alkylierende Cytostatika. Naturwissenschaften 7:184–185

Müller WA (1967) Differenzierungspotenzen und Geschlechtsstabilität der I-Zellen von Hydractinia echinata. Roux Arch Entwicklungsmech 159:412–432

Müller WA (2002) Autoaggressive, multi-headed and other mutant phenotypes in *Hydractinia echinata*(Cnidaria: Hydrozoa). Int J Dev Biol 46:1023–1033

Müller WA, Frank U, Teo R, Mokady O, Guette C, Plickert G (2007) Wnt signaling in hydroid development: Ectopic heads and giant buds induced by GSK-3beta inhibitors. Int J Dev Biol 51:211–220

Müller WA, Leitz T (2002) Metamorphosis in the Cnidaria. Can J Zool 80:755–1771

Müller WA, Plickert G, Berking S (1986) Regeneration in Hydrozoa: distal versus proximal transformation in *Hydractinia*. Roux's Arch Dev Biol 195:513–518

Müller WA, Teo R, Frank U (2004a) Totipotent migratory stem cells in a hydroid. Dev Biol 275:215–224

Müller WA, Teo R, Möhrlen F (2004b) Patterning a multi-headed mutant in *Hydractinia*: enhancement of head formation and its phenotypic normalization. Int J Dev Biol 48:9–15

Neumann R, Schmahl G, Hofmann DK (1980) Bud formation and control of polyp morphogenesis in Cassiopea andromeda (Scyphozoa). In: Tardent P, Tardent R (eds) Developmental and Cellular Biology of Coelenterates. Elsevier, Amsterdam

Nishimiya-Fujisawa C, Sugiyama T (1993) Genetic analysis of developmental mechanisms in hydra. XX. Cloning of interstitial stem cells restricted to the sperm differentiation pathway in *Hydra magnipapillata*. Dev Biol 157:1–9

Okita K, Ichisaka T, Yamanaka S (2007) Generation of germline-competent induced pluripotent stem cells. Nature 448:313–318

Plickert G, Jacoby V, Frank U, Müller WA, Mokady O (2006) Wnt signaling in hydroid development: formation of the primary body axis in embryogenesis and its subsequent patterning. Dev Biol 298:368–378

Plickert G, Kroiher M (1988) Proliferation kinetics and cell lineages can be studied in whole mounts and macerates by means of BrdU/anti-BrdU technique. Development 103:791–794

Plickert G, Kroiher M, Munck A (1988) Cell proliferation and early differentiation during embryonic development and metamorphosis of *Hydractinia echinata*. Development 103: 795–803

Plickert G, Schetter E, Verhey van Wijk N, Schlossherr J, Steinbüchl M, Gajewski M (2003) The role of α-amidated neuropeptides in hydroid development – LWamides and metamorphosis in *Hydractinia echinata*. Int J Dev Biol 47:439–450

Rebscher N, Volk C, Teo R (2008) Vasa protein is a component of the germ plasm in the cnidarian *Hydractinia echinata*. Dev Dyn 237(6):1736–45

Rebscher N, Zelada-Gonzalez F, Banisch TU, Raible F, Arendt D (2007) Vasa unveils a common origin of germ cells and of somatic stem cells from the posterior growth zone in the polychaete *Platynereis dumerilii*. Dev Biol 306:599–611

Sacks PG, Davis LE (1979) Production of nerveless *Hydra attenuata*by hydroxyurea treatments. J Cell Sci 37:189–203

Schmich J, Kraus Y, De Vito D, Graziussi D, Boero F, Piraino S (2007) Induction of reverse development in two marine Hydrozoans. Int J Dev Biol 51(1):45–56

Schmich J, Trepel S, Leitz T (1998) The role of GLWamides in metamorphosis of *Hydractinia echinata*. Dev Genes Evol 208:267–273

Schmid V, Baader C, Bucciarelli A, Reber-Muller S (1993) Mechanochemical interactions between striated muscle cells of jellyfish and grafted extracellular matrix can induce and inhibit DNA replication and transdifferentiation in vitro. Dev Biol 155:483–496

Schmid V, Wydler M, Alder H (1982) Transdifferentiation and regeneration in vitro. Dev Biol 92:76–488

Schmid V, Plickert G (1990) The proportion altering factor (PAF) and the in vitro transdifferentiation of isolated striated muscle of jellyfish into nerve cells. Differentiation 44:95–102

Schmid V, Reber-Müller S (1995) Transdifferentiation of isolated striated muscle of jellyfish in vitro: the initiation process. Sem Cell Biol 6:109–116

Schwarz RS, Hodes-Villamar L, Fitzpatrick KA, Fain MG, Hughes AL, Cadavid LF (2007) A gene family of putative immune recognition molecules in the hydroid *Hydractinia*. Immunogenetics 59:233–246

Seipel K, Yanze N, Schmid V (2004a) The germ line and somatic stem cell gene Cniwi in the jellyfish *Podocoryne carnea*. Int J Dev Biol 48:1–7

Seipel K, Yanze N, Schmid V (2004b) Developmental and evolutionary aspects of the basic helix-loop-helix transcription factors Atonal-like 1 and Achaete-scute homolog 2 in the jellyfish. Dev Biol 269:331–345

Shigeta M, Shibukawa Y, Ihara H, Miyoshi E, Taniguchi N, Gu J (2006) beta1,4-N-Acetylglucosaminyltransferase III potentiates beta1 integrin-mediated neuritogenesis induced by serum deprivation in Neuro2a cells. Glycobiology 16:564–571

Shimizu H, Bode HR (1995) Nematocyte differentiation in hydra: commitment to nematocyte type occurs at the beginning of the pathway. Dev Biol 169:136–150

Shukalyuk AI, Golovnina KA, Baiborodin SI, Gunbin KV, Blinov AG, Isaeva VV (2006) vasa-related genes and their expression in stem cells of colonial parasitic rhizocephalan barnacle *Polyascus polygenea*(Arthropoda: Crustacea: Cirripedia: Rhizocephala). Cell Biol Int 31: 97–108

Strelin GS (1928) Röntgenologische Untersuchungen an Hydren. Wilhelm Roux' Archiv Entwicklungsmechanik Organismen 115:27–51

Sugiyama T, Fujisawa T (1979) Genetic analysis of developmental mechanisms in hydra. VI. Cellular composition of chimera hydra. J Cell Sci 35:1–15

Takahashi T, Koizumi O, Ariura Y, Romanovitch A, Bosch TC, Kobayakawa Y, Mohri S, Bode HR, Yum S, Hatta M, Fujisawa T (2000) A novel neuropeptide, Hym-355, positively regulates neuron differentiation in Hydra. Development 127:997–1005

Teo R, Möhrlen F, Plickert G, Müller WA, Frank U (2006) An evolutionary conserved role of Wnt signaling in stem cell fate decision. Dev Biol 289:91–99

Terada H, Sugiyama T, Shigenaka Y (1988) Genetic analysis of developmental mechanisms in hydra. XVIII. Mechanism for elimination of the interstitial cell lineage in the mutant strain Sf-1. Dev Biol 126:263–269

Teragawa CK, Bode HR (1991) A head signal influences apical migration of interstitial cells in *Hydra vulgaris*. Dev Biol 147:293–302

Thieme C, Hofmann DK (2003) Control of head morphogenesis in an invertebrate asexually produced larva-like bud (*Cassiopea andromeda*; Cnidaria: Scyphozoa). Dev Genes Evol 213:127–133

Torras R, Gonzalez-Crespo S (2005) Posterior expression of nanos orthologs during embryonic and larval development of the anthozoan *Nematostella vectensis*. Int J Dev Biol 49:895–899

Van de Vyver G (1964) Etude histologique du development d'Hydractinia echinata (Flem) Cahiers de Biologie Marine V:295–310

Van de Vyver G (1967) Etude de development embyryonnaire des hydraires athecates (gymnoblastiques) a gonophores. Archives de Biologie (Liege) 78:451–518

Wanek N, Marcum BA, Campbell RD (1980) Histological structure of epithelial hydra and evidence for the complete absence of interstitial and nerve cells. J Exp Zool 212:1–12

Wang YP, Wang ZF, Zhang YC, Tian Q, Wang JZ (2004) Effect of amyloid peptides on serum withdrawal-induced cell differentiation and cell viability. Cell Res 14:467–72

Weis VM, Keene DR, Buss LW (1985) Biology of hydractiniid hydroids. 4. Ultrastructure of the planula of *Hydractinia echinata*. Biol Bull 168:403–418

Weismann A (1883) Die Entstehung der Sexualzellen bei Hydromedusen. Gustav Fischer-Verlag, Jena

Weismann A (1885) Die Continuität des Keimplasmas als Grundlage einer Theorie der Vererbung. Gustav Fischer-Verlag, Jena

Weismann A (1892a) Das Keimplasma. Eine Theorie der Vererbung. Gustav Fischer-Verlag, Jena

Weismann A (1892b) Essays on heredity and kinded biological problems (trans. Poulton EB et al). Clarendon Press, Oxford

Weismann A (1904) Keimplasmatheorie, Regeneration. In: Vorträge über Deszendenztheorie, 2nd edn. Gustav Fischer-Verlag, Jena

Werner B, Hentschel J (1983) Apogamous life cycle of Stephanoscyphus planulophorus. Mar Biol 74:301–304

Wernig M, Meissner A, Foreman R, Brambrink T, Ku M, Hochedlinger K, Bernstein BE, Jaenisch R (2007) In vitro reprogramming of fibroblasts into a pluripotent ES-cell-like state. Nature 448:318–325

Wittlieb J, Khalturin K, Lohmann JU, Anton-Erxleben F, Bosch TC (2006) Transgenic Hydra allow in vivo tracking of individual stem cells during morphogenesis. Proc Natl Acad Sci USA 103:6208–6211

Yanze N, Groger H, Muller P, Schmid V (1999) Reversible inactivation of cell-type-specific regulatory and structural genes in migrating isolated striated muscle cells of jellyfish. Dev Biol 213:194–201

Yu J, Vodyanik MA, Smuga-Otto K, Antosiewicz-Bourget J, Frane JL, Tian S, Nie J, Jonsdottir GA, Ruotti V, Stewart R, Slukvin II, Thomson JA (2007) Induced pluripotent stem cell lines derived from human somatic cells. Science 318:1917–1920

Zeretzke S, Berking S (2002) In the multiheaded strain (mh-1) of *Hydra magnipapillata*the ectodermal epithelial cells are responsible for the formation of additional heads and the endodermal epithelial cells for the reduced ability to regenerate a foot. Dev Growth Differ 44:85–93

Zhang E, Li X, Zhang S, Chen L, Zheng X (2005) Cell cycle synchronization of embryonic stem cells: effect of serum deprivation on the differentiation of embryonic bodies in vitro. Biochem Biophys Res Comm 333:1171–1177

Chapter 4
Stem Cells in Aquatic Invertebrates: Common Premises and Emerging Unique Themes

Yuval Rinkevich, Valeria Matranga and Baruch Rinkevich

Abstract The remarkable ability to generate an embryo from a single fertilized oocyte and to regenerate tissues that are injured or going through natural physiological turnover, is the direct result of stem cells, nature's gift to multicellular organisms. Only recently, studies on marine invertebrates have revealed the diversity of phenomena involved with these cells and their importance to the biology of these organisms. We present an overview on the stem cell biology of four aquatic invertebrate groups: urochordate ascidians, cnidarians, echinoderms and platyhelminthes. While most studies outlined here are based on descriptive data, we found that aquatic invertebrates exhibit multiple cell types with stem cell atributes. Studies revealed that, in contrast to the prevalence of diverse oligopotent and unipotent stem cells in vertebrates, invertebrates appear to display the communal spread of multipotency and pluripotency, with stem cells that give rise to cell lineages characteristic to more than a single germ layer, sometimes with somatic and germ line potentials. The cumulative data further indicate that in contrast to vertebrate systems, stem cells from aquatic invertebrates are disseminated and widespread i.e. not associated with a regulatory microenvironment (niche). We also notice that transdifferentiation is ubiquitous to both anatomically simple and highly evolved invertebrates. These observations delineate common and unique properties to stem cells, posibly tailored, to suit the varied life history and developmental modes characteristic of aquatic invertebrates.

Keywords Ascidian · Cnidarian · Echinoderm · Invertebrate · Platyhelminth · Regeneration · Stem cells · Trans-differentiation

4.1 Introduction: Properties of Stem Cells in Adult Organisms

Early studies (Till and McCulloch 1961; Metcalf and Moore 1971) classified "stem cells" as clonogenic cells, which are capable of both, self-renewal and multilineage

---

Y. Rinkevich (✉)
Institute for Stem Cells Research and Regenerative Medicine, Stanford University School of Medicine, Stanford, CA 94305, USA
e-mail: ryuval@stanford.edu

B. Rinkevich, V. Matranga (eds.), *Stem Cells in Marine Organisms*,
DOI 10.1007/978-90-481-2767-2_4, © Springer Science+Business Media B.V. 2009

differentiation. Subordinate to this terminology, stem cells are responsible for generating somatic tissues during ontogeny (presumably also accountable for astogeny in colonial marine organisms) for constant renewal of the organisms' soma, and are the essential building blocks for urgent needs such as regeneration of damaged or lost parts (reviewed in Weissman 2000). Based on the capabilities of their descendant cells, stem cells are classified as totipotent (able to generate all embryonic and extra-embryonic cell-types), pluripotent (able to differentiate into any of the three germ layers but not to extra-embryonic tissue, such as the placenta), multipotent (able to generate a wide range of cell types in a specific organ, such as hematopoietic stem cells), oligopotent (able to differentiate to a restricted number of cell types), or unipotent (able to differentiate to only one cell type; Wagers and Weissman 2004). Other distinguishing traits of stem cells are their undifferentiated morphology, characterized by a high nucleo-cytoplasmic ratio and small cell size compared to lineage-differentiated progenies (Higuchi et al. 2007), their extreme rarity within resident tissues and their associations with regulatory microenvironments, called niches. Some tissue-specific stem cells had initially been defined by positional or anatomical criteria. These included muscle satellite cells (Mauro 1961), which reside beneath the basal lamina of mature muscle fibers, epidermal stem cells, located predominantly in or near hair follicle's bulge region (Alonso and Fuchs 2003), and intestinal stem cells, residing near the bottom of the intestinal crypts (Bjerknes and Cheng 1999; Spradling et al. 2001). Apparently, the ability to avoid telomere shortening (Vaziri et al. 1994; Allsopp et al. 1995) is required for continuous division and self-renewal and is characteristic of some stem/progenitor cells (Morrison et al. 1996). Asymmetric cell division is another distinct feature of stem cells (reviewed in Moore and Lemischka 2006) in both, developing (reviewed in Betschinger and Knoblich 2004) and adult life (Clevers 2005), when the two daughter cells become different upon the division of the mother cell. Two main modes of asymmetric cell division are described. An intrinsic one, which relies on the asymmetric partitioning of cell components, inherited by only one of the daughter cells, or an extrinsic one, which involves the asymmetric placement of daughter cells relative to external cues. While asymmetric cell division is restricted to stem cell populations, a balance between asymmetric and symmetric cell divisions takes place when stem cells are expanding their pools during development and regeneration (Reviewed in Morrison and Kimble 2006). A special mode of asymmetric cell division (reviewed in Rando 2007) is the so called "immortal strand hypothesis", which proposes that dividing stem cells selectively retain "old" or template DNA strands to prevent accumulation of mutations (Cairns 1975). Data supporting the "immortal strand hypothesis" comes from several model organisms and experimental settings (Potten et al. 1978, 2002; Karpowicz et al. 2005; Smith 2005; Shinin et al. 2006; Conboy et al. 2007), but is not a property of all adult stem cells (Kiel et al. 2007). A similar attribute of some adult stem cells is their ability to retain DNA labels, such as BrdU (Bickenbach 1981) resulting either from the "immortal DNA strand" phenomenon or from extended cell cycle periods compared with their progeny (Tumbar et al. 2004; Cotsarelis et al. 1990; Zhang et al. 2003). At the molecular level, analyses of both global gene expression (Ivanova et al. 2002; Ramalho-Santos et al. 2002) and genetics (Takahashi and Yamanaka 2006; Takahashi et al. 2007; Wernig et al. 2007) of stem cell populations have identified several regulators. However, a "stemness" gene profile ubiquitous to all embryonic and adult stem cells is still elusive (Vogel 2003; Pyle et al. 2004; Lengner et al. 2007).

While much is known on adult stem cells and their properties in vertebrates (especially mammals) and model terrestrial invertebrates (i.e., Drosophila), very little has been learnt on the nature and properties of stem cells in aquatic invertebrates. This chapter will discuss the knowledge gathered on stem cells in four major invertebrate phyla (Cnidaria, Platyhelminthes, Echinodermata, and Tunicata) (see Muller et al., Candia et al., Rychel and Swalla, Ballarin and Manni, Chapters 13, 7, 10, 11 of this book), revealing common characteristics and emerging unique themes of adult stem cells.

4.2 Adult Stem Cells in Tunicates

Urochordate ascidians are valuable for studying the biology of adult stem cells, because stem cells actively participate in a myriad of ascidians' life history traits and ecological phenomena (Rinkevich 2002a; Satoh and Levine 2005; Manni and Burighel 2006). These include aestivation, hibernation, asexual reproduction, age-related growth, genotype-related senescence and death, colonial chimerism, cellular parasitism, and diverse regeneration scenarios (reviewed in Berrill 1951; Rinkevich 2002a; Manni et al. 2007; Ballarin and Manni, Chapter 11 of this book). The importance of exploring stem cells in this phylum is highlighted by the tunicate's key taxonomic position at the origin of the Vertebrata branch (Delsuc et al. 2006). Below, we summarize known and postulated roles of adult stem cells in asexual reproduction and regeneration of ascidians, in colony regression and in somatic and germ cell parasitism.

4.2.1 Stem Cells in Asexual Reproduction

The subphylum Urochordata is composed of three classes: Thaliacea, Larvacea and Ascidiacea. The latter will be discussed here. This class, consists of solitary, social and compound (both are considered colonial) marine organisms, and has long been known for its remarkable budding capacity to form new entities from existing structures. This trait, termed "asexual reproduction", is considered polyphyletic rather than ancestrally acquired (Buss 1999; Cameron et al. 2000), and is carried out by various stem cell systems (reviewed in Berrill 1951). The cumulative data, of which representative examples are highlighted bellow, demonstrate that the Urochordates have acquired at least two independent stem cell systems for budding (epithelial and blood born), which function harmoniously during normal and adverse physiological states.

One common developmental process in botryllid (*Botryllus*, *Botrylloides*), *Symplegma*, *Metandrocarpa*, *Polyzoa* and polystyelid (*Stolonica*, *Distomus* and *Polyandrocarpa*) ascidians is palleal-budding. A palleal bud develops as a double-walled vesicle from the thoracic body region of the parental zooid (Berrill 1941). The outer vesicle is an extension of the parental epidermis, while the inner vesicle develops from the inner (atrial) epithelium and/or hemoblasts recruited from the blood. At an early developmental stage, cells of the inner vesicle display three hallmark stem cell properties: high alkaline phosphatase activity (Akhmadieva et al. 2007), elevated telomerase activity (Laird and Weissman 2004) and are suggested to display a multi-lineage differentiation potential (Kawamura et al. 1988; Kawamura and Nakauchi 1991a; Tiozzo et al. 2005). For example, in botryllid ascidians, cells of the inner vesicle differentiate into most mesodermal and endodermal tissues and organs (Manni et al. 2007). In *Symplegma reptans*, cells of the inner vesicle differentiate into the pharyngeal rudiment (including atrial epithelium, stigmata and endostyle), gut rudiment (including esophagus, stomach, intestine and pyloric caecum) and neural complex (including ciliated duct and neural gland; Kawamura and Nakauchi 1986). In *Polyandrocarpa misakiensis*, buds develop from the participation of both stem cell sources, including epithelial transformation of the hemoblasts. Cells of the inner atrial epithelium enter mitosis with a cycle time of about 12 h, followed by aggregation of hemoblast stem cells in the mesenchymal space to form organ rudiments (Kawamura and Nakauchi 1991a). The contribution of both sources to bud development in *P. misakiensis* was corroborated experimentally in the presence of the mitotic inhibitors colchicine (Kawamura and Nakauchi 1986) or aphidicolin, where only gut rudiments developed from aggregates of blood cells (Kawamura and Nakauchi 1991b).

In the colonial ascidians, *Perophora japonica* and *Ecteinascidia*, animals propagate by a process termed stolonial budding, which takes place at the distal tip of the developing stolon, noted by its high mitotic activity (Koguchi et al. 1993). The stolon is an epidermal outgrowth, inside of which is divided vertically by a single-layered epithelium (called septum), containing the stem cell line for this budding process. These stem cells form most tissues and organs except for the epithelium (Freeman 1964; Mukai et al. 1983). Multilineage contributions (pluripotency) from stem cells was documented experimentally in three closely related species (*Perophora bermudensi*, *P. japonica* and *Ecteinascidia tortugensis*; Freeman 1964, 1970) in which blood cell-injected preparations of allogeneic and xenogeneic combinations rescued budding processes in irradiated colonies. Whenever donor and host were of different species, developed zooids were of the donor type whereas the host cells could only be identified in the vascular system of the zooid (Freeman 1970). Of all blood cell types, only the injection of lymphocytes revived budding activity. Lymphocytes usually represent 0.5% of the total blood cell populations in tunicates, and are highly mitotic (Raftos and Cooper 1991; Rinkevich et al. 2007).

In addition to normal palleal budding, some species of botryllid ascidians (i.e., *Botryllus primigenus*; Oka and Watanabe 1957a), exhibit in a normal healthy state, a distinct budding process termed vascular budding, in which buds are produced inside the vasculature system. Vascular buds emerge from gathering of haemoblasts (n= 15–20) and form the pharyngeal rudiment, gut rudiment, endostyle, and neural complex (Izzard 1973; Nunzi et al. 1979; Kawamura and Nakauchi 1984). Vascular budding is the most tangible example for hemoblasts as totipotent stem cells (Oka and Watanabe 1957a, 1959; Rinkevich et al. 1995, 2007; Voskoboynik et al. 2007). These stem cells are small (3–4 μm), spherical in shape, with a high nucleus to cytoplasm ratio, a well-developed nucleolus and a basophilic, hyaline cytoplasm filled with polysomes (Sabbadin 1955; Oka and Watanabe 1957a; Wright 1981). In *Botryllus primigenus*, these stem cells function in a limited blastogenic phase (from late phase B to early phase C), developing vascular buds in a 10-h period during which the colony and ampullae are at their maximum activity. During this period, tens to hundreds of new buds develop in the most actively growing edges of the colony, at the base of ampullae, about 0.6–0.7 mm from the ampullar tips (Oka and Watanabe 1957a).

4.2.2 Stem Cells in Regeneration

Colonial and solitary ascidians exhibit a wide range of regenerative capabilities, including regeneration of the tunic, mantle, siphons, neuroglandular complex and pharynx in *Ciona intestinalis* (Berrill 1951; Sutton 1953) to whole body regeneration in botryllid ascidians (Oka and Watanabe 1959; Rinkevich et al. 1995, 2007; Voskoboynik et al. 2007). In siphon regeneration of *Ciona intestinalis*, for which only descriptive data is available, the cells responsible for the regeneration of the siphon's connective tissue and muscle are haemoblasts. These cells proliferate, migrate, and differentiate at the regenerative site within four days (Sutton 1953). Not only haemoblasts serve as building block cells. Thoracic tissues of *Ciona intestinalis* and *Styela plicata* regenerate from residual tissues of epidermis and atrial epithelium to form a new mantle and epidermis (Berrill 1951). In *Clavelina lepediformis* and *Archiascidia neapolitana* (Berrill 1951), the new regenerating thoracic emerges from the surviving epidermis, whereas the branchial pharynx, peribranchial sacs and the atrium regenerate from the epicardium, which is a pharyngeal outgrowth. In *Polycarpa tenera* (*P. gracilis*), regeneration of branchial thorax, digestive canal, heart and gonads initiate from folds of the atrial epithelium (Berrill 1951).

Removal of the peripheral circulatory system from a *Botryllus schlosseri* colony results in regeneration within 18–48 h (Voskoboynik et al. 2007; Tiozzo et al. 2008). In *Botrylloides violaceum* (Oka and Watanabe 1959) and *Botrylloides leachi* (Rinkevich et al. 1995, 2007), whole body regeneration (WBR) is induced in vasculature fragments following their separation from the colony (Fig. 4.1A). Even single marginal ampullae containing 100–300 blood cells may regenerate adult zooids within two to three weeks (Rinkevich et al. 1995). WBR develops irrespective to fragment size (number of blood vessels), the colonial blastogenic stage at day of operation, colony size, or location of the amputated vasculature (peripheral, marginal, or from vessels located centrally), but does not develop under normal

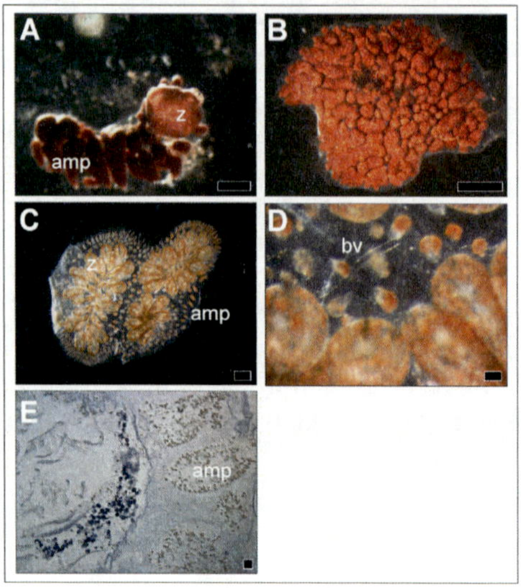

Fig. 4.1 Life history portrailes of colonial ascidians that involve stem cells. (**A**) Whole Body regeneration (WBR) in the colonial ascidian *Botrylloides leachi* from cut-out blood vessels. Experimental separation of marginal blood vessels from the colony leads to regeneration of functional adult zooids within 10–14 days from resident stem cells. (**B**) Colony regression in *Botrylloides leachi* displaying resorped zooids and buds with a "carpet" of dense opaque ampullae and blood lacunae. Upon improved environmental condition, regressed colonies regenerate functional zooids from resident stem cells. (**C**) A cytomictical chimera of two *Botryllus schlosseri* colonies. Following vascular fusion, a common circulatory system is established between the two genetically distinct colonies which commonly leads to somatic and germ-cell parasitism. (**D**) Magnification of C, showing the common vasculature of a cytomictical chimera. (**E**) Piwi positive stem cells within the endostyle niche of *Botrylloides leachi*. Cells of the endostyle niche incorporate into developing and regenerating somatic tissues of botryllid colonies and induce long-term chimerism in transplanted recipient colonies. amp, ampulla; bv, blood vessel; z, zooid. Scale bar represents 500 μm in **A, B** and **C**, 100 μm in **D** and 10 μm in **E**

physiological state of the colony. WBR in *B. leachi* proceeds through a novel mechanism of regeneration, that is depicted by several fundamental criteria, including epimorphic type of restoration, the lack of a regenerative blastema and the generation of a systemic induction in which multiple buds simultaneously develop in newly formed vasculature compartments termed "regeneration niches". Above contradict other documented cases of regeneration in multicellular organisms in which a single restoration center is observed (Rinkevich et al. 2007). In the WBR system, totipotent blood stem cells aggregate within the newly formed compartments and differentiate into all somatic and germ line derivatives, thereby regenerating functionally active adult zooids, capable of sexual and asexual reproduction. A similar stem cell system was documented in several other botryllid ascidians including *Botryllus primigenus* (Oka and Watanabe 1957a), *Botryllus schlosseri* (Voskoboynik et al. 2007) and *Botrylloides violaceus* (Rinkevich, personal observations) and appears to be a conserved feature.

4.2.3 Stem Cells in Colony Regression

Another life history trait of ascidians, which involves stem cells is "colony regression", (Berrill 1951; Nakauchi 1982) termed "aestivation", for colonies that regress following rapid dropping in seawater temperature, in the winter (Bancroft 1903), or "hibernation" (Burighel et al. 1976), for colonies that regress after rapid rise in seawater temperature, in the summer. Colony regression also applies to *Botrylloides* colonies freshly collected from the field, which often resorb all zooids within 24 h after their transfer to the laboratory (Rinkevich et al. 1996). Colony regression entails the rapid involution and resorption of zooids and buds, leaving behind a "carpet" of dense opaque ampullae and blood lacunae (Fig. 4.1B). The vasculature of regressed colonies holds clumps of undifferentiated cells, similar in size and morphology to those of regenerating fragments (Rinkevich et al. 1996). Upon improved environmental condition, regressed colonies resume blastogenesis. It is conceivable that "aestivation" and "hibernations" are two ecological manifestations of the same dormant state, in which colony regeneration appears to come from similar circulating haemoblasts (Rinkevich et al. 1996).

4.2.4 Stem Cells in Somatic and Germ Cell Parasitism

A corpus of reports has documented that resident stem cells in ascidians may parasitize and colonize the animal's soma and germ line following allogeneic fusions (see Voskoboynik et al., Chapter 12 of this book). When peripheral blood vessels (ampullae) of two allogeneic botryllid colonies, in the field or under experimental settings, come into direct contact, they may impinge upon one another at their borders forming cytomictical chimeras (Fig. 4.1C, Rinkevich and Weissman 1987a, b). Allogeneic contacts could also result in inflammatory rejections (Oka and Watanabe 1957b) resulting from allogeneic dissimilarity at a single highly polymorphic FuHC (fusibility/histocompatibility; Weissman et al. 1990) haplotype. Colonies sharing one or two FuHC alleles may undergo vascular fusion and establish a common circulatory system between two genetically distinct colonies (1D, cytomictical chimerism; Oka and Watanabe 1957b; Sabbadin 1962; Scofield et al. 1982). Studies (Sabbadin and Zaniolo 1979; Pancer et al. 1995; Stoner and Weissman 1996; Magor et al. 1999; Stoner et al. 1999; Rinkevich and Yankelevich 2004; Simon-Blecher et al. 2004) revealed that somatic and germ-cell parasitism are common outcomes for cytomictical chimerism. Germ-cell parasitism takes place even in forced chimeras established between allogeneic non-compatible partners (Rinkevich et al. 1998; Simon-Blecher et al. 2004). Germ cell parasitism events are reproducible, exhibit hierarchical organisation, and are sexually inherited, as shown by breeding experiments (Stoner et al. 1999; Rinkevich and Yankelevich 2004; Laird et al. 2005). However, somatic cell parasitism, while reproducible under controlled mariculture conditions and characterized by hierarchical organization, does not reveal the trait of sexual inheritance through a pedigree (Stoner et al. 1999). Therefore, gametic and somatic competitive routes, although reproducible, appear to be unrelated (Stoner et al. 1999; Magor et al. 1999; Rinkevich and Yankelevich 2004). Studies have shown that, in the long term, chimerism might have substantial fitness costs through cell lineage competition and parasitism (Buss 1982; Grosberg and Quinn 1986; Rinkevich and Weissman 1987a, 1992; Sabbadin and Astorri 1988; Magor et al. 1999; Rinkevich and Shapira 1999; Rinkevich 2002a,b). By using microsatellite alleles as polymorphic genetic markers coupled with amplified fragment length polymorphism (AFLP) and seawater temperature as the variable environmental factor, Rinkevich and Yankelevich (2004) documented that the somatic constituent of chimeric zooids may shift from one genotype to another, in accordance with the changes in seawater temperatures. Thus, the variable somatic state of chimerism in the field may carry benefits to the chimeric entity, which at any time, exhibits the best-fitted synergistical combination of its genetic components. It has been also established that stem cell proliferation and chimerism is a dose-dependent phenomenon. Whereas a small number of up to 5000 randomly collected blood cells are sufficient to establish stem cell parasitism, enrichment of cells positive to aldehyde dehydrogenase (ALDH) substrate, which in humans and mice delineates hematopoietic stem cells and progenitors, reduce sufficiently cell numbers needed to establish cell parasitism to as few as 25–50 fractionated cells (Laird et al. 2005). Functional replacement of host colony gametes was also confirmed in germline chimeras three months following transplantation. Self renewal of blood stem cells was documented by infusing whole *Botryllus* blood cells into irradiated *Botrylloides leachi* colonies. These colonies were observed for seven months during which tissue samples were taken from the host *Botrylloides* colonies for genetic evaluation of *Botryllus* genotype with specific microsatellite markers. A 14-fold increase of donor *Botryllus* blood cells was calculated, revealing high rates of cell proliferation and self-renewal within host tissue as a single outbreak event (Simon-Blecher et al. 2004). A niche for somatic stem cells was recently identified in the anerior ventral region of the zooid endostyle (Fig. 4.1E, Voskoboynik et al. 2008). The 10–40 labelled cells taken from the endostyle incorporated into developing and regenerating somatic tissues of *Botryllus* colonies (primary and secondary; body wall and stigmata and epithelial cells of regenerating vasculature) and induced long term chimerism in transplanted recipient colonies (Voskoboynik et al. 2008).

The expression of *vasa* protein (an evolutionarily conserved marker for the germ-line) during ontogeny and colony astogeny has recently been documented in tunicates of the order Pleurogona, in *Botryllus primigenus* (Sunanaga et al. 2006), *Botrylloides violaceus*, *Botryllus schlosseri* (Brown and Swalla 2007) and *Polyandrocarpa misakiensis* (Sunanaga et al. 2007). In these colonial ascidians, *vasa* expression is found specifically in germline cells demarcating the development of ovaries and testis during bud development. In contrast to what was expected from previous experiments (Stoner et al. 1999; Rinkevich and Yankelevich 2004; Laird et al. 2005) no *vasa* positive cells were detected either in the vasculature system of young colonies or in growing buds, indicating that germ line stem cells arise *de-novo* in the developing colonies (Sunanaga et al. 2007). In mature colonies, *vasa* positive cells were found in the developing gonads, but no staining was present in the vasculature of *B. primigenus*. More importantly, during regeneration, vasa positive cells reappeared in the vasculature and within 30 days post regeneration, all regenerating colonies re-expressed the *vasa* transcript (Sunanaga et al. 2006). Therefore, *vasa* may reveal the stage for sequestering the germ cells from a primitive totipotent stem cell.

While the documented contributions of cells with stem cell traits in colonial ascidians are overwhelming, molecular profiling of these stem cell populations is currently minimal. Several reports revealed a close molecular relationship between ascidian and mammalian stem, and progenitor cells including Aldh (aldehyde dehydrogenase), alkaline-phosphatase, pl-10 (a *vasa* family member) and PCNA (proliferating cellular nuclear antigen) (Laird et al. 2005; Rosner et al. 2006; Rinkevich et al. 2007).

4.3 Stem Cells in Echinoderms

The phylum Echinodermata consists of 7000 extant species grouped into five classes: Crinoidea (sea lilies and feather stars), Ophiuroidea (basket stars and brittle stars), Asteroidea (sea stars), Echinoidea (sea urchins, sand dollars, and sea biscuits) and Holothuroidea (sea cucumbers). Commonly, echinoderms exhibit a pentamerous radial symmetry (at some stage of their life), a mesodermal endoskeleton and a vascular water system, which functions in locomotion, respiration, and feeding. Following metamorphosis from a planktonic larva, most echinoderms (except crinoids, whose members are primarily sessile) reach the adult life as benthic slow moving animals (Hyman, 1955). Echinoderms together with hemichordates represent a basal group of non-chordate invertebrate deuterostomes occupying a key position in evolution (Wray and Lowe 2000; Sea Urchin Genome Sequencing Consortium 2006). While not much is known about stem cells in echinoderms, the following will briefly review mechanisms and processes where the contribution of stem cells is evident. These involve larval cloning, adult defense activation and regeneration in larvae and adults.

4.3.1 Larval Cloning

Larval cloning or budding refers to the ability of an echinoderm species to multiply asexually. Mortensen (1921) reported first examples of echinoderm larval cloning, which had been confirmed and have recently been extended to various classes of asteroid, ophiuroid, holothuroid and echinoid larvae (Eaves and Palmer 2003). Larval cloning is not a rare phenomenon; it occurs in about 30–90% of sea star larvae collected from the field and can be abundand, as up to 12% of laboratory-reared sea cucumbers (Bosh et al. 1989; Jaeckle 1994; Vickery et al. 2001b; Knott et al. 2003).

Larval cloning occurs in multiple ways (Knott et al. 2003); at least three different modes of larval cloning were described, depending on the part of the larval body of origin (Jaeckle 1994). The first is by paratomous cloning from posterolateral arms, in which each posterolateral arm is completely modified to produce a secondary larva (Bosh et al. 1989). The second mode involves the autotomy of the anterior portion of the preoral lobe, which first develops a complete digestive system and later assumes the form of a typical bipinnaria larva. The third mode involves budding and the release of blastula or gastrula stage embryos, which resume development (Jaeckle 1994). In bipinnaria larvae of the sea star *Luidia sp.*, larval cloning was documented to occur by fission (Bosh et al. 1989). On the other hand, upon metamorphosis, ophiuroid larvae may leave behind their posterior arms, which swim away and regenerate a complete larva (Balser 1998). It has been proven that clones can grow from different larval body regions, including: arms (sea stars and sea cucumbers; Bosh et al. 1989; Jaeckle 1994; Balser 1998; Vickery and McClintock 2000; Knott et al. 2003), oral hood (Jaeckle 1994; Vickery et al. 2001b; Knott et al. 2003), posterior end (sea stars, sea urchins and sea cucumbers; Vickery et al. 2001b; Eaves and Palmer 2003) and lateral body wall (sea stars and sand dollars; Jaeckle 1994; Eaves and Palmer 2003). Larval clones are usually smaller than primary larvae (Balser 1998). This could be beneficial for survival and be used as an anti predatory strategy, as experimentally demonstrated for sand dollar larvae (*Dendraster excentricus*), which undergo cloning after *in vitro* stimulation with a predator cue (see Fig. 4.2, A–B) (Vaughn and Strathmann 2008).

Asteroid larvae are able to clone by several modes. They can autotomize the preoral lobe, which then regenerates a normal larva, or buds can form from a "site of choice", the posterodorsal, posteroventral, or posterolateral lobes (Bosch et al. 1989; Jaeckle 1994). At first, the lobes swell and then a process similar to gastrulation takes place by forming a blastopore at the underside of the lobe. Next, the bud separates from the parent larva and develops independently, forming a feeding larva (Bosch et al. 1989; Jaeckle 1994; Lacalli 2000). Secondary larvae develop from the reorganization of larval tissues and not from embryonic cells (Balser 1998). In the ophiuroid *Ophiopholis aculeate*, the juvenile, which originates from the primary larva, retains the posterolateral arms after settlement for some period, during which the transverse rods shorten and the posterolateral arm proximal tips make contact with each other. Then, given that the ciliary band remains intact, posterolateral arms are released from the settled juvenile and swim off. The free arms consist of skeletal rods, mesenchymal cells, a ciliated epidermis and the blastocoel of the primary larva's posterolateral arms. Within 24 h of release, the tissue forming the inner part of the arms undergoes a process similar to gastrulation and within 3 or 4 days, the posterolateral arms regenerate the structures typical of an ophiopluteus, including the gut, the coelom and additional feeding arms (Balser 1998). At the time of settlement, the secondary juvenile usually retains the posterolateral arms but, as in primary larva, eventually releases them. The free arms return to the water column and begin a tertiary cycle that results in another planktonic ophiopluteus, producing three clonal organisms from one zygote. At the cellular level, it is presumed that sequential de-differentiation and re-differentiation of larval tissues is

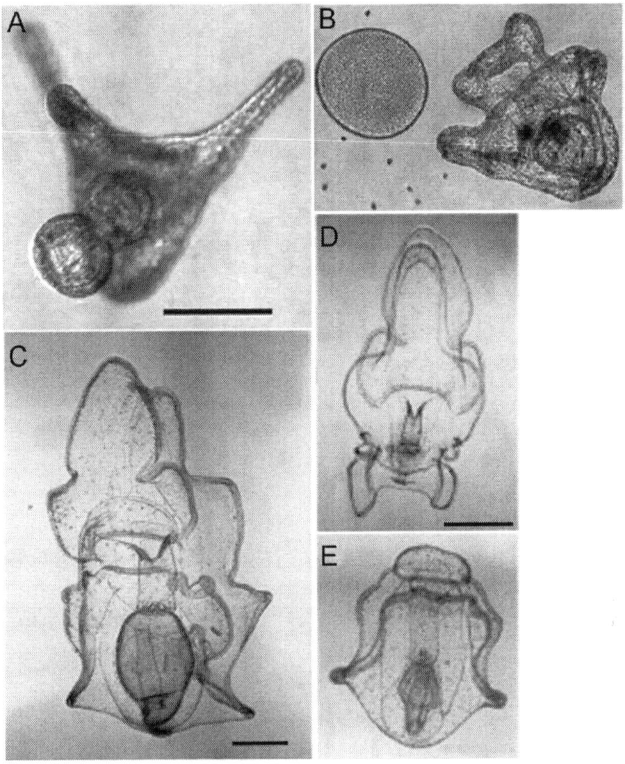

Fig. 4.2 Echinoderm larval cloning and regeneration. (**A, B**) gastrula-like structure and six armed pluteus developed from budding at the aboral surface of a sand dollar primary larva (*Dendraster excentricus*) induced to clone (modified after Vaughn and Strathmann 2008, with permission[1]). (**C–E**) regenerating larva of surgically bidissected sea star (*Luidia foliolata*); missing parts are regenerating in anterior and posterior halves (modified after Vickery et al 2001a, with permission[2])

taking place. The epidermis covering the arms is pulled towards the inner part of the arms and consequently forms the archenteron. It therefore appears that the secondary larva "endodermal" regions grow from the primary larva's differentiated ectoderm, calling for a trans-differentiation process to occur. Within five days, a feeding 4-arm pluteus is formed, morphologically indistinguishable from the primary larva except for having a highly pigmented epidermis, disproportionately long posterolateral arms and a much smaller size (Balser 1998).

---

[1]Reprinted from Vaughn D, Strathmann RR, Predators induce cloning in echinoderm larvae, Fig. 1, p. 1503, (2008), with permission from the America Association for the Advancement of Science.

[2]Reprinted from Vickery et al, Utilization of a novel deuterostome model for the study of regeneration genetics: molecular cloning of genes that are differentially expressed during early stages of larval sea star regeneration, Fig. 1 p 74, (2001a), with permission from Elsevier.

The mesodermal structures, on the other hand, are thought to develop from mesenchyme-like cells and coelomic cavities of the larvae (Jaeckle 1994). An intriguing question relates to the way primary germ cells are formed during larval cloning. Although there is no definite proof so far, three hypotheses are put forward (Balser 1998). The first assumes that the cloned larvae lack the ability to produce germ cells and therefore, they form sterile adults; the second holds that primary germ cells are sequestered from a population of stem cells in the primary larva and then transferred to the secondary (or subsequent) propagule prior to its release from the primary larva. The third proposal suggests that the primary germ cells can arise *de novo* from the cloned larva somatocoelic epithelium. The latter two hypotheses are consistent with the formation of the germ cells in primary larvae either from mesenchyme or from the epithelium of the somatocoel.

Larval cloning by de-differentiation challenges the "set-aside" cell theory (Davidson et al. 1995; Peterson et al. 1997), which postulates that larval body cells are "essentially eutelic" and unable to differentiate into juvenile structures. Set-aside cells, similarly to stem cells, are undifferentiated cells, residing within the larval body. These cells retain indefinite division potential and form tissues of the adult during metamorphosis, but do not contribute to the larva (Cameron et al. 1991; Ransick et al. 1996). Upon specific cues, the set-aside cell populations expand enormously by cell division and then differentiate (Peterson et al. 2000). In sea urchins, set aside cells are believed to grow from four cells (small micromeres) of the 5th cleavage, which eventually proliferate and give rise to the coelomic pouches and later to the adult rudiment (Cameron et al. 1991).

4.3.2 Regeneration in Larval and Adult Echinoderms

The capacity of sea star larvae (*Pisaster ochraceus*) to regenerate and re-form fully functional individuals has been documented following surgical bisection of the larval body. This leads to re-consider the developmental stage at which echinoderm larval cell fates are irreversibly determined (Vickery and McClintock 1998). Another example reveals complete body regeneration following bisection of the planktotrophic larvae of the asteroid *Luidia foliolata* (see Fig. 4.2, C–E) (Vickery et al. 2002).

A number of studies described the regeneration of lost body parts in adult echinoderms (reviewed by Amemiya and Oji 1992; Sánches-Alvarado 2000; García-Arrarás and Greenberg 2001; Candia-Carnevali 2006; San Miguel-Ruiz and García-Arrarás 2007). Regeneration after natural or induced evisceration, autotomy or fission is believed to be triggered by epimorphosis, due to the extensive cell proliferation and differentiation at the cut surface (Candia-Carnevali and Bonasoro 2001) and continues through morphallaxis, characterized by the migration of newly differentiated cells (Mladenov et al. 1989; Moss et al. 1998). However, there is convincing evidence that both mechanisms are functioning and are interchangeable during regeneration processes (García-Arrarás et al. 1998; Candelaria et al. 2006; Candia-Carnevali 2006).

Arm regeneration of the crinoid echinoderm *Antedon mediterranea* has been known as an epimorphic type of regeneration, which involves the reconstruction of several tissue types including muscle, ligaments, branchial nerve, coelomic canals and a feeding organ (ambulacral groove) in a process that takes four weeks (Candia-Carnevali et al. 1989, 1993; Candia-Carnevali and Bonasoro 1994). Two main cellular components contribute to the regeneration process. The blastemal cells and their differentiation lineages are derived from morphologically undifferentiated amoebocytes, while the coelomic cells (coelomocytes) are formed from de-differentiation of the coelomic epithelium and give rise to peritoneocytes, myocytes and wandering coelomocytes of the coelomic canals (Candia-Carnevali et al. 1997). During regeneration, ameobocytes and coelomocytes are actively proliferating in the terminal blastema, the coelomic epithelium of the canals and the cortex of the branchial nerve, which appears to serve as the niche for stem cells (amoebocytes), from which extensive cell migration takes place to intermediate and distal regions (Candia-Carnevali et al. 1998).

Holothuroids (sea cucumbers) represent the best-studied organisms for intestine and Cuverian tubule regeneration following evisceration (Dolmatov 1996; García-Arrarás et al. 1998; Van den Spiegel et al. 2000). Gut regeneration in *Holothuria glaberrima* starts with the disorganization and de-differentiation of the muscle layer. Disorganization is marked by disappearance of cell projections, disruption of intercellular junctions, nucleic activation, and loss of the contractile apparatus within two days. De-differentiated myocytes exhibit a high nuclear/cytoplasmic ratio (Murray and García-Arrarás 2004; Candelaria et al. 2006). After a few days, these cells reacquire muscle markers and form single cell layers (Murray and García-Arrarás 2004). Injecting animals with BrdU showed BrdU-labeled nuclei in muscle cells of 3 week-old regenerates, implying that these cells originated from proliferating precursors (Fig. 4.3, A–D). The peak in muscle precursor proliferation occurred during the second week of regeneration; thus, new muscle cells came from precursor cells that had undergone cell division. During regeneration, myoblasts were suggested to originate from one of the following three cell populations: (1) multipotential stem cells within other tissues, (2) coelomocytes within the coelomic fluid, and (3) cells of the coelomic epithelia (Murray and García-Arrarás 2004). New evidence suggests that regenerating cells (not only muscle cells) originate from coelomocytes in the coelomic fluid (Candia-Carnevali et al. 1997; García-Arrarás et al. 2006; Candia-Carnevali 2006; Pinsino et al. 2007), similarly to muscle cells of the crayfish, which are believed to originate from hemolymph amoebocytes (Uhrík et al. 1989). It is argued that during holothurian longitudinal muscle regeneration, muscle precursors originate from the coelomic epithelium (Dolmatov 1996; Mashanov 2005). Accordingly, mitotic figures are observed within the lining epithelium and the covering peritoneum during regeneration of the pyloric coeca in *Asterias forbesi* (Anderson 1965).

In Cuvierian tubule regeneration of *Holothuria forskali*, BrdU and tritiated thymidine incorporation coupled with electron microscopic observations were used to identify dividing stem cells (Van den Spiegel et al. 2000). All three tissue layers of the regenerates, namely, the mesothelium, the connective tissue layer, and the inner

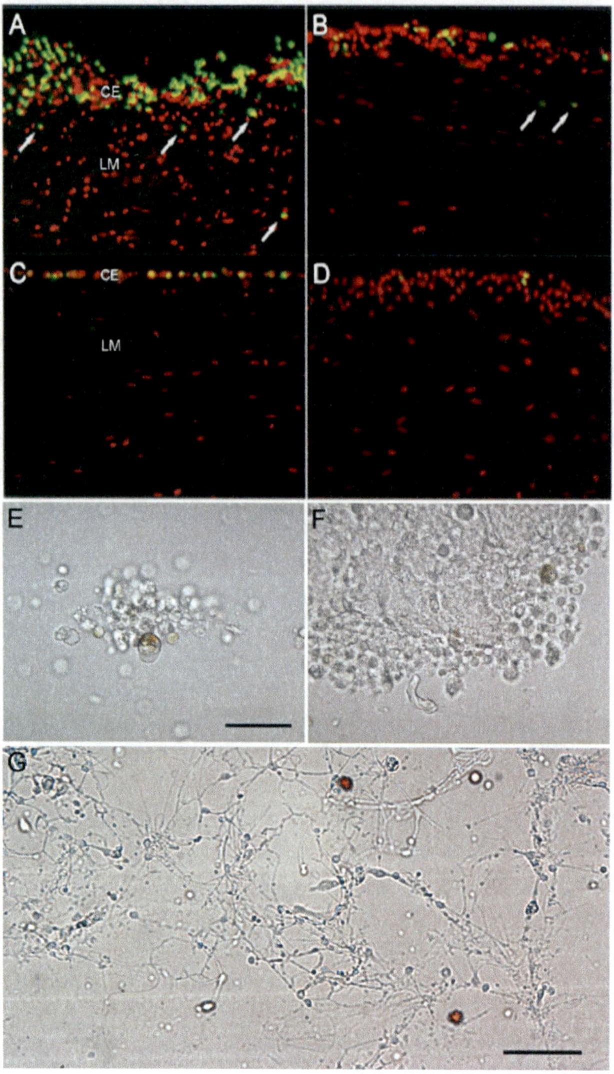

Fig. 4.3 Cell proliferation in holothuroids and echinoids. (**A–D**) BrdU (*green*) and Hoescht dye (*red*) staining in longitudinal muscle (LM) and coelomic epithelia (CE) of regenerating sea cucumber (*Holothuria glaberrima*). Sites proximal to the injury (**A**) and nearby muscle (**B**) are compared to distal sites and control (**D**) (modified after San Miguel-Ruiz and García-Arrarás 2007[3]). (**E–G**) *in vitro* proliferation of coelomocytes from unstimulated sea urchin (*Paracetrotus lividus*) after 1 (**E**), 2 (**F**) and 4 (**G**) days in culture

epithelium tubule, contained dividing cells. Cells of the mesothelium displayed also undifferentiated cell morphology, indicating a contribution to the muscle cell-pool.

---

[3]Reprinted from San Miguel-Ruiz and García-Arrarás, Common cellular events occur during wound healing and organ regeneration in the sea cucumber Holothuria glaberrima, Fig. 4 p 6, (2007).

The enteric nervous system can undergo regeneration within 3–5 weeks. During this process, new neurons are generated by cell division as assessed by BrdU incorporation studies (García-Arrarás et al. 1999). It has been proposed that enteric neurons (except for cathecolaminergic cells) arise either by an epimorphic mechanism in which precursors present in the mesentery divide and then differentiate, or by extensive rearrangement from differentiated tissues (García-Arrarás et al. 2001). Regeneration in crinoids and ophiuroids have been recently described, with a focus on the morphology of stem cells, their proliferation sites and migration paths (Candia et al., Chapter 7 of this book). Generally, the literature confirms the striking plasticity of echinoderms in repair and regenerative processes by both, proliferation of mesenchymal cells in the circulating fluids and transdifferentiation (Candia-Carnevali et al. 1997; Van den Spiegel et al. 2000; García-Arrarás et al. 1998; Dolmatov 1996).

4.3.3 The Echinoid Immune Reservoir: Renewal from Stem Cells

The best studied adult tissue which is able to rejuvenate during adult life is the immune/haematic system (Bryder et al. 2006). Immune functions in echinoderms, are accomplished by effector cells known as coelomocytes, a heterogeneous population of free moving cells found in all coelomic spaces, including the perivisceral coelomic cavities, the water-vascular system and the hemal system (reviewed in Matranga 1996; Gross et al. 1999; Glinski and Jarosz 2000; Smith et al. 2006). They are also present sparsely in the connective tissue and amongst tissues of various organs (Munoz-Chapuli et al. 2005; Pinsino et al. 2007). Classification of coelomic cells, are based on their morphology as observed by optical or electron microscopy (Smith 1981; Smith 2005; Matranga 1996; Matranga et al. 2005). The most abundant cell population (80–85%) are phagocytic cells which undergo a morphological conversion from "petaloid" to "fillopodial" (dendritic-like) phenotype (Edds 1993; Henson et al. 1999; Matranga et al. 2006). Authors have found that minute numbers of small phagocytes, characterized by a larger nucleus/cytoplasm ratio (resembling stem cells) display low phagocytic activity calling for their stemness properties (Gross et al. 1999, 2000). Attempts to establish cultures of propagating cells *in vitro*, aimed to identify progenitor cells of the different morphotypes, produced unfortunate results so far. In fact, after a short period of rapid division (1–2 days), cell clumps usually tend to spread and change their phenotype into a fibroblastoid morphology (Fig. 4.3, E–G).

It has been observed that sea urchin coelomocytes share properties with some lymphocytes, specifically "natural killer" (NK) cells. By flow cytometry, 5–10% of coelomocytes from two genera of sea urchins, *Strongylocentrotus purpuratus* and *Eucidaris tribuloides*, displayed a reactivity for murine monoclonal antibodies to CD56 and CD57 (Koros and Pulsford 1994), as *Lytechinus variegatus*, *Echinus esculentus* and *Arbacia punctulata* (Koros et al. 2002). Interestingly, the CD56 antigen, a specific leucocyte marker, is often associated with cellular pluripotency

(Young et al. 1999). *In vivo* and *in vitro* labelling experiments showed thymidine incorporation in petaloid phagocytes and amoebocytes while circulating in the perivisceral coelomic fluid, thus demonstrating that they can proliferate in circulation (Holland et al. 1965). Evidence for the proliferation of red amoebocytes, which account for 4–5% of total coelomocytes, has also been recorded under various *in vitro* conditions (Matranga et al. 2000, 2002, 2005, 2006; Angelini et al. 2003), as well as in sea urchins collected from the field (Matranga et al., 2000, 2005, 2006; Pinsino et al. 2008). Other examples of rapid renewal of circulating coelomocytes upon stress comes from studies on *Holothuria polii* and *Asterias rubens* with a two-fold rise in the total cell population (mostly phagocytes) within 24 h (Canicattì and D'Ancona 1989; Oweson et al. 2008), observed by incorporation of thymidine and BrdU, respectively. Thus, similar to vertebrate haematopoiesis, echinoderm blood cells may portray several different types of progenitor cells within the coelomic fluid.

The origin of circulating coelomocytes in the adult remains poorly investigated although it is of great interest, given the obvious implications in wound healing and regeneration phenomena (Matranga et al. 2005; Candia-Carnevali 2006). The axial organ has been regarded as the ancestral lymphoid gland since reports described the release of coelomocytes upon injury in echinoids (Millott 1969) and their cellular heterogeneity based on surface lectins (Leclerc et al. 1980; Leclerc and Bajelan 1992). Two sub-populations of cells, adherent and non-adherent cells, resembling mammalian B and T lymphocytes, respectively, have been fractionated from the axial organ of the sea star *Asteria rubens* (Leclerc et al. 1980; Leclerc and Bajelan 1992). The dividing capacity of cells has been assessed by brief exposure to tritiated thymidine, which showed cells retaining the label in the visceral and parietal peritoneum, the epithelium lining all parts of the water vascular system, the axial organ, the Polian vesicles, the haemal strands and the dermal connective tissue of the body wall (Holland et al. 1965). Above may serve as possible sites for coelomocyte production and stem cell reservoirs.

4.3.4 Stemness Genes in Echinoderms

The H2A.Z gene encodes an evolutionary conserved histone variant, whose expression is associated with transcriptional regulation, anti-silencing, silencing and genome stability (Zlatanova and Thakar 2008). Transcripts are enriched in undifferentiated stem cells and decline during their differentiation (Zhong-wei et al. 2001; Ivanova et al. 2002; Easterday et al. 2003; Sato et al. 2003; Faast et al. 2001). H2A.Z was identified in screenings for downstream targets of both the multipotent transcription factors *Oct-4* and *nanog* (Zhong-wei et al. 2001; Boyer et al. 2005). H2A.Z expression is correlated with multipotency in larvae of the polychaete *Hydroides elegans* and the sea urchin *Strongylocentrotus purpuratus* both undergoing indirect development (Arenas-Mena et al. 2007). Undifferentiated precursor cells of various larval organs maintain a strong H2A.Z expression.

Piwi/aragunaute family members serve as epigenetic regulators of stem cells in many systems (Hatfield et al. 2005; Hammond 2006). Their homologues were identified in the sea urchin (Juliano et al. 2006). In accordance with what is known in other systems, sequence analysis revealed that each of the four *Argonautes* examined in *Strongylocentrotus purpuratus* embryos contains conserved residues important for RNAseH activity within the *Piwi* domain (Song and Wessel 2007). In addition, whole-mount *in situ* hybridization showed that all of the important players of the RNAi silencing pathway are abundant in oocytes and eggs, but localize differently (embryonic territories and timing) during development (Song and Wessel 2007).

The sea urchin TRBP (transactivating response RNA binding protein) mRNA accumulates in the coelomic pouches of 72 hr larva, where conserved markers of early primordial germ cell development, such as *vasa* and *nanos* transcripts, are localized (Juliano et al. 2006). In mammalian cells, TRBP is expressed in spermatids where it is involved in translational repression of protamine mRNA translation and regulation in male germ cell differentiation (Lee et al. 1996; Siffroi et al. 2001). *Seawi*, a sea urchin protein member of the *piwi* family (Rodriguez et al. 2005) accumulates within the four micromeres at the vegetal pole of the embryo, those cells that will later give rise to the germ line cells (Juliano et al. 2006). Germ line stem cells homologues were identified in the sea urchin *Strongylocentrotus purpuratus* (Sp) including *boule*, CPEB1, germ cell-less (*gcl*), DMRT, *mago nashi*, MORC, MSY, *ovo*, *nanos2*, *pumilio*, *seawi*, *Sox-E*, *tudor* and *vasa*. All transcripts accumulate uniformly during oogenesis, whereas Sp-*pumilio*, Sp-*tudor*, Sp-MSY and Sp-CPEB1 transcripts are also uniformly distributed during embryonic development. Sp-*nanos2*, Sp-*seawi*, and Sp-*ovo* transcripts, however, are enriched in the vegetal plate of the mesenchyme blastula stage. Sp-*vasa*, Sp-*nanos2*, Sp-*seawi*, and Sp-SoxE transcripts are localized in small micromere descendents at the tip of the archenteron during gastrulation and then enriched in the left coelomic pouch of larvae. The overall results indicate that germ line specific genes are accumulated in the small micromere descendants, thus further supporting the contribution of these cells to the germ line (Juliano et al. 2006). Interstingly, in *S. purpuratus* and *L. variegatus* the *vasa* protein, a DEAD-box RNA helicase that functions in germ line development, accumulates selectively in the 16-cell stage micromeres, and then it is restricted to the small micromeres through gastrulation to larval development (Voronina et al. 2008). In more advanced larvae (approximately 5 days in *S. purpuratus*, 3 days in *L. variegatus*), the *vasa*-positive cells are restricted to the left coelomic sac, the only cells that will give rise to the echinus rudiment and later to the young juvenile. Subseqeuntly, the number of *vasa*-positive cells increases significantly in the left coelomic pouch of larvae and following metamorphosis, *vasa* is expressed in the germ cells of the developing juvenile gonads (Voronina et al. 2008). However, *vasa* expression can be induced in other cell lineages that would normally never express the protein, demonstrating that sea urchins do not have obligate primordial germ cells determined in early development and that *vasa* expression in this embryo is restricted early by translational regulation to the small micromere lineage (Voronina et al. 2008).

It is generally believed that in adult tissues telomerase remains active only in germ cells (Wright et al. 1996) and in those cell types, which require a high proliferative capacity, such as stem cells (Sharpless and DePinho 2004; Zimmermann and Martens 2008). Contrary to vertebrate systems (Forsyth et al. 2002), in echinoderms, telomerase activity is found throughout early stages of development in *L. variegatus* and adult tissues including Aristotle's lantern muscle, ampullae, esophagus, intestine, tube feet, male and female gonads of both *L. variegatus* and *S. franciscanus* coupled to the lack of age-associated telomere shortening (Francis et al. 2006). This, in turn, may represent a continuous need for stem cells during all life history stages of echinoderms.

4.4 Stem Cells in Platyhelminthes

Flatworms are unsegmented triploblastic animals, flattened dorsoventrally, with bilateral symmetry but without circulatory and respiratory systems, an anus and a coelom. A pharynx and a blind gut form the digestive system, whereas a solid mass of tissue, called mesenchyme or parenchyma, fills the space between the epidermis and the gut. The phylum contains both free-living species in fresh or salt water and parasitic species with high tissue plasticity (Reiger 1986). Flatworms exhibit diverse modes of asexual reproduction, such as architomy, in which fission occurs before the formation of new organs, or paratomy (budding) where the development of new organs take place in the presence of the old organs, before fission (Hendelberg and Åkesson 1991; Egger et al. 2007). Both architomy- and paratomy-reproducing groups exhibit pronounced regeneration capabilities (Sánchez Alvarado 2000; Reuter and Kreshchenko 2004; Egger et al. 2007).

4.4.1 The Neoblasts

Studies on free-living members (Class Turbellaria) revealed that planarians' bodies are composed of two main cellular compartments. A functional compartment (70–80% of total cells), made of 12–15 non proliferating differentiated cell types, and a mitotic population of cells (neoblasts), which comprise up to 30% of the total cells in the parenchymal space (Ladurner et al. 2000; Newmark and Sánchez-Alvarado 2000). In *Macrostomum lignano*, the total number of neoblasts compromises approximately 6.5% of all cells, (Bode et al. 2006) and in *Convolutriloba longifissura* neoblasts, over 15% (Gschwentner et al. 2001).

"Neoblasts" is a term often used to indicate small (about 10μm), basophilic cells, round to ovoid in shape, present in *Planaria* and other platyhelminthes. Ultrastructurally, neoblasts consist of a big nucleus with a prominent nucleolus and a scanty rim of cytoplasm filled with free ribosomes, a few mitochondria, and some or no endoplasmic reticulum (Pedersen 1972; Palmberg 1990; Morita 1995; Hori 1997). These stem cells are the only dividing cells in these organisms, and underlie post-embryonic development, homeostasis, starvation, asexual reproduction, and regeneration. Using thymidine labeling experiments, a multilineage contribution of neoblasts was verified into several cell types including epidermal cells, rhabdites, muscle and nerve cells, gastrodermal and germ cells (Thouveny and Tassava 1998; Wikgren et al. 1971; Palmberg and Reuter 1983; Drobysheva 1986, 1988). Cells similar to neoblasts, called "germinative cells", have also been reported in several parasitic Platyhelminthes, and their paths of differentiation to some somatic cell types and germ cells were described (Gustafsson 1976).

Detailed studies, however, revealed that only 3–5% of neoblasts (1–2% of the total cells of the organism) represent true stem cells, whereas the rest are transit amplifying cells or post mitotic cells, which are committed to differentiation (Baguñà and Slack 1981; Saló and Baguñà 1984; Baguñà et al. 1990). Neoblast populations are not homogeneous. In *Schmidtea mediterranea*, a subpopulation of neoblasts was found to arrest in G2 phase of the cell cycle (Baguñà 1976). Saló and Baguñà (1984) identified in *Girardia tigrina* two different populations of fast and slow cycling neoblasts. At least, 2% of all somatic neoblasts in *Macrostomum* sp. do not enter S-phase within 48 hr (Nimeth et al. 2004; Bode et al. 2006) and are considered as a slow-cycling subpopulation. The presence of subtypes of neoblasts had also been supported based on specific staining (Schürmann and Betz 1998), quantitative video microscopy (Behensky and Schürmann 2001) and neoblasts' ultrastructure (Rieger et al. 1999; Higuchi et al. 2007). Recent expression patterns in flatworms of several stem cell markers including *piwi* (Reddien et al. 2005; Rossi et al. 2006) and *nanos* (Sato et al. 2006; Wang et al. 2007; Handberg-Thorsager and Saló 2007) have independently attested to the existence of different somatic and germ-line restricted sub-types.

4.4.2 Neoblasts in Regeneration

Free-living flatworms exercise diverse regeneration capacities, from regeneration of all organs to no regeneration at all (Egger et al. 2007). Such variation in regenerative abilities is associated with the biology of neoblasts, including their exhaustion and multilineage contribution.

Isolated tissue pieces of about 10,000 cells (1/300 of the total body of *Planaria*), anywhere between the head and the tail, will regenerate all somatic and germ cell types of the animal (Montgomery and Coward 1974). Early experiments (Wolff and Dubois 1948) showed that X-irradiation prevents regeneration by causing a progressive disappearance of neoblasts. However, a worm with an unirradiated portion (either shielded region or an unirradiated graft) regenerates fully, even when the healthy portion is away from the amputation site. Furthermore, the time taken for regeneration to begin increases with distance, and negatively correlates with neoblast numbers. Following this evidence of stemness, small fragments of non-irradiated donor tissue grafted into entire irradiated hosts, transformed host into a donor phenotype (Lender and Gabriel 1965; Chandebois 1976; Teshirogi 1967). In pharynx-level amputations of *Macrostomum lignano*, in which all gonadal tissues are removed, complete regeneration of the gonads and their function was demonstrated by breeding experiments in fully regenerated animals (Egger et al. 2006). In similar experiments, Morgan (1902) removed all gonads in *Schmidtea lugubris*, and the animals regenerated to fully functional sexual organisms. More recently, purified neoblasts were tested for their regenerative capabilities by engrafting them into irradiated hosts. In contrast to control irradiated or to engraftment of differentiated cells, only engraftment of enriched neoblasts increased animal's lifespan significantly (Baguñà et al. 1989) and lead, in 20% of the organisms, to complete recovery. Moreover, while injection of neoblasts from the asexual strain into irradiated hosts of the sexual strain rescued the host and transformed it into an asexual individual able to reproduce by fission, the introduction of neoblasts from the sexual strain into irradiated asexual hosts altered the later into individuals capable of mating and laying cocoons (Baguñà et al. 1989).

The most noticeable feature of neoblasts observed by electron microscopy is the germline granule-like "chromatoid body", present also in male and female germ cells of sexual planarians (Coward 1974). During regeneration, the chromatoid bodies decrease in number and size during the process of cell differentiation, and are not visibile in differentiated cells, except of the germ cells (Hori 1982, 1997). Recently, two *vasa* related genes were isolated from the aquatic flatworm *Dugesia japonica* which exhibits a sexually and an asexually reproducing strain. In addition to their expression in ovaries and testes, one of the transcripts was also expressed within neoblasts (Shibata et al. 1999). During regeneration, *vasa* positive cells accumulated into the stump region and later observed accumulating in the growing blastema, followed by their appearance in regenerating brain and pharynx. X-ray irradiation, used to eliminate neoblasts, drastically reduced the number of these *vasa* positive cells.

Several attempts had been made to determine the cellular contribution of neoblasts to maintenance and regeneration in planarians. Previous observations revealed that neoblasts are the only dividing cells in both *Microstomum lineare* (Palmberg and Reuter 1983) and *Convoluta* sp. (Drobysheva 1986). In *Microstomum lineare*, one-hour incubation in tritiated thymidine showed numerous labeled neoblasts, scattered throughout the parenchyma. One hour after cutting, these cells migrated to the wound, and by six hours they clustered and formed the primordia of the brain, the ciliated pits and the pharynx (Palmberg 1986, 1990), followed by differentiation into epithelial, muscle, neuron, rhabdite, and flame cells. BrdU labeled neoblasts in *Macrostomum* sp. were found in two bands along the lateral sides of the body. Pulse-chase experiments revealed that labeled neoblasts migrate to unlabeled areas, differentiate into several somatic cell types and contribute to extensive renewal of epithelial cells and germ cells during normal tissue maintenance (Ladurner et al. 2000) and to blastema formation during regeneration (Newmark and Sánchez-Alvarado 2000). Can a single neoblast contribute to a whole organism, just like a zygote? In the parasitic flatworm *Taenia crassiceps*, individual trypsinated cells from adults injected into the mouse host, develop into complete cysticerci (Toledo et al. 1997). Therefore, in this tapeworm, a single totipotent cell, other than a zygote, is able, in an adequate environment to contribute to a whole animal.

In addition to tissue homeostasis and regeneration, neoblasts appear to regulate other traits in flatworms such as organism growth, lifespan, and rejuvenation. Freshly hatched flatworms are able to regenerate at the same rate as mature adults (Brøndsted 1942, 1969; Hagleithner 1946) resulting from similar distribution and cell cycle patterns of neoblasts (Egger et al. 2006). However, young rhabdocoelates of the genus *Mesostoma* regenerate better than adults (Fulinski 1922) and amputated animals can rejuvenate by regeneration (Haranghy and Balazs 1964; Egger et al. 2006). Specimens of *Macrostomom lignano* amputated 45 times, effectively doubled their lifespan, and outlived non-amputated specimens by 10 months, in culture conditions (Egger et al. 2006). A similar extended lifespan effect was demonstrated in triclads after starvation (Haranghy and Balazs 1964) and in the marine platyhelminth *Macrostomom* sp. (Egger et al. 2006), where following starvation and subsequent feeding, a 55-fold increase in neoblast numbers was observed after six hours (Nimeth et al. 2004).

4.5 Adult Stem Cells in Coelenterates (Cnidaria)

The phylum Cnidaria includes three distinct groups, mostly marine organisms, the anthozoans (corals and sea anemones), scyphozoans and cubozoans (jellyfish), and hydrozoans. Typically, cnidarian animals are diploblastic, signifying a body wall consisting of only two epithelial layers (outer epidermis of an ectodermal origin and an inner gastrodermis of an endodermal origin) separated by an acellular milieu, the mesoglea. The body consists of a single gastrovascular cavity, or coelenteron that can be morphologically simple or highly complex.

Most observations on stem cell functions and behaviors in this phylum were carried on the fresh-water genus *Hydra*, although most hydrozoans are marine animals. *Hydra* is a simple metazoan, abundant in fresh-water ponds and streams, depicting the most basic characteristic morphology of Cnidarians. Studies (reviewed by Bode 1996; Bosch 2007) revealed that the two epithelial layers are maintained, respectively, by two separate epithelial stem cell lines, capable of differentiating into 20–25 different cell types (Campbell 1967a,b). These stem cells enable hydras to bud new polyps and to regenerate when the body wall is injured. Self-renewal capacity of the two epithelial stem cells was experimentally demonstrated more than four decades ago, by using *Hydra* heterocytes (Lenhoff 1965), which contained a mixed cell population from a normal budding strain and a non budding mutant strain. The heterocytes underwent a gradual transformation of phenotype in which the mutant phenotype eventually became dominant in all instances, with the rate depending upon the number and location of mutant cells in the original heterocytes (Lenhoff 1965).

Although the ectodermal and the endodermal epithelial cell lines maintain the two *Hydra* cellular layers, they are not "true" epithelial stem cells, since they exhibit differentiated characteristics of epithelio-muscular cells and/or digestive cells (Bode 1996). However, studies revealed cases of transitions from one epithelial cell to another and regeneration of a complete *Hydra* from a single differentiated somatic cell type (Davis 1970, 1973; Burnett et al. 1973), or the retaining of regeneration capacity in I-cell free strains (Terada et al. 1988). These studies indicate that the ectodermal and the endodermal epithelial cell lines have more stemness capabilities than has been acknolwedged. For example, both epithelial cells in the hydra body column undergo continuously self-renewing mitotic divisions (Dübel et al. 1987) with cell cycle duration of about three days (David and Campbell 1972; Bosch and David 1984). Single GFP-expressing endodermal cells that had been transplanted into wild-type host animals proliferated and actively migrated within the host tissue to distant polyp sites (Wittlieb et al. 2006). Ultimately, this proliferation generated polyps in which the entire endothelium contained the transgene, thus showing that *Hydra* epithelial cells are capable, by successive divisions, to continuously maintain different types of specialized cells such as tentacle or foot specific epithelial cells.

4.5.1 Interstitial Stem Cells

Interstitial stem cells (I-cells) are considered to be the third stem-cell line in the Hydrozoa (not known from Scyphozoa or Cnidaria; Frank et al. 2001; Frank et al., Chapter 3 of this book). I-cells are distributed uniformly along the body column in interstitial spaces of epithelial cells (mostly ectodermal cells), hence their name. I-cells consist of fast cycling, multipotent cells, which contribute to both soma and germ-line, and are probably composed of sub-populations with potencies restricted to somatic and germ cell pathways (Bosch and David 1986; Nishimiya-Fujisawa and Sugiyama 1993). I-cells are continuously dividing in an adult hydra (Campbell and David 1974), with an average cell cycle duration of 24 h. Single cloned I-cells may differentiate into three classes of somatic cell lineages, including secretory cells (gland cells and mucous cells), neurons (several types of sensory cells and ganglion cells), four types of nematocytes and the gametic lineage. The multilineage contribution of I-cells was elucidated by developing I-cell-free animals using hydroxyurea (Sacks and Davis 1979) or nitrogen mustard (Diehl and Burnett 1964), and subsequently by supplementing new I-cells from dissociated normal animals (David and Murphy 1977). Similarly, in colonies of *Hydractinia echinata*, mitomycin elimination of I-cells resulted in the loss of all differentiation derivatives including germ cells. When such animals were repopulated with I-cells from an opposite sex, chimeric entities, with recovered budding activity, emerged (Müller et al. 2004), converting animals to the donor phenotype with sex reversal (Frank et al. 2001).

Despite their uniform morphology, the I-cell population appears to be functionally heterogeneous (David 1973). I-cell lineage probably consists of a mixed population of stem cells with multipotent identities confined to various destinies and cellular fates. Some are restricted to the ectoderm, other intermediates migrate to the endoderm (Smid and Tardent 1986), and others are sequestered into the germ line. This is manifested in "pseudo-epithelial hydra" in which polyps are enriched with a subpopulation of I-cells that differentiate only to the germ-line (Littlefield

1985, 1991 for *Hydra oligactis*; Nishimiya-Fujisawa and Sugiyama 1993, 1995 for *Hydra magnipapillata*) and express the germ-line markers *nanos* (Mochizuki et al. 2000) and *vasa* (Mochizuki et al. 2001). A common characteristic of all three classes of stem cells in *Hydra* is that they are highly regulative during adult life. This is evident during periods of starvation, growth, and budding (Bode et al. 1973), when a constant change of population size and density is taking place.

4.5.2 Stem Cells in Regeneration

Another function of stem cells exemplified throughout this phylum, is the capacity to regenerate missing tissues and germ layers. In *Hydra*, pieces of the body column, representing as little as 2% of tissue mass, or aggregates of dissociated cells isolated anywhere between head and foot, would regenerate an entire animal. Such animals, when fed, grow and reproduce asexually by budding, developing progenies with all cell types (Bode and Bode 1980; Shimizu et al. 1993). These abilities are attributed to epithelial stem cells. When I-cells are removed, the remaining epithelial polyps keep on performing head and foot regeneration and budding (Marcum and Campbell 1978; Sugiyama and Fujisawa 1978). Adult polyps of the box jelly *Carybdea marsupialis* regenerate oral parts following sectioning below the tentacle crown. Furthermore, bisected individuals in the mid-gastric region regenerate both oral and aboral parts. Even polyps sectioned into three fragments of similar size, regenerate the missing fragments (Fischer and Hofmann 2004). Polyps of *Tubularia*, *Aurelia aurata*, *Obelia plana*, *Corymorpha*, *Hydractinia echinata*, *Cordylophora* and *Podocoryne carnea* all show remarkable abilities to regenerate upon fragmentation or cutting (Barth 1940; Müller et al. 1986; Aizu 1967, 1968). In *Corymorpha*, a mixture of cells produced by grinding the body column will form a polyp, if left undisturbed for 3–4 days (Child 1928; Barth 1940). Similar results were obtained with aggregates of *Hydractinia* (Müller et al. 1986). Young polyps of *Hydractinia* excised at the body column, regenerate both oral and aboral ends (Müller et al. 1986).

Such broad regeneration abilities are not restricted to hydrozoans and scyphozoans, but are also present in anthozoans (corals and sea anemones). Sea anemones reproduce asexually by transversal or longitudinal fission and pedal laceration (Polteva and Lenicque 1970; Hand and Uhlinger 1995). Their regenerative abilities were studied in *Actinia*, *Sagartia*, *Nematostella* and *Metridium* (Abeloos 1957, 1961; Lenicque and Féral 1977). When a column is cut transversely at any level, the aboral section regenerates a new oral disc. Pedal disc pieces of any size regenerate into complete new polyps within a few weeks. In these species, regenerative abilities along the oral-aboral axis are not related to fragment size (Hand and Uhlinger 1995) but to the differentiation potential of the regenerative cells. While ectodermal cells preserve total ectodermal potential and proliferate in response to regeneration, endodermal cells are irreversibly differentiated. During regeneration of the pedal disc, the mesenteries degenerate, then de-differentiate to form the endodermal tissues (Lenicque and Féral 1977).

While the regeneration of injured or lost body parts in corals and other sessile cnidarians has been documented as essential for recovery and survival of the animals, practically nothing is known about the roles of stem cells in this process, or even whether they exist at all. For example, the capacity of an injured coral to regenerate from wounds is positively correlated with increasing body size (i.e., Kramarsky-Winter and Loya 2000). This trait may reflect the relative proportion of injured to healthy tissue, the availability of resources derived from adjacent healthy tissue (Oren et al. 1997) and while not studied, may mark the availability of stem cells, their stemness, and numbers. The importance of an "organizing centre" controlling regeneration has also been documented, particularly in solitary fungiid corals. Generally, a mouth must be present in order for these corals to undergo tissue and skeletal regeneration (Kramarsky-Winter and Loya 2000).

4.6 Trans-Differentiation in Aquatic Invertebrates

The differentiation potential of adult stem cells is considered a basic attribute of tissue and organ regeneration in multicellular organisms. However, conversion of pre-existing differentiated cells into new cell lineages, i.e., trans-differentiation, appears to serve as an additional mechanism for generating tissue plasticity, playing a significant role in aquatic invertebrates.

In *Hydra*, different neuron types differentiate according to their location along the body column. The neuropeptide FMRFamide expressing neurons (FLI+ neurons) are found principally in the tentacles and hypostome, but not in the upper or lower body column (Grimmelikhuijzen et al. 1982). Thus, decapitation leads to complete removal of all FLI+ neurons. Treating animals with hydroxyurea, which removes all stem cell and neuron precursors (Sacks and Davis 1979) followed by decapitation, leads to regeneration of the head including FLI+ neurons (Koizumi and Bode 1986). As these cells could not have developed by differentiation from stem cells, they could have been formed by conversion of existing neurons. Similar results were obtained from a subset of neurons positive for a vasoperssin-like immunoreactivity (Koizumi and Bode 1991) and more dramatically, from conversion of ganglion cells into sensory cells (Koizumi et al. 1988). Thus in *Hydra*, either ganglia or sensory neurons, are able to change their phenotype during development and regeneration scenarios through trans-differentiation.

Recently, transgenic GFP *Hydra* has been used to track cells expressing Dickkopf, a negative regulator of Wnt signaling, which is specifically expressed in differentiated zymogen gland cells (Siebert et al. 2008). In normal polyps, GFP positive zymogen cells are located in the gastric region within the endodermal layer, whereas mucous gland cells are located in the head region. During their movement into the head region, zymogen gland cells gradually lose GFP expression and change their phenotype. When transgenic polyps were cut within the boundary of GFP zymogen cells, regenerated animals contained numerous GFP zymogen cells in the head region, which subsequently lost their GFP expression upon differentiation to other cell types. In both scenarios, upon entering the head, eGFP zymogen cells acquired a phenotype of mucous glandular cells and expressed *HyTSR1*, a marker for mucous glandular cells (Siebert et al. 2008).

In the jellyfish *Podocoryne carnea*, isolated mononucleated cross-striated muscle cells are able not only to undergo complete pluripotent trans-differentiation but also to carry out an entire *in vitro* regeneration process (Schmid and Alder 1984). Striated muscle cells explanted from the subumbrella and cultured in the presence of collagenase, form an outer ectodermal layer, an inner endodermal layer separated by a basal lamina, and develop a manubrium (the sexual and feeding organ of the animal) with characteristic cell types including smooth muscle cells, sensory cells and nematocysts (Alder and Schmid 1987). This trans-differentiation process can also be initiated by treatment with extra-cellular matrix (ECM) degrading enzymes or by treatment of the isolated tissue fragments with a monoclonal antibody developed against the ECM of the medusa (Reber-Müller et al. 1994), suggesting that disruption of ECM mediates trans-differentiation and cellular plasticity. Interestingly, the evolutionarily conserved pluripotent stem cell gene *piwi* is expressed in the adult medusa's somatic muscle cells and is upregulated during trans-differentiation of striated muscle cells (Seipel et al. 2004).

The polyp of the hydroid *Aurelia aurita* is free of I-cells. The outermost part of the scyphopolyp is a single layer of ectodermal cells consisting of epithelio-muscular cells and nematocysts, while the innermost endodermal layer forms the wall of the gut and consists of epithelio-muscular and gland cells. Ectoderm fragments separated from the polyp can regenerate into a whole polyp including ectoderm and endoderm (Steinberg 1963). While regeneration of endodermal tissue fragments from the polyp ends in disintegration, tissue fragments taken from the epithelio-muscular (ectoderm) tissue may trans-differentiate into endodermal cells and subsequently regenerate complete polyps (Aizu 1968). Fragmented ectodermal tissue masses become spherical and hollow as of 6–12 h after isolation. A few cells, originating from the epithelio-muscular ectodermal layer, appeare in the hollow space increasing in number until forming a layer of tissue within the ectodermal sphere. This layer forms the wall of the gut. After 4–5 days, a mouth opens, crowned by a few tentacle buds. Similar results were obtained with ectodermal fragments from *Obelia plana* and *Coryne* (Aizu 1967), in which trans-differentiation of endoderm from ectodermal fragments resulted in complete regeneration of a polyp. In flatworms, the interest in trans-differentiation revolves around the origin of neoblasts, which form the blastema. Lang (1912), Steinmann (1925) and more (Coward 1969; Hay and Coward 1975) have suggested that the blastema cells are formed from differentiated cells.

While reports claim for de-differentiation, the terminology and experimental proof discriminating between transdifferentiation and dedifferentiation is ambiguous. Early histological observations showed that during regeneration, germ cells leave the gonads and take part in blastema formation (Gremigni 1981). This was experimentally tested by using *Dugesia lugubrispolychroa* specimens in which somatic cells are triploid whereas male and female germ cells are hexaploid. In sexually mature specimens, regeneration was induced from stumps at the level of gonads.

Karyological analysis of squashed blastemas from stumps devoid of gonads showed that all dividing blastema cells had 12 chromosomes. However, 5% of blastemal cell from testes level stumps had either eight chromosomes or a number of chromosomes varying between 8 and 12, similar to germ cells (Gremigni and Puccinelli 1977). In blastemas formed from stumps at the level of ovaries, 2% of the cells had 24 chromosomes, similar to that of germ cells (Gremigni and Puccinelli 1977). These germ-cell derived cells took part not only in blastema formation but in the regeneration of somatic tissues as well. Karyological analyses showed that some diploid or some hexaploid cells that come from the gonads are present in the fully regenerated cephalic or caudal areas (Gremigni et al. 1980a) and take part in rebuilding somatic tissues. Similar results were obtained by an experimental set-up of prolonged starvation, which led to progressive decrease and loss of gonads. Gonads reappeared in re-nourished animals (Gremigni and Puccinelli 1979).

Planarians also convey trans-differentiation from male to female germ cells. In sexually mature specimens that were cut just behind ovaries, the posterior stumps that possessed testes rebuild anterior fragments with new ovaries including oocytes. The oocytes were karyologically analyzed revealing some with eight bivalents, in addition to the high number of 12 bivalents oocytes (normal situation). All control specimens had the normal karyotype of 12 bivalents. This data suggest migration of male germ cells with eight chromosomes into the blastema during regeneration and their trans-differentiation into female germ cells. Furthermore, cytophotometric analyses of pharynx muscle cell-nuclei from regenerating stumps at the level of testes revealed 5% of pharynx muscle cells had DNA content similar to male germ cells (Gremigni et al. 1980b). Thus in flatworms, both karyological and cytophotometric methods demonstrated trans-differentiation of male germ cells during regeneration into female germ cells or pharyngeal muscle cells.

Trans-differentiation also appeared in echinoderm and tunicate regeneration assays. In eviscerated or experimentally transected holothurian *Eupentacta fraudatrix*, the inner luminal (digestive, endoderm in origin) epithelium of the gut regenerates from anterior and posterior gut rudiments, which extend toward the center until forming a continuous digestive tube. While the posterior gut rudiment originates from cloacal lining epithelium (endoderm in origin), the anterior gut rudiment derives from myoepithelial cells of the mesothelium (mesodermal in origin) by trans-differentiation of both myoepithelial and peritoneal cells (Mashanov et al. 2005). The micro intra-structure of the trans-differentiated cells revealed myoepithelial cells with dense spindle-like myofilaments (Dolmatov 1992; Dolmatov 1996) and peritoneal cells with bundles of intermediate filaments cut into short fragments. In tunicates, trans-differentiation was reported during budding processes (Fujiwara and Kawamura 1992; Kawamura and Fujiwara 1994, 1995). In *Polyandrocarpa misakiensis*, a bud grows from the multipotent atrial epithelium as an evagination from the parental wall, from which all tissues and organs of the adult zooid develop. The cell layer of the atrial epithelium is differentiated structurally and functionally, expressing several differentiation markers, which are lost concomitantly with bud development. Cells of the atrial epithelium express a tissue-restricted antigen, named *Pae 1* (Fujiwara and Kawamura 1992), have a high alkaline-phosphatase activity, and carry orange pigment granules all of which disappear upon differentiation (Kawamura and Fujiwara 1994).

4.7 Summary

While only four invertebrate phyla are considered in this review, we find that aquatic invertebrates from various taxa sustain one or more cell populations, which fulfill the criteria of adult stem cells. These cells exhibit cytological, functional, and genetic properties communal with vertebrate stem cells (summarized in Table 4.1). On the cyto-morphological level, stem cells from aquatic invertebrates are smaller than differentiated progenies, revealing a high nucleo-cytoplasmic ratio, a basophilic staining and a rounded cell appearance (for ascidian hemoblasts Cima et al. 2001; for planarian neoblasts Higuchi et al. 2007; for cnidarian I-cells Holstein and David 1990; for echinoderms Gross et al. 2000). Functional "stemness" features include the power for long-term self-renewal illustrated by their life-long propagation and the ability to differentiate into multiple cell types. In addition, cells with stem cell traits from aquatic invertebrates appear to share the capability of proliferation, differentiation and migration into tissues of the organism throughout adult life. They also share a principal role in homeostasis and numerous developmental and life history traits including growth and regression, asexual reproduction and regeneration of adult tissues and organs (Table 4.1). Common "stemness" genes include the cumulative expressions of Piwi, Vasa, Nanos, PCNA and Aldh within invertebrate and vertebrate stem cells (Table 4.1).

However, in comparison with vertebrate stem cells, adult stem cells in aquatic invertebrates also display a cell strategy outlined by unique features. The first feature deals with their cell-cycle kinetics. While vertebrate stem cells retain DNA label and are considered slow cycling (Bickenbach 1981), cells with stem cell traits from aquatic invertebrates appear to be fast cycling relative to differentiated progenies. The fast cycle rates are responsible for their elimination following irradiation, in ascidians (Freeman 1970; Taneda and Watanabe 1982), cnidarians (Fradkin et al. 1978) and platyhelminthes (Wolff and Dubois 1948; Baguñà et al. 1989). While stem cells in adult vertebrates are present only in extremely low numbers (Moore and Lemischka 2006), they are outnumbered in all aquatic invertebrate taxa, to a degree in which they represent one of the prevalent cell populations of a given tissue/germ layer (i.e. up to 30% in certain platyhelminthes; Ladurner et al. 2000; Newmark and Sánchez-Alvarado 2000). The large proportion of cells with stem cell traits within tissues of aquatic invertebrates is experimentally illustrated in the context of regeneration. In the urochordate *Botrylloides leachi*, any isolated minute peripheral ampullae out of hundreds of ampullae bordering the colony periphery (each possessing as low as 100–200 blood cells including epithelial and tunic cells) has enough stem cells to regenerate a whole organism (Rinkevich et al. 1995, 2007). Similarly, in cnidarians and platyhelminthes, a minimal cell cluster of 270–300 cells,

Table 4.1 Common and unique themes of adult stem cells from four different phyla

Characteristics	Urochordate ascidians	Platyhelminthes	Cnidarians	Echinoderms
Stem cell populations	1. Hemoblast 2. Epithelial	1. Nneoblasts	1. I-cells 2. Epithelial	1. Amoebocytes 2. Coelomocytes
Lineage contributions	Haemoblast: all somatic and germ line derivatives Epithelial: somatic derivatives of all three germ layers	Neoblasts: all somatic and germ line derivatives	1. I-cells: secretory (gland and mucous), neurons (sensory and gangilion), nematocytes and germ line. 2. Epithelial: all ectodermal and endodermal derivatives exept I-cell lineages.	1. Amoebocytes: ligaments, branchial nerve, feeding organ. 2. Coelomocytes: peritoneocytes, myocytes
Distinct germ & somatic stem cell lines	Yes	Yes	Yes	Yes
Migratory	Yes (hemoblasts)	Yes (neoblasts)	Yes (I-cells)	Yes (coelomocytes)
Regeneration	Yes (hemoblasts, epithelial)	Yes (neoblasts)	Yes (I-cells, epithelial)	Yes (amoebocytes, coelomocytes)
Growth/degrowth	Yes (hemoblast)	Yes (neoblasts)	Yes (I-cells, epithelial)	Yes (coelomocytes)
asexual reproduction	Yes (hemoblast)	Yes (neoblasts)	Yes (epithelial)	Yes (epithelial)
"Stemness" genes	1. Hemoblast: alkaline phosphatase, piwi; pcna (unpubl.), aldh. 2. Germ stem cells: vasa, Pl-10 atrial epithelium: pl-10, aldh, alkaline phosphatase	Neoblasts: pcna, vasa, nanos, piwi, bruno-like, pumilio	I-cells: vasa, nanos, piwi	Germ line: vasa, seawi, boule, CPEB1, gcl, DMRT, mago nashi, MORC, MSY, ovo, nanos2, pumilio, sox-E, tudor ependymin

Table 4.1 (continued)

Characteristics	Urochordate ascidians	Platyhelminthes	Cnidarians	Echinoderms
Cell proportion within organisms	Hemoblast 1:100–200	Neoblast 1:1500	I-cells 1:1500	Not determined
Distribution/location	Hemoblast: systemic (circulatory)	Neoblast: systemic (mesenchym)	I-cells: systemic (ectoderm, endoderm)	Coelomocytes: systemic (all coelomic spaces and conective tissues)
Germ cell specification	Epigenesis (budding, regeneration)	Epigenesis (regeneration)	Epigenesis (transdifferentiation)	Epigenesis
Trans-differentiation	Yes Atrial epithelium (ectoderm) to endoderm and mesoderm	Yes Male germ to pharyngeal muscle (germ to soma) male germ to female germ	Yes FMRFamide non-expressing to FMRFamide expressing (FLI− to FLI+ neurons; ectoderm to ectoderm). Ganglion to sensory cells, (ectoderm to zymogen and mucous glandular cells) (endoderm to endoderm). Striated muscle to whole body (ectoderm to endoderm and mesoderm). Outer epithelio-muscular to inner gut epithelium (ectoderm to endoderm)	Yes Coelomic epithelium to luminal gut epithelium (mesoderm to endoderm). Larval epidermis to archenteron (ectoderm to endoderm).

suffices to regenerate an adult organism (Shimizu et al. 1993). Another unique character to stem cells of aquatic invertebrates is their location and distribution within tissues. Vertebrate stem cells are confined to spatio-temporal microenvironment alcoves or “niches”. In contrast, many cells with stem cell traits from aquatic invertebrates, such as the ascidian hemoblasts (Oka and Watanabe 1959; Rinkevich et al. 2007), echinoderm coelomocytes (Candia-Carnevali 2006; Candelaria et al. 2006), cnidarian I-cells (Müller et al. 2004; Bosch 2007) and platyhelminth neoblasts (Ladurner et al. 2000; Newmark and Sánchez-Alvarado 2000) appear to be disseminated throughout the organism’s soma, populating more than a single germ layer (Table 4.1). Another unique “stemness” attribute deals with their multilineage contributions. While vertebrate stem cells exhibit primarily uni- or oligopotent- capabilities, which are restricted to the differentiation of a single germ layer, stem cells from aquatic invertebrates from all four phyla exhibit pluri- and totipotency, with differentiation potential towards cell lineages from more than a single germ layer, including the germ line (Table 4.1). We find that unlike many vertebrate organisms where germ cells are specified during embryonic development (reviewed in Extavour and Akam 2003), an additional epigenetic mechanism exists within the aforementioned phyla where germ cells may arise postembryonically from adult stem cells. Germ cells are formed during sexualisation in planaria (Zayas et al. 2005; Sato et al. 2006) and during budding in colonial ascidians (Sunanaga et al. 2007) as well as in adverse circumstances, such as regeneration in planarian (Shibata et al. 1999; Wang et al. 2007) and ascidian species (Sunanaga et al. 2006; Rinkevich et al. 2007).

An additional property uniform to all four reviewed phyla (Table 4.1) is the use of a trans-differentiation mechanism, which involves the transformation of one cell type into another. Trans-differentiation appears to occur in both developmental and physiological settings, where there is a transient need for a large pool of stem cells. This is especially evident during regeneration of lost bodily parts and during intensive growth and budding in diverse representative species from all examined phyla. These aforementioned “stemness” attributes and differences, at the cellular and organismal levels, may disclose a different strategy employed by adult stem cells to supply the unique demands imposed by life history modes and ecological settings of aquatic invertebrates.

Acknowledgements Authors acknowledge the support of the Marine Genomics Europe Network of Excellence. VM is gratefull to Rosa Bonaventura for preparation of echinoderm figure pannels and to the MoMa ASI project for partial financial support.

References

Abeloos M (1957) Origine de l’organization radiaire dans la régénération des actinies. C R Acad Sci Paris 245:727–729

Abeloos M (1961) Facteurs morphogénétiques dans la régénération des actinies. Ann Fac Sci Marseille 31:11–25

Aizu S (1967) Transformation of ectoderm to endoderm in Obelia. Sci Rep Tohoku Univ Ser V (Biol) 33:375–382

Aizu S (1968) The role of the interstitial cells on regeneration of new endodermal cells in Coryne. Sci Rep Tohoku Univ Ser V (Biol) 34:91–100
Akhmadieva AV, Shukalyuk AI, Aleksandrova YN et al (2007) Stem cells in asexual reproduction of the colonial ascidian *Botryllus tubaratus* (Tunicata: Ascidiacea). Russ J Mar Biol 33: 181–186
Alder H, Schmid V (1987) Cell cycles and in vitro transdifferentiation and regeneration of isolated, striated muscle of jellyfish. Dev Biol 124:358–369
Allsopp RC, Chang E, Kashefi-Aazam M, et al (1995) Telomere shortening is associated with cell division in vitro and in vivo. Exp cell Res 220:194–200
Alonso L, Fuchs E (2003) Stem cells of the skin epithelium. Proc Natl Acad Sci USA Supp 1:11830–11835
Amemiya S, Oji T (1992) Regeneration in sea lilies. Nature 357:546–547
Anderson JM (1965) Studies on visceral regeneration in sea-stars. II. Regeneration of pyloric caeca in Asteriidae, with notes on the source of cells in regenerating organs. Biol Bull 128:1–23
Angelini C, Amaroli A, Falugi C et al (2003) Acetylcholinesterase activity is affected by stress conditions in *Paracentrotus lividus* coelomocytes. Mar Biol 143:623–628
Arenas-Mena C, Wong KS-Y, Arandi-Foroshani NR (2007) Histone H2A.Z expression in two indirectly developing marine invertebrates correlates with undifferentiated and multipotent cells. Evol Dev 9:231–243
Baguñà J (1976) Mitosis in the intact and regenerating planarian *Dugesia mediterranea* n. sp. Mitotic studies during growth, feeding and starvation. J Exp Zool 195:53–64
Baguñà J, Romero R, Saló E et al (1990) Growth, degrowth and regeneration as developmental phenomena in adult freshwater planarians. In: Marthy HJ (ed) Experimental embryology in aquatic plants and animals. Plenum Press, New York, pp 129–162
Baguñà J, Salo E, Auladell C (1989) Regeneration and pattern formation in planarians. III. Evidence that neoblasts are totipotent stem cells and the source of blastema cells. Development 206:73–87
Baguñà J, Slack JM (1981) Planarian neoblasts. Nature 290:14–15
Balser EJ (1998) Cloning by Ophiuroid echinoderm larvae. Biol Bull 194:187–193
Bancroft FW (1903) Aestivation of Botrylloides gascoi Della Valle. Mark Anniversary 8:147–166
Barth LG (1940) The process of regeneration in Hydroids. Biol Rev 15:405–420
Behensky C, Schürmann W (2001) Quantitative analysis of turbellarian cell suspensions by fluorescent staining with acridine orange and video microscopy. Belg J Zool 131:131–136
Berrill J (1951) Regeneration and budding in tunicates. Biol Rev 26:456–475
Berrill NJ (1941) The Development of the Bud in *Botryllus*. Biol Bull 80:169–184
Betschinger J, Knoblich JA (2004) Dare to be different: Asymmetric cell division in *Drosophila*, *C. elegans* and vertebrates. Curr Biol 14:R674–R685
Bickenbach JR (1981) Identification and behavior of label-retaining cells in oral mucosa and skin. J Dent Res 60:1611–1620
Bjerknes M, Cheng H (1999) Clonal analysis of mouse intestinal epithelial progenitors. Gastroenterology 116:7–14
Bode A, Salvenmoser W, Nimeth K et al (2006) Immunogold-labeled S-phase neoblasts, total neoblast number, their distribution, and evidence for arrested neoblasts in *Macrostomum lignano* (Platyhelminthes, Rhabditophora). Cell Tissue Res 325:577–587
Bode HR (1996) The interstitial cell lineage of *Hydra*: A stem cell system that arose early in evolution. J Cell Sci 109:1155–1164
Bode HR, Berking S, David CN et al (1973) Quantitative analysis of cell types during growth and morphogenesis in hydra. Wilhelm Roux's Arch Dev Biol 171:269–285
Bode PM, Bode HR (1980) Formation of pattern in regenerating tissue pieces of *Hydra attenuata*. I. Head-body proportion regulation. Dev Biol 78:484–496
Bosch I, Rivkin RB, Alexander SP (1989) Asexual reproduction by oceanic planktotrophic echinoderm larvae. Nature 337:169–170
Bosch TC, David CN (1986) Male and female stem cells and sex reversal in *Hydra* polyps. Proc Natl Acad Sci USA 83:9478–9482

Bosch TCG (2007) Why polyps regenerate and we don't: Towards a cellular and molecular framework for *Hydra* regeneration. Dev Biol 303:421–433
Bosch TCG, David CN (1984) Growth regulation in *Hydra*: Relationship between epithelial cell cycle length and growth rate. Dev Biol 104:161–170
Boyer L, Lee TI, Cole MF et al (2005) Core transcriptional regulatory circuitry in human embryonic stem cells. Cell 122:947–956
Brøndsted HV (1942) Further experiments on regeneration-problems in planarians. K Dan Videns Selsk Biol Meddr 17:1–27
Brøndsted HV (1969) Planarian regeneration. Pergamon Press, Oxford
Brown FD, Swalla BJ (2007) Vasa expression in a colonial ascidian, *Botrylloides violaceus*. Evol Dev 9:165–177
Bryder D, Rossi DJ, Weissman IL (2006) Hematopoietic stem cells: The paradigmatic tissue-specific stem cell. Am J Pathol 169:338–346
Burighel P, Brunetti R, Zaniolo G (1976) Hibernation of the colonial ascidian *Botrylloides leachi* (Savigny): Histological observations. Bull Zool 43:293–301
Burnett AL, Lowell R, Cyrlin M (1973) Regeneration of a complete Hydra from a single differentiated somatic cell type. In: Burnett AL (ed) Biology of Hydra. Academic Press, New York, pp 225–270
Buss LW (1982) Somatic cell parasitism and the evolution of somatic tissue compatibility. Proc Natl Acad Sci USA 79:5337–5341
Buss LW (1999) Slime molds, ascidians, and the utility of evolutionary theory. Proc Natl Acad Sci USA 96:8801–8803
Cairns J (1975) Mutation selection and the natural history of cancer. Nature 255:197–200
Cameron CB, Garey JR, Swalla BJ (2000) Evolution of the chordate body plan: New insights from phylogenetic analyses of deuterostome phyla. Proc Natl Acad Sci USA 97:4469–4474
Cameron RA, Fraser SE, Britten RJ et al (1991) Macromere cell fates during sea urchin development. Development 113:1085–1091
Campbell RD (1967a) Tissue dynamics of steady state growth in *Hydra littoralis*. I. Pattern of cell divisions. Dev Biol 15:487–502
Campbell RD (1967b) Tissue dynamics of steady state growth in Hydra littoralis. II. Patterns of tissue movement. J Morphol 121:19–28
Campbell RD, David CN (1974) Cell cycle kinetics and development of Hydra attenuata. II. Interstitial cells. J Cell Sci 16:349–358
Candelaria AG, Murray G, File SK et al (2006) Contribution of mesenterial muscle dedifferentiation to intestine regeneration in the sea cucumber *Holothuria glaberrima*. Cell Tissue Res 325:55–65
Candia-Carnevali MD, Bonasoro F (1994) Mechanisms of arm regeneration in *Antedon mediterranea* (Echinodermata, Crinoidea). Anim Biol 9:83–88
Candia-Carnevali MD, Bonasoro F (2001) A microscopic overview of crinoid regeneration. Microsc Res Tech 55:403–426
Candia-Carnevali MD, Bonasoro F, Biale A (1997) Pattern of bromodeoxyuridine incorporation in the advanced stages of arm regeneration in the feather star *Antedon mediterranea*. Cell Tissue Res 289:363–374
Candia-Carnevali MD, Bonasoro F, Patruno M et al (1998) Cellular and molecular mechanisms of arm regeneration in crinoid echinoderms: The potential of arm explants. Dev Genes Evol 208:421–430
Candia-Carnevali MD, Bruno L, Donini DS et al (1989) Regeneration and morphogenesis in the feather star arm. In: Kiortsis V, Koussoulakos S, Wallace H (eds) Recent trends in regeneration research. Nano Asi Series, Plenum Press, New York, London, pp 447–460
Candia-Carnevali MD, Lucca E, Bonasoro F (1993) Mechanisms of arm regeneration in the feather star *Antedon mediterranea*: Healing of wound and early stages of development. J Exp Zool 267:299–317

Candia-Carnevali MDC (2006) Regeneration in Echinoderms: Repair, regrowth, cloning. Invertebr Surviv J 3:64–76

Canicattì C, D'Ancona G (1989) Cellular aspects of *Holothuria polii* immune response. J Invert Pathol 53:152–158

Chandebois R (1976) Histogenesis and morphogenesis in Planarian regeneration. Monogr Dev Biol 11:1–182

Child CM (1928) Axial development in aggregates of dissociated cells from Corymorpha. Physiol Zool 1:419–461

Cima F, Perin A, Burighel P et al (2001) Morpho-functional characterization of haemocytes of the compound ascidian *Botrylloides leachi* (Tunicata, Ascidiacea). Acta Zoologica 82: 261–274

Clevers H (2005) Stem cells, asymmetric division and cancer. Nat Genet 37:1027–1028

Conboy MJ, Karasov AO, Rando TA (2007) High incidence of non-random template strand segregation and asymmetric fate determination in dividing stem cells and their progeny. PLoS Biol 5:E102

Consortium SUGS (2006) The genome of the sea urchin *Strongylocentrotus purpuratus*. Science 314:941–952

Cotsarelis G, Sun TT, Lavker RM (1990) Label-retaining cells reside in the bulge area of pilosebaceous unit: Implications for follicular stem cells, hair cycle, and skin carcinogenesis. Cell 61:1329–1337

Coward SJ (1969) Regeneration in planarians: Some unresolved problems and questions. J Biol Psych 15–19

Coward SJ (1974) Chromatoid bodies in somatic cells of the planarian: Observation on their behavior during mitosis. Anat Rec 180:533–546

David CN (1973) A quantitative method for maceration of hydra tissue. Wilhelm Roux's Arch Dev Biol 171:259–268

David CN, Campbell RD (1972) Cell cycle kinetics and development of *Hydra attenuata* I. Epithelial cells. J Cell Sci 11:557–568

David CN, Murphy S (1977) Characterization of interstitial stem cells in hydra by cloning. Dev Biol 58:372–383

Davidson EH, Peterson KJ, Cameron RA (1995) Origin of bilaterian body plans: Evolution of developmental regulatory mechanisms. Science 270:1319–1325

Davis LE (1970) Cell division during dedifferentiation and redifferentiation in regenerating isolated gastrodermis of Hydra. Exp Cell Res 60:127–132

Davis LE (1973) Ultrastructural changes during dedifferentiation and redifferentiation in regenerating, isolated gastrodermis. In: Burnett AL (ed) Biology of Hydra. Academic Press New York, London, pp 171–219

Delsuc F, Brinkmann H, Chourrout D et al (2006) Tunicates and not cephalochordates are the closest living relatives of vertebrates. Nature 439:965–968

Diehl FA, Burnett AL (1964) The role of interstitial cells in the maintenance of hydra. I. Specific destruction of interstitial cells in normal, asexual, non-budding animals. J Exp Zool 155: 253–260

Dolmatov IY (1992) Regeneration of the aquapharyngeal complex in the holothurian *Eupentacta fraudatrix* (Holothuroidea, Dendrochirota). Monogr Dev Biol 23:40–50

Dolmatov IY (1996) Asexual reproduction, evisceration, and regeneration in holothurians. Russ J Dev Biol 27:256–265

Drobysheva IM (1986) Physiological regeneration of the digestive parenchyma in *Convoluta convoluta* and *Oxyposthia praedator* (Turbellaria, Acoela). Hydrobiologia 132:189–193

Drobysheva IM (1988) An autoradiographic study of the replacement of epidermis in polyclad turbellarians. Fortschr Zool 36:97–101

Dübel S, Hoffmeister SAH, Schaller HC (1987) Differentiation pathways of ectodermal epithelial cells in hydra. Differentiation 35:181–189

Easterday MC, Dougherty JD, Jackson RL et al (2003) Neural progenitor genes. Germinal zone expression and analysis of genetic overlap in stem cell populations. Dev Biol 264: 309–322

Eaves AA, Palmer AR (2003) Widespread cloning in echinoderm larvae. Nature 425:146

Edds KT (1993) Cell biology of echinoid coelomocytes. I. Diversity and characterization of cell types. J Invert Biol 61:173–178

Egger B, Gschwentner R, Rieger R (2007) Free-living flatworms under the knife: Past and present. Dev Genes Evol 217:89–104

Egger B, Ladurner P, Nimeth K et al (2006) The regeneration capacity of the flatworm *Macrostomum lignano*-on repeated regeneration, rejuvenation, and the minimal size needed for regeneration. Dev Genes Evol 216:579–580

Extavour CG, Akam M (2003) Mechanisms of germ cell specification across the metazoans: Epigenesis and preformation. Development 130:5869–5884

Faast R, Thonglairoam V, Schulz TC et al (2001) Histone variant H2A.Z is required for early mammalian development. Curr Biol 11:1183–1187

Fischer AB, Hofmann DK, (2004) Budding, bud morphogenesis, and regeneration in *Carybdea marsupialis* Linnaeus, 1758 (Cnidaria: Cubozoa). Hydrobiologia 530/531:331–337

Forsyth NR, Wright WE, Shay JW (2002) Telomerase and differentiation in multicellular organisms: Turn it off, turn it on, turn it off again. Differentiation 69:188–197

Fradkin M, Kakis H, Campbell RD (1978) Effect of gamma irradiation of hydra: Elimination of interstitial cells from viable hydra. Radiat Res 76:187–197

Francis N, Gregg N, Owenb R et al (2006) Lack of age-associated telomere shortening in long- and short-lived species of sea urchins. FEBS Lett 580:4713–4717

Frank U, Leitz T, Müller WA (2001) The hydroid Hydractinia: A versatile, informative cnidarian representative. Bioessays 23:963–971

Freeman G (1964) The role of blood cells in the process of asexual reproduction in the tunicate *Perphora viridis*. J Exp Zool 156:157–183

Freeman G (1970) The reticuloendothelialystem of tunicates. J Reticuloendothel Soc 7:183–194

Fujiwara S, Kawamura K (1992) Ascidian budding as a transdifferentiation-like system: Multipotent epithelium is not undifferentiated. Dev Growth Differ 34:463–472

Fulinski B (1922) Über das Restitutionsvermögen der Rhabdocoelen. Arch Entwm 51: 575–586

García-Arrarás JE, Estrada-Rodgers L, Santiago R et al (1998) Cellular mechanisms of intestine regeneration in the sea cucumber, *Holothuria glaberrima* Selenka (Holothuroidea: Echinodermata). J Exp Zool 281:288–304

García-Arrarás JE, Greenberg MJ (2001) Visceral regeneration in Holothurians. Microsc Res Tech 55:438–451

García-Arrarás JE, Diaz-Miranda L, Torres II et al (1999) Regeneration of the enteric nervous system in the sea cucumber *Holothuria glaberrima*. J Comp Neurol 406:461–475

García-Arrarás JE, Schenk C, Rodrígues-Ramírez R et al (2006) Spherulocytes in the echinoderm *Holothuria glaberrima* and their involvement in intestinal regeneration. Dev Dyn 235: 3259–3267

Glinski Z, Jarosz J (2000) Immune phenomena in Echinoderms. Arch Immunol Ther Exp (Warsz) 48:189–193

Gremigni V (1981) The problem of cell totipotency, dedifferentiation and transdifferentiation in *Turbellaria*. Hydrobiologia 84:171–179

Gremigni V, Miceli C, Picano E (1980b) On the role of germ cells in planarian regeneration: Cytophotometric analysis of the nuclear Feulgen-DNA content in cells of regenerated somatic tissues. J Embryol Exp Morphol 55:65–76

Gremigni V, Miceli C, Puccinelli I (1980a) On the role of germ cells in planarian regeneration: A karyological investigation. J Embryol Exp Morphol 55:53–63

Gremigni V, Puccinelli I (1977) A contribution to the problem of the origin of blastema cells in planarians: A karyological and ultrastructural investigation. J Exp Zool 199:57–72

Gremigni V, Puccinelli I (1979) Sdifferenziamento e migrazione di cellule germinali in planarie tenute a digiuno: Indagine cariologica. Ric Sci Educ Perm Suppl 6:115–116
Grimmelikhuijzen CJ, Dockray GJ, Schot LP (1982) FMRFamide-like immunoreactivity in the nervous system of *Hydra*. Histochemistry 73:499–508
Grosberg RK, Quinn JF (1986) The genetic control and consequences of kin recognition by the larvae of a colonial marine invertebrate. Nature 322:456–459
Gross PS, Al-Sharif WZ, Clow LA et al (1999) Echinoderm immunity and the evolution of the complement system. Dev Comp Immunol 23:429–442
Gross PS, Clow LA, Smith LC (2000) SpC3, the complement homologue from the purple sea urchin, *Strongylocentrotus purpuratus*, is expressed in two subpopulations of the phagocytic coelomocytes. Immunogenetics 51:1034–1044
Gschwentner R, Ladurner P, Nimeth K et al (2001) Stem cells in a basal bilaterian. Sphase and mitotic cells in Convolutriloba longifissura (Acoela, Platyhelminthes). Cell Tissue Res 304:401–408
Gustafsson MKS (1976) Studies on cytodifferentiation in the neck region of *Diphyllobothrium dendriticum* Nitzsch. 1824 (Cestoda, Pseudophyllidae). Z Parasitenk 50:323–329
Hagleithner G (1946) Das Regenerations problem bei Turbellarien. Dissertation, Innsbruck
Hammond SM (2006) MicroRNAs as oncogenes. Curr Opin Genet Dev 16:4–9
Hand C, Uhlinger KR (1995) Asexual reproduction by transverse fission and some anomalies in the sea anemone *Nematostella vectensis*. Invertebr Biol 114:9–18
Handberg-Thorsager M, Saló E (2007) The planarian nanos-like gene Smednos is expressed in germline and eye precursor cells during development and regeneration. Dev Genes Evol 217:403–411
Haranghy L, Balazs A (1964) Ageing and rejuvenation in planarians. Exp Gerontol 1:77–91
Hatfield SD, Shcherbata HR, Fischer KA et al (2005) Stem cell division is regulated by the microRNA pathway. Nature 435:974–978
Hay ED, Coward SJ (1975) Fine structure studies on the planarian Dugesia. Nature of the 'neoblast' and other cell types in noninjured worms. J Ultrastruct Res 50:1–21
Hendelberg J, Åkesson B (1991) Studies of the budding process in *Convolutriloba retrogemma* (Acoela, Platyhelminthes). Hydrobiologia 227:11–17
Henson JH, Svitkina TM, Burns AR et al (1999) Two components of actinbased retrograde flow in sea urchin coelomocytes. Mol Biol Cell 10:4075–4090
Higuchi S, Hayashi T, Hori I et al (2007) Characterization and categorization of fluorescence activated cell sorted planarian stem cells by ultrastructural analysis. Dev Growth Differ 49:571–581
Holland ND, Phillips JJH, Giese AC (1965) An autoradiographic investigation of coelomocyte production in the purple sea urchin (*Strongylocentrotus purpuratus*). Biol Bull 128: 259–270
Holstein TW, David CN (1990) Cell cycle length, cell size, and proliferation rate in hydra stem cells. Dev Biol 142:392–400
Hori I (1982) An ultrastructural study of the chromatoid body in planarian regenerative cells. J Electron Microsc 31:63–72
Hori I (1997) Cytological approach to morphogenesis in the planarian blastema. . The effect of neuropeptides. J Submicrosc Cytol Pathol 29:91–97
Hyman LH (1955) The Invertebrates Vol 4: Echinodermata. McGraw-Hill, New York, 763p
Ivanova NB, Dimos JT, Schanie C et al (2002) A stem cell molecular signature. Science 298: 601–604
Izzard CS (1973) Development of polarity and bilateral asymmetry in the palleal bud of *Botryllus schlosseri* (Pallas) J Morphol 139:1–26
Jaeckle WB (1994) Multiple modes of asexual reproduction by tropical and subtropical Sea Star larvae: An unusual adaptation for genet dispersal and survival. Biol Bull 186: 62–71

Juliano CE, Voronina E, Stack C et al (2006) Germ line determinants are not localized early in sea urchin development, but do accumulate in the small micromere lineage. Dev Biol 300: 406–415

Karpowicz P, Morshead C, Kam A et al (2005) Support for the immortal strand hypothesis: Neural stem cells partition DNA asymmetrically in vitro. J Cell Biol 170:721–732

Kawamura K, Fujiwara S (1994) Transdifferentiation of pigmented multipotent epithelium during morphallactic development of budding tunicates. Int J Dev Biol 38:369–377

Kawamura K, Fujiwara S (1995) Cellular and molecular characterization of transdifferentiation in the process of morphallaxis of budding tunicates. Semin Cell Biol 6:117–126

Kawamura K, Nakauchi M (1984) Changes of the atrial epithelium during the development of anteroposterior organization in palleal buds of the polystyelid ascidian, *Polyandrocarpa misakiensis*. Mem Fac Sci Kochi Univ (Ser D) 5:15–35

Kawamura K, Nakauchi M (1986) Development of spatial organization in palleal buds of the compound ascidian, *Symplegma reptans*. Biol Bull 171:520–537

Kawamura K, Nakauchi M (1991b) The effect of aphidicolin on morphogenesis of epithelial stem cells during budding in ascidians. Mem Fac Sci Kochi Univ (Ser D) 12:61–65

Kawamura K, Nakauchi N (1991a) Homeostatic integration of stem cell dynamics during palleal budding of ascidians. Zool Sci 8:11–22

Kawamura K, Ohkawa S, Michibata H et al (1988) Autoradiographic studies on cell population kinetics in epithelial and haematopoietic stem cell lines in the process of palleal budding of ascidians. Mem Fac Sci Kochi Univ (Ser D) 9:1–12

Kiel MJ, He S, Ashkenazi R et al (2007) Haematopoietic stem cells do not asymmetrically segregate chromosomes or retain BrdU. Nature 449:238–242

Knott KE, Balser EJ, Jaeckle WB et al (2003) Identification of Asteroid genera with species capable of larval cloning. Biol Bull 204:246–255

Koguchi S, Sugino YM, Kawamura K (1993) Dynamics of epithelial stem cell in the process of stolonial budding of the colonial ascidian, *Perophora japonica*. Mem Fac Sci Kochi Univ (Ser D) 14:7–14

Koizumi O, Bode HR (1986) Plasticity in the nervous system of adult hydra: I. The position-dependent expression of FMRFamide-like immunoreactivity. Dev Biol 116:407–421

Koizumi O, Bode HR (1991) Plasticity in the nervous system of adult hydra: III. Conversion of neurons to expression of a vasopressin-like immunoreactivity depends on axial location. J Neurosci 11:2011–2020

Koizumi O, Heimfeld S, Bode HR (1988) Plasticity in the nervous system of adult hydra: II. Conversion of ganglion cells of the body column into epidermal sensory cells of the hypostome. Dev Biol 129:358–371

Koros AM, Goodwin DG, Siderits RH et al (2002) "Natural killer" (NK) cell antigens CD56, CD57 and others are expressed on breast and lung tumor cells as well as sea urchin coelomocytes. J Biol Regul Homeost Agents 16:173–175

Koros AMC, Pulsford A (1994) Neuroendocrine markers expressed on sea urchin coelomocytes and other immunoregulatory parameters may be used to monitor environmental changes. Ann NY Acad Sci 712:350–353

Kramarsky-Winter E, Loya Y (2000) Tissue regeneration in the coral *Fungia granulosa*: The effect of extrinsic and intrinsic factors. Mar Biol 137:867–873

Lacalli TC (2000) Larval budding, metamorphosis, and the evolution of life-history patterns in echinoderms. Invertebr Biol 119:234–241

Ladurner P, Rieger R, Baguñà J (2000) Spatial distribution and differentiation potential of stem cells in hatchlings and adults in the marine *Platyhelminth macrostomum* sp.: A bromodeoxyuridine analysis. Dev Biol 226:231–241

Laird DJ, De Tomaso AW, Weissman IL (2005) Stem cells are units of natural selection in a colonial ascidian. Cell 123:1351–1360

Laird DJ, Weissman IL (2004) Telomerase maintained in self-renewing tissues during serial regeneration of the urochordate *Botryllus schlosseri*. Dev Biol 273:185–194

Lang A (1912) Über Regeneration bei Planarien. Arch Mikr Anat 79:361–426

Leclerc M, Bajelan M (1992) Homologous antigen for T cell receptor in axial organ cells from the asteroid *Asterias rubens*. Cell Biol 16:487–490
Leclerc M, Brillouet C, Luquet G (1980) The Starfish axial organ: An ancestral lymphoid organ. Dev Comp Immunol 4:605–615
Lee K, Fajardo MA, Braun RE (1996) A testis cytoplasmic RNA-binding protein that has the properties of a translational repressor. Mol Cell Biol 16:3023–3034
Lender TH, Gabriel A (1965) Les néoblastes marqués par Iuridine tritiée migrent et edifient le blasteme de régénération des planaires d'eau douce. C R Hebd Séanc Acad Sci, Paris 260: 4095–4097
Lengner CJ, Camargo FD, Hochedlinger K et al (2007) Oct4 expression is not required for mouse somatic stem cell self-renewal. Cell Stem Cell 1:403–415
Lenhoff HM (1965) Cellular segregation and heterocytic dominance in Hydra. Science 148: 1105–1107
Lenicque PM, Féral JP (1977) Effects of biogenic amines on the regeneration of small pieces of the pedal disc of the sea anemone *Metridium senile* (Linnaeus). Comp Biochem Physiol (C Pharmacol Toxicol Endocrinol) 57:91–93
Littlefield CL (1985) Germ cells in *Hydra oligactis* males. I. Isolation of a subpopulation of interstitial cells that is developmentally restricted to sperm production. Dev Biol 112: 185–193
Littlefield CL (1991) Cell lineages in Hydra: Isolation and characterization of an interstitialstem cell restricted to egg production in *Hydra oligactis*. Dev Biol 143:378–388
Magor BG, De Tomaso A, Rinkevich B et al (1999) Allorecognition in colonial tunicates: Protection against predatory cell lineages? Immunol Rev 167:69–79
Manni L, Burighel P (2006) Common and divergent pathways in alternative developmental processes of ascidians. BioEssays 28:902–912
Manni L, Zaniolo G, Cima F et al (2007) *Botryllus schlosseri*: A model ascidian for the study of asexual reproduction. Dev Dyn 236:335–352
Marcum BA, Campbell RD (1978) Developmental roles of epithelial and interstitial cell lineages in hydra: Analysis of chimeras. J Cell Sci 32:233–247
Mashanov VS, Dolmatov IY, Heinzeller T (2005) Transdifferentiation in holothurian gut regeneration. Biol Bull 209:184–93
Matranga V (1996) Molecular aspects of immune reactions in Echinodermata. In: Müller WEG, Rinkevich B (eds) Invertebrate Immunology Vol 15, Progress in Molecular and Subcellular Biology Series. Springer-Verlag, Heidelberg, pp 235–247
Matranga V, Bonaventura R, Di Bella G (2002) Hsp70 as a stress marker of sea urchin coelomocytes in short term cultures. Cell Mol Biol 48:345–359
Matranga V, Pinsino A, Celi M et al (2006) Impacts of UV-B radiation on short term cultures of sea urchin coelomocytes. Mar Biol 149:24–34
Matranga V, Pinsino A, Celi M et al (2005) Monitoring chemical and physical stress using sea urchin immune cells. Prog Mol Subcell Biol 39:85–110
Matranga V, Toia G, Bonaventura R et al (2000) Cellular and biochemical responses to environmental and experimentally induced stress in sea urchin coelomocytes. Cell Stress Chaperones 5:158–165
Mauro A (1961) Satellite cell of skeletal muscle fibers. J Biophys Biochem Cytol 9:493–495
Metcalf D, Moore MAS (1971) Hematopoietic cells. North Holland Publishing Co., Amsterdam, pp 235–247
Millott N (1969) Injury and the axial organ of echinoids. Experientia 25:756
Mladenov PV, Bisgrove B, Asotra S et al (1989) Mechanisms of arm-tip regeneration in the sea star, *Leptasterias hexactis*. Roux's Arch Dev Biol 189:19–28
Mochizuki K, Nishimiya-Fujisawa C, Fujisawa T (2001) Universal occurrence of the vasa-related genes among meta-zoans and their germline expression in Hydra. Dev Genes Evol 211. 299–308
Mochizuki K, Sano H, Kobayashi S et al (2000) Expression and evolutionary conservation of nanos-related genes in Hydra. Dev Genes Evol 210:591–602

Montgomery JR, Coward SJ (1974) On the minimal size of a planarian capable of regeneration. Trans Am Microsc Soc 93:386–391

Moore KA, Lemischka IR (2006) Stem cells and their niches. Science 311:1880–1885

Morgan TH (1902) Growth and regeneration in *Planaria lugubris*. Arch Entwm 13:179–212

Morita M (1995) Structure and function of the reticular cell in the planarian *Dugesia dorotocephala*. Hydrobiologia 305:189–196

Morrison SJ, Kimble J (2006) Asymmetric and symmetric stem-cell divisions in development and cancer. Nature 441:1068–1074

Morrison SJ, Prowse KR, Ho P et al (1996) Telomerase activity in hematopoietic cells is associated with self-renewal potential. Immunity 5:207–216

Mortensen T (1921) Studies of the Development and Larval Form of Echinoderms. GEC Gad, Copenhagen 266p

Moss C, Hunter J, Thorndyke MC (1998) Pattern of bromodeoxyuridine incorporation and neuropeptide immunoreactivity during arm regeneration in the starfish *Asterias rubens*. Philos Trans R Soc Lond B 353:421–436

Mukai H, Koyama H, Watanabe H (1983) Studies on the reproduction of three species of Perophora (Ascidianea). Biol Bull 164:251–266

Müller WA, Plickert G, Berking S (1986) Regeneration in Hydrozoa: Distal versus proximal transformation in Hydractinia. Dev Genes Evol 195:513–518

Müller WA, Teo R, Frank U (2004) Totipotent migratory stem cells in a hydroid. Dev Biol 275: 215–224

Munoz-Chapuli R, Carmona R, Guadix JA et al (2005) The origin of the endothelial cells: An evo-devo approach for the invertebrate/vertebrate transition of the circulatory system. Evol Dev 7:351–358

Murray G, García-Arrarás JE (2004) Myogenesis during holothurian intestinal regeneration. Cell Tissue Res 318:515–524

Nakauchi M (1982) Asexual development of ascidians: Its biological significance, diversity, and morphogenesis. Am Zool 22:753–763

Newmark PA, Sánchez-Alvarado A (2000) Bromodeoxyuridine specifically labels the regenerative stem cells of planarians. Dev Biol 220:142–153

Nimeth KT, Mahlknecht M, Mezzanato A et al (2004) Stem cell dynamics during growth, feeding, and starvation in the basal flatworm Macrostomum sp. (Platyhelminthes). Dev Dyn 230:91–99

Nishimiya-Fujisawa C, Sugiyama T (1993) Genetic analysis of developmental mechanisms in hydra. XX. Cloning of interstitial stem cells restricted to the sperm differentiation pathway in *Hydra magnipapillata*. Dev Biol 157:1–9

Nishimiya-Fujisawa C, Sugiyama T (1995) Genetic analysis of developmental mechanisms in Hydra. XXII. Two types of female germ stem cells are present in a male strain of *Hydra magnipapillata*. Dev Biol 172324–336

Nunzi MG, Burighel P, Schiaffino S (1979) Muscle cell differentiation in the ascidian heart. Dev Biol 68:371–380

Oka H, Watanabe H (1957a) Vascular budding, a new type of budding in *Botryllus*. Biol Bull 112:225–240

Oka H, Watanabe H (1957b) Colony specificity in compound ascidians as tested by fusion experiments. Proc Japan Acad 33:627–659

Oka H, Watanabe H (1959) Vascular budding in Botrylloides. Biol Bull 117:340–346

Oren U, Rinkevich B, Loya Y (1997) Oriented intra-colonial transport of 14C labeled materials during coral regeneration. Mar Ecol Prog Ser 161:117–122

Oweson C, Sköld H, Pinsino A et al (2008) Manganese effects on haematopoietic cells and circulating coelomocytes of *Asterias rubens* (Linnaeus). Aquat Toxicol 89:75–81

Palmberg I (1986) Cell migration and differentiation during wound healing and regeneration in *Microstomum lineare* (Turbellaria). Hydrobiologia 132:181–188

Palmberg I (1990) Stem cells in microturbellarians-an autoradiographic and immunocytochemical study. Protoplasma 158:109–120

Palmberg I, Reuter M (1983) Asexual reproduktion in *Microstomum lineare* (Turbellaria). I. An autoradiographic and ultrastructural study. Int J Invert Reprod 6:197–206

Pancer Z, Gershon H, Rinkevich B (1995) Cloning of a urochordate cDNA featuring mammalian short consensus repeats (SCR) of complement-control protein superfamily. Comp Biochem Physiol B Biochem Mol Biol 111:625–632

Pedersen KJ (1972) Studies on regeneration blastemas of the planarian Dugesia tigrina with special reference to differentiation of the muscle-connective tissue filament system. Wilhelm Roux Arch Entw Mech 169:134–169

Peterson KJ, Cameron RA, Davidson EH (1997) Set-aside cells in maximal indirect development: Evolutionary and developmental significance. BioEssays 19:623–631

Peterson KJ, Cameron RA, Davidson EH (2000) Bilaterian origins: Significance of new experimental observations. Dev Biol 219:1–17

Pinsino A, Della Torre C, Sammarini V et al (2008) Sea urchin coelomocytes as a novel cellular biosensor of environmental stress: A field study in the Tremiti Island Marine Protected Area Southern Adriatic Sea, Italy. Cell Biol Toxicol Epub 24:541–52

Pinsino A, Thorndyke MC, Matranga V (2007) Coelomocytes and post-traumatic response in the common sea star *Asterias rubens* Cell Stress Chaperones 12:332–342

Polteva DG, Lenicque PM (1970) Effets des neuromédiateurs sur lémbryogénèse somatique chez *Meteridium senile*. Thérapie 25:1067–1081

Potten CS, Hume WJ, Reid P et al (1978) The segregation of DNA in epithelial stem cells. Cell 15:899–906

Potten CS, Owen G, Booth D (2002) Intestinal stem cells protect their genome by selective segregation of template DNA strands. J Cell Sci 115:2381–2388

Pyle AD, Donovan PJ, Lock LF (2004) Chipping away at 'stemness'. Genome Biol 5:235

Raftos DA, Cooper EL (1991) Proliferation of lymphocyte-like cells from the solitary tunicate, *Styela clava*, in response to allogeneic stimuli. J Exp Zool 260:391–400

Ramalho-Santos M, Yoon S, Matsuzaki Y et al (2002) "Stemness": Transcriptional profiling of embryonic and adult stem cells. Science 298:597–600

Rando TA (2007) The immortal strand hypothesis: Segregation and reconstruction. Cell 129: 1239–1243

Ransick A, Cameron RA, Davidson EH (1996) Postembryonic segregation of the germ line in sea urchins in relation to indirect development. Proc Natl Acad Sci USA 93:6759–6763

Reber-Müller S, Ono S, Shimizu O et al (1994) Transdifferentiation of striated muscle of jellyfish to smooth muscle and nerve cells: The role of cell-ECM interactions and carbohydrates revealed by a monoclonal antibody. Differentiation 57:77–87

Reddien PW, Oviedo NJ, Jennings JR et al (2005) SMEDWI-2 is a PIWI-like protein that regulates planarian stem cells. Science 310:1327–1330

Reiger R (1986) Asexual reproduction and the turbellarian archetype. Hydrobiologia 132: 35–45

Reuter M, Kreshchenko N (2004) Flatworm asexual multiplication implicates stem cells and regeneration. Can J Zool 82:334–356

Rieger RM, Legniti A, Ladurner P et al (1999) Ultrastructure of neoblasts in microturbellaria: Significance for understanding stem cells in free-living Platyhelminthes. Invertebr Reprod Dev 35:127–140

Rinkevich B (2002a) The colonial urochordate *Botryllus schlosseri*: From stem cells and natural tissue transplantation to issues in evolutionary ecology. Bio Essays 24:730–740

Rinkevich B (2002b) Germ cell parasitism as an ecological and evolutionary puzzle: Hitchhiking with positively selected genotypes. OIKOS 96:25–30

Rinkevich B, Shapira M (1999) Multi-partner urochordate chimeras outperform two-partner chimerical entities. OIKOS 87:315–320

Rinkevich B, Shlemberg Z, Fishelson L (1995) Whole-body protochordate regeneration from totipotent blood cells. Proc Natl Acad Sci USA 92:7695–7699

Rinkevich B, Shlemberg Z, Fishelson L (1996) Survival budding processes in the colonial tunicate Botrylloides from the Mediterranean Sea: The role of totipotent blood cells. In: Maramorosch K, Loeb MJ, Largo MD (eds) Invertebrate cell culture: Looking towards the 21st century. Society for In Vitro Biology, Raleigh, pp 1–9

Rinkevich B, Weissman IL (1987a) Chimeras in colonial invertebrates: A synergistic symbiosis or somatic-and germ-cell parasitism. Symbiosis 4:117–134

Rinkevich B, Weissman IL (1987b) A long-term study on fused subclones in the ascidian *Botryllus schlosseri*: The resorption phenomenon (Protochordata: Tunicata). J Zool Lond 213: 717–733

Rinkevich B, Weissman IL (1992) Chimeras vs genetically homogeneous individuals: Potential fitness costs and benefits. OIKOS 63:119–124

Rinkevich B, Weissman IL, De Tomaso AW (1998) Transplantation of Fu/HC-incompatible zooids in *Botryllus schlosseri* results in chimerism. Biol Bull 195:98–106

Rinkevich B, Yankelevich I (2004) Environmental split between germ cell parasitism and somatic cell synergism in chimeras of a colonial urochordate. J Exp Biol 207:3531–3536

Rinkevich Y, Paz G, Rinkevich B et al (2007) Systemic bud iduction and retinoic acid signaling underlie whole body rgeneration in the urochordate *Botrylloides leachi*. PLoS Biol 5:900–913

Rodriguez AJ, Seipel SA, Hamill DR et al (2005) Seawi, a sea urchin piwi/argonaute family member is a component of MT-RNP complexes. RNA 11:656–656

Rosner A, Paz G, Rinkevich B (2006) Divergent roles of the DEAD-box protein BSPL10, the urochordate homologue of human DDX3 and DDX3Y proteins, in colony astogeny and ontogeny. Dev Dyn 235:1508–1521

Rossi L, Salvetti A, Lena A et al (2006) DjPiwi-1, a member of the PAZ-Piwi gene family, defines a subpopulation of planarian stem cells. Dev Genes Evol 216:335–346

Sabbadin A (1955) Studies on blood cells in *Botryllus schlosseri* (Pallas). Arch Ital Anat Istol Patol 60:33–67

Sabbadin A (1962) La basi genetiche della capacita di fusione fra colonie in *Botryllus schlosseri* (Ascidiacea). Rend Accad Lincei 32:1031–1035

Sabbadin A, Astorri C (1988) Chimeras and histocompatibility in the colonial ascidian *Botryllus schlosseri*. Dev Comp Immunol 12:737–747

Sabbadin A, Zaniolo G (1979) Sexual differentiseriation and germ cell transfer in the colonial ascidian *Botryllus schlosseri*. J Exp Zool 207:289–304

Sacks PG, Davis LE (1979) Production of nerveless *Hydra attenuata* by hydroxyurea treatments. J Cell Sci 37:189–203

Saló E, Aguñà J (1984) Regeneration and pattern formation in planarians. I. The pattern of mitosis in anterior and posterior regeneration in *Dugesia* (G) *tigrina*, and a new proposal for blastema formation. J Embryol Exp Morphol 83:63–80

Sánchez-Alvarado A (2000) Regeneration in the metazoans: Why does it happen? BioEssays 22:578–590

San Miguel-Ruiz JE, García-Arrarás JE (2007) Common cellular events occur during wound healing and organ regeneration in the sea cucumber Holothuria glaberrima. BMC Dev Biol 7:115

Sato K, Shibata N, Orii H et al (2006) Identification and origin of the germline stem cells as revealed by the expression of nanos-related gene in planarians. Dev Growth Differ 48:615–628

Sato N, Sanjuan IM, Heke M et al (2003) Molecular signature of human embryonic stem cells and its comparison with the mouse. Dev Biol 260:404–413

Satoh N, Levine M (2005) Surfing with the tunicates into the post-genome era. Genes Dev 19:2407–2411

Schmid V, Alder H (1984) Isolated, mononucleated, striated muscle can undergo pluripotent transdifferentiation and form a complex regenerate. Cell 38:801–809

Schürmann W, Betz S (1998) Separation and subtyping of planarian neoblasts by densitygradient centrifugation and staining. Hydrobiologia 383:117–124

Scofield VL, Schlumpberger JM, West LA et al (1982) Protochordate allorecognition is controlled by an MHC-like gene system. Nature 295:499–502

Seipel K, Yanze N, Schmid V (2004) The germ line and somatic stem cell gene Cniwi in the jellyfish *Podocoryne carnea*. Int J Dev Biol 48:1–7

Sharpless NE, DePinho RA (2004) Telomeres, stem cells, senescence, and cancer. J Clin Invest 113:160–168

Shibata N, Umesono Y, Orii H et al (1999) Expression of vasa(vas)-related genes in germline cells and totipotent somatic stem cells of planarians. Dev Biol 206:73–87

Shimizu H, Sawada Y, Sugiyama T (1993) Minimum tissue size required for hydra regeneration. Dev Biol 155:287–296

Shinin V, Gayraud-Morel B, Gomès D et al (2006) Asymmetric division and cosegregation of template DNA strands in adult muscle satellite cells. Nat Cell Biol 8:677–687

Siebert S, Anton-Erxleben F, Bosch TC (2008) Cell type complexity in the basal metazoan Hydra is maintained by both stem cell based mechanisms and transdifferentiation. Dev Biol 313:13–24

Siffroi JP, Pawlak A, Alfonsi MF et al (2001) Expression of the TAR RNA binding protein in human testis. Mol Hum Reprod 7:219–225

Simon-Blecher N, Achituv Y, Rinkevich B (2004) Protochordate concordant xenotransplantation settings reveal outbreaks of donor cells and divergent life span traits. Dev Comp Immunol 28:983–991

Smid I, Tardent P (1986) The potentialities of endoderm interstitial cells in *Hydra attenuata* Pall. Dev Biol 117:672–675

Smith GH (2005) Label-retaining epithelial cells in mouse mammary gland divide asymmetrically and retain their template DNA strands. Development 132:681–687

Smith LC, Rast JP, Brockton V et al (2006) The sea urchin immune system. Invertebr Surviv J 3:25–39

Smith VJ (1981) The echinoderms. In: Ratcliffe NA, Rowley AF (eds) Invertebrate blood cells. Academic Press, New York, pp 513–562

Song JL, Wessel GM (2007) Genes involved in the RNA interference pathway are differentially expressed during sea urchin development. Dev Dyn 236:3180–3190

Spradling A, Drummond-Barbosa D, Kai T (2001) Stem cells find their niche. Nature 414: 98–104

Steinberg MS (1963) The regeneration of whole polyps from ectodermal fragments of scyphostoma larvae of *Aurelia aurata*. Biol Bull 114:337–343

Steinmann P (1925) Das verhalten der Zellen und Gewebe im regenerienden Tricladen Körper. Verh Naturforsch Ges 36:133–162

Stoner DS, Rinkevich B, Weissman IL (1999) Heritable germ and somatic cell lineage competitions in chimeric colonial protochordates. Proc Natl Acad Sci USA 96:9148–9153

Stoner DS, Weissman IL (1996) Somatic and germ cell parasitism in a colonial ascidian: Possible role for a highly polymorphic allorecognition system. Proc Natl Acad Sci USA 93: 15254–15259

Sugiyama T, Fujisawa T (1978) Genetic analysis of developmental mechanisms in Hydra. II. Isolation and characterization of an interstitial cell-deficient strain. J Cell Sci 29:35–52

Sunanaga T, Saito Y, Kawamura K (2006) Postembryonic epigenesis of Vasa-positive germ cells from aggregated hemoblasts in the colonial ascidian, *Botryllus primigenus*. Dev Growth Differ 48:87–100

Sunanaga T, Watanabe A, Kawamura K (2007) Involvement of vasa homolog in germline recruitment from coelomic stem cells in budding tunicates. Dev Genes Evol 217:1–11

Sutton MF (1953) The regeneration of the siphons of *Ciona intestinalis* L. J Mar Biol Assc UK 32:249–268

Takahashi K, Tanabe K, Ohnuki M et al (2007) Induction of pluripotent stem cells from adult human fibroblasts by defined factors. Cell 131:861–872

Takahashi K, Yamanaka S (2006) Induction of pluripotent stem cells from mouse embryonic and adult fibroblast cultures by defined factors. Cell 126:663–676

Taneda Y, Watanabe H (1982) Effects of X-irradiation on colony specificity in the compound ascidian, *Botryllus primigenus* Oka. Dev Comp Immunol 6:665–673

Terada H, Sugiyama T, Shigenaka Y (1988) Genetic analysis of developmental mechanisms in hydra. XVIII. Mechanism for elimination of the interstitial cell lineage in the mutant strain Sf-1. Dev Biol 126:263–269

Teshirogi W (1967) Heteroplastic transplantation in the freshwater planarians *Bdellocephala brunnea* and *Dendrocoelopsis lacteus*. Sci Rep Tohoku Univ Ser V (Biol) 33:383–397

Thouveny Y, Tassava RA (1998) Regeneration through phylogenesis. In: Ferretti P, Geraudie J (eds) Cellular and molecular basis of regeneration: From invertebrates to humans. John Wiley & Sons Ltd, Chichester, UK, pp 9–43

Till JE, McCulloch EA (1961) Direct measurement of the radiation sensitivity of normal mouse bone marrow cells. Radiat Res 14:213–222

Tiozzo S, Christiaen L, Deyts C et al (2005) Embryonic versus blastogenetic development in the compound ascidian *Botryllus schlosseri*: Insights from Pitx expression patterns. Dev Dyn 232:468–478

Tiozzo S, Voskoboynik A, Brown FD et al (2008) A conserved role of the VEGF pathway in angiogenesis of an ectodermally-derived vasculature. Dev Biol 315:243–255

Toledo A, Cruz C, Fragoso G et al (1997) In vitro culture of *Taenia crassiceps* larval cells and cyst regeneration after injection into mice. J Parasitol 83:189–193

Tumbar T, Guasch G, Greco V et al (2004) Defining the epithelial stem cell niche in skin. Science 303:359–363

Uhrík B, Rýdlová K, Zacharová D (1989) The roles of haemocytes during degeneration and regeneration of crayfish muscle fibres. Cell Tissue Res 255:443–449

Van den Spiegel D, Jangoux M, Flammang P (2000) Maintaining the line of defense: Regeneration of Cuvierian tubules in the sea cucumber *Holothuria forskali* (Echinodermata, Holothuroidea). Biol Bull 198:34–49

Vaughn D, Strathmann RR (2008) Predators induce cloning in echinoderm larvae. Science 319:1503

Vaziri H, Dragowska W, Allsopp RC et al (1994) Evidence for a mitotic clock in human hematopoietic stem cells: Loss of telomeric DNA with age. Proc Natl Acad Sci USA 91:9857–9860

Vickery MS, McClintock JB (1998) Regeneration in metazoan larvae. Nature 394:140

Vickery MS, McClintock JB (2000) Effects of food concentration and availability on the incidence of cloning in planktotrophic larvae of the sea star *Pisaster ochraceus*. Biol Bull 199:298–304

Vickery MC, Vickery MS, McClintock JB, et al (2001a) Utilization of a novel deuterostome model for the study of regeneration genetics: molecular cloning of genes that are differentially expressed during early stages of larval sea star regeneration Gene 262:73–80

Vickery MC, Vickery MS, McClintock JB et al (2001b) Regeneration in echinoderm larvae. Microsc Res Tech 55:464–473

Vickery MS, Vickery MCL, McClintock JB (2002) Morphogenesis and organogenesis in the regenerating planktotrophic larvae of asteroids and echinoids. Biol Bull 203:121–133

Vogel G (2003) Stem cells. 'Stemness' genes still elusive. Science 302:371

Voronina E, Lopez M, Juliano CE et al (2008) Vasa protein expression is restricted to the small micromeres of the sea urchin, but is inducible in other lineages early in development. Dev Biol 314:276–286

Voskoboynik A, Simon-Blecher N, Soen Y et al (2007) Striving for normality: Whole body regeneration through a series of abnormal generations. FASEB J 21:1335–1344

Voskoboynik A, Soen Y, Rinkevich Y et al (2008) Identification of the endostyle as a stem cell niche in a colonial chordate. Cell Stem Cell 3:456–64

Wagers AJ, Weissman IL (2004) Plasticity of adult stem cells. Cell 116:639–648

Wang Y, Zayas RM, Guo T et al (2007) Nanos function is essential for development and regeneration of planarian germ cells. Proc Natl Acad Sci USA 104:5901–5906

Weissman IL (2000) Stem cells: Units of development, units of regeneration, and units in evolution. Cell 100:157–168

Weissman IL, Saito Y, Rinkevich B (1990) Allorecognition histocompatibility in a protochordate species: Is the relationship to MHC semantic or structural? Immunol Rev 113:227–241
Wernig M, Meissner A, Foreman R et al (2007) In vitro reprogramming of fibroblasts into a pluripotent ES-cell-like state. Nature 448:318–324
Wikgren BJP, Gustafsson MKS, Knuts GM (1971) Primary anlage formation in diphyllobothriid tapeworms. Parasitol Res 36:131–139
Wittlieb J, Khalturin K, Lohmann JU et al (2006) Transgenic Hydra allow in vivo tracking of individual stem cells during morphogenesis. Proc Natl Acad Sci USA 103:6208–6211
Wolff E, Dubois F (1948) Sur la migration des cellules de régénération chez les planaires. Rev Suisse Zool 55:218–227
Wray GA, Lowe CJ (2000) Developmental regulatory genes and echinoderm evolution. Syst Biol 49:28–51
Wright RK (1981) Urochordates. In: Ratcliffe NA, Rowley AF (eds) Invertebrate blood cells. Academic Press, London, pp 565–626
Wright WE, Piatyszek MA, Rainey WE et al (1996) Telomerase activity in human germline and embryonic tissues and cells. Dev Genet 18:173–179
Young HE, Steele TA, Bray RA et al (1999) Human pluripotent and progenitor cells display cell surface cluster differentiation markers CD10, CD13, CD56, and MHC class-I. Proc Soc Exp Biol Med 221:63–71
Zayas RM, Hernández A, Habermann B et al (2005) The planarian *Schmidtea mediterranea* as a model for epigenetic germ cell specification: Analysis of ESTs from the hermaphroditic strain. Proc Natl Acad Sci USA 102:18491–18496
Zhang J, Niu C, Ye L et al (2003) Identification of the haematopoietic stem cell niche and control of the niche size. Nature 425:836–841
Zhong-Wei D, Hsiao-Chien C, Zhen Y (2001) Identification of putative downstream genes of Oct-4 by suppression-subtractive hybridization. Biochem Biophys Res Commun 282:701–706
Zimmermann S, Martens UM (2008) Telomeres, senescence, and hematopoietic stem cells. Cell Tissue Res 331:79–90
Zlatanova J, Thakar A (2008) H2A.Z: View from the top. Structure 16:166–179

Chapter 5
Stem Cells in Asexual Reproduction of Marine Invertebrates

Helen Nilsson Sköld, Matthias Obst, Mattias Sköld, and Bertil Åkesson

Abstract While sexual reproduction is conserved and almost ubiquitous, asexual reproduction in forms of parthenogenesis or agametic cloning from somatic tissue is less conserved. The phylogeny shows that agametic cloning is widespread but scattered with many different modes for asexual formation of a new animal. This suggests that independent forms of cloning have evolved later from sexual ancestors between and within different phyla. Here, we present an overview of agametic cloning in the marine animal kingdom and discuss molecular and evolutionary aspects of somatic stem cell usage for asexual cloning. The molecular tissue characterizations and the relative role of different stem cells involved in agametic cloning are only at its beginning with whole phyla largely uncovered. An emerging hypothesis is that the first somatic stem cells used in cloning were also able to form a germ-line and that the more limited lineage specific stem cells are derived. We discuss advantages and problems with agametic cloning from somatic tissue and propose that the levels of stem cell potential held in the tissue can have large consequences for the reproductive life cycle strategies and long-term fitness in clonal animals and strains. We finally describe suitable molecular experimental approaches for future research on this topic.

Keywords Agametic cloning · Asexual · Stem cells · Tissue plasticity · Self renewal · Marine invertebrates · Phylogeny · Colonial · Ageing

5.1 Introduction

> *If there were no regeneration there would be no life. If everything regenerated there would be no death. All organisms exist between these two extremes.* Richard Goss (1969)

Life is ensured by solutions for individual longevity and by the generation of new individuals. New individuals can be generated sexually by combining genotypes or

H.N. Sköld (✉)
Marine Ecology, Göteborg University, Kristineberg 566,
SE 450 34, Fiskebäckskil, Sweden
e-mail: helen.skold@marecol.gu.se

B. Rinkevich, V. Matranga (eds.), *Stem Cells in Marine Organisms*,
DOI 10.1007/978-90-481-2767-2_5, © Springer Science+Business Media B.V. 2009

asexually from a single parent. Asexual reproduction can be divided into two fundamentally different processes. First, embryos may be generated by parthenogenesis from oocytes, and second the young or adult body may divide or bud as it grows which is defined as agametic cloning. The fundamental difference between these processes lay in the type of cells contributing to the new animal. While sexually formed embryos derive from heavily selected gametes and gamete fusion processes, agametic progeny form from somatic tissue that may be at a different level of cellular differentiation and physiological age than the pluripotent egg or zygote. Many organisms use both sexual and asexual reproduction in their life cycles. While sexual reproduction is conserved and almost ubiquitous in nature and asexual reproduction also is very common and widespread, asexual reproduction in form of agametic cloning is less conserved and absent in some phyla.

Natural agametic cloning in marine animals is here discussed in the context of stem cell biology, tissue plasticity and evolution. We aim to display and compare the role of stem cells in agametic cloning across the animal diversity with focus on the marine environment. We explain some general patterns and processes and give examples of well studied animal models. We outline the links between regenerative capacity and development as well as asexual reproduction. Finally, we highlight the areas and most importantly the organisms that have stayed unstudied but promise to reveal new insights, models and mechanisms in stem cell research in the future.

5.2 Agametic Cloning and the Role of Stem Cells

5.2.1 Why Study Agametic Cloning?

Stem cell research has been a significant part of experimental biology for more than a hundred years, but has accelerated enormously during the last 30 years. The medical applications are at hand – to improve the regenerative capacity in humans and avoid diseases linked to senility. But also, the reproductive potential of stem cells is debated – to create human clones. Regeneration and especially stem cells research focuses nowadays largely on vertebrates, whole animals or cell lines, but there are diverse phyla of invertebrates, showing extensive regenerative capacity and asexual reproduction by agametic cloning, that virtually are hardly explored.

The work on development has a long and productive tradition, but while embryogenesis is relatively conserved throughout the animal kingdom and is thoroughly investigated, the creation of a new individual by agametic cloning takes other developmental routes and is considerable less explored. Unfortunately, none of the five most used model multicellular animals (mouse, zebrafish, *Xenopus*, *Drosophila*, *Caenorhabditis elegans*) nor the new suggested model deuterostomes with recently sequenced genomes (for a list of sequence projects completed or in progress: http://www.ncbi.nlm.nih.gov/genomes/leuks.cgi), the two *Ciona* species and *Strongylocentrotus purpuratus*, reproduce asexually. Same applies to protocols for rearing asexual animals and maintaining their putative stem cells under *in vitro* conditions. The currently best two model organisms are freshwater protostomes, the planarian *Schmidtea mediterranea* (Sánchez Alvarado and Tsonis 2006) and the cnidarian *Hydra magnipapillata* (Holstein et al. 2003).

Another reason for the limited research in agametic cloning may be that there have not been any obvious implications of this for humans. However, the artificial cloning of the sheep Dolly from adult somatic nuclei did challenge that. The cloning of Dolly tested the paradigm of August Weismann that the soma is non reusable and the germ line the only infinite unit (Weismann 1889; Kirkwood and Austad 2000). In plants, colonial invertebrates and fungi, somatic cells naturally give rise to new individuals in the process of agametic cloning and their phenotypes are in addition highly influenced by the environmental conditions (Buss 1987). Agametic cloning therefore opens up for questions of how plastic somatic cells really are and if/how this may differ in different life forms and at different ages of the organism. There is now a proliferation of reports suggesting that adult cells even in mammals are far more plastic than previously imagined (for examples: Jang et al. 2004; Poulsom et al. 2002). A major, but perhaps not unachievable challenge, has since become to further understand the cellular programs during development and most important to rewind or speed up such programs. One fruitful strategy has been to compare the similarities and differences between embryonic-, germ- or different somatic stem cells, as well as after differentiation to find a stem cell signature. Indeed, recent studies show that it is possible to artificially de-differentiate somatic cells to a pluripotent stem cell state that is phenotypically similar to embryonic stem cells. For mouse, this was done by vector induced expression of the genes Oct3/4, Sox2, c-Myc and Klf4 (Takahashi and Yamanaka 2006) and in humans the genes Oct4, Sox2, Nanog and Lin28 (Yu et al. 2007). De-differentiation is also well described in many invertebrates where it happens spontaneously in the processes of regeneration and agametic cloning (Sánchez Alvarado and Tsonis 2006; Carnevali 2006; Kawamura et al. 2008).

Interestingly, there is a recent understanding of the connection between cellular turnover, somatic stem cell maintenance and lifespan (Rando 2006). Indeed many clonal strains have a very long lifespan and some are even thought of as being potentially immortal. High degree of tissue dynamics and cellular turnover has been discussed as one requirement for their possible escape from senescence (Sánchez Alvarado and Tsonis 2006). It may be that the long-term tissue renewal from adult tissues that takes place during agametic cloning may hold some of the answers and solutions to various fundamental questions like ageing and rejuvenation in addition to pluripotency and de-differentiation. The connection between cellular turnover, somatic stem cell maintenance and lifespan might thus be of special interest for studies on animals with agametic cloning.

In this review, we use a comparative evo-devo approach to address why and how agametic cloning is achieved among marine metazoans, and what we can learn from this process. In the light of this, we have identified some basic questions that remain more or less unclear:

- Are stem cells across the metazoans regulated by the same pathways?
- When and where do stem cells originate – in the animals and in evolution?

- To what extent does asexual morphogenesis recapitulate the ontogeny?
- Why can some animals reproduce by cloning while others cannot?
- What are the signal(s) initiating the formation of the asexual progeny?
- How do cells dedifferentiate to a stem cell state?
- Do somatic stem cells age?
- Do asexual clone lineages age?
- What are the advantages and problems with sexual versus asexual reproduction?

5.2.2 Agametic Cloning – How Does a New Body Form?

Agametic cloning can serve to produce additional individuals for growth, replacing dying individuals in a colony and for generating asexual dispersive propagules (Satoh 1994). Whereas ontogeny through sexual reproduction is relatively conserved throughout the animal kingdom, creation of a new individual by agametic cloning is extremely variable and takes somewhat different developmental routes, a standpoint raised already in 1893 (Weismann 1893). In the filter feeding ascidians, this is particularly obvious because in their agametic cloning, they escape formation of a chordate larvae and metamorphosis, and goes instead directly to form a sessile adult form (Satoh 1994; Rinkevich et al. 2007). A selection of metazoans with agametic cloning is presented in Figs. 5.1 and 5.2.

Formation of the agametic progeny can in principal be interpreted as alterations of the anterior-posterior axis (Fig. 5.3), which is a defining feature of the bilateral symmetry. In annelids and platyhelminths the axis is often subdivided into multiple tandemly arranged axes. It can also grow out as a branch as for budding in colonial ascidians, bryozoans, hydrozoans and corals, or give rise to axes with opposite orientations as the case with some acoel worms. A few animals split longitudinally as discovered for xenoturbella and in an acoel platyhelminth. While the regulatory pathways involved in the anterior-posterior axis formation in the embryo may well be somewhat recapitulated for axis formation in asexual offsprings (see Bely and Wray 2001, as an example), there is still very little information about what cells and what signals, intra- or extra cellular, that initiate the start of it.

Particularly when an animal undergo cloning by fission and subsequent regeneration of the missing body parts, cloning and regeneration seem to be controlled by similar regulatory pathways (Sánchez Alvarado 2000). Regeneration is well investigated and some pathways like BMP, FGF and Wnt are somewhat conserved between animals and tissues (Bannister et al. 2005; Sánchez Alvarado and Tsonis 2006; Stoick-Cooper et al. 2007). The basis of most, but not all, forms of regeneration is a structure known as a blastema and the developmental correlate is a bud, like a limb bud. The blastema is an organizing centre, a place where motile cells gather, proliferate and get their cues of what to do and become (Sánchez Alvarado and Tsonis 2006). In various animals, the correct formation and continuous development of the blastema and regenerate depend on correct innervation (Müller et al. 2003; Carnevali 2006). After the blastema has formed, regeneration follows similar paths to those during development (Sánchez Alvarado and Tsonis 2006). In *Hydra*,

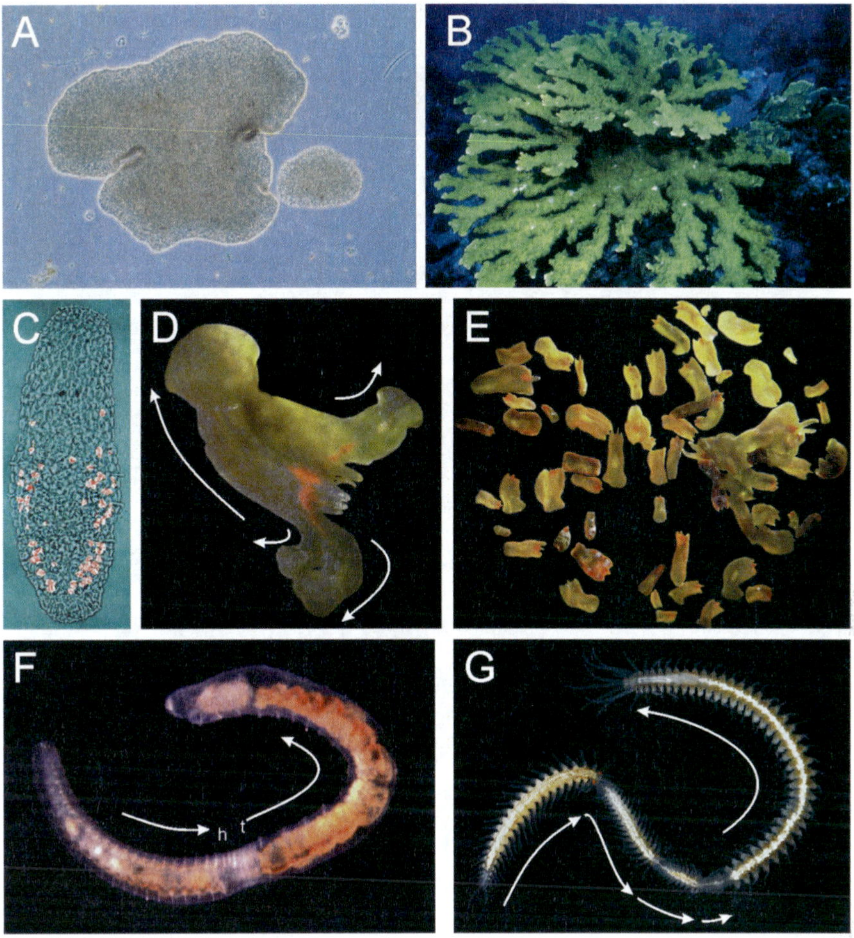

Fig. 5.1 Examples from a range of metazoan phyla containing species which use adult stem cells in agametic cloning. White arrows indicate the a-p-body axes of the clonal progeny. (**A**) Two individuals of *Trichoplax adhaerens* (Placozoa) which have been separated by fission. Photo R. Brons. (**B**) The elkhorn coral *Acropora palmate* (Cnidaria). Cnidarian polyps typically reproduce by budding and often form large colonies consisting of thousands of polyps. Photo National Oceanic and Atmospheric Administration, USA. (**C**) A young specimen of the flatworm *Macrostomum lignano* with marked proliferating stem cells. Photo B. Egger. (**D**) Asexual reproducing individual of *Convolutriloba macropyga* (Acoela), Photo T. Shannon. Three large buds are emerging from the posterior end with a 180° reversed polarity relative to the mother animal. (**E**) Blend of individuals belonging to the genus Convolutriloba (Acoela) performing three different modes of asexual reproduction, (i) transversal fission, in which the posterior fourth of the body detaches to form a new individual, one at a time; (ii) transversal fission with subsequent longitudinal fission, which halves individuals along their a-p-axis; (iii) budding from the posterior part of the mother where the a-p-polarity is 180° reversed relative to the mother animals. Photo T. Shannon. (**F**) *Paranais litoralis* (Oligochaeta, Annelida) at a late stage of paratomic fission. A new tail (t) and a new head (h) are developed at the slight constriction in the middle of worm. The clonal progeny is initially linked with a continuous nervous system, gut, and blood vessels. Photo A. Bely. (**G**) *Myrianida prolifer* (Polychaeta, Annelida). The large individual is an asexual benthic stock individual which develops a posterior chain of sexual individuals (stolons). At a later stage the stolons are released into the water column where they mate. Photo E. Odelgård

Fig. 5.2 Examples from a range of metazoan species including deterostomes which use adult stem cells in agametic cloning. (**A**) Colony of *Pedicellina* sp. (Entoprocta). The individuals remain connected by the stolons. (**B**) *Lophopus crystallinus* (Bryozoa). A young colony with two individuals has developed from a statoblast. (**C**) *Symbion* sp. (Cycliophora). A feeding individual is attached to the hair of the lobster's integument. In the lower part of the trunk a lump of undifferentiated cells continuously develops new feeding structures and other life cycle stages. (**D**) *Xenoturbella* sp. (Xenoturbellida). Few observations on asexual reproduction have been reported. Photo H. Nakano. (**E**) The starfish *Allostichaster polyplax* (Asteroidea, Echinodermata) undergoes fission and regenerates new arms from the disc. (**F**) Young colony of *Polyandrocarpa misakiensis* (Tunicata). Photo K. Kawamura. (**G**) Colony of *Diplosoma listerianum* (Tunicata). The ascidian zooids multiply by budding

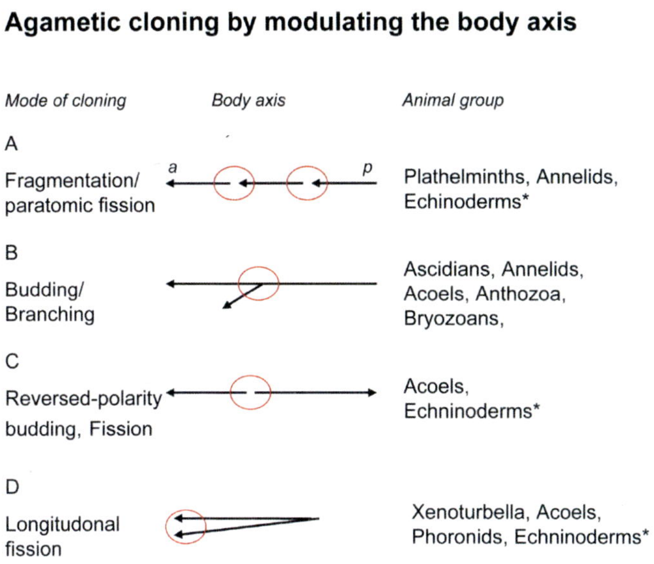

Fig. 5.3 Schematic overview over different modes of body axis multiplication via agametic cloning. The arrows show the direction of the anterior-posterior-body axis during agametic cloning. Circles indicate the initiation site of the asexual progeny. (**A**) A single bilaterian a-p-axis can become fragmented into multiple tandemly repeated axes with the same direction. (**B**) A second a-p-axis can branch off the original axis during budding. (**C**) The original a-p-axis can split into two new axes with reversed orientation. (**D**) The original a-p-axis can split longitudinally. *Echinoderms have a radial symmetrical body axis as adults. Figure provided and modified from A. Bely

regeneration does not involve cell proliferation but instead reorganisation by migration of the existing cells (Sánchez Alvarado 2000; Holstein et al. 2003). Although physical disturbance may trigger asexual propagation and promote clonal structures as for example in the cnidarian *Metridium senile* (Anthony and Svane 1995), a major difference between cloning and regeneration is that cloning is often spontaneously triggered.

At the cellular level, a new organ or body can generally either be formed from one or a group of undifferentiated pluri- or toti potent stem cells, committed cells can de-differentiate to an undifferentiated state where they subsequently redifferentiate and remodel a new organ, or an organ can be reformed by simple mitotic divisions without changing the state of differentiation, like in wound healing. Somatic stem cells in the true sense are those involved in the first process and in invertebrates these cells have been named interstitial cells, neoblasts, hemoblasts or archeocytes (Agata et al. 2006) and they exist in most organisms, at least at the beginning of their embryonic development. These cells are at least pluripotent, potentially immortal and some can also form germ cells, which qualifies them as totipotent (Stoner et al. 1999; Laird et al. 2005; Agata et al. 2006). For simplicity, we call them all here primary stem cells and as such they differ from lineage specific somatic stem cells (also called progenitors) commonly found in vertebrates in their degree of pluripotency and ability for long term renewal (stemness). A number of animal groups have retained the primary stem cells to use for tissue homeostasis, regeneration and/or asexual reproduction. It has further been speculated that the somatic lineage specific stem cells with more limited properties that are found in vertebrates, and perhaps elsewhere, and used for our long-term tissue homeostasis, are derived (Agata et al. 2006).

The special features of stem cells in contrast to other cells are that they are self renewable, proliferative and can give rise to a number of different cell types. The zygote, which may be the prototype of stem cells, is capable of forming all cell types in the animal including the germ line. The derived stem cells seem in many cases, but not all, to have a reduced natural potential when it comes to the degree of pluripotency and long term self renewal (Agata et al. 2006; Rando 2006). At least some of the stemnes need to be involved in clonal propagation, but de-differentiation or growth by simple mitotic division of preexisting cell types, are also potential strategies that may be used (van Bekkum 2004). It may in theory for example be possible to regain toti- or pluripotency in an otherwise renewable system by dedifferentiation into a more uncommitted state.

Noteworthy, some metazoans (nematodes, rotifers, gastrotrichs and insects) not only lack somatic stem cells, but also the number of cells of an adult as well as their fates, are often fixed from the oocyte or zygote stage. Their cells appear not to de-differentiate either. This lack of plasticity correlates with their very limited regenerative capacity as adults, and that they do not reproduce by agametic cloning.

An attempt has been made to summarize the different kinds of animal stem cells, their routes and roles in asexual reproduction and tissue homeostasis (Fig. 5.4).

5.2.3 Constraints for Asexuality and Regeneration

Asexual animals are generally smaller than their sexual species counterparts and the reasons for this could be many. It is well known that animal size increases in species that exercise sexual reproduction due to fecundity advantages of larger animals, especially for females (Andersson 1994). There may also be advantages for asexual species to reduce size or enter the asexual life cycle as physiologically young and therefore small. First, tissue elasticity is a characteristic for fissiparous and regenerating starfish and decreases with increased body size (Marrs et al. 2000). Second, cellular plasticity as well as the amount of lineage specific somatic stem cells diminish with age, at least in deuterostomes like humans (Rando 2006). In modular colonial species like colonial ascidians and bryozoans, an increase in colony size may result in colony regression. Such regression can be circumvented by repeatedly taking cuttings (Bishop et al. 2001; Hughes 2005). Although yet not well investigated, it may be that Kleiber's law (the reduction in metabolic rate with increased body mass) for solitary animals and plants, also applies for colonial species so that larger colonies face metabolic problems (see discussion in Hughes 2005).

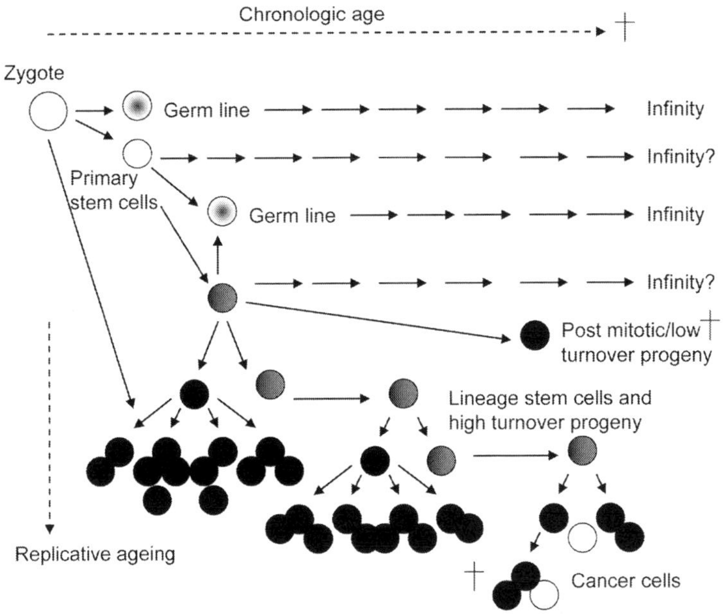

Fig. 5.4 An attempt to demonstrate the contributions of stem cells and lineage specific cells in animal tissue homeostasis. Primary stem cells may be immortal and may contribute to the germ *line*. Depending on the species primary stem cells, lineage specific stem cells or differentiated cells may be involved in agametic cloning and/or tissue regeneration. The immortal germ *line* may come from primary stem cells, as for less complex animals, or somatic cells by epigenetic induction as in many deuterostomes. Somatic lineage specific stem cells are semi-committed with limited degree of stemness and renewal. Commitments and stemness separate in the cellular progeny due to uneven segregation of determinants (*gray shadings*). Animals with less regenerative or nonclonal capacity do either not maintain somatic stem cells or are highly determined and contain many post mitotic cells, as for insects. Clonal or highly regenerative animals have primary stem cells, somatic stemness or less committed cells capable of dedifferentiation. The replicative age may increase the development of cancer (speckled) due to genomic instability. The cross symbilises the death of the soma. *White*: zygote and primary stem cells. *Black*: committed cells. *Shaded gray*: somatic stem cells. *White* with *Black centre*: germ *line*

The capacity of self renewal in stem cells is partly reflected in that they may, in contrast to differentiated cells, be immortal due to long telomeres and sustained telomerase activity, a system highly conserved in the animal kingdom (Traut et al. 2007). In vertebrates, some of the growth promoting hormones, like growth hormone, have positive impact on telomerase activity (Gómez-García et al. 2005), but such hormones decline with age and that could thus influence the adult regenerative capacity. This is not the case for some invertebrates where a sustained high somatic telomerase activity is maintained continuously (Laird and Weissman 2004; Francis et al. 2006; Li et al. 2007). Other mechanisms for long term self renewal, that includes unequal sorting and/or repair of nuclear and mitochondrial DNA as well as damaged proteins, may be crucial as well for maintenance of stem cells (Merok et al. 2002; Rando 2006) and thus the clonal propagation strategy.

Finally, artificial cloning of mammals from somatic nuclei is highly unsuccessful and recent findings implies that epigenetic regulation of the developmental potential in the eggs, like methylation mediated silencing of DNA or other maternally inherited regulators, is lost during development (Yang et al. 2007). Insects have no adult stem cells and their cells are already in embryos highly determined. These factors obstruct natural agametic reproductive strategies in vertebrates and insects (Fig. 5.5).

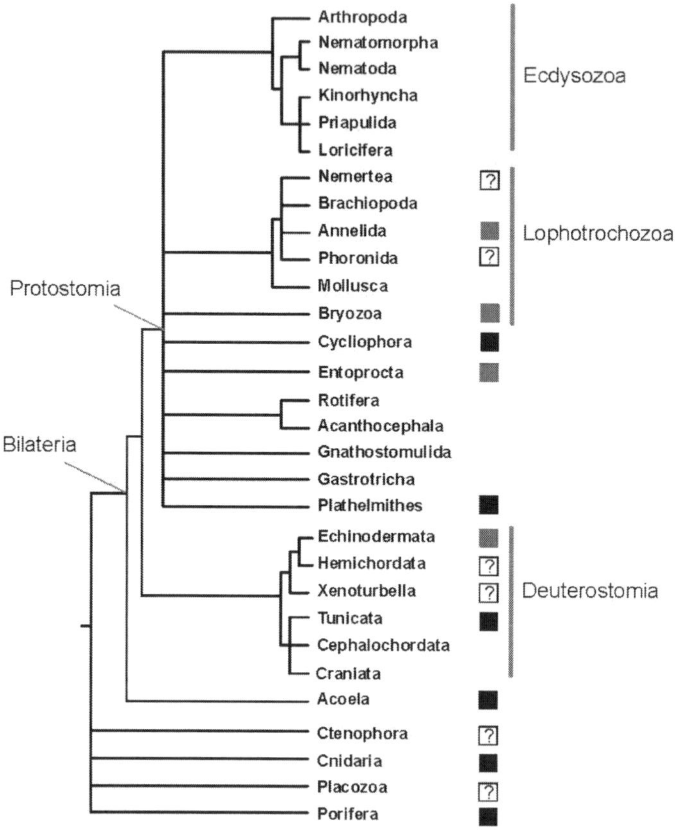

Fig. 5.5 The phylogenetic relationship and agametic cloning capacity among the major metazoan phyla. Taxa performing agametic cloning are appended with a square. *Black squares* indicate the presence of species in which totipotent stem cells have been identified to be active during agametic cloning of adults. *Gray squares* indicate the presence of species with very flexible adult tissue observed during agametic cloning. Some taxa (i.e. annelids) have pluripotent stem cells, while in other groups (i.e. bryozoans and entoprocts), adult cells are able to develop more than one germ layer. *Question marks* point out those groups where it is unknown whether adult stem cells or somatic growth accounts for the potential to reproduce by cloning

5.3 Evolution of Asexual Reproduction

5.3.1 The Origin of Asexual Reproduction and Stem Cells

An important question is if there is a common regulatory pathway for stem cells throughout metazoans. However, if we look within animal groups or phyla and compare the reproductive strategies, there is a noticeable scattered phylogenetic distribution of the current asexual species amongst the obligatory sexual species (Ridley 1996). The asexual species appear out in the branches of the trees, as with the colonial ascidians and sea anemones (Swalla et al. 2000; Geller 2005) but also for clonal echinoderms (Emson and Wilkie 1980). In some starfish, there are even different propagation strategies in different populations within the same species (Sköld et al. 2002). The scattered distribution implies that the colonial and some clonal propagation strategies have evolved independently in different branches as well as within animal groups from obligatory sexual ancestors. The distribution may generally reflect high level of cell and tissue plasticity (Sánchez Alvarado 2000) and shows that the selection pressure to evolve asexuality is very high. One can presume that stem cells are conserved throughout the kingdom, but when considering that a lot of the clonality and colony formation seen today are independently derived from sexual ancestors, this has the consequence that there may not be one major regulatory network controlling agametic cloning or the cells involved in that across the metazoans. Understanding the evolution of the stem cells and reproductive strategies could help clarifying how conserved the stem cell systems are.

It has been suggested that the adult pluri- or totipotent stem cell systems like those found in many invertebrates have evolved as components of clonal asexual reproduction (Agata et al. 2006) and phylogeny reveals that clonality is widespread and indeed a shared basal character down to choanoflagellates (Fig. 5.5; Blackstone and Jasker 2003). The germ line also appears at the base, as a remained undifferentiated and totipotent early cellular progeny (Blackstone and Jasker 2003). In these species where the asexual progeny derives from stem cells with totipotency (Fig. 5.5), ageing may in theory be circumvented (Fig. 5.4; Kirkwood and Austad 2000). Later in evolution, epigenetic induction commonly gives rise to the germ line from somatic cells in a more differentiated animal and this appears in the deuterostome lineage as well as in the phylogenetically more distant myriapods (Blackstone and Jasker 2003; Berekelya et al. 2005). A third type of germ line formation is preformation from an already compartmental oocyte or zygote and here it is unsure if it has evolved late and in a few lines only of both protostomes and some deuterostomes (Extavour and Akam 2003) or if it is more ancient (Blackstone and Jasker 2003). More than one origin of the germ line in evolution is in line with the divergence in the formation of the female meiotic spindles (discussions in Wilson 1928; Sköld et al. 2005). Coloniality exists in fewer phyla (tunicates, cnidarians, entoprocta, ectoprocta and bryozoans) and appears as derived from sexual ancestors (Blackstone and Jasker 2003; Swalla et al. 2000; Geller 2005). In echinoderms and chordates, asexual reproduction and regeneration often involve more than one cell type that dedifferentiates into a more uncommitted state (Satoh 1994; Carnevali 2006). Taken together, at least some cells involved in metazoan agametic cloning may therefore resemble an undifferentiated and totipotent cellular state, but probably not all (Fig. 5.5).

To distinguish the different types of stem cells and their capacities from each other and from other cells, one needs selective markers. The Vasa protein is a major relatively well conserved regulator and marker of the reproductive stem cells, the germ line (Castrillon et al. 2000; Extavour and Akam 2003; Sunanaga et al. 2006; Brown and Swalla 2007). Vasa is not expressed in vertebrate embryonic or lineage-specific somatic stem cells (Castrillon et al. 2000; Deb et al. 2007; Spivakov and Fisher 2007). At least the vertebrate embryonic stem cells seem instead to be maintained by complex regulatory networks and epigenetic modifications that depend on the transcription factors oct3/4, nodal, sox2 (Spivakov and Fisher 2007), but other classic markers are presence of alkaline phosphatase and telomerase (Heins et al. 2004). Interestingly, at least some of the primary stem cells in invertebrates can become germ cells (Laird et al. 2005) and there start to express Vasa (Agata et al. 2006; Sunanaga et al. 2006; Brown and Swalla 2007), like the vertebrate embryonic stem cells also can do (Toyooka et al. 2003). This plasticity supports the hypothesis that the germ line originally evolved from primary stem cells and that the properties if these cells are retained in at least some forms of agametic cloning (Blackstone and Jasker 2003; Agata et al. 2006), but one should be cautious with generalizations given the multiple origins of agametic cloning.

5.3.2 Advantages and Disadvantages with Asexual Reproduction

One of the biggest puzzles in biology has been considered to be why sexual reproduction dominates over asexual reproduction among metazoan taxa. However, in the light of this puzzle, it is surprising that agametic cloning is so common quantitatively and it seems that only some groups have abandoned their potential to use stem cells for either regeneration or reproduction, e.g. vertebrates and the ecdysozoans (Fig. 5.5). Clearly, cloning opens up new reproductive, physiological, morphological and ecological horizons and in phyla which do not undergo agametic cloning, asexual reproduction by parthenogenesis has been invented many times (Hughes 1989).

In the short run, there are some major and direct advantages with agametic cloning as a propagation strategy. First, the offspring is identical to the parent and second fission or budding of the adult body permits rapid colonization even from a single individual. Cloning further save time, risks, and resources that are sexually invested in fertilization and sexual selection. It is generally believed that although sexual animals dilute their genotypes, they gain dispersal and also increased genetic variation in the offspring due to the recombination events both in the gametes as well as during gamete fusion (Weismann 1889; Maynard Smith 1978; Hughes 1989; Ridley 1996). Sexual reproduction may therefore provide short-term advantages by production of offspring with a better fitness than the parents and long-term advantages by being prepared for environmental change due to the increased offspring variation, and there is some experimental data supporting this (see discussion in Agrawal 2006). But on the other hand, armadillos, bryozoans and hymenopterans invest a lot of energy in cloning at the egg or embryonic stage (e.g. polyembryony). In the later two cases large numbers of several hundred individuals are produced and if offspring variation is generally more important than the two fold cost of sex, then this strategy seems odd. Since polyploidy is more common in asexual metazoans than sexual, the genetic diversity may actually not always be so low. It is also evident that many colonial invertebrates, as well as plants, have a high phenotypic plasticity that can be directed by the environment (Buss 1987) and the recent results on transfer of epigenetic information between generations (Bond and Finnegan 2007) suggests that even agametic progenies can be different with time from each others. It has further been argued that recombination by sexual reproduction enables reduction of deleterious mutations (Maynard Smith 1978), but this may also be the case for animals with cloning from somatic tissue where many cells makes up the new animal and where not all contributing cells have the mutation (discussed in Kawamura and Fujiwara 2000). Progeny dispersal does also play a role and can be limited in clonal or colonial species.

At a larger scale, it is apparent that many of the clonal species have appeared quite recently and that the more ancient ones still present more seldom evolve into new species, as for ascidians and anemones (Swalla et al. 2000; Geller 2005). This distribution has been considered as evident for asexual species to be less evolvable and perhaps therefore more prone to extinction in the long run (Ridley 1996). However, it is almost ubiquitous among metazoans that asexual animals, like the ascidians, also involve occasional sexual propagation in their life cycle and then recombination and some dispersal will occur. If the interpretation above is true, the phylogenetic distribution tells us that not only obligatory asexuals, but even colonial species that also occasionally undergo sexual reproduction during their life cycle, may too face long-term problems.

It is further possible that colonial lines that have evolved from sexual ancestors and propagate from physiologically aged somatic tissues and stem cells also undergo ageing. If ageing do occur in these strains, then occasional sexual reproduction would be necessary for a complete rejuvenation. However, if there is no clear distinction between the germ-line and the soma, or the somatic stem cells giving rise to the asexual progeny, then senescence may in theory not occur (Kirkwood and Austad 2000). Unfortunately, there are few experimental studies on ageing in clonal or colonial metazoans (Agrawal 2006), but decline in growth rate is seen in the fissiparous marine oligochaete *Paranius litoralis* and in the rabdocoel *Stenostomum incaudatum* (Martínez and Levinton 1992). Some evidence for colony senescence are from survival rate counting's in the tunicate *Botryllus schlosseri*, the bryozoan *Celleporella hyaline* and the hydroid *Laomedea flexuosa* (Chapter 4 in Hughes 2002). On the other hand, *Hydra* and some clonal lines of the bryozoan *Celleporella hyaline*, the ascidians *Polyandrocarpa mikensis* and *Diplosoma listerianum*, and the polychaets *Zeppelina monostyla*, *Dorvillea bermudensis* and *Dorvillea albomaculata* may be potentially immortal since they have been kept in reproductive isolation for years or decades (Schroeder and Hermans 1975; Åkesson and Rice

1992; Kawamura and Fujiwara 1995b; Martínez 1998; Bishop et al. 2001; Hughes 2005). High degree of tissue dynamics and cellular turnover has been discussed as one requirement for the possible escape from senescence in *Hydra* (Sánchez Alvarado and Tsonis 2006), but what is ten years in culture compared to the life of a tree or a human? More long-term fitness data as well as a better understanding of the stem cells, repair systems and homeostasis in clonal animals and strains are needed. It is noteworthy that *Botryllus schlosseri*, which have totipotent somatic stem cells involved in agametic cloning, still undergoes colony senescence. For further information about the question of senescence in asexual animals, see Chapters 4 and 9 in Hughes (2002).

We conclude that further research of the stem cells and tissues involved in agametic cloning using modern molecular techniques may not only provide novel solutions to adult tissue regeneration and rejuvenation but also new evolutionary models on the putative advantages of sexual versus asexual reproduction.

5.4 A Systematic Overview Over Metazoans Performing Agametic Cloning

Across the animal kingdom and across all levels of animal complexity, it is evident that organisms capable of asexual reproduction are often those with excellent abilities to regenerate, while in those taxa without asexual reproduction the regeneration capacity is generally less pronounced (Fautin 1999). Next, we present a selective overview of what is known about the agametic cloning in some marine animal invertebrates across the animal kingdom. This could serve to reveal potential new model organisms for future studies on stem cells and stemness in asexual reproduction. As we will see, the cellular and molecular mechanisms underlying asexual reproduction are, apart from a hand full of model organisms, yet unstudied for most animals (Fig. 5.5). Hence, it is yet difficult to homologize and compare these processes.

5.4.1 Porifera

Sponges (phylum Porifera) represent one of the oldest metazoan phyla with a very simple body design being constructed with only a few cell types and lacking organs (Funayama et al. 2005). Probably all sponges can regenerate their body and reproduce asexually using stem cells, while some are also able to reassemble their tissue from an unstructured cell mass. The poriferan stem cells are undifferentiated totipotent cells called archeocytes (Borojevic 1966; Weissenfels 1989; Diaz 1977; De Sutter and Van de Vyver 1977; van Bekkum 2004). These cells are evenly distributed in the tissue and used in a number of ways to multiply individuals as for example by budding, fragmentation, or asexual larvae (Ereskovsky and Tokina 2007). Sponges do not have a germ line and hence also the gametes derive from the archeocytes (Witte and Barthel 1994). Many freshwater species and some estuarine/marine sponges use their totipotent archeocytes to develop dormant overwintering stages, so-called gemmules. These are small spherical structures which are highly resistant to freezing and desiccation. The parent sponge usually disintegrates over the winter while the thesocytes (=dormant archeocytes) of the gemmules develop a new individual during spring.

Molecular processes underlying reproduction and regeneration in sponges have been studied and developmental markers identified. The receptor tyrosine kinase (RTKvs) for example is expressed in oocytes and the early larvae while the embryonic development protein (EED) can be found in gemmules (Müller 2006). However, it yet needs to be shown if there are archeocytes that express both markers.

5.4.2 Placozoa

Trichoplax adhaerens is a tiny marine animal that looks more like an amoeba than a metazoan (Fig. 5.1A), consisting of an irregular flattened body with a simple upper and lower epithelium and some loose cells in between. Only five cell types are known from placozoan tissue, not constituting any organs or a basal membrane. Due to their simple organization many features of placozoans are considered to represent the basic and ancestral condition in metazoans and offer thereby a great model for studying metazoan evolution. Placozoans replicate rapidly asexually, forming large clonal assemblages. The most common type is fission into two equal parts, but also multiple fragments can arise (Pearse 1999). Budding from the upper surface of a flattened individual can produce hollow miniature spheres that are able to float in the water column for about a week and after settlement flatten out and restore the upper and lower epithelia. Totipotent stem cells do not seem to occur in these animals, because neither a piece of the body's edge by itself, nor an isolated central part of the body will be able to regenerate. Instead both marginal and central parts of the body must be present to reproduce and regenerate.

5.4.3 Cnidaria

Cnidaria is a likewise ancient phylum with a rather simple body plan possessing great capacity for cloning, regeneration and longevity (Fig. 5.1B; Fautin 1991). The animals are diploblastic with two epithelial layers, the ectoderm and the entoderm, separated by the mesoglea membrane which often contains pluripotent interstitial cells (Bode 1996; Agata et al. 2006). The animals are radial symmetric without cephalisation or a central nervous system. Cnidarians are highly polymorphic and exhibit a bewildering variety of asexual reproductive modes (Miller 1997; Shafir et al. 2001; Fautin 2002) as for example budding, fragmentation, fission, strobilation, asexual larvae and dormant cysts. Physical disturbance may trigger asexual propagation by fragmentation and promote clonal structures, as for the sessile *Metridium senile* (Anthony and Svane 1995) and many corals (Highsmith 1982). The different asexual modes are used to multiply individuals, but also to transform into the next life cycle stage, e.g. from polyp into medusa or from medusa into gamets. Principally most of these processes involve however the formation of a distinct pouch or piece of tissue containing endoderm, ectoderm and the mesoglea membrane. Likewise, the gonophores (gonad carrying organs) are developed by budding from the body wall. Müller et al. (2004) reported that the migratory interstitial stem cells in the hydroid *Hydractina* are totipotent. While the molecular and cytological bases for cloning processes are unstudied in most cnidarians, they are rather well known for the freshwater hydroid *Hydra*. A typical *Hydra* consists of up to 200,000 cells of 20–25 cell types. The interstitial stem cells in hydra are pluripotent but not totipotent as they cannot develop into epidermal and digestive cells (Bode 1996). The dynamic of cloning in *Hydra* seem to depend on the stem cell proliferative activity as these cells continuously produce new tissue and buds. However *Hydra* may not be a well chosen representative for all cnidarians (Carré and Carré 2000) and many processes observed in hydra may not apply to all cnidarians. Piraino et al. (1996) for example, have reported alternative developmental mechanisms in the hydrozoan *Turritopsis nutricula*, where the transformation of a mature medusa into a polyp involves dedifferentiation of fully differentiated cells.

5.4.4 Ctenophores

Some ctenophores are able to regenerate complete individuals, when cut in halves and quarters as for example *Bolinopsis infundibulum*, *Mnemiopsis leidyi* (Matsumoto 1999). Many of the ctenophores in the order Platyctenida are capable of asexual reproduction via fragmentation and budding. Whether these processes involve totipotent adult stem cells or necessitate a piece of all differentiated tissues in the fragments is not studied. However, the fact that regeneration does not occur when *B. infundibulum* is cut into eight equal parts makes it unlikely that totipotent cells are present in ctenophores (Matsumoto 1999).

5.4.5 Platyhelminthes

Plathyhelminthes (flatworms) is a group of worm-shaped, bilateral and acoelomate animals (Figs. 5.1C–E). Flatworms are an extremely diverse group and very likely consist of several subgroups with their own evolutionary history. Asexual reproduction occurs in basal (Acoela, Catenulida, Macrostomorpha) as well as in derived taxa (Tricladida, Neodermata). Some flatworms are among the best studied organisms regarding their potential to regenerate, which is partly because they possess totipotent and highly proliferating adult stem cells, so-called neoblasts (Fig. 5.1C; for an extensive review see Reuter and Kreshchenko 2004). Today we know that neoblasts are probably the only dividing cells in Platyhelminthes (Rieger et al. 1999a,b; Gschwentner et al. 2001). Turbellarians for example can restore the entire individual from a minute piece of tissue, given that neoblasts are present in the fragment. In the parasite *Taenia crassiceps*, single cells of trypsinated cysticerci are able to produce complete cysticerci in a mouse host (Toledo et al. 1997).

Catenulids are reported to have excellent regenerative abilities and undergo paratomic fission (Egger et al. 2007). Acoels have a substantial regenerative capacity and show all major types of asexual reproduction, for example achitomy in *C. longifissura*, paratomy in *Paratomella unichaeta*, and budding in *C. retrogemma*. Triclades show paratomic and architomic fissions, while many of the parasitic flatworms produce multiple individuals from a single egg (polyembryony). Polyembryony is frequently observed also in bryozoans, echinoderms, hymenopterans, and armadillos, but does otherwise not occur often in the animal kingdom. Flatworms use lumps of totipotent and proliferating cells to develop their next stage in the life cycle, as also described for digenetic trematodes (Tyler 1999).

Typical reproductive processes known from flatworms are paratomy and architomy. In paratomy new individuals differentiate in a chain like-fashion from a parent worm before separating from it, while in architomy the body spontaneously fragments and only thereafter individual differentiate from the pieces. Also various ways of budding and fission are known. One example is the acoelous turbellarian, *Convolutriloba retrogemma* described by Hendelberg and Åkesson (1988) that release its progeny by budding from the posterior margin of the body. In early stages, the anlagen of the buds contain symbiotic zoochlorellae algae which serve as food for the developing bud. The subsequently released buds are miniatures of an adult worm. During the release, wounds arise on both animals and sometimes small tissue fragments come loose. These tiny fragments round up to ciliated bodies that eventually they reorganize themselves into very small planula-like worms of which some may grow to adult size (Hendelberg and Åkesson 1988). There are similarities between regeneration and (post)embryonic development. In extreme cases where only a small piece remains for regeneration, resemblance with postembryonic development is particularly striking (Egger et al. 2007). The remarkable variation in budding and fission of the genus *Convolutriloba* and recently the report of successful sexual reproduction in cultures of the described four species (Shannon and Achatz 2007) will make this to a favorite group of new model animals. Here it is thus possible to compare the stem cells that are active in the three modes of development; fission with subsequent regeneration, budding and larvae development.

5.4.6 Nemertea

Nemerteans are unsegmented protostome worms found in marine, terrestrial and freshwater habitats. They have a spiral cleavage, a mesoderm that derives from the 4d cell and a coelomic cavity in their proboscis (Henry and Martindale 1996, 1997). These characters place the nemerteans among other spiralian phyla, for example with mollusks, annelids and plathyhelminths. Some nemertean species of the genus *Lineus* are known to reproduce asexually by spontaneous transversal fragmentation. Each fragment contains a piece of the brain or lateral nerve chord from which it regenerates an entire worm. Small fragments usually encyst and develop within a protective coat (Turbeville 1999). The cytological details and the genetic basis for asexual reproduction in nemerteans have not been studied to date.

5.4.7 Annelids

Among the segmented coelomate worms such as oligochaetes (Fig. 5.1F) and polychaetes (Fig. 5.1G), asexual reproduction is frequent (Schroeder and Hermans 1975). As stated for other taxa, also in annelids the ability to reproduce asexually appears closely connected to the ability to regenerate and this trait seems to have multiple origins within annelids (Berrill 1952). Principally asexual reproduction is achieved by budding or fission into two or more pieces of trunk fragments. This is solved by regeneration of the missing body parts, although new segments can also be formed by remodeling already existing segments. At least regeneration has been shown to be nerve dependent in polychaetes (Müller et al. 2003). In oligochaetes paratomic fission is frequently observed, for example in Naididae and Aeolosomatidae, although *Nais paraguayensis* is an architomic species. Fission or budding likewise occurs in many polychaete families including spionids, cirratulids, syllids, and sabellids (Barnes 1987). However a modified version of trunk fission is known from many polychaetes. Here division occurs into a number of segmented units (schizogenesis) often forming chains of individuals (stolonisation). In these cases only posterior fragments, so called "stolons" acquire the function of pelagic, sexual male and female segments (epitokes) that swarms, leaving behind a benthic asexual "stock" individual that keeps budding further stolons. In the syllid *Autolytus prolifer*, the epitokes can regenerate head structures upon release from the stock and hereby develop into a second generation of sexually mature individuals (Schiedges 1979; Fischer 1999). Asexual polychaetes are very good models and have been kept as clonal strains in laboratories for up to 60 years *Dodecaceria concharum* (Korschelt 1942) and more than 30 years *Dorvillea bermudensis* (Åkesson and Rice 1992; the latter culture is still alive). In the polychaete families listed by Schroeder and Hermans (1975) asexual reproduction alternate with sexual reproduction and often but not always in a regular way. In nature, they are considered to reproduce sexually at some time or another (Jackson 1986).

While the anatomical studies of annelid reproduction and regeneration has a long tradition relatively few molecular studies have been performed to date investigating the mechanisms underlying annelid regeneration and asexual reproduction (Dupin et al. 1991; Bely and Wray 2001; Myohara et al. 2006; Tadokoro et al. 2006). It is not yet clear how potent adult stem cells of annelids are. Recently Paulus and Müller (2006) showed that proliferating cells are present throughout the entire body of the asexually reproducing polychaete *Dorvillea bermudensis*. However, they could not observe the typical stem cell behavior known from flatworm neoblasts, which start to proliferate and move to the regeneration site upon fragmentation. This suggests that new tissue probably arises from already existing tissue. Gibson and Harvey (2000) who studied the morphogenesis during transversal fission in the spionid polychaete *Pygospio elegans* came to a similar conclusion, indicating that the regenerative system in polychates differ from that in flatworms. However, there may be more flexibility in cell fate in some species as Boilly (1967) found evidence in the polychaete *Syllis* that digestive endodermal tissue can be regenerated from ectodermal pharynx tissue. Tadokoro et al. (2006) found germ cell precursors distributed throughout the entire body of the oligochaete *Enchytraeus japonensis*, which migrate during asexual reproduction into regenerating segments to develop the gonads. Annelid stem cell may therefore be pluripotent but not totipotent.

5.4.8 Sessile Filter Feeding Protostomes

Many sessile protostome groups possess extensive abilities of asexual reproduction. These are entoprocts (Fig. 5.2A; Nielsen 2001), bryozoans (Fig. 5.2B; Ryland 1970), cycliophorans (Fig. 5.2C; Funch et al. 2005), and phoronids (Nielsen 2001). Asexual reproduction in phoronids is only known in one species, *Phoronis ovalis* and in this case the adult body splits into two approximately equal pieces which subsequently regenerates the missing parts (Zimmer 1999). Experimentally, this has also been shown for other species, but the gonadal tissue needs to be retained in both fragments. While most phoronids have the ability to regenerate the distal parts of their body which frequently gets lost, *Phoronis ovalis* also reproduces by budding, but it is unknown whether this distal regeneration and budding involves undifferentiated stem cells or dedifferentiation of already specialized tissue (Zimmer 1999). In the two species of cycliophorans known to date, we similarly find that the distal parts of the feeding individual are cast off and replaced in regular intervals (Funch and Kristensen 1995; Obst et al. 2006). For cycliophorans, it is known that the regenerative and reproductive abilities are due to a lump of undifferentiated mesenchymal cells maintained in the trunk of most life cycle stages. Cycliophoran adult stem cells are totipotent as they can replace the feeding structures, give rise to successive live cycle stages, and develop germ cells. The approximately 150 described species of entoprocts have some reproductive traits in common with cycliophorans and phoronids, but some are also colonial. The colonial entoprocts can cast off their distal feeding apparatus and subsequently replace it from a distal bud. Most entoprocts also grow buds from the anterior part of their body. In some stolonate entoprocts encapsulated resting buds (hibernaucula) are formed at stolon tips. These buds start to grow after being exposed to low temperatures. The cytological patterns of bud formation in entoprocts start always with an epidermal proliferation resulting in an evagination that forms the bud primordium. Budding is essentially an ectodermal process, whereas some mesenchyme cells migrate into the bud, while no endodermal cells are involved. According to Nielsen and Jespersen (1997) and Mukai and Makioka (1978, 1980) the buds have an embryonic appearance and consist of large epidermal cells.

Bryozoa is another large and polymorphic group of sessile marine invertebrates with approximately 5000 described species. Almost all bryozoans apply stem cells in order to form modular colonies (Bayer et al. 1994), germinate from overwintering cysts (statoblasts), metamorphose, or propagate via fragmentation (O'Dea 2006). In addition, polyembryony (embryonic fission) is known from cyclostome bryozoans. In most of these processes new tissue is developed from buds containing presumably totipotent stem cells. Principally these buds arise from a primordium developing from a thickening in the body wall and include mainly ectodermal epithel, but may also include mesodermal cells (Nielsen 1971). The modular *Celleporella hyaline* can be maintained long-term in culture by repeatedly taking cuttings (Hughes 2005). However, whether the adult stem cells involved in agametic cloning are really totipotent in that they are able to transform into any germ layer and develop gonad tissue, remains to be proven.

5.4.9 Deuterostome Worms

The benthic marine worm *Xenoturbella* (Fig. 5.2D) grows to about 3 cm in length and inhabits U-shaped burrows in soft bottom sediments (Åkerman 2004). It was first described in 1949 (Westblad 1949) as a delicate, ciliated, marine worm with a simple body plan lacking a through gut, organized gonads, excretory structures and coelomic cavities, and possessing a very simple nervous system that consists of a diffuse nerve net without a brain. Xenoturbella's phylogenetic affinities have long been obscure, but over the last years it became evident that the worm is in fact a deuterostome related to hemichordates and echinoderms (Bourlat et al. 2006). Despite that the life history and ecology of the worm is little studied (Dupont et al. 2007; Bourlat et al. 2008), the basic features of *Xenoturbella* suggests that it has potential for the understanding of deuterostome evolution. The worm seems to process some regenerative potential and may be asexually reproducing through fissions since longitudinal division of the animals into a left and right half have been observed (Israelsson 2006). Undifferentiated cells have been observed inside the gastrodermis (Israelsson 2006), but further cytological processes involved in the possible asexual reproduction of *Xenotubella* are unknown.

The phylum Hemichordata includes two groups of strictly marine invertebrates, the enteropneust worms and the colonial pterobranchs and with approximately 100 described species. In enteropneusts, regeneration and asexual reproduction is described from members of the family Ptychoderidae, i.e. *Balanoglossus capensis* (King 1998), where fragmentation of individuals occurs anterior to the hepatic region of the trunk. But the animals are very fragile and hence fragmentation could also be a result of physical damage. In pterobranchs, asexual reproduction by budding from the zooid base is common and as a result all individuals are interconnected by stolons. In *Cephalodiscus* and *Rhabdopleura*, budding begins when a bulge or evagination arises from the trunk metacoel. After elongation, the anterior end of the bud starts swelling and forms the cephalic shield (proboscis) of the new individual, followed by differentiation of the coelom and then formation of the gut perhaps from the ectoderm (King 1998). To date too little is known about these animals to be able to say whether stem cells are involved in their agametic reproduction.

5.4.10 Echinodermata

The most common form of asexual reproduction in echinoderms is division of the body, usually termed fissiparity (Fig. 5.2E). In seastars and brittlestars, fission usually takes place across the disk dividing the animal in two parts. A few species of seastars are also known to reproduce through autotomy, i.e. nervously mediated self mutilation, regenerating a whole new animal from single arms or even parts of an arm (Emson and Wilkie 1980). Asexual reproduction in adults is not very common in the phylum, and is only known to occur in 21 species of seastars (of ca 1600), 45 brittlestars (of ca 2000) and 8 seacucumbers (of ca 900), so far (Mladenov and Burke 1994). Though fissiparity is uncommon in the phylum, asexual recruitment to populations can be very successful and range widely, and occur in high densities in certain habitats such as temperate intertidal rocky shores and tropical coral reefs among algal turfs and within sponges (e.g. Mladenov and Emson 1988; Barker et al. 1991; Sköld et al. 2002; Conand et al. 2002). Adult fission may also facilitate the introduction of alien species such as the seastar *Asterina burtoni* from the Red Sea that has established clones in the Mediterranean Sea via the Suez channel (Karako et al. 2002).

Echinoderms also reproduce asexually by fragmentation from larval stages at metamorphosis or fission and budding in larvae and have been described in all classes except crinoids (Jaeckle 1994; Balser 1998; Vickery et al. 2001; Eaves and Palmer 2003). The replicate produced is either a differentiated feeding larvae or an earlier embryonic stage that then complete early embryogenesis and form a functional feeding larvae. The relative contribution to recruitment from asexual larval stages is unknown but may well be significant (Bosch et al. 1989). Knott et al. (2003), identified field collected cloning asteroid larvae and concluded that cloning are widespread within Asteroidea and that it occurs regularly and in multiple ways within species that are capable of cloning.

Both exogenous (seasonal changes, environmental stimuli) and endogenous factors (humoral and nervous factors) appears to be involved in regulating asexual reproduction in echinoderms. For most species, the asexual mode is related to small body size, while sexual maturity usually is attained after an individual reaches a certain size. Between habitat variation in levels of asexual and sexual reproduction has also been documented in seastars (Emson and Wilkie 1980; Mladenov 1996). Observations that subtidal populations generally have a larger body size, a lower incidence of fission and larger gonads than intertidal populations, has led to the generality that the asexual phenotype is associated with harsher, less stable resource-poor environments, whereas the sexual phenotype is associated with more favourable, stable and resource-rich environments. In some seastars it has also been observed that fission can be triggered by stress during unintended laboratory holding conditions such as elevated temperature, interruption of seawater circulation, and lack of aeration (Mladenov et al. 1983). These findings and generalisations have raised the question of environmental factors modulating and triggering fission. Few studies have actually tested these observations (Mladenov 1996) but experimental evidences from *Coscinasterias muricata* (Sköld et al. 2002) and larvae of the seastar *Pisaster ochraceus* (Vickery and McClintock 2000) showed that fission is sensitive to starvation and stimulated by high food concentrations.

Adult fission in asteroids, ophiuroids, and holothurians is a slow process taking many minutes to days, and results in irregular wounds and prolonged wound healing (Mladenov and Burke 1994). The following regeneration is slow and involves re-organisation of several old cell structures. Such processes involve extensive migration of cells but can also imply the activity of existing pluripotent stem cells or de-differentiation of tissue into stem cells (Carnevali 2006). Regeneration following autotomy, which usually takes place across pre-defined planes (Wilkie 2001), can also involve stem cells but seem to be a different process as it in contrast to fission involves the formation of an extensively proliferating blastema (Emson and Wilkie 1980; Carnevali 2006), however not in Asteroids (Moss et al. 1998). Both regeneration processes are nerve dependent and involve neurotransmitters and neuropeptides as growth factors (Thorndyke and Carnevali 2001; Carnevali 2006).

5.4.11 Tunicata

Tunicates (Urochordates) are common sessile marine animals with about 2.300 species consisting of four classes, the Thaliacea, Appendicularia, Sorberacea, and Ascidia. Appendicularians (Oikopleura species) do not reproduce asexually, but among the other groups we find extraordinary abilities for asexual reproduction and colony formation (Fig. 5.2F,G), (Satoh 1994). Tunicates are the closest relatives to vertebrates processing agametic cloning (Fig. 5.5). The coloniality appears to have evolved several times from a sexual ancestor (Berrill 1935; Satoh 1994; Swalla et al. 2000) and leads to the direct formation of an adult zooid form, but none of the ascidians are obligatory asexual. Most of the larger tunicates are sexual whereas individual zooids of the colonial tunicates are very small. Ascidians exhibit a bewildering number of ways of budding from nearly any type of tissue. Examples are stroblilation of the abdomen or the postabdomen, herniation of vascular processes extended onto the substrate, accumulation of cell masses in the tunics blood vascular system, or projection of the thoracic wall into the surrounding basal tunic. In addition budding can also give rise to resistant bodies (Nakauchi 1982; Satoh 1994; Newberry 1999) that develop into a new zooid when favourable environmental conditions return. Altogether, there is currently no established general stemness signature for somatic stem cells in ascidians, but cell proliferation, the enzyme aldehyde dehydrogenase, retinoic acid and thyroid hormones are at least involved in budding in more than one species. Elevated activities of telomerase and alkaline phosphatase in early buds are also in alignment with a stem cell state (se markers for human embryonic stem cells: Heins et al. 2004). Berrill (1935) extensively reviewed budding in twenty genera's of ascidians and concluded that although the epidermis of the bud ubiquitously has epidermal origin, the inner tissues of the bud have variable origins.

The thalaceans are pelagic tunicates capable of rapid colonization and some species have successfully been kept in cultures (Gibson and Paffenhöfer 2002). In pyrosomid thaliaceans zooids, a budding site near the heart generates more zooids (Newberry 1999). In salps and doliolids (both Thaliacea) colony formation is extremely complex and can generate polymorphic strands of budded modules that remain for some part of the life cycle connected.

In the cosmopolitan *Botryllus schlosseri*, blood-borne stem cells are involved in budding from the adult body wall or from the distinctive vascular system in absence of zooids (Voskoboynik et al. 2007) and certain blood-born cells can invade other colonies where they differentiate into many different kinds of cells, including germ cells (Laird et al. 2005). These totipotent cells express aldehyde dehydrogenase and telomerase (Laird et al. 2005). In the related *Botrylloides leachi*, budding involves cell proliferation and the morphogene retinoic acid (Ermak 1982; Rinkevich et al. 2007). Alkaline phosphatase, a marker for vertebrate embryonic stem, is expressed in developing buds and in some hematocytes (Akhmadieva et al. 2007). The related *Polyandrocarpa misakiensis* (Fig. 5.2F) develops buds from an evagination of the adult body wall or from stolons. Budding starts with de-differentiation of the pluripotent inner epithelium, called atrial epithelium, in the very early bud. It starts to proliferate to an extent while it protrudes into the bud and differentiates into the inner organs (Kawamura and Nakauchi 1986; Kawamura et al. 1995a,b). The de-differentiation of the atrial epithelium has been demonstrated by tracing the fate of certain differentiation or de-differentiation markers like the TC-14 lectin, TRAMP serine protease, certain atrial epithelium antigens, alkaline phosphatase, aldehyde dehydrogenase and pigment (Kawamura and Fujiwara 1995a; Matsumoto et al. 2001; Kawamura et al. 2008). Early bud morphogenesis is regulated by retinoic acid and thyroxin (Hisata et al. 1998). Some gene sequence data is available from *P. misakiensis* (Kawamura et al. 1998) and the atrial epithelia cells can be cultured *in vitro* where they divide and dedifferentiate (Kawamura and Fujiwara 1995b). One strain of *P. misakiensis* has been kept in culture for more than 20 years (Kawamura and Fujiwara 1995b).

In *Diplosoma listerianum* (Fig. 5.2G) and other didemnids, one bud form the filtering apparatus and another bud form the stomach region. Both grow out from the intestine area and appear to involve the for *Didemnids* specific epicardia structures that protrudes into the buds (Berrill 1935). When kept in reproductive isolation, no oocytes are formed and several colony lines have been kept growing in reproductive isolation since 10 years back (J. Bishop, personal communication) but to prevent colony senescence, the colony needs to be cut back from time to time. Some gene sequence information in form of ESTs is available for *D. listerianum*. The related *Clavelina gemmae*, produce star-shaped buds that originate from a basal vessel and once buds are developed, they detach and disperse into the water (Turon 2005). Similarly, species of *Perophora* bud from vascular stolons and this seems to involve mitotic blood born lymphocyte- like cells (Freeman 1964). In regressing colonies however, survival budding occurs without mitotic growth and instead by recycling of existing cells or material, a matter little understood (Barth and Barth 1966). Budding in *Perophora* is stimulated by thiourea and inhibited by thyroid hormones (Fukumoto 1971).

5.5 Future Perspectives

5.5.1 Summary

When summarizing what we know about agametic cloning and the stem cells involved, it is apparent that the knowledge is based on findings from a few species and phyla only. Altogether, animal size and age, tissue elasticity, and cellular or nuclear commitment, plasticity and repair systems, may, together with evolutionary constraints and selection pressures, influence and limit the choice of reproductive strategy, asexual through gametic or agametic cloning and/or sexual. In evolution, the basic metazoans emerge as clonal, acolonial and with initiation of germ line from undetermined primary stem cells. The colonial strategy is derived from sexual ancestors and appears independently in different branches, as the case for also some clonal animals. Generally, the ability to reproduce by cloning is reflected in an extensive regenerative capacity and it is clear that the widespread distribution of asexual reproduction as well as the many different modes of clonality within phyla reflect high levels of cell and tissue plasticity. This plasticity is however still very little characterised and understood. In addition, when considering that a lot of the asexual reproduction seen today is derived, this has the consequence that there is probably no universal regulatory network or stem cell system used in cloning across the metazoans. To better understand the stem cells and issues concerned with asexual reproduction or cloning, both evolutionary and comparative experimental approaches are necessary. We believe that further research of the stem cells and tissues involved in agametic cloning using modern molecular techniques may not only provide novel solutions to adult tissue regeneration and long term rejuvenation but also new evolutionary models. Some suggested experimental approaches are described in the following section and involve analyses of cell proliferation, different aspects of self renewal, stem cell markers, gene silencing and regulators of developmental pathways.

5.5.2 Experimental Strategies

Cellular proliferation is a key characteristic of tissue renewal that should be distinguished from cell migration or transformation without cell proliferation. In fact, regeneration, which in some but not all cases show similarities to asexual reproduction, depend often on cell migration and de-differentiation (Holstein et al. 2003; Carnevali 2006; Sánchez Alvarado and Tsonis 2006). High cellular turnover is considered crucial for the proposed infinite asexual propagation capacity of *Hydra* and planarians (Holstein et al. 2003; Sánchez Alvarado and Tsonis 2006).

The ability of self renewal in stem cells is in addition to mitosis also reflected in that they are in contrast to other somatic cells immortal or at least long-lived. Immortality at the cellular level is achieved by maintained chromosome telomeres and active telomerase (Blackburn 1991). Telomerase and telomere length are therefore potential prerequisites for self renewal during agametic cloning, as indicated for the colonial ascidian *Botryllus schlosseri* (Laird and Weissman 2004). Other issues to consider for long term maintained cellular function and renewal are mechanisms for nuclear and mitochondrial DNA repair, protein damage and handling of oxidative stress (Rando 2006). This should be especially important for stem cells with high proliferation rates. One interesting mechanism to maintain healthy DNA could be for the stem cells to selectively retain the DNA template and sort the DNA copies to the daughter cells (Merok et al. 2002). The long-lived asexual animals and strains can hold information about novel mechanisms for renewal.

The degree of pluripotency is probably reflected in what the role of the stem cell is and this in turn may be reflected in its developmental and evolutionary origin. A primary or an embryonic stem cell with capacity to form the germ line may be immortal while lineage specific stem cells seem to have less potential (Rando 2006). For the clonal animal, this means that if it uses totipotent primary stem cells then the lineage may not age, while if it uses a somatic tissue with no connection to the germline occasional sexual reproduction will be needed for a complete rejuvenation. It is thus important to be able to distinguish between different kinds of stem cells and their degree of potential. Established stem cell markers like Vasa, Sox2, Oct3/4, Nanog, alkaline phosphatase activity, telomerase activity and ultimately also graft experiments, should be tested comparatively for many asexual animals.

Pluripotency in the germ line is maintained by total suppression of mRNA transcription (Berekelya et al. 2005). Transcription from DNA can be down-regulated by histone modifications or methylation of the DNA and appears to be important also for embryonic stem cell maintenance (Spivakov and Fisher 2007). It is important for regulating early invertebrate embryogenesis (Maharajan et al. 1986) and seems also to be involved in regulating asexual budding in *Hydra* (De Petrocellis et al. 1986). RNA silencing of genes is also important for stem cell maintenance for many organisms (Seto et al. 2007) and the highly conserved piwi protein that is part of the silencing machinery is expressed in the neoblast stem cells of planarians (Reddien et al. 2005). Further research on the silencing machinery it self as well as identification of what genes in the stem cells which are silenced or down-regulated by DNA modifications are wanted not only for agametic cloning but for stem cells in general including germ cells. Since there is a large lack of *in vitro* invertebrate cell culture procedures with the exception for eggs and early embryos (but see, Kawamura and Fujiwara 1995b), marine invertebrate eggs can here complement studies on cultured stem cells.

Finding the regulatory pathways involved in asexual morphogenesis is also important. Many of the factors that interfere with formation of asexual progeny are also major players in embryogenesis, but one must be aware of that many different solutions may be involved in agametic cloning given its high diversity in the animal kingdom. Nevertheless, retinoic acid is a well known morphogen that stimulates cellular differentiation and is used in leukemia therapies as a cancer stem cell differentiation agent (Sell 2004). It induces multiple buds as well as interferes with bud morphogenesis in ascidians (Rinkevich et al. 2007; Kawamura et al. 2008). In Cnidarians and vertebrates, modulation of the WNT/GSK/beta catenin signaling cascade similarly results in the formation of secondary heads along the body axis

(Fagotto et al. 1997; Müller et al. 2004; Broun et al. 2005). It is however currently unknown if factors inducing multiple budding stimulate stem cell proliferation or promotes their differentiation. If we also consider agametic cloning as a secondary axis formation, signals involved in axis formation in the embryo are additional candidates to investigate. This would however benefit greatly if a larger quantity of gene sequence information was available for some different asexual animals.

Acknowledgments We are grateful for comments and suggestions from B. Rinkevich during the preparation of this chapter. We thank our colleagues and collaborators for contributing with images. The work is supported by the Swedish Taxonomy Initiative (contract grant no. dha 15/06 1.4), Royal Swedish Academy of Sciences and the Network of Excellence 'Marine Genomics Europe', contract GOCE-04-505403 to M.O., and Swedish Research Council (Natural and Engineering Sciences), Royal Swedish Academy of Sciences and Magnus Bergvalls Stiftelse to H.S.

References

Agata K, Nakajima E, Funayama N, Shibata N, Saito Y, Umesono Y (2006) Two different evolutionary origins of stem cell systems and their molecular basis. Semin Cell Dev Biol 17(4):503–509

Agrawal AF (2006) Evolution of sex: Why do organisms shuffle their genotypes? Curr Biol 16:R696–R704

Åkerman M (2004) On the behaviour and ecology of *Xenoturbella* sp. Master Thesis in Marine Ecology, Göteborg University, Sweden

Åkesson B, Rice SA (1992) 2 New Dorvillea Species (Polychaeta, Dorvilleidae) with obligate asexual reproduction. Zoologica Scripta 21 (4):351–362

Akhmadieva AV, Shukalyuk AI, Aleksandrova YN, Isaeva VV (2007) Stem cells in asexual reproduction of the colonial ascidian *Botryllus tubaratus* (Tunicata: Ascidiacea) Russian. J Mar Biol 33:181–186

Anthony KRN, Svane I (1995) Effects of substratum instability on locomotion and pedal laceration in *Metridium senile* (Anthozoa: Actiniaria). Mar Ecol Prog Ser 124:171–180

Andersson M (1994) Sexual selection. Princeton University Press, Princeton

Balser EJ (1998) Cloning by ophiuroid echinoderm larvae. Biol Bull 194 (2):187–193

Bannister R, McGonnell IM, Graham A, Thorndyke MC, Beesley PW (2005) Afuni, a novel transforming growth factor-beta gene is involved in arm regeneration by the brittle star *Amphiura filiformis*. Dev Genes Evol 215 (8):393–401

Barker MF, Scheibling R, Mladenov P (1991) Seasonal changes in population structure of the fissiparous asteroids *Allostichaster insignis* (Farquhar) and *Coscinasterias calamaria* (Gray). Echinoderm Res 191–196

Barnes RD (1987) Invertebrate Zoology 5th Edition. Saunders College Publishing, Philadelphia

Barth LG, Barth LJ (1966) A study of regression and budding in *Perophora viridis*. J Morphol 118:451–460

Bayer MM, Cormack RM, Todd CD (1994) Influence of food concentration on polypide regression in the marine bryozoan *Electra-Pilosa* (L) (Bryozoa, Cheilostomata). J Exp Mar Biol Ecol 178 (1):35–50

Bely AE, Wray GA (2001) Evolution of regeneration and fission in annelids: Insights from engrailed- and orthodenticle-class gene expression. Development 128 (14):2781–2791

Berekelya LA, Ponomarev MB, Mikryukov AA, Luchinskaya NN, Belyavsky AV (2005) Molecular mechanisms of germ line cell determination in animals. Mol Biol 39:572–584

Berrill NJ (1935) Studies in tunicate development. IV. Asexual reproduction. Philos Trans R Soc Lond B 225:327–379

Berrill NJ (1952) Regeneration and budding in worms. Biol Rev 27:401–438

Bishop JDD, Pemberton AJ, Sommerfeldt AD, Wood CA (2001) Laboratory studies of mating in the aplousobranch *Diplosoma listerianum*. In: Sawada H, Lambert CC, Yokosawa H (eds) Biology of Ascidians. Springer-Verlag, Tokio

Blackburn EH (1991) Structure and function of telomeres. Nature 350:569–573

Blackstone NW, Jasker BD (2003) Phylogenetic considerations of clonality, coloniality and mode of germline development in animals. J Exp Zool 29 (7B):35–47

Bode HR (1996) The interstitial cell lineage of hydra: A stem cell system that arose early in evolution. J Cell Sci 109:1155–1164

Boilly B (1967) Regeneration of gut in pharyngeal region of *Syllis amica* Quatrefages (Polychete Annelides). Cahiers de Biologie Marine 8 (3):221–228

Bond DM, Finnegan EJ (2007) Passing the message on: Inheritance of epigenetic traits. Trends Plant Sci 12:211–216

Borojevic R (1966) Tude experimentale de la diffrenciation des cellules de l'ponge au cours de son dvelopment. Dev Biol 14:130–153

Bosch I, Rivkin RB, Alexander SP (1989) Asexual reproduction by oceanic planktotrophic echinoderm larvae. Nature 4337 (6203):169–170

Bourlat SJ, Juliusdottir T, Lowe CJ, Freeman R, Aronowicz J, Kirschner M, Lander ES, Thorndyke M, Nakano H, Kohn AB, Heyland A, Moroz LL, Copley RR, Telford M (2006) Deuterostome phylogeny reveals monophyletic chordates and the new phylum Xenoturbellida. Nature 444 (7115):85–88

Bourlat S, Nakano H, Akerman M, Telford M, Thorndyke M, Obst M (2008) Feeding ecology of *Xenoturbella bocki* (phylum Xenoturbellida) revealed by genetic barcoding. Mol Ecol Notes 8 (1):18–22

Broun M, Gee L, Reinhardt B, Bode HR (2005) Formation of the head organizer in hydra involves the canonical Wnt pathway. Development 132 (12):2907–2916

Brown FD, Swalla BJ (2007) Expression of vasa in a colonial ascidian, *Botrylloides violaceus*. Evol Dev 9:165–177

Buss LW (1987) The evolution of individuality. Princeton University Press, Princeton

Carnevali MDC (2006) Regeneration in Echinoderms: Repair, regrowth, cloning. Invert Surv J 3:64–76

Carré D, Carré C (2000) Origin of germ cells, sex determination, and sex inversion in medusae of the genus Clytia (Hydrozoa, leptomedusae): The influence of temperature. J Exp Zool 287 (3):233–242

Castrillon DH, Quade BJ, Wang TY, Quigley C, Crum CP (2000) The human VASA gene is specifically expressed in the germ cell lineage. Proc Natl Acad Sci USA 97:9585–9590

Conand C, Uthicke S, Hoareau T (2002) Sexual and asexual reproduction of the holothurian Stichopus chloronotus (Echinodermata): A comparison between La Reunion (Indian Ocean) and east Australia (Pacific Ocean). Invert Reprod Dev 41(1–3):235–242

Deb KD, Jayaprakash AD, Sharma V, Totey S (2007) Embryonic stem cells: From markers to market. Rejuvenation Res 11:19–37

De Petrocellis L, Maharajan V, De Petrocellis B, Minei R (1986) Bud induction in decapitated Hydra attenuata by 5-azacytidine: A morphological study. J Embryol Exp Morphol 93: 105–119

De Sutter D, Van de Vyver G (1977) Aggregative properties of different cell types of the fresh-water sponge *Ephydatia fluviatilis* isolated on ficoll gradients. Wilhelm Roux's Archiv 182:329–346

Diaz JP (1977) Transformation histologiques et cytologiques posttraumatiques chez la demosponge *Suberites massa*, Nardo. Bull Museum Natural History (Paris) 445:375–396

Dupin F, Coulon J, Le Parco Y, Fontes M, Thouveny Y (1991) Formation of the extracellular matrix during the epimorphic anterior regeneration of *Owenia fusiformis*: Autoradiographical and in situ hybridization studies. Intern J Dev Biol 35:109–119

Dupont S, Wilson K, Obst M, Sköld H, Nakano H, Thorndyke MC (2007) Marine ecological genomics: When genomics meets marine ecology. Mar Ecol Prog Ser 332:257–273

Eaves AA, Palmer AR (2003) Reproduction-Widespread cloning in echinoderm larvae. Nature 425(6954):146

Egger B, Gschwentner R, Rieger R (2007) Free-living flatworms under the knife: Past and present. Dev Genes Evol 217 (2):89–104

Emson RH, Wilkie IC (1980) Fission and autotomy in echinoderms. Ocean Mar Biol Ann Rev 18:155–250

Ereskovsky AV, Tokina DB (2007) Asexual reproduction in homoscleromorph sponges (Porifera; Homoscleromorpha). Mar Biol 151 (2):425–434

Ermak TH (1982) The renewing cell populations of ascidians. Am Zool 22:795–805

Extavour CG, Akam M (2003) Mechanisms of germ cell specification across the metazoans: Epigenesis and preformation. Development 130:5869–5884

Fagotto F, Guger K, Gumbiner BM (1997) Induction of the primary dorsalizing center in *Xenopus* by the Wnt/GSK/beta-catenin signaling pathway, but not by Vg1, Activin or Noggin. Development 124 (2):453–460

Fautin DG (1991) Developmental Pathways of Anthozoans. Hydrobiologia 216:143–149

Fautin DG (1999) Cnidaria. In: Knobil E, Neill JD (eds) Encyclopedia of reproduction, vol. 1. Academic Press, San Diego, pp 645–653

Fautin DG (2002) Reproduction of Cnidaria. Canadian Journal of Zoology-Revue Canadienne de Zoologie 80 (10):1735–1754

Fischer A (1999) Reproductive and developmental phenomena in annelids: A source of exemplary research problems. Hydrobiologia 402:1–20

Francis N, Gregg T, Owen R, Ebert T, Bodnar A (2006) Lack of age-associated telomere shortening in long- and short-lived species of sea urchins. FEBS Lett 580:4713–4717

Freeman G (1964) The role of blood cells in the process of asexual reproduction in the tunicate *Perophora viridis*. J Exp Zool 156:157–183

Fukumoto M (1971) Experimental control of budding and stolon elongation in *Perophora orientalis*, a compound ascidian. Dev Growth Diff 13:73–88

Funayama N, Nakatsukasa M, Hayashi T, Agata K (2005) Isolation of the choanocyte in the fresh water sponge, *Ephydatia fluviatilis* and its lineage marker, Ef annexin. Dev Growth Diff 47 (4):243–253

Funch P, Kristensen RM (1995) Cycliophora is a new phylum with affinities to Entoprocta and Ectoprocta. Nature 378 (6558):711–714

Funch P, Sorensen MV, Obst M (2005) On the phylogenetic position of Rotifera – have we come any further? Hydrobiologia 546:11–28

Geller JB (2005) Fission in sea anemones: Integrative studies of life cycle evolution. Integr Comp Biol 45:615–622

Gibson GD, Harvey JML (2000) Morphogenesis during asexual reproduction in *Pygospio elegans* Claparede (Annelida, Polychaeta). Biol Bull 199 (1):41–49

Gibson DM, Paffenhöfer GA (2002) Asexual reproduction of the doliolid, *Dolioletta gegenbauri Uljanin* (Tunicata, Thaliacea). J Plankton Res 24:703–712

Gómez-García L, Sánchez FM, Vallejo-Cremades MT, de Segura IA, del Campo Ede M (2005) Direct activation of telomerase by GH via phosphatidylinositol 3'-kinase. J Endocrinol 185:421–428

Goss R (1969) Principles of regeneration. Academic Press, New York

Gschwentner R, Ladurner P, Nimeth K, Rieger R (2001) Stem cells in a basal bilaterian – S-phase and mitotic cells in *Convolutriloba longifissura* (Acoela, Platyhelminthes). Cell Tissue Res 304 (3):401–408

Heins N, Englund MC, Sjöblom C, Dahl U, Tonning A, Bergh C, Lindahl A, Hansson C, Semb H (2004) Derivation, characterization, and differentiation of human embryonic stem cells. Stem Cells 22:367–376

Hendelberg J, Åkesson B (1988) *Convolutriloba retrogemma gen. et sp.n.*, a turbellarian (Acoela, Platyhelminthes) with reversed polarity of reproductive buds. Fortschritte der Zoologie 36: 321–327

Henry JQ, Martindale MQ (1996) The establishment of embryonic axial properties in the nemertean, Cerebratulus lacteus. Dev Biol 180 (2):713–721

Henry JQ, Martindale MQ (1997) Regulation and the modification of axial properties in partial embryos of the nemertean, Cerebratulus lacteus. Dev Genes Evol 207 (1):42–50

Highsmith RC (1982) Reproduction by fragmentation in corals. Mar Ecol Prog Ser 7:207–226

Hisata K, Fujiwara S, Tsuchida Y, Ohashi M, Kawamura K (1998) Expression and function of a retinoic acid receptor in budding ascidians. Dev Genes Evol 208:537–546

Holstein TW, Hobmayer E, Technau U (2003) Cnidarians: An evolutionarily conserved model system for regeneration? Dev Dyn 226:257–267

Hughes RN (1989) A functional biology of clonal animals. Chapman Hall, London, New York

Hughes RN (2002) Reproductive biology of invertebrates. Vol XI, Progress asexual reproduction. John Wiley & Sons, Chichester, UK

Hughes RN (2005) Lessons in modularity: The evolutionary ecology of colonial invertebrates. Sci Mar 69:169–179

Israelsson O (2006) Observations on some unusual cell types in the enigmatic worm *Xenoturbella* (phylum uncertain). Tissue Cell 38 (4):233–242

Jackson JBC (1986) Distribution and ecology of clonal and aclonal benthic invertebrates. In: Jackson JBC, Buss LW, Cook RE (eds) Population biology and evolution of clonal organisms. Yale University Press, New Haven London, pp 297–355

Jaeckle WB (1994) Multiple-modes of asexual reproduction by tropical and subtropical sea star larvae – an unusual adaptation for genet dispersal and survival. Biol Bull 186 (1):62–71

Jang YY, Collector MI, Baylin SB, Diehl AM, Sharkis SJ (2004) Hematopoietic stem cells convert into liver cells within days without fusion. Nat Cell Biol 6:532–539

Karako S, Achituv Y, Perl-Treves R, Katcoff D (2002) *Asterina burtoni* (Asteroidea; Echinodermata) in the Mediterranean and the Red Sea: Does asexual reproduction facilitate colonization? Mar Ecol Prog Ser 234:139–145

Kawamura K, Nakauchi M (1986) Mitosis and body patterning during morphollactic development of palleal buds in ascidians. Dev Biol 116:39–50

Kawamura K, Fujiwara S (1995a) Cellular and molecular characterization of transdifferentiation in the process of morphallaxis of budding tunicates. Semin Cell Biol 6:117–126

Kawamura K, Fujiwara S (1995b) Establishment of cell lines from multipotent epithelial sheet in the budding tunicate, *Polyandrocarpa misakiensis*. Cell Struct Funct 20:97–106

Kawamura K, Hayata D, Fujiwara S, Yubisui T (1998) Serine protease inhibitors expressed in the process of budding of tunicates as revealed by EST analysis. J Biochem 124:1004–1012

Kawamura K, Fujiwara S (2000) Advantage or disadvantage: Is asexual reproduction beneficial to survival of the tunicate, *Polyandrocarpa misakiensis*. Zool Sci 17:281–291

Kawamura K, Sugino Y, Sunanaga T, Fujiwara S (2008) Multipotent epithelial cells in the process of regeneration and asexual reproduction in colonial tunicates. Dev Growth Diff 50:1–11

King GM (1998) Reproduction in the Hemichordata. In: Knobil E, Neill JD (eds) Encyclopedia of reproduction vol. 2 Academic Press, San Diego, pp 599–603

Kirkwood TBL, Austad SN (2000) Why do we age? Nature 409:233–237

Knott KE, Balser EJ, Jaeckle WB, Wray GA (2003) Identification of asteroid genera with species capable of larval cloning. Biol Bull 204 (3):246–255

Korschelt E (1942) Weiteres über die Dauer der ungeschlechtlichen Fortpflanzung des Ctenodrilus monostylus. Zoologischer Anzeiger 137:162–166

Laird DJ, Weissman IL (2004) Telomerase maintained in self-renewing tissues during serial regeneration of the urochordate *Botryllus schlosseri*. Dev Biol 273:185–194

Laird DJ, De Tomaso AW, Weissman IL (2005) Stem cells are units of natural selection in a colonial ascidian. Cell 123:1351–1360

Li Y, Yates JA, Chen JJ (2007) Identification and characterization of sea squirt telomerase reverse transcriptase. Gene 400:16–24

Maharajan P, Maharajan V, Branno M, Scarano E (1986) Effects of 5 azacytidine on DNA methylation and early development of sea urchins and ascidia. Differentiation 32:200–207

Maynard SJ (1978) The evolution of sex. Cambridge University Press, Cambridge

Marrs J, Wilkie IC, Sköld M, Maclaren WM, McKenzie JD (2000) Size-related aspects of arm damage, tissue mechanics, and autotomy in the starfish *Asterias rubens*. Mar Biol 137: 59–70

Martínez DE (1998) Mortality patterns suggest lack of senescence in hydra. Exp Gerontol 33 (3):217–225

Martínez DE, Levinton JS (1992) Asexual metazoans undergo senescence. Proc Natl Acad Sci USA 89 (20):9920–9923

Matsumoto GI (1999) Phylum Ctenophora. In: Knobil E, Neill JD (eds) Encyclopedia of reproduction, vol 1. Academic Press, San Diego

Matsumoto J, Nakamoto C, Fujiwara S, Yubisui T, Kawamura K (2001) A novel C-type lectin regulating cell growth, cell adhesion and cell differentiation of the multipotent epithelium in budding tunicates. Development 128:3339–3347

Merok JR, Lansita JA, Tunstead JR, Sherley JL (2002) Cosegregation of chromosomes containing immortal DNA strands in cells that cycle with asymmetric stem cell kinetics. Cancer Res 62:6791–6795

Miller KJ (1997) Gentic structure of black coral populations in New Zealand fiords. Mar Ecol Prog Ser 161:12–132

Mladenov PV (1996) Environmental factors influencing asexual reproductive processes in echinoderms. Oceanol Acta 19 (3–4):227–235

Mladenov PV, Burke RD (1994) Echinodermata: Asexual propagation. In: Adiyodi KG, RG Adiyodi (eds) Reproductive biology of invertebrates Vol VI, Part B, Asexual propagation and reproductive strategies. Oxford IBH Publishing Co Pvt Ltd, New Delhi Bombay Calcutta, pp 339–383

Mladenov PV, Emson RH, Colpit LV, Wilkie IC (1983) Asexual reproduction in the West Indian brittle star *Ophiocomella ophiactoides* (H. L. Clark) (Echinodermata: Ophiuroidea). J Exp Mar Biol Ecol 72:1–23

Mladenov PV, Emson RH (1988) Density, size structure and reproductive characteristics of fissiparous brittle stars in algae and asponges: Evidence for interpopulational variation in levels of sexual and asexual reproduction. Mar Ecol Prog Ser 42:181–194

Moss C, Hunter AJ, Thorndyke MC (1998) Patterns of bromodeoxyuridine incorporation and neuropeptide immunoreactivity during arm regeneration in the starfish *Asterias rubens*. Philos Trans R Soc Lond B 353:421–436

Mukai H, Makioka T (1978) Studies on regeneration of an Entoproct, *Barentsia-Discreta*. J Exp Zool 205 (2):261–275

Mukai H, Makioka T (1980) Some observations on the sex-differentiation of an Entoproct, *Barentsia-Discreta* (Busk). J Exp Zool 213 (1):45–59

Müller MCM, Berenzen A, Westheide W (2003) Experiments on anterior regeneration in *Eurythoe complanata* ("Polychaeta", Amphinomidae): Reconfiguration of the nervous system and its function for regeneration. Zoomorphology 122:95–103

Müller WA, Teo R, Frank U (2004) Totipotent migratory stem cells in a hydroid. Dev Biol 275 (1):215–224

Müller WEG (2006) The stem cell concept in sponges (Porifera): Metazoan traits. Semin Cell Dev Biol 17 (4):481–491

Myohara M, Niva CC, Lee JM (2006) Molecular approach to annelid regeneration: cDNA subtraction cloning reveals various novel genes that are upregulated during the large-scale regeneration of the oligochaete, Enchytraeus japonensis. Dev Dyn 235 (8):2051–2070

Nakauchi M (1982) Asexual development of ascidians: Its biological significance, diversity and mophogenesis. Am Zool 22:753–763

Newberry AT (1999) Tunicata. In: Knobil E, Neill JD (eds) Encyclopedia of reproduction. Academic Press, San Diego

Nielsen C (1971) Entoproct life-cycles and the entoproct/ectoproct relationship. Ophelia 9:209–341

Nielsen C (2001) Animal Evolution: Interrelationships of the living phyla, 2 edn. Oxford University Press, Oxford
Nielsen C, Jespersen A (1997) Entoprocta. In: Harrison FW (ed) Microscopic anatomy of invertebrates, vol. 13. Wiley-Liss, New York, pp 13–43
Obst M, Funch P, Kristensen RM (2006) A new species of Cycliophora from the mouthparts of the American lobster, *Homarus americanus* (Nephropidae, Decapoda). Org Divers Evol 6 (2): 83–97
O'Dea A (2006) Asexual propagation in the marine bryozoan *Cupuladria exfragminis*. J Exp Mar Biol Ecol 335 (2):312–322
Paulus T, Müller MCM (2006) Cell proliferation dynamics and morphological differentiation during regeneration in *Dorvillea bermudensis* (Polychaeta, Dorvilleidae). J Morphol 267 (4):393–403
Poulsom R, Alison MR, Forbes SJ, Wright NA (2002) Adult stem cell plasticity. J Pathol 197 (4):417–565
Pearse VB (1999) Placozoa. In: Knobil E, Neill JD (eds) Encyclopedia of reproduction, vol. 3. Academic Press, San Diego, pp 898–901
Piraino S, Boero F, Aeschbach B, Schmid V (1996) Reversing the life cycle: Medusae transforming into polyps and cell transdifferentiation in *Turritopsis nutricula* (Cnidaria, Hydrozoa). Biol Bull 190 (3):302–312
Rando TA (2006) Stem cells, ageing and the quest for immortality. Nature 441:1080–1086
Reddien P.W, Oviedo NJ, Jennings JR, Jenkin JC, Sánchez Alvarado A (2005) SMEDWI-2 is a PIWI-like protein that's regulates planarian stem cells. Science 310:1327–1330
Reuter M, Kreshchenko N (2004) Flatworm asexual multiplication implicates stem cells and regeneration. Canadian J Zool 82:334–356
Ridley M (1996) Evolution, 2nd edn. Blackwell Science, Oxford
Rieger RM, Gehlen M, Haszprunar G, Holmlund M, Legniti A, Salvenmoser W, Tyler S (1999a) Laboratory cultures of marine Macrostomida (Turbellaria). Fortschritte der Zoologie 36:523
Rieger RM, Legniti A, Ladurner P, Reiter D, Asch E, Salvenmoser W, Schürmann W, Peter R (1999b) Ultrastructure of neoblasts in microturbellaria: Significance for understanding stem cells in free-living Platyhelminthes. Invert Reprod Dev 35 (2):127–140
Rinkevich Y, Paz G, Rinkevich B, Reshef R (2007) Systemic bud induction and retinoic acid signaling underlie whole body regeneration in the urochordate *Botrylloides leachi*. PLoS Biol 5 (4):e71
Ryland JS (1970) Bryozoans. Hutchinson University Library, London
Sánchez Alvarado A (2000) Regeneration in the metazoans: Why does it happen? BioEssays 22:578–590
Sánchez Alvarado A, Tsonis PA (2006) Bridging the regeneration gap: Genetic insights from diverse animal models. Nat Rev Gen 7:873–884
Satoh N (1994) Developmental biology of ascidians. Cambridge University Press, Cambridge
Schiedges KL (1979) Reproductive-biology and ontogenesis in the polychaete genus autolytus (Annelida, Syllidae) – Observations on Laboratory-Cultured Individuals. Mar Biol 54:239–250
Schroeder PC, Hermans CO (1975) Annelida: Polychaeta. In: Giese AC, Pearse JS (eds) Reproduction of marine invertebrates, vol III. Academic Press, New York, San Francisco, London, pp 1–213
Sell S (2004) Stem cell origin of cancer and differentiation therapy. Critical Rev Oncol Hematol 51:1–28
Shafir S, Van Rijn J, Rinkevich B (2001) Nubbing of coral colonies: A novel approach for the development of island broodstocks. Aquarium Sci Conserv 3:183–190
Shannon T, Achatz JG (2007) *Convolutriloba macropyga* sp nov, an uncommonly fecund acoel (Acoelomorpha) discovered in tropical aquaria. Zootaxa 1525:1–17
Seto AG, Kingston RE, Lau NC (2007) The coming of age for Piwi proteins. Mol Cell 26:603–609
Sköld HN, Komma DJ, Endow SA (2005) Assembly pathway of the anastral *Drosophila* oocyte meiosis I spindle. J Cell Sci 118:1745–1755

Sköld M, Barker MF, Mladenov PV (2002) Spatial variability in sexual and asexual reproduction of the fissiparous seastar *Coscinasterias muricata*: The role of food and fluctuating temperature. Mar Ecol Prog Ser 233:143–155

Spivakov M, Fisher AG (2007) Epigenetic signatures of stem-cell identity. Nat Rev Gen 8:263–271

Stoick-Cooper CL, Moon RT, Weidinger G (2007) Advances in signaling in vertebrate regeneration as a prelude to regenerative medicine. Genes Dev 21 (11):1292–315

Stoner DS, Rinkevich B, Weissman IL (1999) Heritable germ and somatic cell lineage competitions in chimeric colonial protochordates. Proc Natl Acad Sci USA 96:9148–9153

Sunanaga T, Saito Y, Kawamura K (2006) Postembryonic epigenesis of Vasa-positive germ cells from aggregated hemoblasts in the colonial ascidian, *Botryllus primigenus*. Dev Growth Diff 48:87–100

Swalla BJ, Cameron CB, Corley LS, Garey JR (2000) Urochordates are monophyletic within the deuterostomes. Syst Biol 49:52–64

Tadokoro R, Sugio M, Kutsuna J, Tochinai S, Takahashi Y (2006) Early segregation of germ and somatic lineages during gonadal regeneration in the annelid *Enchytraeus japonensis*. Curr Biol 16 (10):1012–1017

Takahashi K, Yamanaka S (2006) Induction of pluripotent stem cells from mouse embryonic and adult fibroblast cultures by defined factors. Cell 126:663–676

Thorndyke MC, Carnevali MDC (2001) Regeneration neurohormones and growth factors in echinoderms. Canadian Journal of Zoology-Revue Canadienne De Zoologie 79 (7):1171–1208

Toledo A, Cruz C, Fragoso G, Laclette JP, Merchant MT, Hernandez M, Sciutto E (1997) In vitro culture of *Taenia crassiceps* larval cells and cyst regeneration after injection into mice. J Parasitol 83 (2):189–193

Toyooka Y, Tsunekawa N, Akasu R, Noce T (2003) Embryonic stem cells can form germ cells in vitro. Proc Natl Acad Sci USA 100:11457–11462

Traut W, Szczepanowski M, Vitkova M, Optitz C, Marec F, Zrzavý J (2007) The telomere repeat motif of basal Metazoa. Chromosome Res 15:371–382

Turbeville JM (1999) Nemertea. In: Knobil E, Neill JD (eds) Encyclopedia of reproduction, vol 3. Academic Press, San Diego, pp 341–349

Turon X (2005) A new mode of colony multiplication by modiefied budding in the ascidian *Clavelina gemmae* n. sp. (Clavelinidae). Invert Biol 124:273–283

Tyler S (1999) Platyhelminthes. In: Knobil E, Neill JD (eds) Encyclopedia of reproduction. Academic Press, San Diego, pp 901–908

van Bekkum DW (2004) Phylogenetic aspects of tissue regeneration: Role of stem cells – A concise overview. Blood Cells Mol Dis 32 (1):11–16

Vickery MS, McClintock JB (2000) Effects of food concentration and availability on the incidence of cloning in planktotrophic larvae of the sea star *Pisaster ochraceus*. Biol Bull 199 (3): 298–304

Vickery MCL, Vickery MS, Amsler CD, McClintock JB (2001) Regeneration in echinoderm larvae. Microsc Res Techn 55 (6):464–473

Voskoboynik A, Blecher N, Sonen Y, Rinkevich B, De Tomaso AW, Ishizuka KJ, Weissman IL (2007) Striving for normality: Whole body regeneration through a series of abnormal zooidal generations. FASEB J 21:1335–1344

Weismann A (1889) Essays upon heredity and kindred biological subjects. Oxford University Press, Oxford

Weismann A (1893) The Germ-Plasm: A theory of heredity. Charles Scribner's Sons, New York

Westblad E (1949) *Xenoturbella bocki* n.g., n. sp. a peculiar, primitive Turbellarian type. Arkiv für Zoologi 1:11–29

Wilkie IC (2001) Autotomy as a prelude to regeneration in echinoderms. Microsc Res Techn 55 (6):369–396

Wilson EB (1928) The cell in development and heredity. Macmillan, New York

Weissenfels N (1989) Biologie und Mikroskopische Anatomie der Süsswasserschwämme (Spongillidae). Fischer, Stuttgart

Witte U, Barthel D (1994) Reproductive cycle and oogenesis of *Halichondria panicea* (Pallas) in Kiel Bight. In: Van Soest RWM, Van Kempen TMG, Braekman JC (eds) Sponges in time and space. Balkema, Rotterdam, pp 389–393
Yang X, Smith SL, Tian XC, Lewin HA, Renard JP, Wakayama T (2007) Nuclear reprogramming of cloned embryos and its implications for therapeutic cloning. Nat Gen 39:295–302
Yu J, Vodyanik MA, Smuga-Otto K, Antosiewicz-Bourget J, Frane JL, Tian S, Nie J, Jonsdottir GA, Ruotti V, Stewart R, Slukvin II, Thomson JA (2007) Induced pluripotent stem cell lines derived from human somatic cells. Science 318 (5858):1917–1920
Zimmer R (1999) Phoronida. In: Knobil E, Neill JD (eds) Encyclopedia of Reproduction. Academic Press, San Diego

Chapter 6
Neuroimmune Chemical Messengers and Their Conservation During Evolution

George B. Stefano, Michel Salzet, and Enzo Ottaviani

Abstract Cellular communication is mediated primarily by chemical signal molecules. This suggests that during the course of evolution, organisms in which this form of communication developed have greatly increased their chances of survival, ensuring that this trait passes on to their descendents. Interestingly, for the most part, these signaling molecules and their apparent systems/mechanisms have remained relatively intact during evolution. However, this principle of conservation does not preclude events that may lead to an old signal system being used in a new functional capacity. A classical example of this phenomenon can be observed in the immune and neuroendocrine systems. They share a common pool of identical molecules, i.e., opioids, involved in the maintenance of homeostasis, which occurs in both invertebrates and vertebrates. Specifically, many of these protein molecules, active as chemical messengers, are also derived from larger polypeptide gene products and classified into three families: the proopiomelanocortin (POMC), the proenkephalin and the prodynorphin families. In marine mussels the immunocytes produce and react to opioid peptides, demonstrating autocrine and paracrine signaling. Under stressful stimuli immune system alteration occurs, in part mediated by opioid signals, coupling these processes and demonstrating neuroimmune-regulatory phenomena. Additionally, both immune and nervous systems contain mammalian-cytokine-like molecules, which also interact with the endogenous opioid system. Recent data have demonstrated the presence of novel opiate receptors on human multi-lineage progenitor cells. Interestingly, these same receptors are found on molluscan neural cells, suggesting their early evolutionary origins and conservation. The only data on molluscan stem cells are the presence of CD14- and CD-34-like molecules on prohemocyte membranes. These data suggest that the highly sophisticated mammalian immune and neuroendocrine systems had their origins in their invertebrate counterparts.

Keywords Molluscs · Stem cells · Opioids · POMC · CRH

E. Ottaviani (✉)
Department of Animal Biology, University of Modena and Reggio Emilia, Via Campi 213/D, 41100 Modena, Italy
e-mail: enzo.ottaviani@unimore.it

B. Rinkevich, V. Matranga (eds.), *Stem Cells in Marine Organisms*,
DOI 10.1007/978-90-481-2767-2_6, © Springer Science+Business Media B.V. 2009

6.1 Introduction

Many of the smaller mammalian protein molecules active in chemical communication appear to be derived from larger polypeptide gene products (Udenfriend and Kilpatrick 1984; Udenfriend and Meienhofer 1984; Pasternak 1988). For many years, opioid peptides were classified into three families: the proopiomelanocortin (POMC), the proenkephalin (proenk) and the prodynorphin (prodyn) families. The peptides of the POMC family are derived from a single macromolecular precursor. POMC is an interesting molecule because its processed products include hormones such as adrenocorticotropin (ACTH), LPH, α-MSH (melanocyte stimulating hormone), and β-MSH as well as the opioid peptide, β-endorphin (Udenfriend and Meienhofer 1984). β-endorphin possesses a very strong analgesic effect and contains the Met-enkephalin sequence at its N-terminal. In many animals, POMC appears to be processed to different final products in the anterior and in the neurointermediate lobes of the pituitary gland; it is possible that POMC serves different functions in different tissues. The absence of a basic paired amino acid after the Met-enkephalin in β-endorphin makes it an unlikely precursor for Met-enkephalin. In addition, immunocytochemistry studies have shown that β-endorphin and enkephalin do not coexist in the same regions of the brain.

The ratio of Met-enkephalin sequence to Leu-enkephalin sequence in proenk is 6:1. On the other hand, prodyn contains three copies of Leu-enkephalin sequence, but no Met-enkephalin sequence (Stefano and Salzet 1999). Similar to proenk, prodyn in the tissues appears to be processed first to various dynorphins, which have potent opioid activities and contain the Leu-enkephalin sequence. Here again, Leu-enkephalin may only be one of the "intended" products of prodyn.

Eukaryotic genes tend to have numerous copies of certain exons, as is the case of methionine enkephalin in proenk. Thus, this phenomenon, following enzymatic cleavage, may yield multiple copies of a signal molecule, indicating its significance. These repetitions of important exons may also be regarded as a mechanism to insure signal fidelity in older organisms. Again, if an important signal molecule is going to be used during an organism's life span, precautions have to be incorporated into stabilizing the molecule.

Other, equally important phenomena, are the mechanisms of simultaneous expression of the signal molecule and the receptor in different cell types. It is known that the same signal system can be used in different ways in the same animal and different phyla. What causes the precise expression of both complementary systems? This dual expression certainly suggests the existence of a functional interaction between these two "separate" aspects of the same signal system. We surmise that since both "up" and "down" regulation of a receptor population can be regulated by the concentration of the signal molecule, the signal molecule itself can induce the presence of its receptor in a distant cell.

With these considerations in mind, we now review the scientific literature for a determination of whether many mammalian signal molecules first arose before

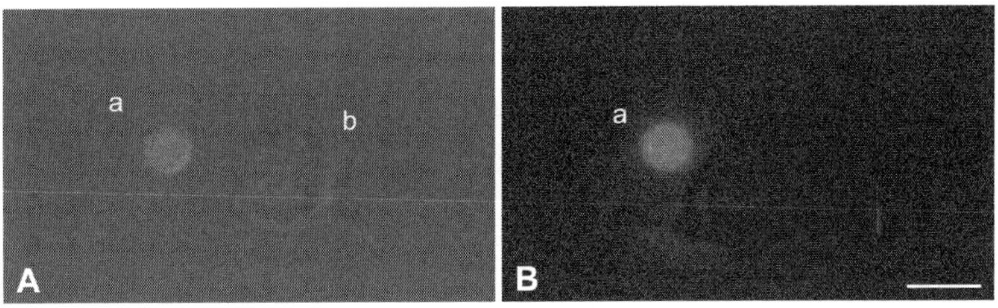

Fig. 6.1 Immunofluorescence procedure with (**A**) and without (**B**) phase-contrast. *M. galloprovincialis* prohemocyte positive (a) and hemocyte (b) negative to anti-CD34 monoclonal antibody. Bar, 10 μm

divergence of protostomian and deuterostomian. Most of the data reported in this chapter derive from studies on the *Mytilus edulis* and *Mytilus galloprovincialis*.

6.1.1 The Model System

The bivalves *M. edulis* and *M. galloprovincialis* show a central nervous system composed of three pairs of ganglia: cerebral, visceral and pedal (for details see Stefano et al. 1990). The cerebral ganglia are located around the animal's esophagus, while the pedal one is found in the mussel foot lying between the retractor muscle and the digestive gland and consists of two symmetrical ganglia connected by a small commissure. The immune system possesses a well-developed capability to discriminate between self and non-self and the recognition is mainly carried out by both the cellular and humoral component of the hemolymph. With regards to the former, in *M. edulis* two cell types have been reported: agranular cells (hyalinocytes) and granular cells (granulocytes) (Renwrantz 1990), while in *M. galloprovincialis* only one type in two different stages, young or old, is present. Furthermore, in *M. galloprovincialis*, a prohemocyte (Ottaviani et al. 1998a) positive to anti-CD34 (Fig. 6.1) (unpublished data) was observed. The circulating cells (hemocytes) are involved in immune and neuroendocrine responses, such as chemotaxis, phagocytosis, cytotoxicity and stress responses (Ottaviani et al. 1998a; Renwrantz 1990; Malagoli et al. 2003; Malagoli and Ottaviani 2005).

6.1.2 Molluscan Stem Cells

As far as the presence of CD34-like molecules is concerned, the prohemocytes show an immunophenotype of hemopoietic stem cells. CD34 is a surface glycophosphoprotein expressed on human lymphohematopoietic stem and progenitor cells (Krause et al. 1996). Immunoreactive CD-34 molecules were found also in other invertebrates such as the clam *Tapes philippinarus* (Cima et al. 2000), kuruma prawn

Penaeus japonicus (Rojtinnakorn et al. 2002), the grasshopper *Euprepocnemis shirakii* (Lim et al. 2004) and the ascidian *Botryllus schlosseri* (Ballarin and Cima 2005). In the mollusc *Planorbarius corneus* a marker for the monocyte-derived multipotential cells, such as CD14, was detected (Franceschi et al. 1991; Seta and Kuwana 2007). In this regard, distant stem cells in the molluscan nervous system have not been reported. In *Drosophila melanogaster* the neuronal precursor cells are the neuroblasts in the central nervous system and the sensory organ precursor cells in the peripheral nervous system (Bellaïche and Schweisguth 2001; Harrison and Harrison 2006). In this regard, Scadden (2006) reported that the regulation of stem cells depends on their interaction with a specialized microenvironment or niche. In *D. melanogaster* it was found that the lymph gland represents the hemopoietic organ and the posterior signalling centre is a hemopoietic niche essential for the maintenance of the blood cell precursors (Mandal et al. 2007).

6.2 Invertebrate Opioid Precursors

The presence of biologically active neuropeptides in invertebrates, which are comparable to those of vertebrates, has been known for a considerable period of time (Scharrer 1967,1978; Frontali and Gainer 1977; Haynes 1980). However, detailed information on a specific class of these peptides, the endogenous opioid peptides, is almost exclusively confined to the mammalian nervous system. The upsurge of interest in the diverse roles and modes of operation of these molecules, including immune actions, has sparked a search for their evolutionary history. While several reports on the occurrence of endogenous opioid peptides in submammalian vertebrates have been available (Audigier et al. 1980), comparable data in invertebrates are still emerging. They consist of its demonstration in certain invertebrate ganglia of either opioid peptides or their specific receptor sites (Leung and Stefano 1987). Additionally, ACTH and β-endorphin-like amino acid sequences were detected immunologically in protozoa as part of a high-molecular-weight macromolecule (LeRoith et al. 1982).

In regard to invertebrates, the presence of an opiate receptor mechanism in the central nervous system of the marine mussel, *M. edulis*, was first suggested by a rise in ganglionic dopamine levels following intracardiac administration of exogenous Met- and Leu-enkephalin, an effect reversible by naloxone (Stefano and Catapane 1979; Stefano et al. 1982). The first actual demonstration of high affinity opiate binding sites in an invertebrate ganglion was accomplished by Stefano and colleagues (Stefano et al. 1980) in *M. edulis*. The biochemical characteristics of this system, analyzed in detail by Kream and colleagues (Kream et al. 1980), have been found to parallel those of mammalian systems. Early indications to specific binding sites in insects are the studies by Pert and Taylor (1980) and by Edley et al. (1982), who showed that tissue suspensions prepared from *Drosophila* heads avidly bind (3H) Leu-enkephalin and the opioid ligand (3H)-diprenorphine. Specific high affinity binding sites for a synthetic enkephalin analog, D-Ala2-Met5-enkephalinamide

(DAMA), were demonstrated in the cerebral ganglia and midgut of the insect Leucophaea (Stefano et al. 1989c). The results strongly suggest the presence, in *Mytilus*, of opioid receptors that are confined to certain areas of the nervous tissue. Once more, these opioid receptors were found to resemble those described in mammalian systems.

6.2.1 Proenkephalin

In the past, Stefano and colleagues have biochemically sequenced Met- and Leu-enkephalin as well as Met-enkephalin-Arg-Phe from *M. edulis* neural tissues (Leung and Stefano 1984; Makman and Stefano 1984). These signaling molecules were later isolated and sequenced in arthropods (Luschen et al. 1991; Rothe et al. 1991), annelids (Laurent and Salzet 1996b) and the mollusc *Lymnaea stagnalis* (Ewandinger et al. 1996) thereby providing evidence for the presence of an invertebrate proenk-like molecule.

This study has demonstrated that invertebrate proenk is quite similar to that found in mammals (Udenfriend and Kilpatrick 1984). In this context, we have established the presence of proenk in two representative invertebrates, namely in the leech *Theromyzon tessulatum*, and in the marine mussel *M. edulis* (Salzet and Stefano 1997a). This opioid precursor was found in the animal's immunocytes. The structure of the leech proenk material demonstrates considerable amino acid sequence similarity with amphibian proenkephalin (26.2%). *M. edulis* proenk exhibits a higher sequence identity with human and guinea pig proenk (39% and 50%, respectively). This proenk contains Met- and Leu-enkephalin in a ratio of 3/1 in *M. edulis* and 1/2 in the leech. They also possess Met-enkephalin-Arg-Gly-Leu and Met-enkephalin-Arg-Phe that are flanked by dibasic amino acid residues, showing cleavage sites. Furthermore, using both sequence comparison and a specific antiserum raised against bovine proenk A (209–237), the enkelytin peptide, FAEPLPSEEEGESYSKEVPEMEKRYGGFM, was identified in invertebrate proenk and it exhibited a sequence identity of 98% with mammalian enkelytin (Goumon et al. 1996). This demonstration of proenk in invertebrates supports the observations of the previous studies (Stefano and Leung 1984; Leung and Stefano 1984; Luschen et al. 1991; Rothe et al. 1991; Ewandinger et al. 1996; Laurent and Salzet 1996b) that identified proenk-derived peptides as free signaling molecules since they are flanked by basic amino acid residues in the precursor, showing, as in mammals, that they are products of enzymatic processing. In this regard, the difficulty in obtaining these pentapeptides in invertebrate tissues is due to the presence of proteolytic enzymes, i.e., neutral endopeptidase (Shipp et al. 1990; Turner et al. 1994; Laurent and Salzet 1996b; Salzet et al. 1995).

Of equal importance is the phenomenon of having multiple copies of a repeating sequence in a precursor, i.e., Met-enkephalin (Udenfriend and Meienhofer 1984). This occurrence has been regarded as a simple amplification mechanism, arising from gene crossover. This is demonstrated by the leech having two copies of Met-enkephalin, the mussel three and bovine adrenal proenk four. From this, we may conclude that Met-enkephalin is singularly important in opioid processing. This is also surmised by the presence of delta-1 and -2 binding sites on mammalian and invertebrate tissues (Stefano et al. 1989a,b, 1992a, 1996).

The presence of the enkelytin (Goumon et al. 1996) in invertebrate proenk with a nearly perfect sequence match to that found in bovine chromaffin cells (98%; [Goumon et al. 1996]) further supports the hypothesis that these molecules first evolved in simpler animals. Indeed, enkelytin, with its high antibacterial activity (Strub et al. 1996) further associates opioid peptides with immune related activities (Stefano et al. 1998a). We speculate that immune signaling or alerting may lead to enhanced proenk proteolytic processing freeing both opioid peptides and enkelytin (Stefano et al. 1998a). In this scenario, the opioid peptides would stimulate immunocyte chemotaxis and phagocytosis as well as the secretion of mammalian-like cytokines (Stefano et al. 1996). During this process, the simultaneously liberated enkelytin would attack bacteria immediately, allowing time for the immune stimulating capabilities of opioid peptides to manifest itself. This hypothesis is further supported by the presence of specific Met-enkephalin receptors on these cells (Liu et al. 1996b). Interestingly, this same scenario may occur in neural tissues (Stefano 1989; Sonetti et al. 1994) given the presence of glial cell types, i.e., microglia. Thus, it appears that many of the mammalian molecular and cellular survival strategies first appeared in organisms that evolved, at least, 500 million years ago.

6.2.2 Prodynorphin

A mammalian prodyn-derived peptide, α-Neo-endorphin, was purified from *Theromyzon tessulatum* central nervous system (CNS) and suckers (Laurent and Salzet 1996b; Salzet et al. 1996), suggesting the presence of a larger precursor peptide similar to prodyn of vertebrates. In mammals, processing of prodyn yielded a number of bioactive peptides including Leu-enkephalin, Neo-endorphins (α, ß), and dynorphins (A, B) (Patey and Rossier 1986).

Sequence alignment of the entire prodyn opioid precursor with vertebrate prodyn revealed a 28.8% sequence identity with rat, and 22% with the human and pig (Civelli et al. 1985). In leech prodyn, α-Neo-endorphin was found at position 67–76 and it exhibited a 100% sequence identity with the respective mammalian material. Dynorphin A-like material at 93–105 exhibited a 50% sequence identity and dynorphin B-like material at 106–117 exhibited a 76.6% sequence homology with its mammalian counterpart. Although the α-Neo-endorphin was identical to the one found in vertebrate, the dynorphins were slightly shorter. The amount of Leu-enkephalin is similar to that found in vertebrates, i.e., 3. Moreover, the C-terminus of leech prodyn was similar to that of vertebrates whereas the N-terminus was shorter. This explains the difference in mass observed between the leech (14291 Da), rat (23386 Da) (Civelli et al. 1985), pig (28616 Da) (Horikawa et al. 1983) and human prodyn (28385 Da) (Kakidani et al. 1982), suggesting that these additions occurred later in evolution. As with the proenk-derived peptides, the leech prodyn-derived peptides were found in positions flanked by basic amino acids, indicating cleavage sites. Furthermore, the N-terminus of leech prodyn exhibited a 54.5% sequence homology with that of rat (Civelli et al. 1985). In addition, Leu-enk, α-Neo-endorphin, dynorphin A, and dynorphin B were present at the C-terminal side of the protein.

We have also characterized a prodyn molecule in hemocytes of the free-living bivalve mollusc *M. edulis* (Stefano et al. 1998a). The ca. 16 kDa protein was purified by cut-off filtration pre-purification, anti-leucine-enkephalin affinity column separation followed by reversed-phase HPLC. Its primary sequence was determined by Edman degradation, endoproteinase Glu-C digestion and CNBr treatment. *Mytilus* prodyn contained, α-Neo-endorphin, dynorphin-A and dynorphin-B at the C-terminus, exhibiting 100%, 70.5%, and 85% sequence identity with the rat prodyn-derived counterparts, respectively. The number of leucine-enkephalins in this precursor was identical to that found in vertebrates. *M. edulis* prodyn was distinguished from the earlier described leeches in its longer N-terminus. Additionally, sequence comparison established the existence of an orphanin FQ-like peptide, exhibiting 50% sequence homology with that found in mammals (Stefano et al. 1998a). This was the first report of the complete biochemical characterization of a prodyn in a non-parasitic invertebrate and mollusc.

6.2.3 POMC

Duvaux-Miret and colleagues (Duvaux-Miret et al. 1990) demonstrated the presence of β-endorphin and of a POMC-related gene in *Schistosoma mansoni*. Dot blots of cercarial genomic DNA, hybridized with two oligonucleotide probes complementary to highly conserved POMC sequences, showed a POMC-related gene in this trematode. Northern blot analysis of adult worm RNA indicated that this gene was actively transcribed and β-endorphin, ACTH, and α-MSH were detected in all developmental stages of the parasite by radioimmunoassay. Furthermore, *S. mansoni* secreted ACTH-like and β-endorphin-like peptides into its incubation medium (Duvaux-Miret et al. 1992a,b). This study constituted the first expression of a POMC-related gene transcribed in an invertebrate.

The second report was by Salzet and colleagues (Salzet and Stefano 1998) who sequenced a mammalian-like POMC, and six of its derived peptides, including ACTH and MSH, in the immune tissues of the leech *T. tessulatum*. Of the six peptides, three showed high sequence similarity to their vertebrate counterparts, namely, Met-enkephalin, α-MSH and ACTH (100, 84.6 and 70% respectively) whereas γ-MSH, β-endorphin and γ-LPH exhibited only 45, 20 and 10% sequence identity. No dibasic amino acid residues were found at the C-terminus of the γ- and β-MSH peptides. In contrast, the leech α-MSH was flanked at its C-terminus by the Gly-Arg-Lys amidation signal. ACTH and CLIP were also C-terminally flanked by dibasic amino acid residues. The coding region of leech POMC was also reported by RT-PCR using degenerated oligonucleotide primers (Salzet et al. 1997).

By *in situ* hybridization using a digoxigenin-labelled human DNA probe, phagocytic hemocytes of *M. galloprovincialis* were found to express POMC-mRNA. Moreover, immunocytochemistry has identified POMC-derived peptides such as ACTH, α-MSH and β-endorphin. In this context, an ACTH receptor-like mRNA (Franchini et al. 1994; Ottaviani et al. 1998c).

Stefano and Salzet (1999) reported that *M. edulis* hemocytes also contain a mammalian–like POMC. Of the six peptides found in this opioid precursor, methionine-enkephalin, γ-MSH, α-MSH and ACTH exhibited 100, 100, 90 and 74% sequence identity, respectively. The β-endorphin-like and γ-LPH-like molecules exhibited only 25% and 10% sequence identity. Dibasic amino acid residues are found at the C-terminus of MSH and ACTH, indicating cleavage sites. The α-MSH was flanked at the C-terminus by Gly-Arg-Lys, representing the amidation signal. ACTH and CLIP were also C-terminally flanked by dibasic amino acid residues. Of interest is the fact that the methionine enkephalin present in the "β-endorphin-like peptide" is not flanked by basic amino acids, as in mammals, suggesting that it may be present functionally inactive. Taken together, the results from parasites and a free-living mollusc demonstrate conclusively that, without the presence of vertebrate "contamination", POMC and many of the derived bioactive peptides, i.e., α-MSH, are present in invertebrates. Furthermore, regarding their function, i.e., immune regulatory actions, they appear to be conserved as well (Stefano et al. 1996; Salzet et al. 1997).

6.3 Opioid Processing

Since precursor processing involves enzymes, the presence of specific enzymes becomes important. A quick examination of the literature reveals the presence of many types of enzymes in both vertebrates and invertebrates, some of which are important in processing neuropeptides, e.g., neutral endopeptidase (NEP) and angiotensin converting enzyme (ACE) (Hughes et al. 1990; Shipp et al. 1990; Turner et al. 1994; Laurent and Salzet 1995; Laurent and Salzet 1996a; Laurent and Salzet 1996b). Neutral endopeptidase 24.11 (NEP, CD10, CALLA, enkephalinase) is found on the surface of granulocytes and invertebrate immunocytes (Shipp et al. 1990). In invertebrates, opioid precursor molecules are also found in the hemolymph (Salzet et al. 1997; Salzet and Stefano 1997a,b).

The enzyme NEP appears to be quite important. For example, it may not only be responsible for cleaving the precursor proenk or POMC/ACTH but the active processed peptides as well, i.e., Met-enkephalin and MSH, respectively (Stefano et al. 1991a, 1996; Duvaux-Miret et al. 1992a; Smith 1992). Furthermore, NEP deactivates the effects of heterologous cytokines. Indeed, NEP inhibits the PDGF-AB- and TGF-β1-induced shape changes in *M. galloprovincialis* hemocytes (Caselgrandi et al. 2000). This is a marvelously sophisticated processing mechanism because the same enzyme breaks down the precursor to generate active peptides, then goes on to inactivate the same active molecules producing inactive products. This represents a multidimensional process that requires less DNA "message" since the same enzyme performs these tasks. Furthermore, in some cases, the actual inactive products may act as competitive inhibitors to limit further the activity of the prime enzyme, adding another degree of microenvironmental control. This has been observed in our laboratory by NEP processing of Met-enkephalin-Arg-Phe (Stefano and Scharrer 1991).

We surmise that in invertebrates the outcome of an immune proinflammatory response depends on the precise time-dependent build-up of enzyme levels followed by their dissipation and resulting in lack of immunocyte recruitment due to diminished peptide presence that also stops further immunocyte stimulation, i.e., cytokine secretion. Clearly, in this scenario, cascading immune responses can be better understood it is important to realize that invertebrate immune/defense systems have been utilizing these processes before vertebrates came into being (Stefano et al. 1996,1998d,e; Salzet et al. 1997; Sonetti et al. 1997; Salzet and Stefano 1997a,b).

6.4 Corticotropin-Releasing Hormone (CRH)

CRH-like molecules were found in hemocytes of different invertebrates, including *M. galloprovincialis* (Ottaviani et al. 1998b), and in CRH receptor subtype (CRH-R1)- and (CRH-R2)-like mRNAs. Furthermore, CRH induced hemocyte shape changes via protein kinase (PK) A, PKC and PKB/Akt and its effect needs the synergic action of two second messengers, cAMP and IP3 (Malagoli et al. 2000) in order to be functional.

The presence of CRH, ACTH, catecholamines (see below) and cytokines (see below) in the same phagocytic hemocyte is not a chance. We found that all these molecules are involved in the neuroendocrine response, i.e., stress response (Ottaviani and Franceschi 1996). Indeed, as in vertebrates, the invertebrate response shows the same key mediator molecules and the cascade follows the same order and pattern, i.e., CRH>ACTH>catecholamines. It should be underlined that this invertebrate scenario did not need of all the vertebrate organs (hypothalamus, pituitary and adrenal glands), but is all concentrated in the phagocytic hemocyte. Another important similarity with vertebrates is represented by the involvement of molecules of immune system, such as cytokines (Ottaviani and Franceschi 1996; Ottaviani et al. 2004). The incubation of the molluscan hemolymph with different mammalian cytokines such as IL-2, IL-1α and TNF-α provoked the release of biogenic amines (Ottaviani et al. 2004). However, such release was higher when the single cytokines were incubated alone with respect to the incubation of the single cytokines plus CRH. By immunocytochemical and cytofluorimetric approaches it has been found that cytokines and CRH recognize a common epitope on molluscan immunocytes suggesting the presence of an ancestral receptor capable of binding these molecules.

These findings allow us to surmise that invertebrate cytokine receptors show a certain degree of promiscuity. In this context, it should be underlined that the structure of mammalian cytokine receptors are characterized by multi-subunits and the same subunit is shared by different cytokine receptors, such as IL-2, IL-4 and IL-7 (Kondo et al. 1993; Noguchi et al. 1993; Russell et al. 1993; Taga and Kishimoto 1992).

6.5 Monoamines

It is widely known that monoamine, i.e., indoleamine (serotonin), catecholamine, signaling emerges in invertebrates with dopamine, the major molecule used in neural systems (Stefano and Aiello 1975; Hiripi and Stefano 1980; Stefano et al. 1976, 1978; Stefano 1982; Vehovszky and Salanki 1983; Hiripi et al. 1985; Kandel et al. 2000; Nagy and Hiripi 2002). However, there even are some reports that in long-lived invertebrates norepinephrine is always present at low levels (Stefano 1990). Importantly, despite numerous attempts, our laboratory could not identify epinephrine in invertebrate or plant tissues. However, one report identified this chemical messenger in a mollusc, via HPLC via electrochemical detection (Ottaviani et al. 1988).

Recently, we have demonstrated that dopamine and its precursors serve also as endogenous morphine precursors (Zhu et al. 2005b,c; Zhu and Stefano 2004) in invertebrates, further highlighting catecholaminergic significance while providing proof for opiate alkaloid biosynthesis in invertebrate animal tissues.

6.6 Endogenous Morphine

6.6.1 μ_3 Binding

Based on the historical conviction that morphine was not present endogenously in animals, studies on the pharmacologic properties of morphine and morphine-like substances had focused on the effects of exogenous opiates, a family of important analgesic and antinociceptive drugs. This focus changed following the discovery that morphine binds to the same receptors used by endogenous opioid peptides (Lord et al. 1977). An important step forward was the demonstration of endogenous opiates in various vertebrate tissues, including the nervous system (Gintzler et al. 1976; Gintzler et al. 1978). Most of our knowledge of opioid receptors is derived from studies of these receptors in the nervous system. These receptors were classified as delta, mu and kappa, and heterogeneity was shown for each of these major subtypes (Clark et al. 1989; Pasternak 1986, 1988; Rothman et al. 1990; Mattia et al. 1992; Stefano et al. 1992a; Traynor and Elliott 1993). Amino acid sequences were obtained from cloning of certain delta, mu and kappa (neuronal) opioid receptor subtypes (Evans et al. 1992; Kieffer et al. 1992; Chen et al. 1993; Yasuda et al.

1993). Another mu opiate receptor subtype was demonstrated by Stefano et al. (1993), namely μ_3. The μ_3 receptor differed from previously described neuronal opioid receptor subtypes in that it exhibited essentially no, or exceedingly low, affinity for naturally occurring endogenous opioid peptides, or their analogues (Stefano 1999). However, the opiate alkaloid binding that was present was naloxone sensitive, demonstrating its opiate receptor properties (Stefano et al. 1996). Additionally, and of crucial importance, these properties corresponded to effects of opiates, i.e., morphine, on immunocytes that were not mimicked by opioid peptides. In contrast, each of the other opioid receptor subtypes bound at least one of the endogenous opioid peptides with high affinity. Furthermore, certain opiate alkaloids, benzomorphans and other drugs bound to classical opioid receptors but not to the μ_3 receptor (Cruciani et al. 1994; Makman et al. 1995a,b; Stefano et al. 1993, 1995a). Binding sites for this novel morphine (opiate alkaloid-selective, opioid peptide-insensitive) receptor, designated μ_3, were first reported to be present in human peripheral blood monocytes and in invertebrate immunocytes and later on other cell types (Cruciani et al. 1994; Dobrenis et al. 1995; Makman et al. 1995a; Stefano et al. 1993, 1995a; Stefano and Scharrer 1996). The newly discovered opioid peptides endomorphin-1, -2 and orphanin FQ did not bind to this opiate receptor subtype (Rialas et al. 1998) and neither did the potent mu opiate agonist fentanyl (Dobrenis et al. 1995; Makman et al. 1995a; Stefano et al. 1995a), but 6-Glucuronide, not the 3-glucuronide metabolite of morphine, did bind to the μ_3 receptor. Classical opioid receptors are linked to trimeric G proteins that in turn modulate Ca^{++} and K^+ channels, adenylyl cyclase, and probably other signal transduction systems (Pasternak 1988). In this regard, the μ_3 receptor is also linked to G protein, based on guanine nucleotide effects on agonist binding to the receptor (Makman et al. 1995a).

6.6.2 *Nitric Oxide (NO) and Morphine*

The literature documents an association of nitric oxide (NO) with morphine actions. Peripheral morphine analgesia involves NO-stimulated increases in intracellular cGMP (Ferreira et al. 1991). NO also was associated with antinociception (Przewlocki et al. 1993) as well as tolerance and dependence (Majeed et al. 1994). In addition, the morphine-induced suppression of splenic lymphocyte proliferation was shown to involve NO (Fecho et al. 1994). Morphine and NO were linked to gastrointestinal regulation (Gyires 1994; Stefano et al. 2004). Furthermore, morphine, not opioid peptides, stimulates constitutive NO release in macrophages, granulocytes, various types of human and rat endothelial cells, invertebrate neurons and immunocytes and in rat median eminence fragments, all in a naloxone antagonizable manner (Bilfinger et al. 1997, 1998; Sawada et al. 1997; Stefano et al. 1996, 1997, 1998b,c; Prevot et al. 1998; Rialas et al. 1998; Stefano 1998). These data suggest that the μ_3 receptor is coupled to constitutive NO release in these cells.

Receptor Transcript

5'atacaccaagatgaagactgccaccaacatctacattttcaaccttgctctggcagatgccttagccaccagtac
cctgcccttccagagtgtgaattacctaatgggaacatggccatttggaaccatcctttgcaagatagtgatctccata
gattactataacatgttcaccagcatattcaccctctgcaccatgagtgttgatcgatacattgcagtctgccaccctgt
caaggccttagatttccgtactccccgaaatgccaaaattatcaatgtctgcaactggatcctctcttcagccattggtc
ttcctgtaatgttcatggctacaacaaaatacaggcaaggttccatagattgtacactaacattctctcatccaacctg
gtactgggaaaacctgctgaagatctgtgttttcatcttcgccttcattatgccagtgctcatcattaccgtgtgctatgga
ctgatgatcttgcgcctcaagagtgtccgcatgctctctggctccaaagaaaaggacaggaatcttcgaaggatca
ccaggatggtgctggtggtggtggctgtgttcatcgtctgctggactcccattcacatttacgtcatcattaaagccttgg
ttacaatcccagaaactacgttccagactgtttcttggcacttctgcattgctctaggttacacaaacagctgcctcaac
ccagtcctttatgcatttctggatgaaaacttcaaacgatgcttcagagagttctgtatcccaacctcttccaacattga
gcaacaaaactccactcgaattcgtcagaacactagagaccacccctccacggccaatacagtggatagaacta
atcatcagaattattatataattcatagatgttgctgcaataccccctcttatttctcaaaagccagtcttgctctggttctg**t**
gattaaagagagagggtgagtgccttgcccactgtggtcatggatgcaagatattcacagaaaattagcatcatag
aaaaaaaannnaaaaaaaaaaaaaaaaaancatgtcggccgcctcggccaaacatcgggtcgagcatgcat
ctagggcggccaattccgcccctctcccccccngcnntttc**cacaccgaggagtccagtttgtgcaagacacc**
cagcggaaccaaaacccatcgtggtatgtgaatcgaagtcatcataaaggtgacccttctgtctgtaaga
ttttaatttaagcatatatttatgacctcaacaaagacgaaccatcttttgttaattcaccgtagtaacacataa
agttatgctacctctgatcaaag-3'

MOR-1: 1________________________________-2162
MOR-3: 1__________ _____ - - - - ________1338

Fig. 6.2 μ3 cDNA sequence (1338 bp). The underlined sequence represents the novel 263 bp segment, and the bold letters represent *Homo sapiens* Oprm 3′UTR DNA sequence (nucleotide position 1625–1829 of Oprm). From the expression of the mu opiate receptor (MOR)-1 and MOR-3, it can be observed that the latter is a truncated version of the MOR-1, which we now speculate is a 6-transmembrane receptor (Kream et al. 2007) (Modified from Cadet et al. 2003a)

6.6.3 Cloning of μ_3

In order to identify μ_3 at the molecular level, a human testis cDNA library was screened using a mu opiate receptor (MOR)-1 gene-specific probe and a 1338 base pair clone was identified (Fig. 6.2; Cadet et al. 2003a). Sequencing of this clone and subsequent analysis (NCBI Blast software) showed that the clone exhibits 100% identity to mu1 in the center and conserved region, but is truncated at the 5′-end (position 503 of mu1 mRNA; missing several hundred nucleotides). In addition, the 3′-end of the new clone contains the 3′-end of the mu1 receptor, followed by a new fragment of 263 bases, and then a 202 bp fragment of the 3′-end of the mu1 gene untranslated region (Cadet et al. 2003a).

6.6.4 Endogenous Morphine

The demonstration of endogenous opiates, i.e., morphine, codeine, in various vertebrate tissues, including the nervous system (Gintzler et al. 1976; Goldstein et al. 1985; Oka et al. 1985; Cardinale et al. 1987; Donnerer et al. 1986, 1987; Stefano and Scharrer 1994), is quite important for establishing the significance of the μ_3 opiate receptor subtype. There is a body of evidence that shows opiate alkaloids such as morphine, morphine-3- and 6- glucuronide, as well as the morphine putative precursor molecules (thebaine, salutaridine, norcocolarine, reticuline, tetrahydropapoverine (THP) and codeine) exist in vertebrates (Donnerer et al. 1986; Lee and Spector 1991; Epple et al. 1994; Zhu et al. 2003). In invertebrates, specifically *M. edulis*, the presence of morphine, morphine 6-glucuronide, morphine 3-glucuronide, codeine, THP and reticuline were also reported (Stefano et al. 1993, 2000b; Goumon et al. 2001; Zhu et al. 2001, 2002). Endogenous opiate levels can be induced to change following stimulation (Stefano et al. 2000b). Morphine was also found in human plasma (Brix-Christensen et al. 1997; Cadet et al. 2003b), suggesting a hormonal action with immune, vascular and gut tissues as targets (Stefano et al. 2002, 2004). Additionally, in recent studies sources of contamination for these compounds were examined, i.e., rat food, supplies, etc, and not found (Zhu et al. 2003). Besides these biochemical studies, immunocytochemical localization of a morphine-like material was reported in neural and immune tissues (Bianchi et al. 1993, 1994; Liu et al. 1996a; Esch et al. 2004) as well as in invertebrate tissues (Stefano et al. 1993). Taken together, these reports suggest that animals appear to have the ability to synthesize opiate alkaloids. In earlier reports, we formulated a hypothesis stating that endogenous morphine, used as either a hormone or neurotransmitter down regulates immune, vascular, neural and gut tissues under normal and following trauma situations (Stefano and Scharrer 1994; Stefano et al. 1996, 2000b, 2004; Esch et al. 2004; Guarna et al. 2005). Our recent work has provided compelling *prima facie* evidence that chemically authentic morphine is endogenously synthesized by diverse animal cellular systems, including invertebrate neural tissues, from L-tyrosine-derived small molecules within a strikingly similar biochemical pathway to that described in opium poppy (Zhu et al. 2005a,b; Kream and Stefano 2006a,b). Furthermore, these reports have empirically demonstrated a functional linkage of *de novo* morphine synthesis and its evoked release upon physiological demand. These accumulated data suggest that low steady-state levels of morphine in many, if not all, mammalian organ systems, indicate dynamic utilization and turnover of releasable cellular pools of morphine, thereby lending support to its essential role as an autocrine/paracrine factor devoted to hierarchal integration of cellular function (Stefano et al. 2000a,b). In general, it appears that morphinergic signaling occurs in plants, invertebrates and human tissues, demonstrating its importance via its conservation during evolution.

6.7 Cytokines

Invertebrate ganglia, immunocytes, and microglia contain interleukin-like (IL)-1- and IL-6 and -10 signaling molecules (Beck and Habicht 1986; Hughes et al. 1990, 1991a; Stefano 1992; Paemen et al. 1992; Stefano et al. 1992b; Hughes and Chin 1994; Scharrer et al. 1996). Based on these findings, it one can surmise that an interleukin-like molecule secreted from these invertebrate cells may have the ability to interact with monoamine chemical messengers. Sawada and colleagues, as well as others, demonstrated that mammalian IL-1 and IL-2 and -4 have the ability to alter invertebrate neural ion channels in a stereoselect manner, further straightening the hypothesis that these immunocyte-derived molecules can alter neural activities as well as stimulate them (Fig. 6.3), including tumor necrosis like molecules (Hughes et al. 1990, 1991a,b,c, 1994; Sawada et al. 1991; Stefano et al. 1991b, 1999; Szucs et al. 1992a,b; Rozsa et al. 1997). PDGF-AB and TGF-β1-like molecules as well as PDGF-AB receptor-α- and β- and TGF-β-receptor (type II)-like molecules were demonstrated in hemocytes of *M. galloprovincialis* (Franchini et al. 1996;

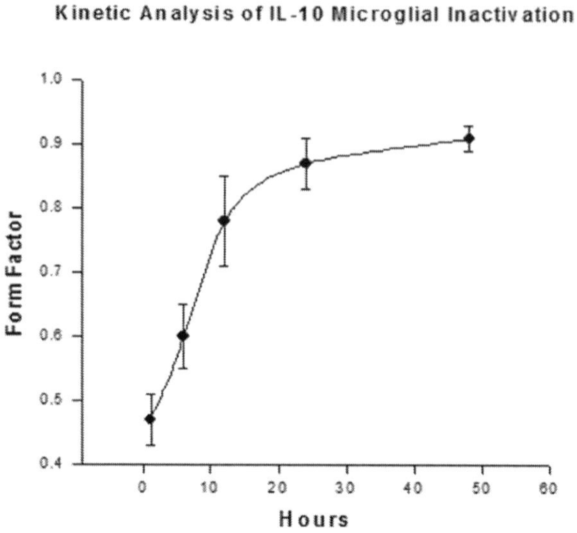

Fig. 6.3 IL-10 (100 U/ml) slowly induces active (mobile and ameboid) microglia to become round and inactive. The experiments were replicated 4 times. The mean value of these trials is represented along with the standard deviation. Form factor (ff) is equal to $4 \times \pi \times$ area/perimeter^2. It was demonstrated that the lower the form factor the more active the cell (mobile) is, whereas a ff of 1, indicative of a round shape, the cells are immobile and inhibited from becoming mobile (Stefano et al. 1989a,b). Statistical significance ($P < 0.01$) was determined by a one-tailed students t-test comparing the ff at 1 h with that obtained at 24 h. In control samples 46 + 5.9% of the cells were in the ameboid-active conformation. IL-10 reduced this level of activation to 13.7 + 2.5%

Kletsas et al. 1998). These cytokines affect molluscan hemocytes chemotaxis (i.e., the expression of cell migration) in a dose-correlated fashion and phagocytosis in a species-specific manner (Ottaviani et al. 1997). With regards cell shape changes (i.e., the expression of cell motility) both the cytokines provoke cellular shape changes in hemocytes via interactions with respective receptors and the extracellular signals are transduced along the phosphoinositide signaling pathway (Kletsas et al. 1998). The stimulatory effect of PDGF-AB is independent of the influx of extracellular Ca^{2+}, while that of TGF-β1 is Ca^{2+}-dependent (Ottaviani et al. 1997).

6.8 Specific Regulatory Processes

There is a growing body of evidence demonstrating that morphine influences ACTH processing in vertebrates and invertebrates (Stefano et al. 1996; Stefano and Smith 1996; Sonetti et al. 2005). This is especially important since it is a naturally occurring signal molecule found in human plasma and invertebrate hemolymph (Stefano et al. 1993, 1995b; Liu et al. 1996b). In yet another mechanism associated with ACTH, morphine, in a dose-dependent manner, and again by way of

NO increases leech processing of POMC as noted by higher hemolymph levels of α-MSH and ACTH (Salzet et al. 1997). In *M. edulis* we also demonstrate that morphine stimulates the processing of ACTH (1–39) to MSH (1–13) by NEP as determined by phosphoramidon inhibition. The ability of morphine to enhance enzyme levels has also been noted in other studies using mammalian and human tissues (Malfroy et al. 1978; Stefano et al. 1996). The mechanism for this morphine action, based on these reports, is by increasing the processing of the precursor or stimulating the release of the precursor or both.

The significance and specificity of opiate molecules in these studies are enhanced by the observation that LPS stimulation results in ACTH (1–24) in the hemolymph, indicating that other enzymatic processes can occur by way of different signaling molecules. Furthermore, ACTH (1–24) processing occurred by an enzymatic process independent of NEP, i.e., renin-type enzyme (Salzet and Stefano 1998). Taken together, as in mammals, differential processing of ACTH occurs in invertebrates. Additionally, invertebrate immunocytes are capable of displaying different responses to ACTH fragments, including those of *M. edulis* (Genedani et al. 1993), further supporting the differential processing pattern and its potential significance as a meaningful event.

6.9 The Hemocyte as "Immune-Mobile Brain"

From the data reported in this chapter, the phagocytic hemocyte represents a model to unravel the predicted common origin of immune and neuroendocrine systems (Ottaviani and Franceschi 1997; Ottaviani et al. 2007). Indeed, invertebrate hemocytes contain different bioactive molecules involved in immune-neuroendocrine responses. From the original idea coined by Blalock (Blalock 1984) and Blalock and Smith (Blalock and Smith 1985) for the mammalian lymphocytes in which are combined both immune and neuroendocrine properties, we extend this concept to our ancestral invertebrate hemocytes and in this perspective, we suggested the concept of "immune-mobile brain" for those phagocytic cells (Ottaviani et al. 1991a,b) in order to describe the complex situation where mobile cells are able to recognize a variety of stimuli and to set up a complex response in which primitive, but very efficient, forms of immune and neuroendocrine responses are intermixed.

6.10 Human Stem Cells with Invertebrate Promise

Recently, programmed expression of opioid peptide and opioid receptor genes in cultured neural progenitor cells, at various stages of differentiation, was demonstrated (Ventura and Maioli 2000; Kim et al. 2006). Human multi-lineage progenitor cells (MLPC) derived from post-partum umbilical cord blood have recently been established as a high resolution model (Collins 2006) for studying biochemical and molecular mechanisms underlying differentiation of multi-potent progenitor cells into clonal cell lines (e.g., adipocytes, osteoblasts, myocytes, vascular endothelial cells, neurons, astrocytes, and oligodendrocytes). Because MLPC are non-transformed, and non-immortalized, their potential for both proliferation and differentiation into phenotypically distinct clonal lines is temporally defined by the complex chemical profile of their respective microenvironments (Horowitz et al. 2002).

Ongoing studies from our group evaluated whether a μ_3 opiate receptor/NO coupled regulatory pathway exists in MLPC prepared from umbilical cord blood (Cadet et al. 2007). Real-time PCR analysis of extracted RNA from undifferentiated human MLPC indicated selective expression of a μ_3-like opiate receptor-encoding RNA, in the absence of traditional μ_1 opioid receptor-encoding RNA expression. Pharmacological analyses provided confirmatory evidence of functional μ like opiate receptor/NO coupling via morphine-evoked real-time release of NO into the bath medium from primary cultures of undifferentiated human MLPC, blockable by the opiate alkaloid antagonist L-naloxone or the constitutive NO synthase inhibitor N(G)-nitro-L-arginine methyl ester. Complementary microarray analysis of extracted RNA indicated that traditional mu, delta, and kappa opioid receptor gene expression is not detected in both undifferentiated and differentiated MPLC. Chemical differentiation of MLPC into neuronal progenitor cells effected significant phenotypic expression of a variety of neurally-associated genes (Cadet et al. 2007). Our data provide compelling evidence in support of both the evolutionary primacy and primordial regulatory role of μ_3 opiate receptor/NO in embryogenesis. Its presence, as discussed earlier in invertebrates, strongly demonstrates the conservation of this signaling during evolution.

6.11 Conclusion

In all probability, the chemical messenger systems observed in this select review appear to have originated in "simple" animals. The same sequential order of the derived peptides in POMC and their flanking by dibasic amino acids is of noteworthy significance in this regard. Therefore, aside from their historical origin, it may be more correct to refer to these mammalian signal molecules as invertebrate-like. We surmise that the reason these molecules were maintained during evolution is because conformational "matching" of molecules within each of their respective signal system is difficult and time consuming to achieve. The conformational complexity and rigidity of the "match" among the sequential components for a given signal system, i.e., enzymes, receptors, would thus seem to exert a determining influence during evolution to maintain the conformational integrity of the signal system (Makman and Stefano 1984; Stefano 1986). Thus, "ancient" communication systems, e.g., opioids, would tend to remain relatively intact in increasingly complex animal phyla, especially the structure or conformation of the bioactive portions of the molecules themselves.

The list of "mammalian type" signaling molecules in "simpler" organisms is steadily growing. In prokaryotes, chorionic gonadotropin-like material has been detected (Acevedo et al. 1978). In protozoans, not only have similar mammalian type neuropeptides been detected, but opioids have been shown to alter feeding behavior, an effect inhibited by naloxone (Josefsson and Johansson 1979). This evidence indicates that signal effector and receptor communication mechanisms may be present in unicellular organisms, suggesting the origin of signal systems may have occurred during prokaryotic development. Indeed, many of these systems could have started out as intracellular communication mechanisms associated with energy metabolism modulation coupled to longevity advances.

References

Acevedo HF, Sliifkin M, Pouchet GR, Pardo M (1978) Immunohistochemical localization of a choriogonadotropin-like protein in bacteria isolated from cancer patients. Cancer 41:1217–1229

Audigier Y, Deprat AM, Cros J (1980) Comparative study of opiate and enkephalin receptors on lower vertebrates. Comp Biochem Physiol 67:191–194

Ballarin L, Cima F (2005) Cytochemical properties of *Botryllus schlosseri* haemocytes: Indications for morpho-functional characterisation. Eur J Histochem 49:255–264

Beck G Habicht GS (1986) Isolation and characterization of a primitive interleukin-1-like protein from an invertebrate, *Asterias forbesi*. Proc Natl Acad Sci USA 83:7429–7433

Bianchi E, Alessandrini C, Guarna M, Tagliamone A (1993) Endogenous codeine and morphine are stored in specific brain neurons. Brain Res 627:210–215

Bianchi E, Guarna M, Tagliamonte A (1994) Immunocytochemical localization of endogenous codeine and morphine. Adv Neuroimmunol 4:83–92

Bilfinger TV, Fimiani C, Stefano GB (1998) Morphine's immunoregulatory actions are not shared by fentanyl. Int J cardiol 64:61–66

Bilfinger TV, Hartman A, Liu Y, Magazine HI, Stefano GB (1997) Cryopreserved veins used for myocardial revascularization: A 5 year experience and a possible mechanism for their increased failure. Ann Thoracic Surg 63:1063–1069

Blalock JE (1984) The immune system as a sensory organ. J Immunol 132:1067–1070

Blalock JE, Smith EM (1985) The immune system: Our mobile brain? Immunol Today 6:115–117

Bellaïche Y, Schweisguth F (2001) Lineage diversity in the Drosophila nervous system. Curr Opin Gen Dev 11:418–423

Brix-Christensen V, Tonnesen E, Sanchez RG, Bilfinger TV, Stefano GB (1997) Endogenous morphine levels increase following cardiac surgery as part of the antiinflammatory response? Intern J Cardiol 62:191–197

Cadet P, Mantione KJ, Stefano GB (2003a) Molecular identification and functional expression of μ3, a novel alternatively spliced variant of the human mu opiate receptor gene. J Immunol 170:5118–5123

Cadet P, Zhu W, Mantione K, Rymer M, Dardik I, Reisman S, Hagberg S, Stefano GB (2003b) Cyclic exercise induces anti-inflammatory signal molecule increases in the plasma of Parkinson's patients. Int J Mol Med 12:485–492

Cadet P, Mantione KJ, Zhu W, Kream RM, Sheehan M, Stefano GB (2007) A functionally coupled μ3-like opiate receptor/nitric oxide regulatory pathway in human multi-lineage progenitor cells. J Immunol 179:5839–5844

Cardinale GJ, Donnerer J, Finck AD, Kantrowitz JD, Oka K, Spector S (1987) Morphine and codeine are endogenous components of human cerebrospinal fluid. Life Sci 40:301–306

Caselgrandi E, Kletsas D, Ottaviani E (2000) Neutral endopeptidase-24.11 (NEP) deactivates PDGF- and TGF-beta-induced cell shape changes in invertebrate immunocytes. Cell Biol Int 24:85–90

Chen Y, Mestek A, Liu J, Hurley JA, Yu L (1993) Molecular cloning and functional expression of a m-opioid receptor from rat brain. Mol Pharmacol 44:8–12

Cima F, Matozzo V, Marin MG, Ballarin L (2000) Haemocytes of the clam Tapes philippinarum (Adams & Reeve, 1850): Morphofunctional characterisation. Fish Shellfish Immunol 10: 677–693

Civelli O, Douglass J, Goldstein A, Herbert E (1985) Sequence and expression of the rat prodynorphin gene. Proc Natl Acad Sci USA 82:4291–4295

Clark JA, Liu L, Price M, Hersh BS, Edelson M, Pasternak GW (1989) Kappa opiate receptor multiplicity: Evidence for two U50, 488- sensitive K1 subtypes and a novel K3 subtype. J Pharmacol Exp Therap 251:461–468

Collins DP (2006) Isolation and characterization of umbilical cord blood-derived multipotent stem cells arising from an adherent CD45+/CD34+ cell subset. (Paper presented at the 4th Ann Int Umbilical Cord Blood Transplant Symp)

Cruciani RA, Dvorkin B, Klinger HP, Makman MH (1994) Presence in neuroblastoma cells of a m3 receptor with selectivity for opiate alkaloids but without affinity for opioid peptides. Brain Res 667:229–237

Dobrenis K, Makman MH, Stefano GB (1995) Occurrence of the opiate alkaloid-selective m3 receptor in mammalian microglia, astrocytes and kupffer cells. Brain Res 686:239–248

Donnerer J, Cardinale G, Coffey J, Lisek CA, Jardine I, Spector S (1987) Chemical characterization and regulation of endogenous morphine and codeine in the rat. J Pharmacol Exp Therap 242:583–587

Donnerer J, Oka K, Brossi A, Rice KC, Spector S (1986) Presence and formation of codeine and morphine in the rat. Proc Natl Acad Sci USA 83:4566–4567

Duvaux-Miret O, Dissous C, Guatron JP, Pattou E, Kordon C, Capron A (1990) The helminth *Schistosoma mansoni* expresses a peptide similar to human beta-endorphin and possesses a POMC-related gene. New Biol 2:93–99

Duvaux-Miret O, Stefano GB, Smith EM, Capron A (1992a) Neuroimmunology of host parasite interactions: Proopiomelanocortin derivedpeptides in the infection by *Schistosoma mansoni*. Adv Neuroimmunol 2:297–311

Duvaux-Miret O, Stefano GB, Smith EM, Mallozzi L, Capron A (1992b) Proopiomelanocortin-derived peptides as tools of immune evasion for the human trematode *Schistosoma mansoni*. Acta Biol Hung 43:281–286

Edley SM, Hall L, Herkenham M, Pert CB (1982) Evolution of striated opiate receptors. Brain Res 249:184–188

Epple A, Nibbio B, Spector S, Brinn J (1994) Endogenous codeine: Autocrine regulator of catecholamine release from chromaffin cells. Life Sci 54:695–702

Esch T, Guarna M, Bianchi E, Zhu W, Stefano GB (2004) Commonalities in the central nervous system's involvement with complementary medical therapies: Limbic morphinergic processes. Med Sci Monit 10:MS6–MS17

Evans CJ, Keith DE Jr, Morrison H, Magendzo K, Edwards RH (1992) Cloning of a delta opioid receptor by functional expression. Science 258:1952–1955

Ewandinger NM, Ridgway RL, Syed NI, Lukowiak K, Bulloch AGM (1996) Identification and localization of a (Met5)-enkephalin-like peptide in the mollusc, *Lymnaea stagnalis*. Brain Res 737:1–15

Fecho K, Maslonek KA, Coussons-Read ME, Dykstra LA, Lysle DT (1994) Macrophage-derived nitric oxide is involved in the depressed concanavalin A responsiveness of splenic lymphocytes from rats administered morphine in vivo. J Immunol 152:5845–5851

Ferreira SH, Duarte ID, Lorenzetti BB (1991) The molecular mechanism of action of peripheral morphine analgesia: Stimulation of the cGMP system via nitric oxide release. Eur J Pharmacol 201:121

Franceschi C, Cossarizza A, Monti D, Ottaviani E (1991) Cytotoxicity and immunocyte markers in cells from the freshwater snail *Planorbarius corneus* (L.) (Gastropoda pulmonata): Implications for the evolution of natural killer cells. Eur J Immunol 21:489–493

Franchini A, Fontanili P, Ottaviani E (1994) Expression of pro-opiomelanocortin (POMC)-mRNA in phagocytic hemocytes of *Mytilus galloprovincialis*. In: Argano R, Cirotto C, Grassi Milano E, Mastrolia L (eds) Contributions to animal biology. Halocynthia Assoc, Palermo, pp 233–236

Franchini A, Kletsas D, Ottaviani E (1996) Immunocytochemical evidence of PDGF- and TGF-beta-like molecules in invertebrate and vertebrate immunocytes: An evolutionary approach. Histochem J 28:599–605

Frontali N, Gainer H (1977) Peptides in invertebrate nervous system. In: Gainer H (ed) Peptides in neurobiology. Plenum Press, New York, pp 259–271

Genedani S, Bernardi M, Ottaviani E, Franceschi C, Leung MK, Stefano GB (1993) Differential modulation of invertebrate hemocyte motility by CRF, ACTH, and its fragments. Peptides 15:203–206

Gintzler AR, Gershon MD, Spector S (1978) A nonpeptide morphine-like compound: Immunocytochemical localization in the mouse brain. Science 199:447–448

Gintzler AR, Levy A, Spector S (1976) Antibodies as a means of isolating and characterizing biologically active substances: Presence of a non-peptide morphine-like compound in the central nervous system. Proc Natl Acad Sci USA 73:2132–2136

Goldstein A, Barrett RW, James IF, Lowney LI, Weitz C, Knipmeyer LI, Rapoport H (1985) Morphine and other opiates from beef brain and adrenal. Proc Natl Acad Sci USA 82: 5203–5207

Goumon Y, Casares F, Zhu W, Stefano GB (2001) The presence of morphine in ganglionic tissues of *Modiolus deminissus*: A highly sensitive method of quantitation for morphine and its derivatives. Mol Brain Res 86:184–188

Goumon Y, Strub JM, Moniatte M, Nullans G, Poteur L, Hubert P, Van Dorsselaer A, Aunis D, Metz-Boutigue MH (1996) The C-terminal bisphosphorylated proenkephalin-A-(209–237)-peptide from adrenal medullary chromaffin granules possesses antibacterial activity [published erratum appears in Eur J Biochem (1996) 237(3):883]. Europ J Biochem 235:516–525

Guarna M, Ghelardini C, Galeotti N, Stefano GB, Bianchi E (2005) Neurotransmitter role of endogenous morphine in CNS. Med Sci Monitor 11:RA190–RA193

Gyires K (1994) The role of endogenous nitric oxide in the gastroprotective action of morphine. Eur J Pharmacol 255:33–37

Harrison SM, Harrison DA (2006) Contrasting mechanisms of stem cell maintenance in Drosophila. Semin Cell Dev Biol 17:518–533

Haynes LW (1980) Peptide neuroregulation in invertebrates. Progr Neurobiol 15:205–223

Hiripi L, Salanki J, Stefano GB, Assanah P (1985) Heavy metal pollution influences serotonin level and dopamine-stimulated adenylate cyclase activity in the CNS of molluscs. In: Salanki J (ed) Symposia Biologica Hungarica, vol 29. Academy Hungarica Press, Budapest, pp 401–412

Hiripi L, Stefano GB (1980) Dopamine inhibition of tryptophane hydroxylase in molluscan nervous tissue homogenates: Evidence for intracellular site of action. Life Sci 27: 1205–1209

Horikawa S, Takai T, Toyosata M, Takahashi H, Noda M, Kakidani H, Kubo T, Hirose T, Inayama S, Hayashida H, Miyata T (1983) Isolation and structural organization of the human preproenkephalin B gene. Nature 306:611–614

Horowitz D, Callahan JF, Pelus LM, Fukuda S, King AG (2002) Inhibition of hematopoietic progenitor cell growth by Tyr-MIF, an endogenous opiate modulator, and its degradation products. Int Immunopharmacol 2:721–730

Hughes TK, Cadet P, Rady PL, Tyring SK, Chin R, Smith EM (1994) Evidence for the production and action of IL-10 in pituitary cells. Cellular Mol Neurobiol 14:59–69

Hughes TK, Chin R (1994) Interactions of neuropeptides and cytokines. In: Scharrer B, Smith EM, Stefano GB (eds) Neuropeptides and immunoregulation. Springer-Verlag, Berlin, pp 101–119

Hughes TK, Chin R, Smith EM, Leung MK, Stefano GB (1991a) Similarities of signal systems in vertebrates and invertebrates: Detection, action, and interactions of immunoreactive monokines in the mussel, *Mytilus edulis*. Adv Neuroimmunol 1:59–70

Hughes TK, Smith EM, Cadet P, Sinisterra JI, Leung MK, Shipp MA, Scharrer B, Stefano GB (1990) Interaction of immunoactive monokines (IL-1 and TNF) in the bivalve mollusc *Mytilus edulis*. Proc Natl Acad Sci USA 87:4426–4429
Hughes TK, Smith EM, Stefano GB (1991b) Detection of immunoreactive Interleukin-6 in invertebrate hemolymph and nervous tissue. Progr Neuroimmune Endocrinol 4:234–239
Hughes TKJ, Smith EM, Barnett JA, Charles R, Stefano GB (1991c) LPS stimulated invertebrate hemocytes: A role for immunoreactive TNF and IL-1. Dev Comp Immunol 15:117–122
Josefsson JO, Johansson P (1979) Naloxone reversible effect of opioid on pinocytosis in *Amoeba proteus*. Nature 78:283–292
Kakidani H, Furutani Y, Takahashi H, Noda M, Morimoto Y, Hirose T, Asai M, Inayama S, Nakanishik S, Numa S (1982) Cloning and sequence analysis of cDNA for porcine beta-neo-endorphin/dynorphin precursor. Nature 298:245–249
Kandel ER, Schwartz JH, Jessell TM (2000) Principles of Neural Science. McGraw-Hill/Appleton & Lange, United States
Kieffer BL, Befort K, Gaveriaux-Ruff CE, Hirth CG (1992) The d-opioid receptor: Isolation of a cDNA by expression cloning and pharmacological characterization. Proc Natl Acad Sci USA 89:12048–12052
Kim E, Clark AL, Kiss A, Hahn JW, Wesselschmidt R, Coscia CJ, Belcheva MM (2006) Mu- and kappa-opioids induce the differentiation of embryonic stem cells to neural progenitors. J Biol Chem 281:33749–33760
Kletsas D, Sassi D, Franchini A, Ottaviani E (1998) PDGF and TGF-beta induce cell shape changes in invertebrate immunocytes via specific cell surface receptors. Eur J Cell Biol 75: 362–366
Kondo M, Takeshita T, Ishii N, Nakamura M, Watanabe S, Arai K, Sugamura K (1993) Sharing of the interleukin-2 (IL-2) receptor gamma chain between receptors for IL-2 and IL-4. Science 262:1874–1877
Krause DS, Fackler MJ, Civin CI, May WS (1996) CD34: Structure, biology, and clinical utility. Blood 87:1–13
Kream RM, Sheehan M, Cadet P, Mantione KJ, Zhu W, Casares F, Stefano GB (2007) Persistence of evolutionary memory: Primordial six-transmembrane helical domain mu opiate receptors selectively linked to endogenous morphine signaling. Med Sci Monit 13:SC5–SC6
Kream RM, Stefano GB (2006a) De novo biosynthesis of morphine in animal cells: An evidence-based model. Med Sci Monit 12:RA207–RA219
Kream RM, Stefano GB (2006b) Morphine synthesis in animals (Editorial). Med Sci Monit 12:ED1–ED2
Kream RM, Zukin RS, Stefano GB (1980) Demonstration of two classes of opiate binding sites in the nervous tissue of the marine mollusc *Mytilus edulis*. Positive homotropic cooperativity of lower affinity binding sites. J Biol Chem 255:9218–9224
Laurent V, Salzet M (1995) Isolation of a renin-like enzyme from the leech *Theromyzon tessulatum*. Peptides 16:1351–1358
Laurent V, Salzet M (1996a) Identification and properties of an angiotensin-converting enzyme in the leech *Theromyzon tessulatum*. Peptides 17:737–745
Laurent V, Salzet M (1996b) Metabolism of enkephalins in head membranes of the leech Theromyzon tessulatum by peptidases. Regul Pept 65:123–131
Lee SC, Spector S (1991) Don't use changes in endogenous morphine and codeine contents in the fasting rat. J Pharmacol Exp Therap 257:647–652
LeRoith D, Trudgill DL, Roth J, Shiloach J, Lewis ME, Pert CB, Kriger DT (1982) Corticotropin and b-endorphin-like materials are native to unicellular organisms. Proc Natl Acad Sci USA 79:2086–2090
Leung MK, Stefano GB (1984) Isolation and identification of enkephalin in pedal ganglia of *Mytilus edulis* (mollusca). Proc Natl Acad Sci USA 81:955–958
Leung MK, Stefano GB (1987) Comparative neurobiology of opioids in invertebrates with special attention to senescent alterations. Progr Neurobiol 28:131–159

Lim JY, Lee BH, Kang SW, Wago H, Han SS (2004) Association of reticular cells with CD34+/Sca-1+ apoptotic cells in the hemopoietic organ of grasshopper, *Euprepocnemis shirakii*. J Insect Physiol 50:657–665

Liu Y, Bilfinger TV, Stefano GB (1996a) A rapid and sensitive quantitation method of endogenous morphine in human plasma. Life Sci 60:237–243

Liu Y, Shenouda D, Bilfinger TV, Stefano ML, Magazine HI, Stefano GB (1996b) Morphine stimulates nitric oxide release from invertebrate microglia. Brain Res 722:125–131

Lord JAH, Waterfield AA, Hughes J, Kosterlitz HW (1977) Endogenous opioid peptodes: Multiple agonists and receptors. Nature 267:495–499

Luschen W, Buck F, Willig A, Jaros PP (1991) Isolation, sequence analysis, and physiological properties of enkephalins in the nervous tissue of the shore crab, *Carcinus maenas* L. Proc Natl Acad Sci USA 88:8671–8675

Majeed NH, Przewlocka B, Machelska H, Przewlocki R (1994) Inhibition of nitric oxide synthase attenuates the development of morphine tolerance and dependence in mice. Neuropharmacology 33:189–192

Makman MH, Bilfinger TV, Stefano GB (1995a) Human granulocytes contain an opiate receptor mediating inhibition of cytokine-induced activation and chemotaxis. J Immunol 154: 1323–1330

Makman MH, Dvorkin B, Stefano GB (1995b) Murine macrophage cell lines contain m3-opiate receptors. Eur J Pharmacol 273:R5–R6

Makman MH, Stefano GB (1984) Marine mussels and cephalopods as models for study of neuronal aging. In: Mitchell DH, Johnson TE (eds) Invertebrate models in aging research. CRC Press Inc, Boca Raton, Florida, pp 165–190

Malagoli D, Franchini A, Ottaviani E (2000) Synergistic role of cAMP and IP(3) in corticotropin-releasing hormone-induced cell shape changes in invertebrate immunocytes. Peptides 21: 175–182

Malagoli D, Gobba F, Ottaviani E (2003) Effects of 50-Hz magnetic fields on the signalling pathways of fMLP-induced shape changes in invertebrate immunocytes: The activation of an alternative "stress pathway". Biochim Biophys Acta 1620:185–190

Malagoli D, Ottaviani E (2005) Cytotoxicity as a marker of mussel health status. J Mar Biol Ass UK 85:359–362

Malfroy B, Swerts JP, Guyon A, Roques BP, Schwartz JC (1978) High affinity enkephalin degrading peptidase in brain is increased after morphine. Nature 276:523–526

Mandal L, Martinez-Agosto JA, Evans CJ, Hartenstein V, Banerjee U (2007) A Hedgehog- and Antennapedia-dependent niche maintains Drosophila haematopoietic precursors. Nature 446:320–324

Mattia A, Farmer SC, Takemori AE, Sultana M, Portoghese PS, Mosberg HI, Bowen WD, Porreca F (1992) Spinal opioid delta antinociception in the mouse mediation by a 5'-NTII-sensitive delta receptor subtype. J Pharmacol Exp Therap 260:518–525

Nagy L, Hiripi L (2002) Role of tyrosine, DOPA and decarboxylase enzymes in the synthesis of monoamines in the brain of the locust. Neurochem Int 41:9–16

Noguchi M, Nakamura Y, Russell SM, Ziegler SF, Tsang M, Cao X, Leonard WJ (1993) Interleukin 2 receptor gamma chain: A functional component of the interleukin-7 receptor. Science 262:1877–1880

Oka K, Kantrowitz JD, Spector S (1985) Isolation of morphine from toad skin. Proc Natl Acad Sci USA 82:1852–1854

Ottaviani E, Caselgrandi E, Bondi M, Cassarizza A, Monti D, Franceschi C (1991a) The "immune-mobile brain": Evolutionary evidence. Adv Neuroimmunol 1:27–39

Ottaviani E, Caselgrandi E, Franchini A (1988) Epinephrine investigation in the snail brain of *Helicella virgata* (Gastropoda, pulmonata). Comp Biochem Physiol C Toxicol Pharmacol 89:267–269

Ottaviani E, Cossarizza A, Ortolani C, Monti D, Franceschi C (1991b) ACTH-like molecules in gastropod molluscs: A possible role in ancestral immune response and stress. Proc Biol Sci R Soc 245:215–218

Ottaviani E, Franceschi C (1996) The neuroimmunology of stress from invertebrates to man. Progr Neurobiol 48:421–440
Ottaviani E, Franceschi C (1997) The invertebrate phagocytic immunocyte: Clues to a common evolution of immune and neuroendocrine systems. Immunol Today 18:169–174
Ottaviani E, Franchini A, Barbieri D, Kletsas D (1998a) Comparative and morphofunctional studies on *Mytilus galloprovincialis* hemocytes: Presence of two aging-related hemocytes stages. Ital J Zool 65:349–354
Ottaviani E, Franchini A, Franceschi C (1998b) Presence of immunoreactive corticotropin-releasing hormone and cortisol molecules in invertebrate haemocytes and lower and higher vertebrate thymus. Histochem J 30:61–67
Ottaviani E, Franchini A, Hanukoglu I (1998c) In situ localization of ACTH receptor-like mRNA in molluscan and human immunocytes. Cell Mol Life Sci 54:139–142
Ottaviani E, Malagoli D, Franceschi C (2007) Common evolutionary origin of the immune and neuroendocrine systems: Ten years from circumstantial evidences to in silico approaches. Trends Immunol 28:497–502
Ottaviani E, Malagoli D, Franchini A (2004) Invertebrate humoral factors: Cytokines as mediators of cell survival. Progr Mol Subcell Biol 34:1–25
Ottaviani E, Sassi D, Kletsas D (1997) PDGF- and TGF-beta-induced changes in cell shape of invertebrate immunocytes: Effect of calcium entry blockers. Eur J Cell Biol 74:336–341
Paemen LR, Porchet-Hennere E, Masson M, Leung MK, Hughes TK, Stefano GB (1992) Glial localization of interleukin-1a in invertebrate ganglia. Cell Mol Neurobiol 12:463–472
Pasternak GW (1986) Multiple μ opiate receptors: Biochemical and pharmacological evidence for multiplicity. Biochem Pharmacol 35:361–364
Pasternak GW (1988) The opiate receptors. Humana Press, New Jersey
Patey G, Rossier J (1986) Decouverte, anatomie et biosynthese des differentes familles de peptides opioides endogenes. Ann Endocrinologie (Paris) 47:71–87
Pert CB, Taylor D (1980) Type 1 and type 2 opiate receptors: A subclassification scheme based upon GTP's differential effects on binding. In: Way EL (ed) Endogenous and exogenous opiate agonists and antagonists. Pergamon Press, New York, pp 87–94
Prevot V, Rialas C, Croix D, Salzet M, Dupouy JP, Puolain P, Beauvillain JC, Stefano GB (1998) Morphine and anandamide coupling to nitric oxide stimulated GnRH and CRF release from rat median eminence: Neurovascular regulation. Brain Res 790:236–244
Przewlocki R, Machelska H, Przewlocka B (1993) Inhibition of nitric oxide synthase enhances morphine antinociception in the rat spinal cord. Life Sci 53:PL1–PL5
Renwrantz L (1990) Internal defense system of *Mytilus edulis*. In: Stefano GB (ed) Neurobiology of *Mytilus edulis*. Studies in Neuroscience. Manchester University Press, Manchester, pp 256–275
Rialas C, Bilfinger TV, Salzet M, Stefano GB (1998) Endomorphin 1 and 2 do not interact with the m3 opiate receptor subtype. Acta Pharmacol Sinica 19:403–407
Rojtinnakorn J, Hirono I, Itami T, Takahashi Y, Aoki T (2002) Gene expression in haemocytes of kuruma prawn, *Penaeus japonicus*, in response to infection with WSSV by EST approach. Fish Shellfish Immunol 13:69–83
Rothe H, Luschen W, Asken A, Willig A, Jaros PP (1991) Purified crustacean enkephalins inhibits release of hyperglycemic hormone in the crab, *Carcinus maenas* L. Comp Biochem Physiol C Toxicol Pharmacol 99:57–62
Rothman RB, Bykov V, DeCosta BR, Jacobson AE, Rice KC, Brady LS (1990) Interaction of endogenous opioid peptides and other drugs with four kappa binding sites in guinea pig brain. Peptides 11:311–331
Rozsa KS, Rubakhin SS, Szucs A, Hughes TK, Stefano GB (1997) Opposite effects of IL-2 and -4 on GABA-induced inward currents in dialyzed Lymnaea neurons. Gen Pharmacol 29:73–77
Russell SM, Keegan AD, Harada N, Nakamura Y, Noguchi M, Leland P, Friedmann MC, Miyajima A, Puri RK, Paul WE (1993) Interleukin-2 receptor gamma chain: A functional component of the interleukin-4 receptor. Science 262:1880–1883

Salzet M, Bulet P, Verger-Bocquet M, Malecha J (1995) Isolation and structural characterization of enkephalin-related peptides in the brain of the Rhynchobdellid leech *Theromyzon tessulatum*. FEBS Lett 357:187–191

Salzet M, Cocquerelle C, Verger-Bocquet M, Pryor SC, Rialas CM, Laurent V, Stefano GB (1997) Leech immunocytes contain proopiomelanocortin: Nitric oxide mediates hemolymph POMC processing. J Immunol 159:5400–5411

Salzet M, Stefano GB (1997a) Invertebrate proenkephalin: Delta opioid binding sites in leech ganglia and immunocytes. Brain Res 768:224–232

Salzet M Stefano GB (1997b) Prodynorphin in invertebrates. Mol Brain Res 52:46–52

Salzet M, Stefano GB (1998) Evidence for an invertebrate neuroendocrine system: Neuropeptide processing in leech-host communication. Res Trends 5:85–98

Salzet M, Verger-Bocquet M, Bulet P, Beauvillain JC, Malecha J (1996) Purification, sequence analysis and cellular localization of a prodynorphin-derived peptide related to the a neo-endorphin in the rhynchobdellid leech *Theromyzon tessulatum*. J Biol Chem 271:13191–13196

Sawada M, Hara N, Maeno T (1991) Ionic mechanism of the outward current induced by extracellular ejection of interleukin-1 onto identified neurons of Aplysia. Brain Res 545:248–256

Sawada M, Ichinose M, Stefano GB (1997) Nitric oxide inhibits the dopamine-induced K+ current via guanylate cyclase in Aplysia neurons. J Neurosci Res 50:450–456

Scadden DT (2006) The stem-cell niche as an entity of action. Nature 441:1075–1079

Scharrer B (1967) The neurosecretory neuron in neuroendocrine regulatory mechanisms. Am Zool 7:161–168

Scharrer B (1978) Peptidergic neuron: Facts and trends. Gen Comp Endocrinol 34:50–62

Scharrer B, Paemen LR, Smith EM, Hughes TK, Liu Y, Pope M, Stefano GB (1996) The presence and effects of mammalian signal molecules in immunocytes of the insect *Leucophaea madarae*. Cell Tissue Res 283:93–97

Seta N, Kuwana M (2007) Human circulating monocytes as multipotential progenitors. Keio J Med 56:41–47

Shipp MA, Stefano GB, D'Adamio L, Switzer SN, Howard FD, Sinisterra JI, Scharrer B, Reinherz E (1990) Downregulation of enkephalin-mediated inflammatory response by CD10/neutral endopeptidase 24.11. Nature 347:394–396

Smith EM (1992) Hormonal activities of cytokines. Chem Immunol 52:154–172

Sonetti D, Ottaviani E, Bianchi F, Rodriquez M, Stefano ML, Scharrer B, Stefano GB (1994) Microglia in invertebrate ganglia. Proc Natl Acad Sci USA 91:9180–9184

Sonetti D, Ottaviani E, Stefano GB (1997) Opiate signaling regulates microglia activities in the invertebrate nervous system. Gen Pharmacol 29:39–47

Sonetti D, Peruzzi E, Stefano GB (2005) Endogenous morphine and ACTH association in neural tissues. Med Sci Monit 11:MS22–MS30

Stefano GB (1982) Comparative aspects of opioid-dopamine interaction. Cell Mol Neurobiol 2:167–178

Stefano GB (1986) Comparative opioid and related neuropeptide mechanisms. CRC Press, Boca Raton, Florida

Stefano GB (1989) Role of opioid neuropeptides in immunoregulation. Progr Neurobiol 33: 149–159

Stefano GB (1990) Norepinephrine: Presence and interaction with endogenous biogenic amines. In: Stefano GB (ed) Neurobiology of *Mytilus edulis*. Manchester University Press, Manchester, pp 93–103

Stefano GB (1992) Invertebrate and vertebrate immune and nervous system signal molecule commonalities. Cell Mol Neurobiol 12:357–366

Stefano GB (1998) Autoimmunovascular regulation: Morphine and anandamide stimulated nitric oxide release. J Neuroimmunol 83:70–76

Stefano GB (1999) The μ3 opiate receptor subtype. Pain Forum 8:206–209

Stefano GB, Aiello E (1975) Histoflourescent localization of serotonin and dopamine in the nervous system and gill of *Mytilus edulis* (Bivalvia). Biol Bull 148:141–156

Stefano GB, Cadet P, Scharrer B (1989a) Stimulatory effects of opioid neuropeptides on locomotory activity and conformational changes in invertebrate and human immunocytes: Evidence for a subtype of delta receptor. Proc Natl Acad Sci USA 86:6307–6311

Stefano GB, Cadet P, Sinisterra JI, Charles R, Barnett JA, Kuruvilla S, Aiello E (1990) Functional neural anatomy of *Mytilus edulis*: Monoaminergic and opioid localization. In: Stefano GB (ed) Neurobiology of *Mytilus edulis*. Manchester University Press, Manchester, pp 38–56

Stefano GB, Cadet P, Sinisterra JI, Scharrer B (1991a) Comparative aspects of the response of human and invertebrate immunocytes to stimulation by opioid neuropeptides. In: Stefano GB, Florey E (eds) Comparative aspects of neuropeptide function. University of Manchester Press, Manchester, pp 329–334

Stefano GB, Catapane EJ (1979) Enkephalin increases dopamine levels in the CNS of a marine mollusc. Life Sci 24:1617–1622

Stefano GB, Catapanem EJ, Aiello E (1976) Dopaminergic agents: Influence on serotonin in the molluscan nervous system. Science 194:539–541

Stefano GB, Digenis A, Spector S, Leung MK, Bilfinger TV, Makman MH, Scharrer B, Abumrad NN (1993) Opiate-like substances in an invertebrate, an opiate receptor on invertebrate and human immunocytes, and a role in immunosuppression. Proc Natl Acad Sci USA 90: 11099–11103

Stefano GB, Goumon Y, Bilfinger TV, Welters I, Cadet P (2000a) Basal nitric oxide limits immune, nervous and cardiovascular excitation: Human endothelia express a mu opiate receptor. Progr Neurobiol 60:513–530

Stefano GB, Goumon Y, Casares F, Cadet P, Fricchione GL, Rialas C, Peter D, Sonetti D, Guarna M, Welters I, Bianchi E (2000b) Endogenous morphine. Trends Neurosci 9:436–442

Stefano GB, Hartman A, Bilfinger TV, Magazine HI, Liu Y, Casares F, Goligorsky MS (1995a) Presence of the μ3 opiate receptor in endothelial cells: Coupling to nitric oxide production and vasodilation. J Biol Chem 270:30290–30293

Stefano GB, Hiripim L, Catapane EJ (1978) The effect of short- and long-term temperature stress on serotonin, dopamine and neuroepinephrine metabolism in molluscan ganglia. J Thermal Biol 3:79–83

Stefano GB, Kahoud J, Hughes J (1999) Inhibition of microglial egress in excised ganglia by human interleukin 10: Implications for its activity in invertebrates. Acta Biol Hungarica 50:247–256

Stefano GB, Kream RM, Zukin RS (1980) Demonstration of stereospecific opiate binding in the nervous tissue of the marine mollusc *Mytilus edulis*. Brain Res 181:445–450

Stefano GB, Leung MK (1984) Presence of met-enkephalin-Arg6-Phe7 in molluscan neural tissues. Brain Res 298:362–365

Stefano GB, Leung MK, Bilfinger TV, Scharrer B (1995b) Effect of prolonged exposure to morphine on responsiveness of human and invertebrate immunocytes to stimulatory molecules. J Neuroimmunol 6:175–181

Stefano GB, Leung MK, Zhao X, Scharrer B (1989b) Evidence for the involvement of opioid neuropeptides in the adherence and migration of immunocompetent invertebrate hemocytes. Proc Natl Acad Sci USA 86:626–630

Stefano GB, Melchiorri P, Negri L, Hughes TK, Scharrer B (1992a) (D-Ala2)-Deltorphin I binding and pharmacological evidence for a special subtype of delta opioid receptor on human and invertebrate immune cells. Proc Natl Acad Sci USA 89:9316–9320

Stefano GB, Paemen LR, Hughes TK Jr (1992b) Autoimmunoregulation: Differential modulation of CD10/neutral endopeptidase 24.11 by tumor necrosis factor and neuropeptides. J Neuroimmunol 41:9–14

Stefano GB, Salzet B, Fricchione GL (1998a) Enkelytin and opioid peptide association in invertebrates and vertebrates: Immune activation and pain. Immunol Today 19:265–268

Stefano GB, Salzet B, Rialas CM, Pope M, Kustka A, Neenan K, Pryor SC, Salzet M (1997) Morphine and anandamide stimulated nitric oxide production inhibits presynaptic dopamine release. Brain Res 763:63–68

Stefano GB, Salzet M (1999) Invertebrate opioid precursors: Evolutionary conservation and the significance of enzymatic processing. Int Rev Cytol 187:261–286

Stefano GB, Salzet M, Bilfinger TV (1998b) Long-term exposure of human blood vessels to HIV gp120, morphine and anandamide increases endothelial adhesion of monocytes: Uncoupling of Nitric Oxide. J Cardiovasc Pharmacol 31:862–868

Stefano GB, Salzet M, Magazine HI, Bilfinger TV (1998c) Antagonist of LPS and IFN-g induction of iNOS in human saphenous vein endothelium by morphine and anandamide by nitric oxide inhibition of adenylate cyclase. J Cardiovasc Pharmacol 31:813–820

Stefano GB, Salzet-Raveillon B, Salzet M (1998d) *Mytilus edulis* hemocytes contains pro-opiomelanocortin: LPS and morphine stimulate differential processing. Molecular Brain Res 63:340–350

Stefano GB, Salzet-Raveillon B, Salzet M (1998e) *Mytilus edulis* hemolymph contain prodynorphin. Immunol Lett 63:33–39

Stefano GB, Scharrer B (1991) A possible immunoregulatory function for [Met]-enkephalin-Arg6-Phe7 involving human and invertebrate granulocytes. J Neuroimmunol 31:97

Stefano GB, Scharrer B (1994) Endogenous morphine and related opiates, a new class of chemical messengers. Adv Neuroimmunol 4:57–68

Stefano GB, Scharrer B (1996) The presence of the m3 opiate receptor in invertebrate neural tissues. Comp Biochem Physiol C Toxicol Pharmacol 113:369–373

Stefano GB, Scharrer B, Leung MK (1989c) Neurobiology of opioids in *Leucophaea maderae*. Cockroaches as models for neurobiology: Applications in biomedical research. CRC Press Inc, Boca Raton, Florida

Stefano GB, Scharrer B, Smith EM, Hughes TK, Magazine HI, Bilfinger TV, Hartman A, Fricchione GL, Liu Y, Makman MH (1996) Opioid and opiate immunoregulatory processes. Crit Rev Immunol 16:109–144

Stefano GB, Smith EM (1996) Adrenocorticotropin, a central trigger in immune responsiveness: Tonal inhibition of immune activation. Med Hypotheses 46:471–478

Stefano GB, Smith EM, Hughes TK (1991b) Opioid induction of immunoreactive interleukin-1 in *Mytilus edulis* and human immunocytes: An interleukin-1-like substance in invertebrate neural tissue. J Neuroimmunol 32:29–34

Stefano GB, Zhu W, Cadet P, Mantione K (2004) Morphine enhances nitric oxide release in the mammalian gastrointestinal tract via the m3 opiate receptor subtype: A hormonal role for endogenous morphine. J Physiol Pharmacol: Official J Polish Physiol Soc 55:279–288

Stefano GB, Zhu W, Cadet P, Mantione K, Bilfinger TV, Bianchi E, Guarna M (2002) A hormonal role for endogenous opiate alkaloids: Vascular tissues. Neuroendocrinol Lett 23:21–26

Stefano GB, Zukinm RS, Kream RM (1982) Evidence for the presynaptic localization of a high affinity opiate binding site on dopamine neurons in the pedal ganglia of *Mytilus edulis*. J Pharmacol Exp Therap 222:759–764

Strub JM, Goumon Y, Lugardon K, Capon C, Lopez M, Moniatte M, Van Dorsselaer A, Aunis D, Metz-Boutigue MH (1996) Antibacterial activity of glycosylated and phosphorylated chromogranin A-derived peptide 173–194 from bovine adrenal medullary chromaffin granules. J Biol Chem 271:28533–28540

Szucs A, Stefano GB, Hughes TK, Rozsa KS (1992a) Modulation of voltage-activated ion currents on identified neurons of *Helix pomatia* L. by interleukin-1. Cell Mol Neurobiol 12: 429–438

Szucs A, Stefano GB, Hughes TK, Rozsa KS (1992b) Modulation of voltage-activated ion currents on identified neurons of Helix pomatia L. by interleukin-1. Cell Mol Neurobiol 12:429–438

Taga T, Kishimoto T (1992) Cytokine receptors and signal transduction. FASEB J 6:3387–3396

Traynor JR, Elliott J (1993) d-Opioid receptor subtypes and cross-talk with μ-receptors. Trends Pharmacol Sci 14:84–86

Turner A, Leung MK, Stefano GB (1994) Peptidases of significance in neuroimmunoregulation. In: Scharrer B, Smith EM, Stefano GB (eds) Neuropeptides in neuroimmunology. Springer-Verlag, Berlin, pp 152–169

Udenfriend S, Kilpatrick DL (1984) Proenkephalin and the products of its processing: Chemistry and biology. In: Udenfriend S, Meienhofer J (eds) The Peptides, vol 6. Academic Press, San Diego, pp 26–69

Udenfriend S, Meienhofer J (1984) Opioid peptides: Biology, chemistry and genetics. Academic Press, San Diego

Vehovszky A, Salanki J (1983) Pharmacological characterization of postsynaptic potentials evoked in the bimodal pacemaker neuron of *Helix pomatia* L. Acta Physiol Hungarica 62:35–46

Ventura C, Maioli M (2000) Opioid peptide gene expression primes cardiogenesis in embryonal pluripotent stem cells. Circulation Res 87:189–194

Yasuda K, Raynor K, Kong H, Breder CD, Takeda J, Reisine T, Bell GI (1993) Cloning and functional comparison of κ and δ-opioid receptors from mouse brain. Proc Natl Acad Sci USA 90:6736–6740

Zhu W, Baggerman G, Goumon Y, Casares F, Brownawell B, Stefano GB (2001) Presence of morphine and morphine-6-glucuronide in the marine mollusk *Mytilus edulis* ganglia determined by GC/MS and Q-TOF-MS. Starvation increases opiate alkaloid levels. Mol Brain Res 88:155–160

Zhu W, Cadet P, Baggerman G, Mantione KJ, Stefano GB (2005a) Human white blood cells synthesize morphine: CYP2D6 modulation. J Immunol 175:7357–7362

Zhu W, Ma Y, Cadet P, Yu D, Bilfinger TV, Bianchi E, Stefano GB (2003) Presence of reticuline in rat brain: A pathway for morphine biosynthesis. Mol Brain Res 117:83–90

Zhu W, Ma Y, Stefano GB (2002) Presence of isoquinoline alkaloids in molluscan ganglia. Neuroendocrinol Lett 23:329–334

Zhu W, Mantionem KJ, Shen L, Cadet P, Esch T, Goumon Y, Bianchi E, Sonetti D, Stefano GB (2005b) Tyrosine and tyramine increase endogenous ganglionic morphine and dopamine levels in vitro and in vivo: CYP2D6 and tyrosine hydroxylase modulation demonstrates a dopamine coupling. Med Sci Monitor 11:BR397-BR404

Zhu W, Mantione KJ, Shen L, Stefano GB (2005c) In vivo and in vitro L-DOPA exposure increases ganglionic morphine levels. Med Sci Monit 11:MS1-MS5

Zhu W, Stefano GB (2004) Reticuline exposure to invertebrate ganglia increases endogenous morphine levels. Neuroendocrinol Lett 25:323–330

Chapter 7
Regenerating Echinoderms: A Promise to Understand Stem Cells Potential

M. Daniela Candia-Carnevali, Michael C. Thorndyke, and Valeria Matranga

Abstract The potential for regenerating tissues, organs and body parts, even the ability to reconstruct virtually a complete animal from a body fragment, is expressed to a maximum extent in echinoderms which provide fantastic and tractable models for the study of regeneration. Regenerative processes are common in all classes of the phylum, even though specific capabilities differ remarkably between the classes, depending on individual potential of morphogenetic and histogenetic plasticity at tissue and most of all at cellular levels. These phenomena, particularly in adults, imply the existence of stem cells which can be present in the circulating fluids or in the tissues in the form of resident cells, ready to be recruited in the repair and regenerative processes that follow traumatic or self-induced damage. In spite of the impressive effectiveness of their regenerative processes, only a few model systems for the study of regeneration have been developed in echinoderms, each model being unique for its specificity and versatility, and useful for unravelling peculiar aspects of the phenomenon. In addition, larvae of all classes display a unique capacity for rapid regeneration regardless of their developmental stage, showing an unexpected plasticity in terms of processes and mechanisms closely related to events of asexual reproduction and cloning. On the basis of their regenerative potential echinoderms can provide a broad range of valuable new deuterostome models for the study of regeneration genetics, with potential applications in vertebrate regeneration. Since the complexity of the echinoderm genome, as exemplified by the sea urchin genome project, indicates that echinoderms share at least 70% of their proteins with mankind, we shall consider how this provides an important tool kit to aid our understanding of the phenomenon as well as support the development of realistic methods to pursue tissue and organ regeneration in humans.

Keywords Regenerative Potential · Differentiation · Plasticity · Stemness · Progenitor cells

---

M.D. Candia-Carnevali (✉)
Dipartimento di Biologia, Università degli Studi di Milano, Via Celoria 26, 20133 Milano, Italy
e-mail: daniela.candia@unimi.it

B. Rinkevich, V. Matranga (eds.), *Stem Cells in Marine Organisms*,
DOI 10.1007/978-90-481-2767-2_7, © Springer Science+Business Media B.V. 2009

7.1 Introduction

The potential for repairing and replacing cells, tissues, organs and body parts, commonly called regeneration potential, is considered a primitive attribute of life shared by all the organisms, even though expressed to a different extent, and essential for survival of both individual and whole species. Speaking of animals, there is a remarkable variability in terms of post-traumatic response and capabilities of morphological and functional recovery, not only between unrelated taxa, but also between closely related species, and even between organs and parts of the same individual (Candia-Carnevali and Bonasoro 2001a; Carlson 1997; Sánchez-Alvarado and Tsonis 2006). In addition, the regenerative potential can vary a lot according to the life stage (embryonic, larval, adult) and the age of the individual, being higher in larval tissues and organs in comparison with adults and in many cases expressed only in the early embryonic stages (mammals).

It is now clear that, in contrast to old traditional views, regeneration is not a unique prerogative of simple and primitive animals, but is a common and widespread phenomenon through phylogeny, its quite heterogeneous distribution from the lowest to the highest phyla being independent of their organization and complexity level (Candia-Carnevali and Bonasoro 2001a; Ferretti and Géraudieu 1997; Thouveny and Tassava 1997). Actually, the regenerative capabilities appear to depend upon the individual potential for histogenetic and morphogenetic plasticity expressed in terms of recruitment of adult pluripotential cells (true stem cells and/or dedifferentiated reprogrammed cells), cell potential for proliferation and migration, supply of specific regulatory/trophic factors, and finally expression or re-expression of specific parts of the developmental program in adult organisms (Goss 1992). Despite evident species-specific differences, reserve progenitor cells appear to be present in all organisms, and differentiated tissues (such as liver or muscles) can be replaced *de novo* even in adult mammals, at least to a certain degree. The role of the pluripotential cells is not restricted to the embryonic development, but is expressed in the whole life cycle in the processes of tissue homeostasis and, most of all, of tissue regeneration leading to the recover/replacement of damaged body organs, till the reformation of entire individuals (cloning). The reconstruction of complex body parts requires the coordinated activation of complicate genetic cascades (networks), involving different cell types which must interpret complex arrays of signals and then translate these instructions into spatially and temporally appropriate morphogenetic responses. In this view, regeneration can not be considered an accelerated version of embryogenesis, but rather a well distinct and more complicate process, occurring in a complex organism, at embryonic, larval or adult stage, in the context of established and differentiated tissues.

The crucial point of the "progenitor" elements involved in regenerative processes, in terms of cell recruitment, sources and fate, and totipotentiality, pluripotentiality or unipotentiality, is a central problem, particularly relevant for its topical interest. The identification of the cellular and molecular events induced by wounding and repair in regenerating animals can significantly help to understand the conditions and factors that allow developmental fields to be established *de novo* in terminally differentiated tissues.

7.2 Fundamental Aspects

Regeneration is a common phenomenon in all the animals, but for the reasons mentioned above, the study of the regeneration-competent cells can not be standardized, but requires an appropriate study and a differential approach according to the organism, its life stage and age, and to the type of response (traumatic or self-induced), the extent of the process (repair, regrowth, cloning) and finally the specific conditions, intrinsic (physiological) and extrinsic (environmental) in which it takes place. In regeneration an anatomically defined part of the organism, small or large, is reformed after its loss or severe injury and the new cells, starting from somatic cells, develop in an established context of mature tissues and differentiated cells in individuals (adult or larval) well characterized anatomically and functionally. From invertebrates to humans the main questions concerning regeneration are always related to fundamental cellular and molecular aspects, particularly in terms of: (1) responsible cells, specifically which are the origin and the differentiation or dedifferentiation/redifferentiation pathway of those cells employed in the repair and reconstruction of new structures; (2) involved factors, namely which factors induce activation of "dormant" populations of adult stem cells or reversal/re-entry of differentiated cells into the cell cycle, and, importantly, which factors regulate end of growth and beginning of differentiation or redifferentiation at the right place and time ensuring the re-establishment of structure and function; (3) involved genes, those which are activated and/or reactivated upon regeneration and their networks (Carlson 1997). In spite of the wide choice of potential models for studying regeneration, it has been explored in detail only in a few animals and there are still wide surprising gaps in our knowledge of the basic regenerative/developmental mechanisms, in particular in relation to responsible cells and molecules involved. On the basis of these premises, the problem of stem cell biology in regenerative processes, which obviously appears to be of topical impact for its possible broad implications in both basic and applied research, urgently deserve to be extensively and comparatively studied in appropriate experimental models not so commonly employed but potentially very amenable for this approach. In fact they can offer useful alternatives to traditional embryonic models for exploring the specific problem of the "progenitor cells", their potential of stemness and clonogenicity, and the specific factors involved in regulating their activation, proliferation and differentiation plasticity.

7.3 Regeneration in Echinoderms

The regenerative potential finds its maximum expression in echinoderms which provide fantastic and tractable experimental models for the study of regeneration. Regenerative processes are common in all classes, even though specific capabilities differ between the groups, going from localized processes of tissue repair to replacement of lost body parts/organs (arms, viscera) and even to complete re-growth of whole individuals from body fragments. The extent of these phenomena contradict Goss's paradigm about the "non-regenerability" of the truly vital structures

(Goss 1965, 1969), since vital organs such as the whole visceral mass can be easily and completely regenerated after evisceration. Through the phylum, regeneration is largely a predictable event and in most cases follows autotomic self-induced mutilations (see for a review Wilkie 2001, 2005). The impressive effectiveness of these processes, which can be easily reproduced in the laboratory, mimicking the natural conditions, has allowed development of a few established model systems for the study of regeneration.

The regenerative processes of arms or viscera in crinoids and holothuroids can provide valuable and flexible experimental models for studying regeneration from the whole organism level up to the tissue, cellular and molecular level, focussing on the plasticity of the regeneration-competent cells and the possible factors, particularly neurotrophic, involved in regulating their recruitment, proliferation, differentiation and de-differentiation (Garcia-Arraras and Greenberg 2001; Mashanov and Dolmatov 2004; Odintsova et al. 2005; Suárez-Castillo et al. 2004). In particular, the integrated comparative analysis carried out in crinoids by means of parallel *in vitro* and *in vivo* experimental approaches throws light on still unexplored cellular aspects, showing an unexpected plasticity. Ophiuroids and asteroids have been also frequently used to investigate adult growth and differentiation rates and to identify stem-cell niches during regeneration (Mladenov et al. 1989; Moss et al. 1998; Thorndyke et al. 2001). In both groups studies on regeneration following accidental loss, autotomy or experimentally-induced trauma led to the understanding of the first repair phases, cell recruitment areas and involvement of presumptive stem cells. On occasions, autotomy can be envisaged as a form of asexual reproduction, given a non permissive environment. Although regeneration is not a major occurrence in echinoids, regeneration of their spines, pedicellariae and even test has been described allowing comparison of the mechanisms and processes operating, especially at the cellular level (Bonasoro et al. 2004; Dubois and Ameye 2001). A major focus has been addressed to the early regeneration stages following the wound healing process, with special emphasis on the immune cells involved (Matranga et al. 2005; Ottaviani and Franceschi 1997). As in all the other animals, regeneration in echinoderms must be regarded as a post-embryonic developmental process: repair and re-growth processes can involve in fact both the adult and the larval stages (Eaves and Palmer 2003), where they show an unexpected plasticity and are closely related to spectacular events of asexual reproduction and cloning.

Thus the cellular aspects of regeneration deserve to be explored in both adults and larvae following as much as possible complementary approaches. The regenerative process, in any case, implies the existence of stem cells, present in the circulating fluids or in the tissues in the form of resident cells, ready to be recruited in the repair and regenerative processes that follow traumatic or self-induced damage. A thorough analysis of the regenerative phenomena can hopefully contribute (1) to provide a new cellular insight of the striking regenerative potential of echinoderms; (2) to unravel the crucial problems of the expression/suppression of the regenerative potential in the animal phylogeny. The stimulation of regeneration of new tissues from old tissues *in vivo* constitutes the emerging and exciting interdisciplinary field of "regenerative biology" (Pearson 2001; Stocum 2001) whose wide potential of expansion is strictly dependent on the knowledge of the fundamental mechanisms at the cellular and molecular levels in the animal models endowed with remarkable regenerative capacities.

The experimental use of echinoderms has two additional relevant advantages: (1) they are animal models amenable to experimental manipulation which do not present any ethical or normative problems; (2) the close phylogenetic relationship and the established gene conservation between echinoderms and vertebrates (the sea urchin genome, recently sequenced, shows 70% homologies with human genome - see Science 2006), allows not only the successful employment of protocols and molecules commonly used in vertebrate models, including mammals, but implies also the potential for applicability and transferability to biomedicine.

7.3.1 Experimental Models

This section provides a comparative account of stem cell regenerative potential in echinoderms by briefly reviewing analogous regeneration events in established models which are rather different in terms of both macroscopic processes and range of histogenetic potential and differentiation plasticity of the responsible cells. Since each model is unique for its specificity and versatility, each one can be useful for contributing data to the puzzle of stem cell biology, so incomplete as far as marine invertebrates are concerned, and providing new cellular insights into the echinoderm "recipe" for a successful regeneration. Several echinoderm models have been employed to analyse the molecular and cellular bases of adult regeneration. These models are used for different purposes. The first model is the sea urchin, whose whole genome has been recently released. The sea urchin enables us to analyse echinoderm gene structure and organisation (in other words it is possible to assess the degree of conservation of echinoderm gene sequences). However, this group has limited regeneration capabilities. Regeneration of their appendices (spines and pedicellariae, Dubois and Ameye 2001) and of test portions (Bonasoro et al. 2004) have been described allowing comparison of the mechanisms and processes operating, especially at the cellular level. After an initial wound-healing phase, regeneration takes place through the recruitment of progenitor cells (undifferentiated or differentiated, depending on the case) accumulating in the lesion region. In brief, authors propose that regeneration can proceed through two different processes, morphallactic or epimorphic.

The second and very tractable experimental model is the brittle star. Brittle stars are very abundant and because marine predators eat brittle star arms regularly, brittle stars use most of their energy to regenerate predated arms within a couple of weeks. This regeneration time frame is appropriate for laboratory experimentation because it allows the rapid collection of arms at key stages of the regeneration process (e.g. stages where cell proliferation, differentiation, functional recovery of the new tissues occur). In some ophiuroids, for example, there is an apparent trade-off between growth and differentiation according the amount of tissue lost (Dupont and Thorndyke 2006). In this way there is a clear plasticity in the regulation of regeneration that allows a switch between differentiation or cell proliferation that is determined by local signals. In particular, it is also clear that in brittlestars the radial nerve cord plays a key-role in regeneration and can undergo substantial cell proliferation (Thorndyke et al. 2001) raising the possibility of new adult neurons arising from a resident stem cell population in the adult nerve cord. With the cloning of the BMP proteins we now have the ability of study this phenomenon in the brittle star at the gene level for the first time.

A third model is the common starfish. Starfish regenerate at a much slower rate than brittle stars, but they are larger and so allow dissection of the nerve cords and easy extraction of the coelomic cells thought to be involved in regeneration. Recently, in the attempt to identify the first intervening mechanisms taking place after experimentally-induced autotomy, *Asterias rubens* sea stars were analysed looking for putative stem cells and dedifferentiation process. A time-dependent increase in the number of circulating cells as well as a iper-reactivity of the coelomic epithelium with coelomocyte specific monoclonal antibodies was demonstrated (Pinsino et al. 2007; Oweson et al. 2008. The results obtained so far reinforced the notion that coelomocytes proliferate in the coelomic epithelium, as reported by earlier studies (Bossche and Jangoux 1976) and confirmed results obtained in crinoids by BrdU incorporation (see below, Candia-Carnevali et al. 1995; Candia-Carnevali and Bonasoro 2001b). Because the coelomic and nervous systems are implicated in regeneration, these are the focus for an extensive analysis of gene networks, known in other animals to play a role in one or both systems as well as in embryogenesis, stem cell biology and cell lineage regulation.

In any case, since crinoids are the best explored echinoderm model in terms of overall regeneration mechanisms (see for a review Candia-Carnevali 2006; Candia-Carnevali and Bonasoro 2001b; Thorndyke and Candia-Carnevali 2001), they can be considered the most appropriate and representative echinoderm model for a cellular approach specifically addressed to explore the problem of the regeneration-competent cells and their stemness, in terms of origin and derivation (stem cells or dedifferentiated cells), activities (proliferation and/or migration), plasticity and pluripotential/unipotential for differentiation (derived cellular phenotypes).

7.3.2 Crinoids

Among echinoderms crinoids are well known for their spectacular regenerative potential, well documented in the adult. In all crinoids regenerative processes are quite common and are utilized to replace the loss of internal and external organs. Feather stars extensively employ regeneration to reconstruct both external appendices (arms, pinnules and cirri) and internal organs (digestive apparatus, gonads, entire visceral mass), which can be frequently lost following traumatic injury, predation or spontaneous autotomy (Perrier 1873; Minckert 1905; Reichensperger 1912). On the whole, regeneration appears to be a quick and effective process when the mutilation involves a vital organ or the injury is particularly traumatic. In contrast, it seems to be a slower and less effective phenomenon when the part involved is not so indispensable for survival (Reichensperger 1912). Interestingly, there is no apparent correlation between availability/assimilation of food and regenerative capabilities which seem to be comparably rapid and effective in both well-fed and starved animals (Candia-Carnevali and Bonasoro 2001b). Although less documented, regeneration events are present and relevant also in the larvae (Runnström 1915, 1925; Barbaglio et al. in preparation).

7.3.2.1 The Arm Experimental Model

Arm regeneration in *Antedon mediterranea* is the best established experimental model, successfully employed in both old classical studies (Minckert 1905; Perrier 1873; Reichensperger 1912) and recent comprehensive studies (Candia-Carnevali et al. 1993, 1995, 1996, 1998; Candia-Carnevali and Bonasoro 2001a,b; Candia-Carnevali 2005, 2006; Patruno et al. 2002, 2003; Thorndyke and Candia-Carnevali 2001) for exploring regeneration in all its aspects from the macroscopic to the molecular level.

These regenerative processes can be easily reproduced in the lab mimicking the autotomy conditions and amputating the arms at the level of the autotomy plane (sutures). Serial experiments of arm regeneration at different stages were carried out following pseudo-autotomic amputations or after traumatic mutilations, in standard environment or in the presence of specific compounds. A well established experimental strategy was to use in parallel normal regenerating arms and their respective amputated arm segments (*explants*) maintained in living conditions. Different types of isolated explants have been employed to test their capabilities to undergo extensive repair and regenerative processes in parallel with their donor arms. Pioneering experiments were addressed to understand both mechanisms and pattern of the regenerative processes under standard conditions (Candia-Carnevali et al. 1993, 1995, 1997). Comparison between the regenerative processes of standard regenerating arms and their respective *explants* (Candia-Carnevali and Bonasoro 2001b; Candia-Carnevali et al. 1998) was useful for testing the arm regenerative potential in terms of autonomy of resources and control, and highlights that, beside general similarities in the basic regenerative processes, there are some meaningful differences in terms of mechanisms employed and cellular/tissue elements involved. The regenerative potential, mechanisms and pattern have been also explored and compared with regard to aberrant regenerations resulting from arms deliberately subjected to traumatic mutilations which do not reproduce autotomy (Candia-Carnevali and Bonasoro 2001b). Finally, the regenerative response was also employed in applied research as a new valuable model for ecotoxicological studies addressed to the effects of exposure to specific classes of environmental contaminants (EDs, *endocrine disrupters*- Candia-Carnevali 2005; Sugni et al. 2007) well-known for their dramatic effects on developmental and reproductive processes. In fact, the regenerative phenomena, which reproduce developmental processes in adult animals, are regulated by endocrine and neurohumoral mechanisms and are

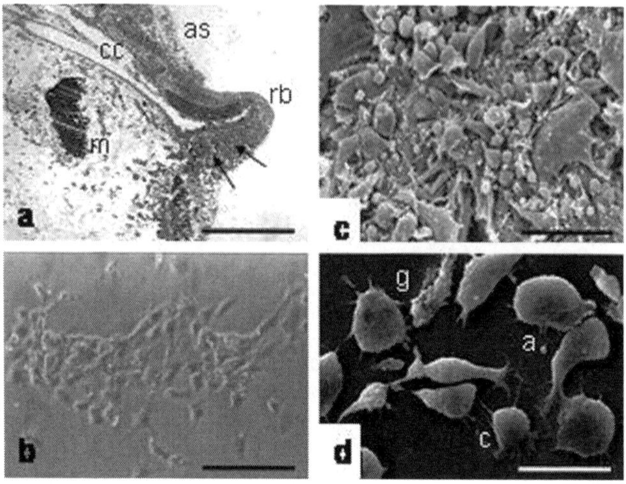

Fig. 7.1 Progenitor cells in crinoids and their contribution to regenerative processes: *in vivo* and *in vitro* micrographs. (**a**) Light microscopic view of *in vivo* sample. Regenerating arm of *A. mediterranea* in sagittal histological section at 72 h post amputation. An evident regenerative blastema (rb) is regrowing on the amputation surface (as). The coelomic canals (cc) are involved in active phenomena of cell migration and proliferation providing new cells (*arrows*) to the blastema. *Bar*, 50 μm; (**b**) Scanning electron microscopic view of *in vivo* sample. Amputation surface of *A. mediterranea* at 24 h post amputation. The wound is covered by a cycatricial layer formed by migratory cells mixed to fibrous material and cellular debris. A number of small roundish cells (coelomocytes) and elongated ameboid elements (amoebocytes) are recognizable. *Bar*, 25 μm; (**c**) Contrast-phase microscopic view of *in vitro* sample. Primary cell culture from *A. mediterranea* arm explants at 4 days of culture. The cells are actively migrating and proliferating. The different cytotypes can not be discriminated by this technique. *Bar*, 20 μm; (**d**) Scanning electron microscopic view of *in vitro* sample. Primary cell culture from *A. mediterranea* arm explants at 4 days of culture. Different cytotypes can be recognized: coelomocytes (c), amoebocytes (a), granulocytes (g). *Bar*, 10 μm

very sensitive to the presence of chemicals in the environment, particularly to persistent organic micropollutants with pseudo-hormonal activity (Candia-Carnevali. 2005).

The arm regeneration process turned out to be a typical blastemal regeneration in which the new structures develop from a recognizable blastemal bud (Fig. 7.1a). The main responsible elements for both repair and regenerative processes are migratory progenitor cells, plausibly pluripotent, actively proliferating in the presence of presumptive regulatory factors (Fig. 7.1b,c,d). However, all the phases of arm regeneration are clearly implying the prompt intervention of more than one type of migratory cells, which appear to be differently involved in migration, proliferation and differentiation processes. Therefore, priority problems to answer are those related to: (i) the characterization of the regeneration-competent cytotypes and their plasticity potential, and (ii) the identification of the putative factors (neurotrophic or non-neurotrophic) involved in regulating their recruitment, proliferation, differentiation and de-differentiation.

7.3.2.2 The Visceral Experimental Model

In all crinoids, in spite of the apparent complexity of the organs and tissues involved, the loss of the visceral mass can be easily repaired by prompt regeneration (Clark 1921; Hyman 1955). Current research (Dolmatov et al. 2003; Mozzi et al. 2004, 2006) is actually focusing on the overall process of visceral regeneration (involving gut and associated tissues and organs) in *A. mediterranea*, a phenomenon which was explored in the past in the historical study by Dendy (1886).

Experiments of complete regeneration of viscera were carried out in serial experiments of regeneration at different stages. The basic mechanisms and pattern of the regenerative processes in standard conditions have been established in experimental regenerations of different stages following evisceration (Mozzi et al. 2006), and also compared in preliminary transplantation experiments. The preliminary results collected so far show that visceral regeneration is a very rapid and effective process during which a small gut, functionally and anatomically complete, is reformed *de novo* in a loose context of new tissues. In terms of responsible cellular elements, the migratory cells usually employed in regeneration (see arm regeneration) are involved. In addition, recent experiments on transplantation of tegmen and related viscera between different individuals also show an unexpected plasticity and adaptability of tissues and organs in response to particularly critical and traumatic conditions (Mozzi et al. 2004).

7.3.2.3 Larval Regeneration

There are only a few available data related to processes of partial regeneration of body parts following traumatic amputations in larval stages (Runnström 1915, 1925). Recent preliminary results clearly indicate that also in this group the larval regenerative potential is relevant (Barbaglio et al. in preparation): regeneration of larval body parts post-traumatic amputations was preliminary explored in both mobile and sessile stages and provided indication on the potential of cellular plasticity and clonal reproduction in crinoid post-embryonic stages.

7.3.3 Experimental Approaches

In the crinoid regeneration model (arm or visceral), the problem of the progenitor cells, their regenerative plasticity, and the specific factors involved in regulating their activation, proliferation, differentiation, or in reprogramming their differentiation, appears to be a crucial node which deserves to be explored by both an *in vivo* and *in vitro* approach. The main cellular aspects of the regenerative processes were therefore approached by parallel *in vivo* and *in vitro* studies by employing standard methods of basic and advanced microscopy (optical and confocal microscopy, transmission and scanning electron microscopy, immunocytochemistry, immunohistochemistry), specific methods for monitoring cell proliferation, basic techniques of biochemistry and molecular biology, specific and original protocols for cell cultures of marine invertebrates. The comparison of results derived from different approaches in the same animal model was useful for verifying mechanisms and determining specificity of roles and implications of regeneration-competent cells.

7.3.3.1 *In vivo* Experiments

The mechanisms of regeneration, the histogenetic potential of regeneration-competent cells, their times and modalities of recruitment and their relative differential contribution in the regenerative processes were investigated *in vivo* by employing different regenerating systems. The obtained data were derived from an integrated approach which utilizes experimentally induced regenerations (standard or abnormal) obtained in significantly different experimental conditions, including extreme mutilations or transplantations.

The *in vivo* experiments were very successful in terms of cellular approach. The following goals were obtained: (1) the different cytotypes were characterized (Figs. 7.1c and 7.2a,c,e,g,i); (2) their main activities and capabilities were comparatively explored and defined (migration, proliferation, potential for differentiation); (3) specific neurothrophic and non-neurothrophic factors were identified; (4) specific genes involved were identified, with particular reference to a gene well known for its role in controlling cellular regenerative plasticity in other animals (BMP2/4).

7.3.3.2 *In vitro* Experiments

The employment of the crinoid experimental model turned out to be a successful strategy and provided a significant improvement also in terms of an *in vitro* approach (Parma et al. 2006). The experimental method employing cell cultures of marine invertebrate models such as echinoderms should be considered an appropriate and potentially powerful tool to address the wide problem of the biology of stem cells. However, until now this approach encountered severe limits in echinoderms due to actual difficulties to obtain long-term cultures, so the positive results obtained so far are particularly relevant. The following goals were obtained: (1) mixed long-term cell cultures from *explants* and regenerating arms were developed and stabilized in optimized experimental conditions determined for the first time (Fig. 7.1b); (2) the different cell types were characterized and compared to the *in vivo* phenotypes (Figs. 7.1d and 7.2b,d,f,h,j); (3) the growing potential and the properties of the cultured cells were explored and their main activities and capabilities defined (migration, proliferation, potential for differentiation). Preliminary but not completely satisfactory results were also obtained by employing visceral or larval tissues as primary source for cell cultures.

7.3.4 Regeneration-Competent Cells

According to recent results (for a review see Candia-Carnevali and Bonasoro 2001b; Candia-Carnevali 2006; Mozzi et al. 2006; Parma et al. 2006), in crinoid echinoderms at least four different cell types (Figs. 7.1c,d and 7.2a–h), namely

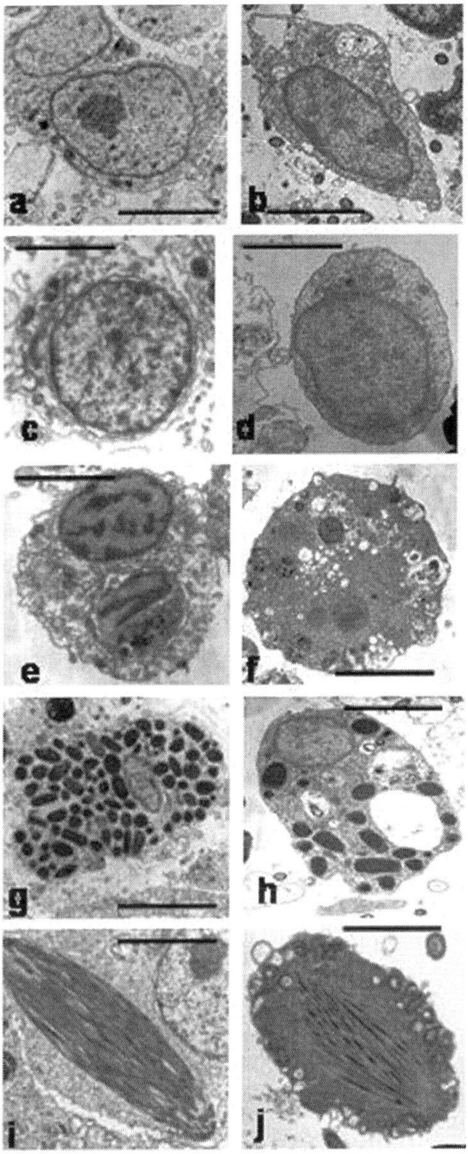

Fig. 7.2
Regeneration-competent cells in crinoids: Trasmission electron micrographs of the involved cytotypes compared *in vivo* (*left column*) and *in vitro* (*right column*). The cells present in the cultures correspond to those recruited during regeneration as shown by the histological specimens, and display the same ultrastructural distinctive features. (**a**) Amoebocyte. *In vivo* view. *Bar*, 1.5 μm; (**b**) Amoebocyte. *In vitro* view. *Bar*, 1.5 μm; (**c**) Coelomocyte. *In vivo* view. *Bar*, 1 μm; (**d**) Coelomocyte. *In vitro* view. *Bar*, 1 μm; (**e**) Phagocyte. *In vivo* view. *Bar*, 2 μm; (**f**) Phagocyte. *In vitro* view. *Bar*, 4 μm; (**g**) Granulocyte. *In vivo* view. *Bar*, 6 μm; (**h**) Granulocyte. *In vitro* view. *Bar*, 3 μm; (**i**) Dedifferentiating myocyte. *In vivo* view. *Bar*, 1.5 μm; (**j**) Dedifferentiating myocyte. *In vitro* view. *Bar*, 1.5 μm

amoebocytes, celomocytes, phagocytes and granulocytes, are involved in arm or visceral regeneration, according to the traditional terminology proposed by Reichensperger (1912). In addition there is evidence that, according to the specific model and experimental conditions, terminally differentiated cells can be recruited (Fig. 7.2i,j) and partially or completely re-programmed, thus re-differentiating and contributing to the formation of new tissues. These different cells can be well defined and characterized by their morphology, ultrastructure and immunocytochemistry and were shown to give a differential contribution according to the different phases (repair phase, early regenerative phase, advanced regenerative phase).

7.3.4.1 Migratory Cells

Some cytotypes can be clearly correlated to specific roles, whereas other elements potentially represent true adult stem cells and it is of fundamental importance to understand if one or more cell types exhibit at least two fundamental stem cell properties: self renewal and totipotentiality or pluripotentiality. The bulk of the results obtained so far in crinoids show that, in terms of specific cellular contribution, most of the migratory cell types involved in regeneration are morphologically undifferentiated elements which were previously identified as amoebocytes and coelomocytes (Fig. 7.2a–d). This is confirmed by the parallel analysis of the *explants*, where these cytotypes are involved in effective migration and proliferation activities, evident in both directions, proximal and distal, according to the phase. In particular, during the repair phase, they are typically involved in bidirectional phenomena, whereas, during the regenerative phase, and in accord with what happens for the blastemal growth, they occur preferentially in the proximal-distal direction (Candia-Carnevali et al. 1998).

Amoebocytes (Fig. 7.2a,b) are resident stem cells (RSC) which are stored in the arm around the brachial nerve in form of aggregates of satellite cells and able to be activated/recruited in response to injury signals and to migrate actively through the tissues (Candia-Carnevali 2006; Candia-Carnevali and Bonasoro 2001b). RSC usually appear to be small cells *in vivo*, diplaying an elongated shape and characterized by a large nucleolate euchromatic nucleus surrounded by scarce granular cytoplasm. They are employed during the overall regeneration period, even though their contribution appear to be especially essential during the repair and the early regenerative phase. In addition, they are endowed with ameboid movements and migrate along the brachial nerve towards the amputation area. Following RSC fate upon arm regeneration, available evidence suggests that, at least *in vivo*, after their migration towards the wound area, they undergo an extensive local proliferation. Thus RSC give a fundamental contribution first to the formation of the cicatricial layer (Fig. 7.1c) and then to the blastema growth. In fact, they appear to be the progenitor cells of the blastemal cells and consequently of all the blastema derived differentiated cells (Candia-Carnevali et. al. 1995, 1997, 1998). The blastema growth is therefore the result of the migratory/proliferative activity of these presumptive stem cells (Fig. 7.1a). When observed *in vitro*, RSC morphology confirms what seen *in vivo*: in particular, their elongated shape evidenced by SEM images (Fig. 7.1b,d) and the presence of many dynamic phyllopodia spreading around.

Coelomocytes (Fig. 7.2c,d) represent circulating stem cells (CSC), freely moving in the coelomic fluids and produced by the proliferating coelomic epithelium (coelothelium). Apart from their typical roundish shape, they are morphologically comparable to RSC. CSC do not represent reserve stem cells already present in the tissues, but are rather pluripotential elements produced by the continuous turnover of the coelothelium (Candia-Carnevali et al. 1995, 1997, 1998). Since they derive directly from coelothelial proliferation, this implies that the coelothelial cells (peritoneocytes) are plausibly partially involved in both dedifferentiation or transdifferentiation phenomena. Also CSC contribute to all the regeneration process phases. They migrate massively along the coelomic canals towards the amputation site during the overall regeneration period. Their contribution to regenerating tissues seems to be restricted to the coelomic compartments. In fact, once migrated into the wound area, they form first typical clots which seal the injured coelomic canals and later give rise to the re-growth of the coelomic components. The latter develop in parallel, but apparently independently, to the blastema itself. So CSC can be considered the progenitor cells of all the coelom-derived differentiated cells, including perytoneocytes, myocytes, neurons of the plexus and free coelomocytes, although a possible extra-coelomic contribution to tissue re-growth can not be excluded. *In vitro*, CSC show a roundish rather regular shape (Fig. 7.1c,d) and are often involved in proliferation phenomena. It is relevant that in the long-term cultures the majority of the cells belong to CSC cytotype. A relevant question is whether RSC and CSC actually represent two distinct types of progenitor cells, or they should be simply regarded as two morphotypes of the same pluripotent progenitor cell type. The origin of the RSC is possibly coelomic, as that of the CSC, originating from the proliferation activity of coelothelial components at the central axial organ level. In any case, the re-growth processes are supported by extensive cell cycle activity related to these two progenitor cytotypes, as evidenced by BrdU incorporation studies showing sites of extensive cell proliferation in the satellite cells around the brachial nerve, in the blastema and in the coelomic epithelium (Candia-Carnevali et al. 1995, 1997, 1998).

As mentioned above, other types of migratory cells significantly involved in regeneration are phagocytes and granulocytes, which certainly can not be interpreted as presumptive progenitor cells but represent separate classes of differentiated migratory cells.

Phagocytes (PC) are obviously recognizable for their large phagosomes (Fig. 7.2e,f), different in size and contents, and are involved in defense mechanisms. As expected PC are especially numerous at the level of the amputation region during the repair phase (Candia-Carnevali et al. 1993). As far as the *in vitro* approach is concerned, PC are present in the cultures even to a more relevant extent and their phagosomes tend to form huge cytoplasmic inclusions (Fig. 7.2f).

Granulocytes (GC) are cells considered to be exclusive of crinoids (*wanderzellen*, Reichensperger 1912; Smith 1981) and recognizable for their abundant cytoplasmic granules (Fig. 7.2 g,h). Their large basophilic and electron-dense granules represent a source of putative growth factors (see below), presumably involved in RSC activation. GC are randomly scattered in all tissues, with a specific preferential distribution all around the brachial nerve and are activated by traumatic events. During the repair phase they migrate along the nerve or inside the coelomic canals towards the amputation region where their chromatophilic granules are discharged by exocytosis (Candia-Carnevali et al. 1993). Free intact granules can also be frequently detected in the tissue context all around the amputation site. *In vitro*, GC are present to a different extent (Fig. 7.1d), particularly in few days cultures (up to 1 week). Their morphology and large size are always well recognizable among other cytotypes. PC and GC are recruited at the beginning of the regeneration process and, although their number appreciably varies according to the experimental model, their involvement is always related and limited to the repair phase, in both arm and visceral regeneration. PC and GC do not show an up-regulated cell cycle activity as their overall number progressively decreases during the advanced regeneration phases. In particular, an exaggerated number of GC during the early/proper regenerative stages can be considered an evident indication of an occurring histological anomaly as in the case of aberrant regeneration and can be always correlated to significant developmental dysfunctions in terms of growth, morphogenesis and differentiation of cells and tissues (Candia-Carnevali and Bonasoro 1995; Candia-Carnevali 2005; Sugni et al. 2008).

7.3.4.2 Dedifferentiated Cells

Dedifferentiation is particularly evident in myocytes belonging to the arm muscle bundles close to the amputation area, and, to a minor extent, in other mesodermal cell types (sclerocytes, fibrocytes) (Candia-Carnevali et al. 1993; Candia-Carnevali and Bonasoro 2001b; Candia-Carnevali 2006).

Dedifferentiating myocytes (DDM), in progressive stages of dedifferentiation (Fig. 7.2i,j), characterized by evident signs of disorganization in their contractile apparatus, can be involved in active migration through coelomic fluids or loose connective tissue. They are observed in the regenerating samples only occasionally or quite frequently, depending on individual variability and experimental conditions. Dedifferentiation occurs in parallel to a conspicuous rearrangement of the pre-existing muscle bundles which are involved in dynamic processes of cell rearrangement/recycling and/or dedifferentiation (Fig. 7.1a). Comparison between regenerating samples obtained using different experimental conditions (normal regenerating arms, regenerating samples exposed to pollutants, *explants*, etc.) highlights that there are general similarities in the basic regenerative processes but meaningful differences related to cellular/tissue elements involved. Specifically, the contribution of dedifferentiated cell types to the regenerating tissues appears to be quite significant in extreme experimental conditions, e.g. in both treated samples and explants: in these latter, in fact, it was observed that the arm muscle bundles undergo a particularly massive rearrangement and myocytes are extensively employed for cell recruitments. Actually, it is not clear if these processes represent dedifferentiation/redifferentiation processes of myocytes to give rise directly to new populations of migrating coelomocytes, or if it they are part of more complicated processes mediated by phagocytes in which myocytes undergo degeneration and contribute indirectly to form new progenitor cell. Maybe both phenomena can occur. In any case, whichever mechanism is involved, direct or indirect, it is significant that these processes of rearrangement/dedifferentiation at the level of the muscles are closely associated to massive phenomena of cell migration occurring in the adjacent coelomic canals, the migratory elements often including well recognizable semi-dedifferentiated myocytes. *In vitro*, DDM are particularly numerous in the first days of culture, during which their contractile activity can be recorded (unpublished). However, after the first week in culture, they lose both their elongated shape and contractile activity and tend to progressively acquire undifferentiated morphology (Fig. 7.2j).

In addition, mainly in the explants and the pollutant-exposed regenerating samples, a certain rearrangement can be detected in other differentiated cells of the stump tissues, with particular reference to dedifferentiating sclerocytes (DDS) of the endoskeletal tissue and fibrocytes (DDF) of the connective tissues (Candia-Carnevali et al. 1998, Candia-Carnevali 2005). However, in these cases, tissue rearrangement mainly involves obvious degeneration phenomena which lead to an appreciable vacuolization/vesiculation of both extracellular matrix and cells. These tissue/cell turn-over processes are always associated with a marked and unusual PC and GC presence indicating that connective tissue can be actually employed as a secondary indirect source rather than a primary producer of undifferentiated cells. BrdU experiments do not show any significant proliferation activity in all these cases of tissue rearrangement and cell dedifferentiation.

7.4 Cellular Mechanisms of Regeneration

In terms of cellular mechanisms there is evidence that in all cases of blastemal regeneration typical epimorphic processes can involve not only a local proliferation of undifferentiated cells but also a significant contribution of morphallactic reorganization processes, depending on the specific need and individual availability of suitable cells (Candia-Carnevali and Bonasoro 2001a,b). On the other hand, the morphallactic rearrangement appears to be massive in extreme conditions (*explants* and exposed specimens), whereas it is maintained to low basal levels in standard conditions, where regeneration is mainly involving undifferentiated cells (RSC and CSC), without any relevant reorganization and contribution from the muscles or any other differentiated tissue of the stump (Candia-Carnevali et al. 1993, 1995, 1997).

In other cases of non-blastemal regeneration (e.g. visceral regeneration) new tissues and organs develop mainly as a result of extensive migration and transdifferentiation phenomena, whereas proliferation seems to play only a minor role. In these cases transdifferentiation phenomena of coelothelial cells into other epithelial cytotypes are particularly evident. The extensive employment of cell migration in regenerative development and reconstruction of new tissues from the remnants of old tissues can thus be interpreted as other obvious signs of morphallaxis. Interestingly, in experiments of visceral transplantations in crinoids, the mechanisms involved appear to be comparable to a certain extent to those occurring in visceral regeneration. Namely, all the migratory cells described above contribute to repair/regeneration and are engaged in a mutual cell exchange between donor and acceptor tissues (Mozzi et al. 2004). Dedifferentiation and transdifferentiation processes have also been described. These results show a striking potential of cell plasticity and tissue histocompatibility in crinoids and confirm their remarkable repair/regenerative capabilities.

7.5 Molecules, Genes, Expression Patterns

The nervous system with its ubiquitous distribution and differentiated components (ectoneural, entoneural, hyponeural) plays a crucial role in regeneration. This is mainly due to its striking capacity of self regenerating, to its pilot action as a promoter/inducer of the overall regenerative processes, and to its contribution in terms of release of neurothrophic factors (Candia-Carnevali et al. 1989, 1996; Thorndyke and Candia-Carnevali 2001; Patruno et al. 2002, 2003). Thus, a number of studies have directed the search for molecules involved in regenerative development towards neurohumoral factors with paracrine or autocrine action. Among those are: neurotransmitters, particularly monoamines such as dopamine and serotonin; neuropeptides, such as Substance-P, SALMFamide 1 (S1), SALMFamide 2 (S2); nerve-derived growth factors, particularly TGF-ß and related peptides (BMP), NGF, FGF-2 (see for a review Thorndyke and Candia-Carnevali 2001). As an example, studies on the expression of the TGF-β1 and TGF-β-type II receptor show different localizations and levels during crinoid regeneration (Patruno et al. 2002). Thus, it is not surprising that migratory cytotypes are immuno-reactive to antibodies specific for any of these factors, with a cell type specific correspondence. In fact, RST and CSC are reactive to dopamine and serotonin antibodies, respectively, whereas GC granules to Substance-P and TGF-β1 (Thorndyke and Candia-Carnevali 2001)

In terms of genes regulating/involved in regeneration, results related to the cloning of native growth factors confirm the implications of these molecules for regenerative processes (Patruno et al. 2003): in fact, genes identified so far are new members of the TGF-ß superfamily. In crinoids *AnBMP2/4* which shows a sequence similarity with other echinoderm and human BMPs, has been found in regenerates. According to its expression pattern a plausible role has been inferred in the specification of migratory stem cells, blastemal growth and skeletogenic tissue differentiation (Patruno et al. 2003). The bone morphogenetic protein/transforming growth factor-β (BMP/TGFβ)-signaling pathway has been shown to function in regeneration also in brittlestars (Bannister et al. 2005, 2008). It is relevant to remind that, in general, the active gradient established by BMP ligands is considered as one of the main factors responsible for wound healing (O'Kane and Ferguson 1997) or generating the positional information during development (Graff 1997). So it becomes of great interest to establish whether or not the same gene regulatory pathways are deployed in both adult and embryonic/larval development and regeneration. In ophiuroids, *AFUNI* was identified as the first gene implicated and cloned in regenerating brittlestars. *In situ* hybridisation analysis showed that it is expressed in coleomocytes and in different sites of regenerating arms, suggesting that it may be involved in more than one process during regenerative development (Bannister et al. 2005). This finding reinforces the notion that the coelomic network plays a key role in adult echinoderms regeneration acting as a centralized structure for the recruitment of progenitor cells of any type. It has been suggested that BMP proteins act like morphogens, specifying and directing cell fates in a concentration-dependent manner (O'Kane and Ferguson 1997; Graff 1997). An interesting question is which effects would be produced on the regeneration process by experimentally induced *BMPs* over-expression or supression. Results from functional knockout will probaly reveal the true function of BMPs. Over-expression and supression of a gene can be achieved through different methods. The mRNA injection approach has already been successfully applied to other echinoderms, like the sea urchin, *Paracentrotus lividus* (Zito et al. 2003) and it is currently being implemented in the brittlestar *Amphiura filiformis* (unpublished).

A very promising perspective to the genetic approach to echinoderm regeneration was recently opened by the sequencing of the sea urchin genome. The sea urchin genome project (see Science 2006) indicates that echinoderms share at least 70% of their proteins with humans. Clearly this provides an important tool kit to aid our understanding of regeneration and stem cell biology as well as to support the development of realistic methods to pursue tissue and organ regeneration in humans. In some case more genes of a particular type were found in the urchin genome compared to humans and one group that is very relevant for stem cell studies are those related to the immune system. Humans have both innate and acquired immunity systems and one surprise was the finding that the sea urchin has some of the genes of the acquired immunity system, whereas genes from the innate immunity group are greatly expanded (Hibino et al. 2006). From this point of view, it will be very useful to explore the homologues of key stem cell marker genes and other pathways involved in stem cell self-renewal such as STATs (Signal Transducers and Activators of Transcription) that are transcription factors phosphorylated by JAK kinases in response to cytokine activation. These, together with LIF (leukaemia inhibitory factor), BMPs and stem cell markers such as Oct4, Nanog and Rex 1 with the GSK pathways (glycogen synthase kinase) are all central to our understanding of cell renewal and stem cell biology in echinoderms. Moreover, the massive expansion of genes associated with the immune system and hence the coelomocytes of echinoderms are likely to provide a rich source of inspiration and material for future research. It might be predicted that the amazing powers of regeneration seen in echinoderms might be a reflection of this increase in relevant gene families.

7.6 Concluding Remarks and Future Perspectives

The bulk of the results obtained so far in the different groups contribute to throw light on the possible contribution of stem cells in echinoderm regeneration. In terms of general mechanisms, echinoderm regeneration employs both epimorphic and morphallactic processes. In the first case new tissues arise from undifferentiated cells (stem cells or dedifferentiated cells) which form a blastema, i.e. a discrete centre of proliferative activity giving rise to all the new structures. In the second case the regenerative processes imply substantial rearrangement of the old structures, involving less extensive and more localized proliferation, and the responsible cells derive from existing tissues by dedifferentiation, transdifferentiation and/or migration. Often both types of processes give a differential contribution to regeneration. In the light of what seen above, it is possible to suggest that echinoderms can utilize possible alternative mechanisms of regeneration which imply a modulated differential recruitment of progenitor cells, including presumptive stem cells. In terms of cellular plasticity, in fact, it is evident that echinoderms can employ very flexible mechanisms which involve both the direct recruitment of undifferentiated stem cells and the indirect recruitment of differentiated cells, via partial or complete reversal of their differentiation pathway.

In particular in crinoids, the *in vivo* and *in vitro* results clearly show that the success of the whole complex regenerative phenomenon with all its variety of variables, intrinsic or extrinsic, appear to depend mainly by the extremely broad adaptability/versatility of the progenitor cells involved. This implies their origin and derivation (true stem cells or reprogrammed dedifferentiated cells), activity (proliferation and/or migration), plasticity and pluripotential/unipotential for differentiation (derived cellular phenotypes), and their striking flexibility to respond to specific factors involved in regulating their activation and recruitment. Therefore regeneration-competent cells deserve to be explored in detail by further extensive *in vivo* and *in vitro* studies in the crinoid and in other promising echinoderm experimental models, also in view of future successful applications. The accurate identification of adult stem cells and the possibility to develop specific stem cell cultures represent in fact priority targets which could have relevant implications. As shown by recent published data (Bulgakov et al. 2002; Chia and Xing 1996; Matranga et al. 2005; Weissman 2000), the experimental approach employing cell cultures of invertebrate deuterostomian models (such as echinoderms or protochordates) should be considered an appropriate and potentially powerful tool to address the wide problem of the biology of stem cells. The study of the mechanisms regulating specification of pluripotent cells in vertebrates represents, in particular, a promising and expanding area of applied research: in this perspective it seems to be very timely and appropriate to enlarge the interest to deuterostomian relatives, such as echinoderms. However, untill now this approach encountered severe limits in echinoderms due to actual difficulties to obtain long-term cultures, as in other marine invertebrates (Rinkevich 2005). The employment of the crinoid experimental model turned out to be a successful strategy and provided a significant improvement to the *in vitro* approach (Parma et al. 2005, 2006; Di Benedetto et al. PhD Thesis in preparation). The encouraging results obtained so far, which showed the way to the isolation and *in vitro* expansion of cell cultures from regenerating crinoids, offer a promising basis for the success of the *in vitro* approach, relatively new for invertebrate models. Our preliminary integrated analysis opens new perspectives to the study of stem cell biology, throwing light on still unexplored cellular aspects of the regenerative phenomena, with particular reference to progenitor (stem) cells and their regenerative plasticity, and represents an important basis to set up a suitable parallel study in other echinoderm classes.

In terms of future prospects, current *in vivo* and *in vitro* experiments need to address the following: (1) establishment of long-term cultures of specific cell clones; (2) definition of their activities, capabilities (migration, proliferation, apoptosis, clonogenecity) and response to modulator signal molecules (mitogen, morphogen, chemo-attracting agents); (3) determination of their potential for differentiation plasticity (dedifferentiation, transdifferentiation, redifferentiation), particularly the response to different growth and reprogramming factors; (4) identification of the role played by specific involved genes and their expression pattern, with particular reference to some sets of genes well known for their key-role in controlling cellular regenerative plasticity in other animals (Odelberg 2004).

A comprehensive analysis of changes in gene expression during the regeneration process can be accomplished using the micro- or macro-array technology. Thus, gene expression patterns associated with regeneration could be interfaced with parallel studies to determine the cellular pathways involved in wound healing, rapid cell proliferation, differentiation of new tissues and formation of new adult structures. These studies include the determination of temporal and spatial sequence of specific cell types using *in situ* hybridization or localisation of gene products with antibodies.

The target will be to obtain a comprehensive overview of the signal transduction pathways with which BMPs and other specific stemness genes may be involved and the neurotrophic ligands and receptors in echinoderm models then compare these proteins and/or their encoding genes over a wide phylogenetic range in order to gain insight into potential common pathways in chordates, especially mammals.

Acknowledgements This work was carried out thanks to the support of a number of specific projects financed through the years by CNR, MURST (COFIN 2003, 2006) and University of Milano (FIRST) programs. VM is grateful for the partial financial support of the Marine Genomics Europe Network of Excellence and MoMa ASI project. The first author is particularly grateful to the following colleagues for their indispensable help and valuable collaboration during the research work: Alice Barbaglio, Anna Biressi, Francesco Bonasoro, Giulio Cossu, Cristiano Di Benedetto, Suzanne Denis Donini, Daniela Mozzi, Lorenzo Parma, Marco Patruno, Michela Sugni.

References

Bannister R, McGonnell IM, Graham A, Thorndyke MC, Beesley PW (2005) Afuni, a novel transforming growth factorbeta gene is involved in arm regeneration by the brittle star Amphiura filiformis. Dev Genes Evol 215:393–401

Bannister R, McGonnell IM, Graham A, Thorndyke MC, Beesley PW (2008) Coelomic expression of a novel bone morphogenetic protein in regenerating arms of the brittle star Amphiura filiformis. Dev Genes Evol 218:33–38

Bonasoro F, Ferro P, Di Benedetto C, Sugni M, Mozzi D, Candia-Carnevali MD (2004) Regenerative potential of echinoid test. In: Heinzeller T, Nebelsick JH (eds) Echinoderms: Munchen. Taylor & Francis Group, London, pp 97–103

Bossche JP, Jangoux M (1976) Epithelial origin of starfish coelomocytes. Nature 261:227–228

Bulgakov VP, Odintosova NA, Plotnikov SV, Kislev KV, Zacharov VE, Zhuravlev YN (2002) Gal4-gene-dependent alterations of embryo development and cell growth in primary culture of sea urchins. Mar Biotech 4:480–486

Candia-Carnevali MD, Bruno L, Donini DS et al (1989) Regeneration and morphogenesis in the feather star arm. In: Kiortsis V, Koussoulakos S, Wallace H (eds) Recent trends in regeneration research. Nano Asi Series, Plenum Press, New York, London, pp 447–460

Candia-Carnevali MD, Lucca E, Bonasoro F (1993) Mechanisms of arm regeneration in the feather star Antedon mediterranea: Healing of wound and early stages of development. J Exp Zool 267:299–317

Candia-Carnevali MD, Bonasoro F, Lucca E, Thorndyke MC (1995) Pattern of cell proliferation in the early stages of arm regeneration in the feather star Antedon mediterranea. J Exp Zool 272:464–474

Candia-Carnevali MD, Bonasoro F, Invernizzi R, Lucca E, Welsch U, Thorndyke MC (1996) Tissue distribution of monoamine neurotransmitters in normal and regenerating arrns of the feather star Antedon mediterranea. Cell Tissue Res 285:341–352

Candia-Carnevali MD, Bonasoro F, Biale A (1997) Pattern of bromodeoxyuridine incorporation in the advanced stages of arm regeneration in the feather star Antedon mediterranea. Cell Tissue Res 289:363–374

Candia-Carnevali MD, Bonasoro F, Patruno M et al (1998) Cellular and molecular mechanisms of arm regeneration in crinoid echinoderms: The potential of arm explants. Dev Genes Evol 208:421–430

Candia-Carnevali MD, Bonasoro F (2001a) A microscopic overview of crinoid regeneration. Microsc Res Tech 55:403–426

Candia-Carnevali MD, Bonasoro F (2001b) Introduction to the Biology of Regeneration in Echinoderms. Microsc Res Tech 55:365–368

Candia-Carnevali MD (2005) Regenerative response and Endocrine Disrupters in crinoid Echinoderms: An old experimental model, a new ecotoxicological test. In: Matranga V (ed) Echinodermata. Progr Mol Subcell Biol (Marine Molecular Biotechnology), vol 39. Springer-Verlag, Heidelberg, pp 167–198

Candia-Carnevali MDC (2006) Regeneration in Echinoderms: Repair, regrowth, cloning. Invertebr Surviv J 3:64–76

Carlson BM (1997) Development and regeneration, with special emphasis on the amphibian limb. In: Ferretti P, Géraudie J (eds) Cellular and molecular basis of regeneration: From Invertebrates to Humans. John Wiley and Sons, Chichester, New York, pp 45–61

Chia FS, Xing J (1996) Echinoderm coelomocytes. Zool Stud 35:231–254

Clark AH (1921) A monograph of the existing crinoids. Vol 1. The comatulids. Part 1. Bull US Nat Mus 82:1–389

Dendy A (1886) On the regeneration of the visceral mass in Antedon rosaceus, vol I. Stud. Biol. Lab. Owens College, Manchester, p 299

Dolmatov I, Bonasoro F, Ferreri P, Candia-Carnevali MD (2003) Visceral regeneration in the crinoid Antedon mediterranea. In: Feral JP (ed) Echinoderm Research 2001. Balkema, Rotterdam, pp 215–220

Dubois P, Ameye L (2001) Regeneration of spines and pedicellariae in echinoderms: A review. Micr Res Tech 55:427–437

Dupont S, Thorndyke MC (2006) Growth or differentiation? Adaptive regeneration in the brittlestar Amphiura filiformis. J Exp Biol 209:3873–3881

Eaves AA, Palmer AR (2003) Widespread cloning in echinoderm larvae. Nature 425:146

Ferretti P, Géraudieu J (1997) Preface. In: Ferretti P, Géraudie J (eds) Cellular and molecular basis of regeneration: From Invertebrates to Humans. John Wiley and Sons, Chichester, New York, pp XIX–XX

Garcia-Arraras JE, Greenberg MJ (2001) Visceral regeneration in holothurians. Micr Res Tech 55:438–451

Goss RJ (1965) Mammalian regeneration and its phylogenetic relationships. In: Kiortsis V, Trampush HAL (eds) Regeneration in animals and related problems. North-Holland publishing Company, Amsterdam, pp 33–38

Goss RJ (1969) Principles of regeneration. Academic Press, New York and London

Goss RJ (1992) The evolution of regeneration: Adaptive or inherent? J Theor Biol 159:241–260
Graff JM (1997) Embryonic patterning: To BMP or not to BMP, that is the question. Cell 89: 171–174
Hibino T, Loza-Coll M, Messier C et al (2006) The immune gene repertoire encoded in the purple sea urchin genome. Dev Biol 300:349–365
Hyman HL (1955) The Invertebrates: Echinodermata, vol IV. McGraw-Hill, New York, Toronto, London, p 763
Mashanov VS, Dolmatov IY (2004) Functional morphology of the developing alimentary canal in the holothurian Eupentacta fraudatrix (Holothuroidea, Dendrochirota). Acta Zool 85:29–39
Matranga V, Pinsino A, Celi M, Natoli A, Bonaventura R, Schroder HC, Muller WEG (2005) Monitoring chemical and physical stress using sea urchin immune cells. In: Matranga V (ed) Echinodermata. Springer, Heidelberg, pp 85–110
Minckert W (1905) Über Regeneration bei Comatuliden nebst Ausführungen über die Auffassung und Bedeutung der Syzygieen. Arch f Naturg Jahrg 71:163–244
Mladenov PV, Bisgrove B, Asotra S, Burke RD (1989) Mechanisms of arm-tip regeneration in the sea star, Leptasterias hexactis. Roux's Arch Dev Biol 189:19–28
Moss C, Hunter J, Thorndyke MC (1998) Pattern of bromodeoxyuridine incorporation and neuropeptide immunoreactivity during arm regeneration in the starfish Asterias rubens. Phil Trans R Soc London B 353:421–436
Mozzi D, Ferreri P, Petrillo P, Dolmatov IY, Bonasoro F, Candia-Carnevali MD (2004) Graft and regeneration of viscerain the crinoid Antedon mediterranea. In: Heinzeller T, Nebelsick JH (eds) Echinoderms: München. Taylor & Francio Group, London, pp 135–139
Mozzi D, Dolmatov I, Bonasoro F, Candia-Carnevali MD (2006) Visceral regeneration in the crinoid Antedon mediterranea: Basic mechanisms, tissues and cells involved in gut regeneration. Cent Eur J Biol 1(4):609–635
Odintsova NA, Dolmatov IYu, Mashanov VS (2005) Regenerating holothurian tissues as a source of cells for long-term cell cultures. Mar Biol 146:915–921
Odelberg SJ (2004), Unravelling the molecular basis for regenerative cellular plasticity. Plos Biol 2:1068–1071
O'Kane S, Ferguson MW (1997) Transforming growth factor beta s and wound healing. Int J Biochem Cell Biol 29:63–78
Ottaviani E, Franceschi F (1997) The invertebrate phagocytic immunocyte: Clues to a common evolution of immune and neuroendocrine systems. Immunol Today 18:169–174
Oweson C, Sköld H, Pinsino A, Matranga V, Hernroth B (2008) Manganese effects on haematopoietic cells and circulating coelomocytes of Asterias rubens (Linnaeus). Aquat Toxicol 89:75–81
Parma L, Di Benedetto C, Denis Donini S, Cossu G, Candia-Carnevali MD (2006) Primary cell culture in crinoid echinoderms: Exploring the plasticità potential of regenerative competent cells. Proceedings of 3rd Eur Conf on Regeneration, EMBO Conference, 2006, p 21
Patruno M, Smertenko A, Candia-Carnevali MD, Bonasoro F, Beesley PW, Thorndyke MC (2002) Expression of TGF-B-like molecules in normal and regenerating arms of the crinoid Antedon mediterranea: Immunocytochemical and biochemical evidence. Proc Roy Soc Lond B 269:1741–1747
Patruno M, McGonnell IM, Graham A, Beesley P, Candia-Carnevali MD, Thorndyke M (2003) AnBMP2/4 is a new member of the TGF-b superfamily isolated from a crinoid and involved in regeneration. Proc Roy Soc Lond B 270:1341–1347
Pearson H (2001) The regeneration gap. Nature 414:388–390
Perrier E (1873) L'anatomie et la régénération des bras de la comatula. Arch Zool Exp Gen 2:29–86
Pinsino A, Thorndyke MC, Matranga V (2007) Coelomocytes and post-traumatic response in the common sea star Asterias rubens. Cell Stress Chaperones 12:332–342
Reichensperger A (1912) Beiträge zur Histologie und zum Verlauf der Regeneration bei Crinoiden. Ztschr Wiss Zool 101:1–69
Rinkevich B (2005) Marine invertebrate cell cultures new millennium trends. Mar Biotechnol 7:429–439

Runnström J (1915) Analytische Studien über die Seeigelentwicklung.II. Arch EntwMec Org 41:1–56
Runnström J (1925) Zur Experimentaellen Analyse Der Entwiglung Von Antedon. Wilh Roux Arch Entw Mech Org 105:63–113
Sánchez Alvarado A, Tsonis PA (2006) Bridging the regeneration gap: Genetic insights from diverse animal models. Nat Rev Genet 7:873–884
Sea Urchin Genome Sequencing Consortium (2006) The genome of the sea urchin Strongylocentrotus purpuratus. Science 314:941–952
Smith VJ (1981) The echinoderms. In: Ratcliffe NA, Rowley AF (eds) Invertebrate blood cells. Academic Press, New York, pp 513–562
Stocum DL (2001) Stem cells in regenerative biology and medicine. Wound Repair Regen 9: 429–442
Suárez-Castillo EC, Medina-Ortíz WE, Roig-López JL, García-Arrarás JE (2004) Ependymin, a gene involved in regeneration and neuroplasticity in vertebrates, is overexpressed during regeneration in the echinoderm Holothuria glaberrima. Gene 334:133–143
Sugni M, Mozzi D, Barbaglio A, Bonasoro F, Candia-Carnevali MD (2007) Endocrine disrupting compounds and echinoderms: New ecotoxicological sentinels for the marine ecosystem. Ecotoxicology 16:95–108
Sugni M, Manno V, Barbaglio A, Mozzi D, Bonasoro F, Tremolada P, Candia-Carnevali MD (2008) Echinoderm regenerative response as a sensitive ecotoxicological test for the exposure to endocrine disrupters: Effects of p,p'DDE and CPA on crinoid arm regeneration. Cell Biol Toxicol 24:573–586
Thorndyke MC, Candia-Carnevali MD (2001) Regeneration neurohormones and growth factors in echinoderms. Can J Zool 79:1171–1208
Thorndyke MC, WC Chen, Beesley PW, Patruno M (2001) Molecular approach to echinoderm regeneration. Micr Res Tech 55:474–485
Thouveny Y, Tassava RA (1997) Regeneration through phylogenesis. In: Ferretti P, Géraudie J (eds) Cellularand molecular basis of regeneration: From Invertebrates to Humans. John Wiley and Sons, Chichester, New York, pp 9–43
Weissman IL (2000) Stem Cells: Units of Development, Units of regeneration, and Units in Evolution. Cell 100:157–168
Wilkie IC (2001) Autotomy as a prelude to regeneration in echinoderms. Micr Res Tech 55: 369–396
Wilkie IC (2005) Mutable collagenous tissue: Overview and biotechnological perspective. In: Matranga V (ed) Echinodermata. Progress in Molecular and Subcellular Biology (Marine Molecular Biotechnology), vol 39. Springer-Verlag, Heidelberg, pp 221–250
Zito F, Costa C, Sciarrino S, Poma V, Russo R, Angerer L, Matranga V (2003) Expression of univin, a TGF-B growth factor, requires ectoderm-ECM interaction and promotes skeletal growth in the sea urchin embryo. Dev Biol. 246:217–227

Chapter 8
Secondary Mesenchyme Cells as Potential Stem Cells of the Sea Urchin Embryo

Francesca Zito and Valeria Matranga

Abstract In the last years, stem cells have been the subject of great interest. Although considerable progress has been made in this field, the signaling pathways and related molecules controlling stem cells behaviour still remain unclear. Among marine invertebrates, echinoderms have the attracting ability to regenerate parts of their bodies, involving stem cells recruitment, and thus providing themselves as excellent models for studies on stem cells in adult organisms. On the contrary, hardly any research focused on embryonic stem cells has been performed using echinoderm embryos, although they have been utilised extensively and with noteworthy results, for example, for studies on basic developmental biology. Indeed, the great amount of data accumulated over the years, the availability of new genomic and proteomic research tools, together with the advantage of experimental manipulation support the sea urchin embryo as a good candidate for detailed studies on embryonic stem cells. Here, we review fundamental findings concerning the two distinct populations of mesodermal cells in the sea urchin embryo, discussing the possibility to identify some of them as embryonic stem cells. In particular, secondary mesenchyme cells (SMCs) are a heterogeneous population of cells with several different fates and behaviours. Taken together, a number of evidence indicates that SMCs function as multipotent stem cells, thus sharing some features with vertebrate embryonic stem cells, including the neural crest cells. Our aim is to address new viewpoints for forthcoming studies on SMCs as well as to open new directions for research on fundamental mechanisms of stem cell biology exploiting the sea urchin embryo as a model system.

Keywords Echinoderms · Neural crest cells · Pigment cells · Skeletogenesis · Transfating

---

F. Zito (✉)
Consiglio Nazionale delle Ricerche, Istituto di Biomedicina e Immunologia Molecolare "Alberto Monroy", Via Ugo La Malfa 153, 90146 Palermo, Italy
e-mail: zito@ibim.cnr.it

B. Rinkevich, V. Matranga (eds.), *Stem Cells in Marine Organisms*,
DOI 10.1007/978-90-481-2767-2_8,

8.1 Introduction

One of the most fascinating and promising areas of modern science is the biology of stem cells.

Research in this important field is providing a considerable amount of data, which are advancing our knowledge about mechanisms of action of stem cells as well as about essential developmental processes. Thus, the results obtained so far offer a great promise for cell-based therapies able to treat diseases (namely the regenerative therapies) and concurrently provide an incredible opportunity to connect both applied and basic disciplines (Denham et al. 2005; Zandonella 2005). Stem cells are defined as cells that have a unique ability to both self-renew and produce newly differentiated cells playing essential roles, ranging from embryonic development and organogenesis (embryonic stem cells) to tissue regeneration (adult stem cells) (Zhan 2008). Studies on diverse systems, from insects to mammals, have shown the occurrence of evolutionarily conserved mechanisms, both at cellular and molecular levels, which regulate stem cell maintenance and self-renewal. This is a crucial point, meaning that results obtained from studies on less evolved organisms can be noticeably applied to highly evolved animals. Furthermore, the exploitation of simple animal models, which are particularly well suited for analyses on their gene expression and in which stem cells can be easily identified, may accelerate understanding of the gene regulatory pathways governing stem cell biology. Among marine invertebrates, echinoderms have a unique position in the evolutionary tree, since they are classified as deuterostomes, that is to say a "sister" group to chordates (Bromham and Degnan 1999). Although members of the echinoderms and chordates phyla appear so morphologically dissimilar, both fossil records and molecular data strongly support the notion that they are close enough to make genetic and biological comparisons worthwhile (Smith 2004; Bromham and Degnan 1999). In the stem cell biology field, echinoderms have been successfully used as model systems for studying regenerative processes occurring in adult organisms because of their ability to regenerate body parts lost following autotomy (self-induction) or after traumas caused by predation or environmental accidents. On the contrary, echinoderms have never been considered as a valuable model for any research focused on embryonic stem cells, although their embryos have been studied to a great extent. In particular, the sea urchin embryo has been known for its versatility and suitability since the end of the 19th century, when classical embryologists performed the very first studies on the basic mechanisms of embryo development (Driesch 1892; Hörstadius 1939). Since then the sea urchin embryo has been widely used, especially thanks to: (1) its rather rapid embryogenesis, which takes a few days; (2) the reasonable simplicity in its shape and organization and (3) its transparency, which simplify experimental manipulations and observations. Our knowledge of the molecular basis of developmental processes of the sea urchin embryo has advanced considerably in the past 30 years. Embryonic cell lineage studies (significant also in modern stem cell research) in combination with high-throughput molecular biology technologies led to the design of a complex gene regulatory network, which is gradually evolving in its complexity (Davidson et al. 2002a,b). In addition, the sequencing of the entire sea urchin genome has been accomplished, which represents the first full sequenced genome of a non-chordate deuterostome (Sea Urchin Genome Sequencing Consortium 2006). Even with such remarkable improvements, to date the stem cell biology remains an unexplored topic in this system. Indeed, the available information together with new genomic and proteomic research tools turn the sea urchin embryo into an appealing model system for undertaking detailed studies on embryonic stem cells.

In this chapter, we will first review a few basic notions concerning the sea urchin embryo developmental biology. Then, we will focus on the phenotypic and molecular analyses performed on the two main mesodermal populations among which embryonic stem cells are included, emphasizing those that in our opinion are the most relevant aspects to compare sea urchin stem cells with vertebrate ones, in particular with the neural crest cells. Thus, vertebrate stem cell signalling pathways will be described whenever common aspects with sea urchin stem cells emerge. Our aim is to suggest a new viewpoint for forthcoming studies on these mesenchymal cells and to open new directions for future research on fundamental principles of stem cell biology using the sea urchin embryo as a model system.

8.2 A Brief Overview of the Sea Urchin Development

The sea urchin embryo is one of the best studied among marine invertebrate animals. The developmental mode can be different among diverse sea urchin species, although they are similar as adults. In general, shortly after fertilization, the zygote starts a number of cell divisions entering in the cleavage stage, which gives rise to a definite number of unequally divided cells, namely mesomeres, macromeres and micromeres (Fig. 8.1A,B). In a few hours a blastula is formed, consisting of one layer of epithelial cells surrounding a fluid-filled cavity, the blastocoel (Fig. 8.1C). A fate map showing different embryonic territories of gene expression has been accurately described by the late blastula stage (see review by Davidson et al. 1998). This includes the following territories, from the vegetal to the animal pole: the small micromeres descendants, from which part of coelomic pouches arise; the large micromeres that are the primary mesenchyme cells (PMCs) precursors; the vegetal plate territory, from which both the secondary mesenchyme cells (SMCs) and endoderm arise; the oral and aboral ectoderm (Fig. 8.1B,C; Davidson et al. 1998; Angerer and Angerer 2000; Ettensohn and Sweet 2000). Then, a highly dynamic period of development follows, which is known as gastrulation, during which three germ layers are formed, namely ectoderm, endoderm and mesoderm, and the overall body plan is established (Fig. 8.1D–H). During this stage, the morphology of the embryo is dramatically reorganized by means of massive cell migrations, which lead to a relatively complex shape from a very simple starting form. In indirect developing species, a typical pluteus larva is formed (Fig. 8.1I), which, under the appropriate environmental conditions, feeds and undergoes a radical metamorphosis within few weeks (Burke 1982; Pearse and Cameron 1991; Miller and Emlet 1999; Yokota et al. 2002). By contrast, in species showing a direct development, a non-feeding larva

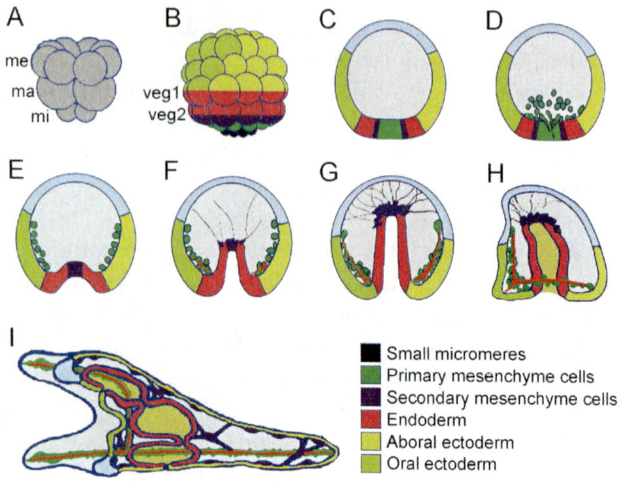

Fig. 8.1 Diagram of sea urchin embryo development and fate map showing different embryonic territories. (**A**) 16 blastomeres; (**B**) 64 blastomeres; (**C**) blastula; (**D**) mesenchyme blastula; (**E**) early gastrula; (**F**) middle gastrula; (**G**) late gastrula; (**H**) prism; (**I**) pluteus. me, mesomeres; ma, macromeres; mi, micromeres

forms, which undergoes metamorphosis in few days (Wray and Raff 1989; Wray 1996). Among the numerous differences between indirect and direct developmental modes, one concerns the mesoderm differentiation (see details in the following sections).

8.3 Sea Urchin Embryo Mesoderm

The mesoderm of the indirect-developing sea urchin embryo consists of two populations of mesenchymal cells, PMCs and SMCs, which differ with respect to lineage, developmental fates, time of epithelial-mesenchymal transition (EMT) and ingression into the blastocoel. In the following, we will review the main notions accumulated over the last 30 years about indirect-developing sea urchin embryo mesoderm, as we believe that embryonic stem cells can be identified and studied in this territory.

8.3.1 Primary Mesenchyme Cells

The morphogenetic behaviour of PMCs and the cellular events leading to skeleton formation has been described to a great extent (see reviews by Decker and Lennarz 1988; Hardin 1996). PMCs originate from the four large micromeres appearing at the vegetal pole of the 32-cell stage embryo and are their unique descendants (Fig. 8.1A,B). Before gastrulation, PMCs undergo EMT, a complex phenomenon involving changes in both cell adhesion behaviour and gene expression (Fink and McClay 1985; Wu and McClay 2007), and ingress into the blastocoel, where they

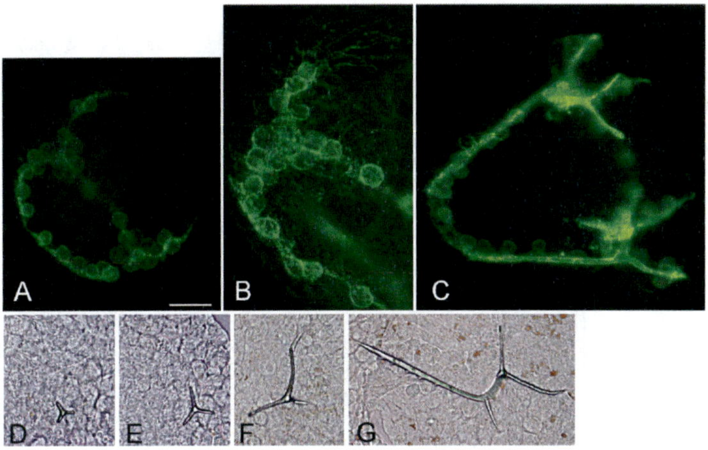

Fig. 8.2 Primary mesenchyme cells organization and skeleton formation in *P. lividus* embryos. (**A–C**) whole-mount embryos labelled with a monoclonal antibody against msp130 (kindly provided by D.R. McClay); (**D–G**) bright-field of squashed embryos at different stages to follow triradiate spicule rudiment growth. (**A**) early gastrula embryo; (**B**) higher magnification of part of an early gastrula embryo; (**C**) pluteus; (**D**) early gastrula stage; (**E**) middle gastrula stage; (**F**) late gastrula; (**G**) prism. *Bar*, 20 μm

migrate by means of numerous filopodia (Fig. 8.1C,D) (Gustafson and Wolpert 1963; Malinda et al. 1995; Miller et al. 1995). Progressively PMCs gather round to form a characteristic ring-like pattern beneath the equator of the embryo towards the vegetal pole; eventually their filopodia fuse to form a syncytial network which connects the cells by means of long cables (Fig. 8.2A–C) (Okazaki 1975; Hodor and Ettensohn 1998). PMCs begin to synthesise two triradiate rods, called spicules, consisting of magnesian calcite (calcium carbonate containing small amounts of magnesium carbonate) and occluded specific matrix proteins. The elongation and branching of spicules in the three directions (X, Y, Z) will give rise to an elaborate, species-specific patterned larval skeleton (Fig. 8.2D–G) (Okazaki 1975; Decker and Lennarz 1988; Beniash et al. 1997; Wilt 2002, 2005).

The early determination of micromeres to differentiate exclusively as skeletogenic cells has been extensively demonstrated, starting from the pioneering observations of Okazaki (1975; reviewed by Davidson et al. 1998). These progenitor cells undergo only a few rounds of divisions, the last one occurring soon before or just after their descendants (PMCs) ingression into the blastocoel, depending on the species (Stephens et al. 1986; Ettensohn and Malinda 1993). The final number of PMCs in the 4-armed pluteus is about 32 in *Paracentrotus lividus* and *Strongylocentrotus purpuratus* (Zito et al. 2003; Harkey et al. 1992) or 64 in *Lytechinus variegatus* (Ettensohn 1990).

Several genes encoding PMC-specific proteins have been cloned and shown to play important roles in the biomineralization process, including the spicule matrix proteins SM30, SM50, PM27, SM37 and the cell surface protein MSP130 (see reviews by Wilt 1999, 2002; Wilt et al. 2003; Livingston et al. 2006). Although the cell-autonomous competence to produce a skeleton, PMCs require several types of cues, including axial, temporal and scalar signals, provided by the overlying ectoderm and the apical extra cellular matrix (ECM) in order to synthesise a normal sized and patterned skeleton (Ettensohn and Malinda 1993; Armstrong and McClay 1994; Peterson and McClay 2003; Zito et al. 2003; Kiyomoto et al. 2004). Few of such signals have been identified, including the growth factors univin, VEGF and FGF (Zito et al. 2003; Duloquin et al. 2007; Röttinger et al. 2008). These growth factors seem to be independent and not functionally redundant and thus each of them seems required for controlling skeleton morphogenesis.

8.3.2 Secondary Mesenchyme Cells

Since PMCs are early determined to differentiate towards a single cell population and undertake only few rounds of cell cycles, they do not have the appropriate features to be considered as embryonic stem cells. However, the description of their features and functions has been instrumental for the understanding of some SMCs traits and performances, which will be described in details in the following sections. It should be highlighted that SMCs constitute a distinct population of mesodermal cells, ontogenetically distinct from PMCs, and display a number of characteristics typical of embryonic stem cells, as described in the following.

8.3.2.1 SMC Lineage Specification

SMCs derive from the veg2 tier of blastomeres, which forms at the 64-cell stage, after the 6th cleavage (Fig. 8.1B) (Cameron et al. 1991; Ruffins and Ettensohn 1993, 1996). These blastomeres are also the precursors of endodermal cells as the segregation of the two territories, mesoderm and endoderm, occurs later in development (Davidson et al. 1998). During the late blastula stage, the veg2 blastomere descendants form the vegetal plate, which consists of a disc of epithelial cells that initiate to invaginate at the onset of gastrulation, after PMCs ingression (Fig. 8.1C–E). By labelling single cells of the vegetal plate with fluorescent dyes and scoring pluteus embryos for labelled cells, Ruffins and Ettensohn (1993, 1996) found that SMC precursors are localized in the most central region of the vegetal plate, whereas endoderm precursors are arranged concentrically around them and in close proximity to ectodermal precursors (Fig. 8.1C). Both SMCs specification, arising by the early cleavage stage, and mesoderm-endoderm segregation, occurring by the late blastula stage, are mostly dependent upon an inductive signal. The latter is mainly transmitted via the Delta/Notch signalling pathway from micromeres and their progeny, although this is not the only signaling pathway involved (Khaner and Wilt 1991; Sherwood and McClay 1999, 2001; Sweet et al. 1999, 2002; McClay et al. 2000). The range of Delta actions is limited to the adjacent cells: thus, only those veg2 cells receiving a Delta signal from the nearby micromeres activate the Notch pathway and, as a consequence, they are specified as SMCs (Sweet et al.

2002). It is well known that the Notch intercellular signaling pathway is evolutionarily conserved in both vertebrate and invertebrate embryos and is the most commonly used signaling pathway among many different cell-to-cell communication systems. Numerous studies have demonstrated Notch fundamental role in various stages of cell maturation, either promoting stem cell self-renewal or terminal cell differentiation, depending on the environmental context in which cells are located (see reviews by Chiba 2006; Molofsky et al. 2004). Similarly, other signaling pathways, such as Wnt/β-catenin and Hedgehog, appear to be involved in the regulation of stem cell self-renewal depending on the environmental context too (Kleber and Sommer 2004; Molofsky et al. 2004). Gain and loss-of-function studies established crucial roles for the Wnt signaling pathway in the early neural crest development, both for its progenitors proliferation and induction of differentiation (Wu et al. 2003; Raible 2006). In the sea urchin embryo, many of the components of the Wnt/β-catenin and Hedgehog signaling pathways have been found. Cell transplantation and gene expression studies showed that micromeres are activated by β-catenin to produce an inductive ligand (i.e. Delta) for SMC specification (McClay et al. 2000). Moreover, Wnt8 has been identified as one key element acting downstream in this signaling pathway for the regulation of SMC specification (Wikramanayake et al. 2004; Minokawa et al. 2005). On the contrary, contrasting data have been reported concerning Hedgehog expression in different sea urchin species. It appears to be expressed in pigment cell precursors in *S. purpuratus*, in small micromeres in *Hemicentrotus pulcherrimus* and in endoderm cells in *L. variegatus*, but its functional role has not yet been established in any of the mentioned species (Egana and Ernst 2004; Hara and Katow 2005; Walton et al. 2006). In conclusion, the presence of components of these signaling pathways in sea urchin embryo is very intriguing, and stands for the first evidence on a hypothetical correspondence between SMCs and vertebrate stem cells.

While a number of data clarified the regulation of SMC specification, the processes leading to the differentiation of each set of SMC-derived cells have been studied but with limited results. In some cases, the morphogenetic behaviour of SMC-derived cells is poorly understood, probably because of the insufficient amount of markers specific for each single SMC population. It is unclear whether SMC precursors specific for any single population do exist or not. If such precursor cells exist, it has to be determined when they are specified, how many of them are present and how many times they divide before turning into fully differentiated cells. To date, we have hardly few answers to these questions as only some regulatory mechanisms of pigment cell lineage development have been elucidated. In the following, we will discuss details of what is known and about speculations on possible mechanisms involved.

8.3.2.2 SMCs Differentiation

As a general rule, SMCs are expected to differentiate into four types of non-skeletogenic mesoderm cells (Ettensohn and Ruffins 1993): the freely migrating pigment (Gibson and Burke 1985) and blastocoelar cells (Tamboline and Burke

1992), as opposed to the coelomic pouches and circumesophageal muscle cells, which do not show extensive migratory activity (Ishimoda-Takagi et al. 1984; Burke and Alvarez 1988). Recent studies, performed on different sea urchin embryo species, have reported evidences for the presence of at least three new SMC-derived cells randomly distributed in the blastocoel, and a fourth new type showing skeletogenic fate (described in detail in the following sections).

Some authors attempted to determine the relationship among the different SMC-derived cells during the presumed period of SMC specification in the *H. pulcherrimus* embryo (Tokuoka et al. 2002). Their results suggest that the specification period of pigment and blastocoelar cells is partially overlapping, and it can be confined around the ninth cleavage, with slight variations depending on the species. In contrast, coelomic pouches and circumesophageal muscle cells seem to be specified at much later stages and independent of the previously quoted cell types. Similar studies remain to be addressed in other sea urchin species.

8.3.2.3 Pigment Cells

Among SMC-derived populations, pigment cells are the most numerous cells in pluteus larvae and, in general, are the first to become recognizable from the late gastrula stage onwards. Indeed, these cells become easily detectable by the presence of red pigment granules in their cytoplasm (Monroy et al. 1951; Gustafson and Kinnander 1956; Chaffee and Mazia 1963), which accumulate as the cells disperse throughout the ectoderm and assume elongated and branched shapes, with two to three pseudopodia (Fig. 8.3). The cytoplasmic granules store carotenoids and naphthoquinone compounds (Chaffee and Mazia 1963; Griffiths 1965; Ryberg and Lundgren 1979), which have been suggested to function in body colouring, phototropism and photo-protection, thus for the defence of larval ectoderm (Matsuno and Tsushima 2001; Smith et al. 2006), and to have effective

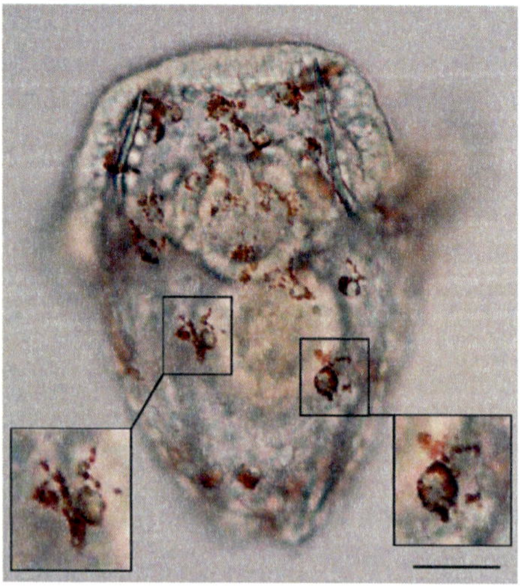

Fig. 8.3 Pigment cells in *P. lividus* pluteus embryos. Open squares encircle branched pigment cells. *Bar*, 20 μm

antibacterial properties (Service and Wardlaw 1984). The ability of phagocytosis showed by pigment cells suggests their participation in wound healing in the larva (Hibino et al. 2006).

Pigment cells have the capability to invade ectodermal epithelium with amoeboid movements as a consequence of the rapid extension and contraction movements of their pseudopodia (Gibson and Burke 1987). Because of their localization, pigment cells were originally thought to be of ectodermal origin (Young 1958), but now it is widely accepted that they are derived from a population of SMCs (Gibson and Burke 1985; Ettensohn and Ruffins 1993).

Among SMC populations, only the pigment cell lineage has been significantly studied and the regulatory processes leading their specification have been elucidated (see review by Materna and Davidson 2007). One of the pigment cell-specific genes, a transcription factor identified and referred to as *glial cell missing (gcm)*, has been identified as a direct target of the Notch signaling pathway, and its expression seems required for pigment cells specification (Ransick and Davidson 2006). *Gcm* is expressed early in development in the veg2 blastomeres and remains detectable all over gastrulation in pigment cell precursors (Ransick et al. 2002). Its function is to positively regulate the expression of genes coding for echinochrome bio-synthetic enzymes (Ransick and Davidson 2006). *Gcm* is a conserved gene, since it is found from Drosophila, where it was first isolated, to mammals, although it appears to have different functions in different animals (Akiyama et al. 1996). However, despite the differences, the interesting aspect to be highlighted is that *gcm* has a common role as a master regulatory gene and is regulated by the Notch signaling pathway in each animal system analysed to date.

On the basis of the data reported so far, the presence of cytoplasmic pigments, the morphology of the cell and its migration ability altogether support a speculative similarity between sea urchin pigment cells and vertebrate melanocytes. Although these latter cells arise from the neural crest (which is of ectodermal origin), melanocytes are migratory cells, as they move towards several sites in the embryo, for example the basal layer of the epidermis or the hair bulb, where they differentiate into pigment-producing cells. A number of factors are involved in controlling the balance between self-renewal and differentiation of melanocyte stem cells in the hair follicle bulge, including Notch and Wnt signaling molecules (see reviews by Sommer 2005; Delfino-Machín et al. 2007). In particular, the expression or the absence of the transcription factors *Pax3*, *Sox10* and *Mitf* is particularly critical for the regulation of melanocyte stem cells gene expression (Vance and Goding 2004). However, until now none of the melanocyte-specific markers have been found expressed in sea urchin pigment cells. Surprisingly, the only one identified in the sea urchin genome, *Mitf*, seems to be expressed in PMCs rather than in SMCs (Howard-Ashby et al. 2006).

8.3.2.4 Blastocoelar Cells

Distinct populations of migrating SMCs are the blastocoelar cells, which have been initially recognized by their fibroblast-like morphology (Dan and Okazaki

1956; Okazaki 1975) and by immuno-staining procedures (Tamboline and Burke 1992), while their biochemical and molecular properties remain still unknown. Blastocoelar cells leave the tip of the invaginating archenteron during gastrulation and remain localized within the blastocoel cavity. There, they spread and form a network of cells connected by their cytoplasmic filopodia, which surrounds the gut and localizes along the skeletal rods and within the arms of the pluteus larva (Tamboline and Burke 1992). To date, rather few studies have explored the role of blastocoelar cells in the embryo and, as a consequence, few possible functions have been suggested. The localization of a number of these cells associated with the skeletal rods suggested they might play a role in the growth of the larval skeleton (Shoguchi et al. 2002). Blastocoelar cells are also thought to be part, together with pigment cells, of the immune system of the embryo, since they seem capable of phagocytosing yeast injected into the blastocoelar cavity and of clearing of bacteria contamination (Silva 2000; Smith et al. 2006; Hibino et al. 2006).

Currently, the insufficient amount of specific molecular markers for blastocoelar cells makes the study of their cell lineage quite difficult.

8.3.2.5 Muscle Cells

Different kinds of mononucleated muscle cells have been described in sea urchin larvae, although they have been studied in greater details in other classes of echinoderms, as for example the sand dollar *Dendraster excentricus* (Burke 1981) and the starfish *Pisaster ochraceus* (Crawford and Martin 1998). By histological and ultrastructural methods, it has been possible to distinguish smooth muscle cells localized around the esophagus, whose elongated pseudopods form contractile fibers and therefore referred to as circumesophageal cells. Others are the striated myoepithelial cells placed in the cardiac sphincter, while cells placed in the pyloric and anal sphincters appear similar to muscle cells surrounding stomach or intestine found in vertebrates (Burke 1981). All these cells segregate from the two coelomic pouches that form at the late gastrula stage and can be detected by immunofluorescence staining with anti-tropomyosin (Ishimoda-Takagi et al. 1984) or anti-actin antibodies (Burke and Alvarez 1988).

Interestingly, all sea urchin embryonic muscle cells express the receptor for the fibroblast growth factor (FGFR1) (McCoon et al. 1998). The finding that detectable FGFR1 protein levels are found in muscle cells just before their migration from the tip of the archenteron, suggested that in the sea urchin embryo muscle differentiation, but not muscle commitment, requires some kind of FGF-mediated signaling (McCoon et al. 1998). In particular, the expression timing of FGFR is consistent with its possible role in the migration of muscle cell precursors from the archenteron, and in the proliferation or extension of cytoplasmic processes.

In general, the myogenic development seems to follow a conserved pathway, since embryonic muscle progenitors have been identified in a variety of vertebrate species (see review Figeac et al. 2007). Several molecular markers and few genes controlling muscle stem cells specification or their entry into the differentiation process have been so far identified in vertebrate and invertebrate species, i.e. the transcription factors *Pax3*, *Pax7* and *twist* (Figeac et al. 2007; Leptin 1991). Very recently, a sea urchin ortholog of the *twist* transcription factor has been identified in the *L. variegatus* species (Wu et al. 2008). It appears to be mainly an essential regulator of the skeletogenic gene regulatory network, but it also seems involved in SMC formation (Wu et al. 2008). Interestingly, the functional knockdown of twist, besides the impairment of PMCs development, reduces the number of pigment cells and compromise muscle development. Unfortunately, the last observations have not been further investigated by the authors.

8.3.2.6 Coelomic Pouches Cells

The coelomic pouches cells are clustered in two structures evaginating from the tip of the archenteron during the mouth formation (Gustafson and Wolpert 1963). In indirect developers, the left coelomic pouch begins to proliferate after the onset of larval feeding, and forms the adult rudiment, which gives rise to the juvenile sea urchin at metamorphosis (Burke 1982; Pearse and Cameron 1991; Miller and Emlet 1999).

Coelomic pouches include all the progeny of the small micromeres, as demonstrated by staining for 5-bromodeoxyuridine (Tanaka and Dan 1990). The small micromeres are candidate primordial germ cells in sea urchin, although this assumption has not been definitely demonstrated yet (Pehrson and Cohen 1986; Juliano et al. 2006; Voronina et al. 2008). However, the germ cell lineage will not be further discussed here, since additional information can be found in other chapters of the same book.

8.3.2.7 Other SMC-derived Cells

Among blastocoelar cells, a distinct cell type has been characterized by the expression of a serotonin receptor at late developmental stages, i.e. from the prism to the pluteus stage (Katow et al. 2004). By immuno-localization studies, the authors were able to describe the appearance and spatial organization of these cells, which form a blastocoelar network. They have been called "serotonin receptor cells" (SRCs) and have been proposed as new members of the SMC lineage (Katow et al. 2004, 2007). The presence of short fibers extending from the network through the ciliary band ectoderm, or closely associated with the basal surface of the circumesophageal-muscles, strongly suggests that the SRC network might function as a mediator of serotonin signaling, for the neural regulation of ciliary beating or muscle contraction. In agreement, several studies on echinoderm larvae (Asteroids, Ophiuroids and Holothuroids) have demonstrated the presence of a serotonergic system, constituted by ectodermal sensory cells forming the apical organ. From this structure, neurites emerge to innerve the ciliary band, the antero-lateral arms and the esophagus (Bisgrove and Burke 1986; Byrne et al. 2007; Yaguchi et al. 2000; Nakajima et al. 2004; Hirokawa et al. 2008; Bishop and Burke 2007).

Studies on *L. variegatus* embryo reported the transient expression of snail in a population of SMC precursors, including two clusters of mesenchyme cells near the presumptive arm buds at the early prism stage and isolated cells at the pluteus stage. On the basis of the fate mapping of SMCs and double-label immuno-staining with antibodies to snail and to msp130, a PMC specific marker, the presence of a new non-skeletogenic population of SMCs has been suggested (Hardin 1996; Hardin and Illingworth 2006). *Snail* is a transcription factor found and characterized in both protostomes and deuterostomes, and appears to have a crucial role in the development of mesoderm and neural crest by controlling EMTs in chordate embryos (Manzanares et al. 2001; Morales and Nieto 2004). It should be recalled that, in the sea urchin embryo, both PMC and SMC mesodermal populations undergo an EMT at different times during development. Thus, it is intriguing that some mesenchymal cells in the two phyla both express snail, emphasizing the importance of the Snail gene family in mesenchyme differentiation among deuterostomes.

Additional new type of SMCs seems present in the sea urchin embryo, but, since these cells express an unknown gene with no significant similarity with other known genes, for the moment it seems impossible to hypothesize any functional role for them in the embryo and will not be further described (Shoguchi et al. 2002).

8.3.2.8 SMC-derived Skeletogenic Cells and the Transfating Response

A number of studies using surgical and molecular manipulations have revealed that, although skeletogenesis is normally carried out only by PMCs, almost every lineage of the sea urchin embryo has the capacity to activate the skeletogenic gene program, which is a consequence of the developmental plasticity typical of this embryo. Actually, the removal of micromeres at the 16-cell stage results in the transformation of macromere-derived cells into skeletogenic phenotypes (Hörstadius 1939; Sweet et al. 1999). Besides, treatment of isolated animal blastomeres with LiCl (Livingston and Wilt 1989), or their exposure to micromeres (Minokawa et al. 1997), enables them to activate a skeletogenic gene program. In addition, many cells of the vegetal region possess the same ability even after the onset of gastrulation. Thus, the microsurgical removal of PMCs induces some SMCs to differentiate as skeletogenic cells (see details in the following; Ettensohn and McClay 1988) or the combined removal of PMCs from the blastocoel and SMCs from the tip of the elongating archenteron leads some presumptive endodermal cells to switch their fate towards a skeletogenic one (McClay and Logan 1996). To date, these events have been interpreted as the reaction to specific experimental conditions of some cells, which would have never produced a skeleton spontaneously and which are unnaturally induced to change their cellular fate. It seems that, in normal development, SMCs are inhibited to activate the skeletogenic program probably by direct contacts with PMCs' filopodial extension (Ettensohn 1992; Miller et al. 1995). The process leading SMCs to change their cellular fate towards a skeletogenic one, studied at first in *L. variegatus* (Ettensohn and McClay 1988) and recently in *P. lividus* (Kiyomoto et al. 2007), has been conventionally termed as SMC "transfating" or "conversion". It is associated with the *de novo* expression of biomineralization-related genes and PMC-specific transcription factors, and generates the migration of "transfated" cells and their clustering at the PMC-specific sites. A detailed study on gene regulation operating during "transfating" has been reported (Ettensohn et al. 2007). The authors showed that the key components in the regulation of the skeletogenic gene program in the PMC lineage play also an essential role in the regulative pathway operating in "transfating" cells. On the basis of evolutionary studies performed on various echinoderm species, the SMCs' skeletogenic potential has been explained as remains of an ancient program of skeletogenesis, which is not used in extant sea urchin embryos in natural conditions (Ettensohn et al. 2007; see details in the following paragraphs). The evolutionary studies on the echinoderm phylum have been simplified by the possibility to distinguish among embryos showing indirect, direct and intermediate development. Interestingly, this echinoderm feature allowed, for instance, the further characterization of mesenchyme cell lineages in these marine invertebrates. In typical indirect-developing sea urchins, such as *L. variegatus*, *S. purpuratus*, *P. lividus*, *H. pulcherrimus*, where embryo growth proceeds through a feeding larva which undergoes metamorphosis, two populations of embryonic mesenchyme cells develop separately, PMCs and SMCs. On the contrary, direct-developing sea urchins, whose embryos bypass the formation of a larva and turn into juveniles in 3 days without feeding, such as *Heliocidaris erythrogramma*, possess only one embryonic mesenchyme cell population (Wray and McClay 1988; Wray and Raff 1989). In intermediate development, such as in the sand dollar *Peronella japonica*, embryos develop into pluteus larvae that undergo metamorphosis without feeding (Okazaki and Dan 1954) and two mesenchymal cell lines have been distinguished (Yajima 2007a). The "early mesenchyme cells" are micromere descendants involved only in larval skeletogenesis, while the "late mesenchyme cells" are non-micromere descendants involved in adult skeletogenesis. Recently, studies on larval and adult rudiment skeletogenesis occurring in Japanese indirect-developing embryos have been addressed and new interesting data have been reported (Yajima and Kiyomoto 2006; Yajima 2007b). Indeed, it has been surprisingly shown that some SMCs have also a "natural" skeletogenic fate, which becomes manifest at late developmental stages (Yajima 2007b). Therefore, while PMCs are definitely the only cells that synthesize skeleton matrix proteins of the early sea urchin larva (4-armed pluteus), SMCs seem responsible for the formation of the late developing larva skeleton (from the 6-armed pluteus on), reminding the "late mesenchyme cells" observed in *P. japonicus*. All together, these findings agreeably account for the "SMCs' potential" towards a skeletogenic differentiation revealed only under experimental conditions, as reported in old papers. Therefore, it seems necessary a re-evaluation and re-interpretation of previous results concerning SMCs, so far ascribed to generate all types of mesoderm cells, except for the skeletogenic ones. According to Yajima, an important question still remains concerning "late skeletogenic cells" origin, and, for the first time (in the sea urchin embryo literature), the possible involvement of pluripotent stem cells is suggested.

In agreement, some results obtained in our laboratory by microsurgical experiments using *P. lividus* embryos call for a role of embryonic stem cells in the sea urchin embryo. Briefly, microsurgical removal of PMCs resulted in the development of a typical pluteus larva, showing a normally elongated and patterned skeleton, but with a higher number of cells expressing surface PMC-specific antigens, i.e.

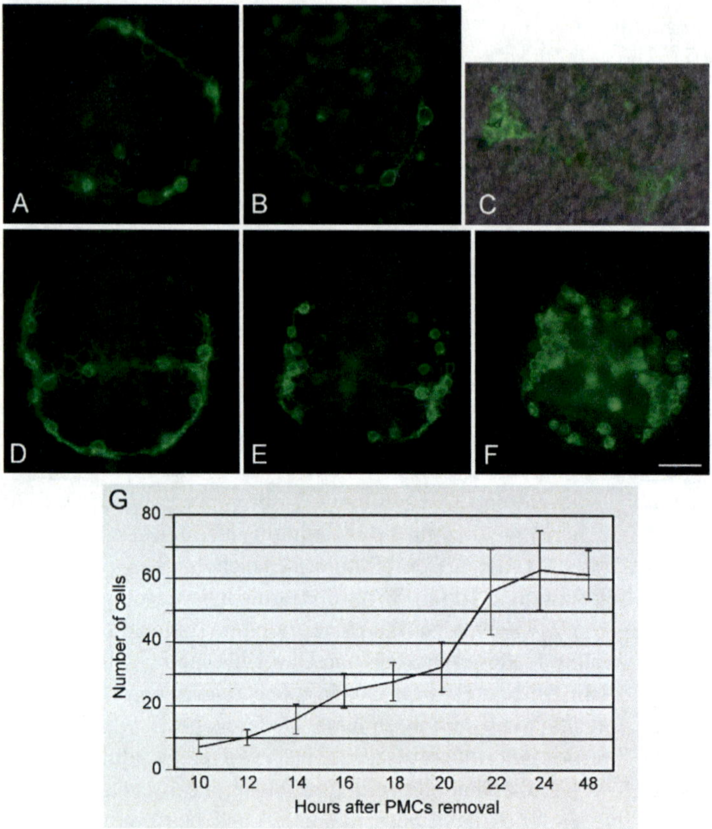

Fig. 8.4 Spatial organization of "transfated" cells in *P. lividus* embryos at different hours after PMCs removal, labelled with a monoclonal antibody against msp130. (**A,B**) and (**D–F**) whole-mount embryos; (**C**) higher magnification of labelled whole-mount embryo merged with corresponding bright-field area. (**A**) 10 h; (**B,C**) 12 h; (**D**) 18 h; (**E**) 20 h; (**F**) 24 h; (**G**) histogram showing number of WGA-positive cells during development at different hours after PMCs removal. *Bar*, 20 μm

WGA (wheat germ agglutinin) and msp130, than those observed in normal development (Kiyomoto et al. 2007). A time course of the so called "transfating" process in *P. lividus* resulted in a linear increase in the appearance of WGA-positive cells together with their arrangement in the typical ventro-lateral clusters, between 10 and 20 h after PMCs removal (Fig. 8.4A–E,G). About 22 h after PMCs removal, the number of WGA/msp130-positive cells associated with the skeleton doubled with respect to that of PMCs present in normal embryos, and such condition remained later on (compare Figs. 8.4F,G with 8.2A). Concurrently, the number of SM30-expressing cells remained almost equal to that of PMCs expressing SM30 in normal embryos (Kiyomoto et al. 2007). Our results appeared rather surprising and in disagreement with the reported and commonly accepted notion that the replacement of missing PMCs is always quantitative, which means that the number of SMCs switching fate relies on the number of PMCs experimentally removed by microsurgery (Ettensohn and McClay 1988). In the attempt to explain the reason for the doubling of WGA/msp130-positive cells in *P. lividus* species, we proposed two likely hypotheses. First, bona fide "transfated" SMCs should be considered only those cells expressing the three skeletal markers (WGA-binding and msp130 proteins and SM30 gene), in agreement with the findings accumulated from different species. Then, in accordance with this assumption, the doubled cells expressing only two of those markers, i.e. WGA and msp130, which are characteristic of early skeletogenesis, should be considered as partially "transfated" cells, which will later, or never, switch completely to a skeletogenic phenotype. An alternative hypothesis is that SMCs "transfate" within 20 h from PMCs removal and, for a still unknown reason, divide in the next few hours, doubling their total number. This hypothesis would account for the very short interval (about 2 h) before we observed cells doubling.

According to the previously reported data showing an intrinsic SMCs late skeletogenic fate (Yajima 2007b), a new hypothesis to explain our results may take into account the involvement of multipotent stem cells. This hypothesis is based on a likely hierarchical model for SMC lineage segregation, according to which progressive fate restrictions of multipotent stem cell give rise to a number of precursors committed to at least seven different SMC-derived phenotypes, namely pigment cells, blastocoelar cells, serotonin receptor cells, snail-positive cells, coelomic pouches cells, muscle cells and late skeletogenic cells (Fig. 8.5). For the moment, the existence of different precursors for blastocoelar, serotonin receptor and snail-positive cells or the presence of a common precursor for the three cell types cannot be verified (see dotted arrows in Fig. 8.5). On the contrary, the presence of a common precursor for pigment and blastocoelar cells, as well as for coelomic pouches and muscle cells, seems established to some extent (Fig. 8.5). Nevertheless, a noteworthy consequence of this model is that the so called "transfated" SMCs actually are an independent cell population, the "late skeletogenic cells", that differentiate ahead of time when are stimulated by particular conditions, such as PMCs removal. In agreement with this hypothesis, we didn't observe any reduction in the number of pigment cells after PMCs removal in *P. lividus* (Fig. 8.6A,B; Kiyomoto et al. 2007), thus excluding that these cells are the "transfated" cells as, on the contrary, reported for other species. In fact, it seems that in *L. variegatus* some pigment cell precursors retain the potential for skeletogenic differentiation, since a reduction in the number of these cells has been documented in PMC-removed embryos (Ettensohn and Ruffins 1993). For the moment, we do not have any explanation for the doubling in the number of "late skeletogenic cells" observed in *P. lividus*.

8.4 SMCs as Embryonic Stem Cells

On the basis of what has been discussed so far and rather in agreement with what suggested by Katow (2005), we believe that sea urchin SMCs share a number of features with vertebrate neural crest cells, despite the fundamental difference in the

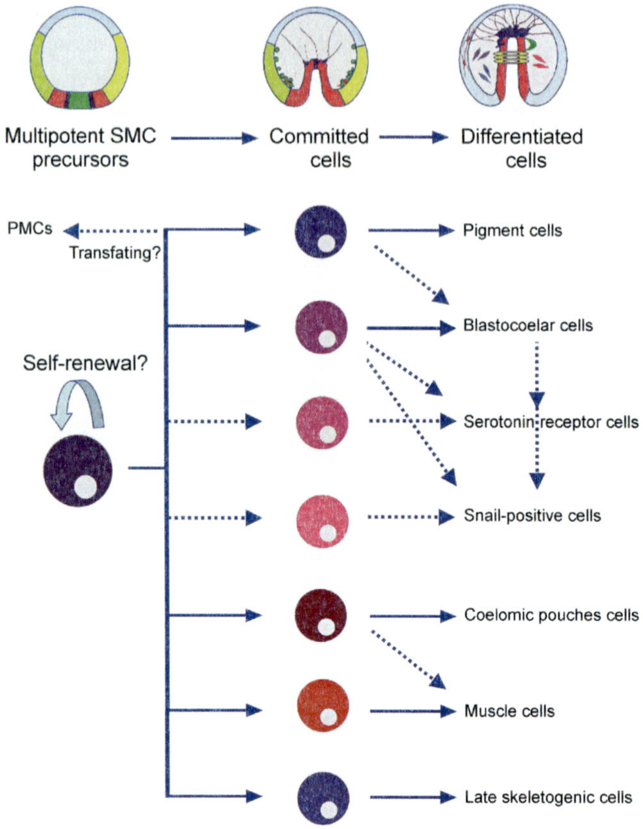

Fig. 8.5 Model of multipotent SMCs specification and SMC-derived cells differentiation (see text for description). Dotted arrows indicate questionable possibilities

germ layer where the two cell populations originate from (mesoderm and ectoderm, respectively) (Fig. 8.7). Because of the different origin of the two cell populations, it doesn't seem appropriate, and it is not our intent, to propose SMCs as the "sea urchin neural crest cells", but we simply think about the neural crest cells as a

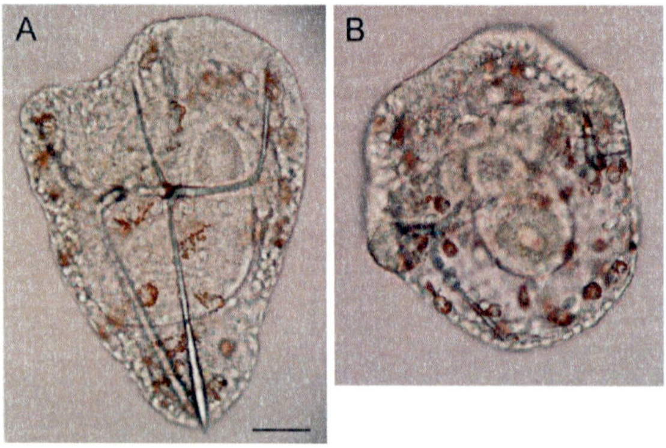

Fig. 8.6 Normal (**A**) and PMC-less (**B**) *P. lividus* embryos, 24 h after PMCs removal. *Bar*, 20 μm

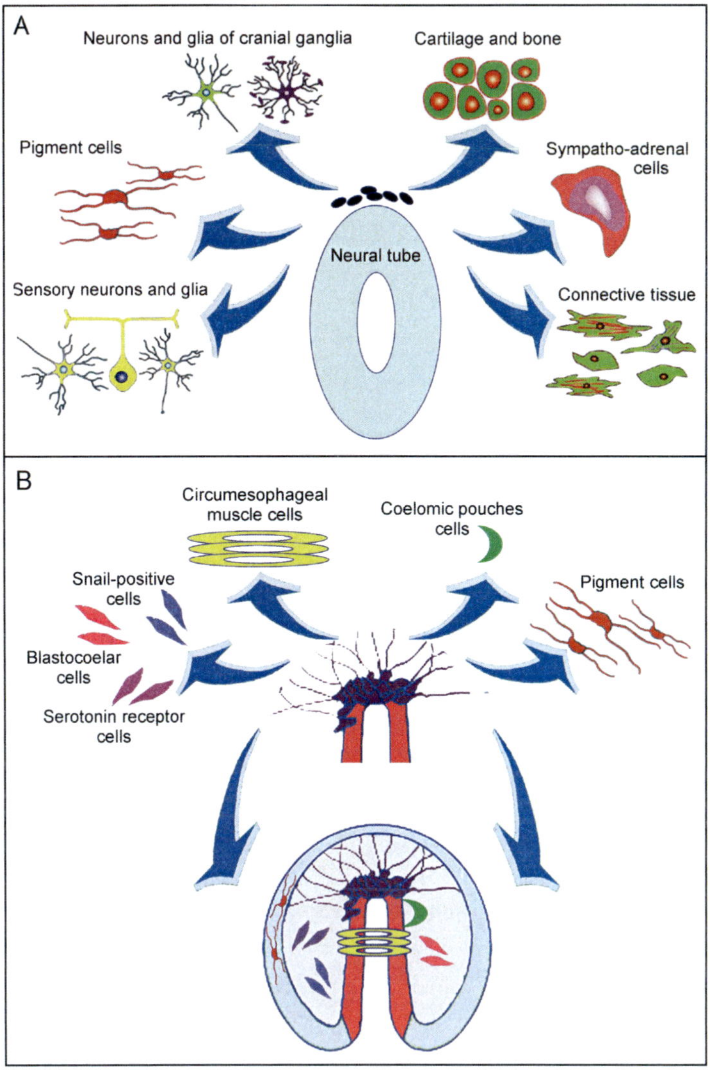

Fig. 8.7 Schematic drawing of neural crest (**A**) and SMCs (**B**) derivatives (see text for description)

good example of vertebrate embryonic stem cells, with a number of interesting well-defined characteristics. Indeed, the neural crest is a not-permanent structure formed by a transient heterogeneous population of stem cells, which are characterized by migratory capabilities for the colonization of various regions of the embryo (see reviews by Fuchs and Sommer 2007; Delfino-Machín et al. 2007). The neural crest cells contribute to a variety of neural and non-neural cell types, such as peripheral neurons (including serotonin receptor cells) and glial cells (Schwann and satellite cells), some endocrine cells (in thyroid and adrenal glands), mesenchymal cells in head and neck (chondrocytes, osteoblasts, odontoblasts, adipocytes, muscle cells) and pigment cells (melanocytes) (Fig. 8.7A). It is generally acknowledged a progressive fate restriction during neural crest development, by which the multipotent capacity of stem cells tends to become increasingly limited over time (see reviews by Weston 1991; Le Douarin and Dupin 2003). These same features can be identified in sea urchin SMCs: they are multipotent cells giving rise to many different types, such as pigment cells, muscle cells, serotonin receptor cells, blastocoelar cells, coelomic pouch cells, skeletogenic cells, with the capacity to migrate inside the embryo before their final differentiation (Fig. 8.7B). As neural crest cells, SMCs arise by delamination from the edge of an epithelial invagination, which for these cells is the elongating archenteron, undergoing an EMT (Fig. 8.7B). As genetic information accumulates, the similarity appears more than phenomenal. In vertebrates, induction and specification of the neural crest is regulated by an interplay of signals, belonging to various signal transduction pathways, including Notch, Wnt, FGF and BMP, released by the surrounding non-neural ectoderm and mesoderm layers (reviewed by Meulemans and Bronner-Fraser 2004; Lee et al. 2004). As a general concept, data on stem cells biology indicate that diverse signaling pathways interact with each other during stem cell development and the resulting signaling output is therefore determined by the concerted action of multiple factors. This implies that each "signaling pathway" has to be considered as part of a more complex "signaling network", meaning that stem cells have to integrate all the signals they receive with their intrinsic properties so to guarantee the generation of both new stem and differentiated cells. In sea urchin embryo, several studies have demonstrated the mutually dependent interactions between Notch and Wnt signalling pathways in early germ layer specification and, in particular, in SMCs specification (see review by Angerer and Angerer 2003). As a consequence of the proposal of SMCs as embryonic stem cells in the sea urchin embryo, a main question arises: do SMCs actively divide, as a typical stem cell would do? By general agreement, stem cells are proliferating precursor cells whose fundamental property, besides the ability to generate a large number of cells committed to further differentiation, is their capacity for self-renewal, reproducing themselves without losing their developmental potentiality. Stem cells can accomplish these two tasks by either asymmetric or symmetric cell divisions, and the equilibrium between these two modes of division is controlled by developmental and environmental signals (details concerning this issue can be found in the review by Morrison and Kimble 2006). As a general rule, in vertebrates, the progression of cells through the cell cycle is regulated by a complex network of kinase activities, a central role played by two groups of evolutionarily conserved molecules, namely cyclins and cyclin-dependent kinases (Cdks) (Obaya and Sedivy 2002; Le Douarin and Dupin 2003). The complete known repertoire of genes associated with the cell cycle control is conserved also in the sea urchin, although with some diversities with respect to the vertebrate one (Fernandez-Guerra et al. 2006). Before the onset of the sequenced genome, however, the sea urchin had been crucially utilised for cell cycle studies. Actually, the first cyclin had been identified exactly in the *Arbacia punctulata* species (Evans et al. 1983). Other cyclins and Cdks have been identified later on and their role in the first mitotic cycles of the early sea urchin embryo has been extensively studied (Geneviere-Garrigues et al. 1995; Moreau et al. 1998). Unfortunately, there is no evidence of a mitotic activity of SMCs yet, thus it remains to be determined whether Cdks are involved in the regulation of SMCs cell cycle.

Another aspect of stem cell biology worthy of attention is the concept of niche, which emerged as a major mechanism of stem cell regulation. The term is referred to a specialized microenvironment with both anatomic and functional implications, i.e. it is a site where stem cells are often located, and provides a set of signals, such as growth factors and extracellular matrix molecules, controlling stem cell self-renewal and responsiveness to differentiation cues (Ohlstein et al. 2004).

In the sea urchin embryo, the blastocoelic cavity might be recognized as the niche for SMCs, since it is where these cells reside during gastrulation. It should be recalled that the blastocoelic cavity is filled with a fluid, containing an indefinite number of undetermined growth factor-like molecules and ECM proteins, thus providing the necessary environment for sea urchin stem cells.

8.5 Concluding Remarks

The molecular tools available nowadays allow several studies on gene expression patterns in a variety of stem cell systems, such as neural crest cells and mouse embryonic stem cells, making possible the description of a number of gene regulatory networks involved in the control of stem cell pluripotency (Meulemans and Bronner-Fraser 2004; Zhou et al. 2007; Niwa 2007). Despite the great amount of molecular data collected so far, many aspects of the basic mechanisms regulating the maintenance of pluripotency of stem cells as well as the beginning of differentiation await to be elucidated. As a result of what discussed in this chapter, we think that the sea urchin embryo could be a useful model system also in the field of stem cell biology. Indeed, for further understanding of the processes controlling SMCs migration, homing and differentiation, detailed studies focused on the identification of stem cell-specific markers and directed at deciphering the role of genetic networks and molecular pathways involved in these processes are necessary. One of the strategies of choice to identify stem cell lineages is still the use of specific antibodies, either monoclonal or polyclonal, which allow the assignment of the "stemness" phenotype by the analysis of surface molecules and intracellular markers (Cai et al. 2005). By this strategy, a good number of molecules have been found to serve as specific stem cell molecular markers by means of their unique pattern and timing of expression, rather than by their function, which sometimes remains unknown. Of course, no-one marker is uniquely sufficient for the identification of a particular stem cell, but rather a critical amount of some of them is required to sustain stem cell pluripotency. For example, human and mouse embryonic stem cells can be characterized by high expression levels of Oct3/4 and Nanog, two transcription factors which are down-regulated as cells differentiate *in vitro* and *in vivo* (Niwa et al. 2000; Cavaleri and Scholer 2003).

Unfortunately, few markers of the typical "stemness" phenotype are currently known in the sea urchin but, certainly, a good number of them are waiting to be identified and characterized in the near future. Thus, with the availability of specific markers, an *in vitro* approach could be taken into account. In fact, it was the opportunity to culture individual neural crest cells *in vitro* and to treat them with soluble growth factors that led to the characterization of the extracellular signals involved in stem cell development. Then, it seems useful to device a method for the isolation of multipotent SMCs from the tip of the invaginating archenteron and to culture them *in vitro*, supposing that the right conditions of culture are found (Fig. 8.8). Such

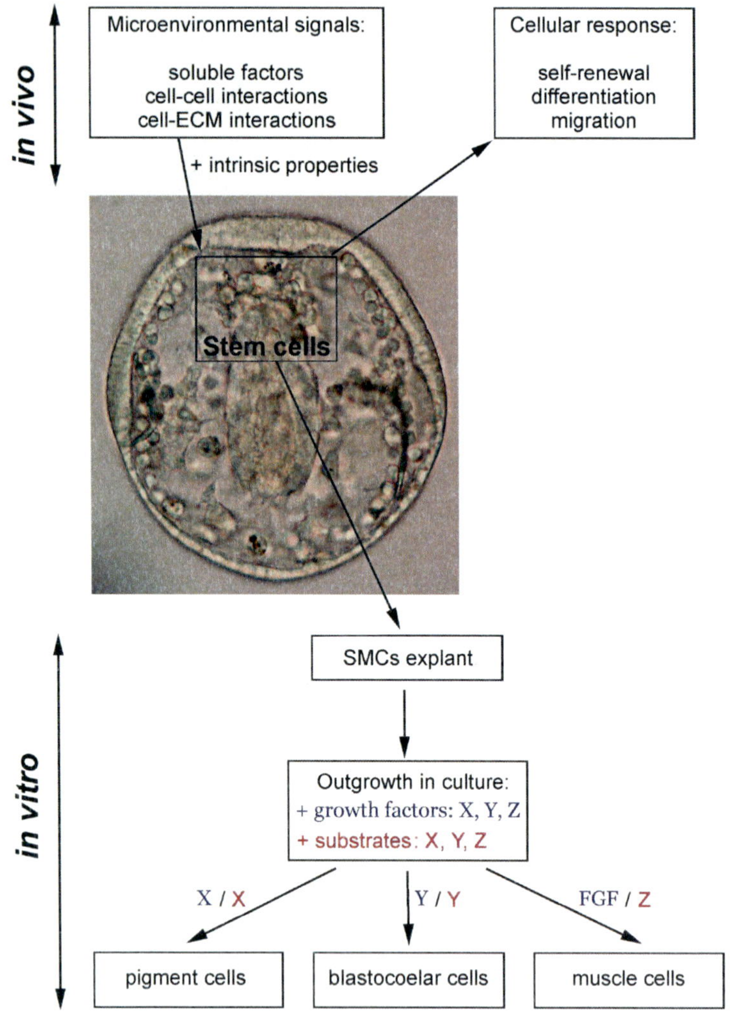

Fig. 8.8 Schematic drawing of supposed factors regulating *in vivo* and *in vitro* SMCs fate in sea urchin embryo

an *in vitro* approach would highlight the variety of progenitors present in SMC population (see Fig. 8.5) and allow the identification of the extracellular signals controlling their development.

Taken together, growing evidence point to the statement that sea urchin embryo has much to teach us yet, as it provides a valuable physiological *in vivo*, and probably *in vitro*, context in which to investigate the complexities of stem cell biology.

Acknowledgments We are grateful to Maria Byrne and Hideki Katow for their critical reading of a draft of this manuscript. We warmly thank all the other members of the group for their help and support. F.Z. wishes to express her gratitude to a few colleagues of IBIM for their continuous encouragement, helpfulness and friendliness. Authors are grateful for the partial financial support of the Marine Genomics Europe Network of the Excellence and MoMe ASI project.

References

Akiyama Y, Hosoya T, Poole AM, Hotta Y (1996) The gcm-motif: A novel DNA-binding motif conserved in Drosophila and mammals. Proc Natl Acad Sci USA 93:14912–14916

Angerer LM, Angerer RC (2000) Animal-vegetal axis patterning mechanisms in the early sea urchin embryo. Dev Biol 218:1–12

Angerer LM, Angerer RC (2003) Patterning the sea urchin embryo: Gene regulatory networks, signaling pathways, and cellular interactions. Curr Top Dev Biol 53:159–198

Armstrong N, McClay DR (1994) Skeletal pattern is specified autonomously by the primary mesenchyme cells in sea urchin embryos. Dev Biol 162:329–338

Beniash E, Aizenberg J, Addadi L, Weiner S (1997) Amorphous calcium carbonate transforms into calcite during sea urchin larval spicule growth. Proc R Soc Lond B Biol Sci 264: 461–465

Bisgrove BW, Burke RD (1986) Development of serotonergic neurons in embryos of the sea Urchin, *Strongylocentrotus purpuratus*. Dev Growth Diff 28:569–574

Bishop CD, Burke RD (2007) Ontogeny of the holothurian larval nervous system: Evolution of larval forms. Dev Genes Evol 217:585–592

Bromham LD, Degnan BM (1999) Hemichordate and deuterostome evolution: Robust molecular phylogenetic support for a hemichordate plus echinoderm clade. Evol Dev 1:166–171

Burke RD (1981) Structure of the digestive tract of the pluteus larva of *Dendraster excentricus* (Echinodermata: Echinoida). Zoomorphology 98:209–225

Burke RD (1982) Echinoderm metamorphosis: Retraction and resorption of larval tissues. In: Lawrence JM (ed) International Echinoderms Conference. Tampa Bay, Balkema, Rotterdam, pp 513–518

Burke RD, Alvarez CM (1988) Development of the esophageal muscles in embryos of the sea urchin *Strongylocentrotus purpuratus*. Cell Tissue Res 252:411–417

Byrne M, Nakajima Y, Chee FC, Burke RD (2007) Apical organs in echinoderm larvae: Insights into larval evolution in the Ambulacraria. Evol Dev 9(5):432–445

Cai J, Olson JM, Rao MS, Stanley M, Taylor E, Ni HT (2005) Development of antibodies to human embryonic stem cell antigens. BMC Dev Biol 5:26

Cameron RA, Fraser SE, Britten RJ, Davidson EH (1991) Macromere cell fates during sea urchin development. Development 113:1085–1091

Cavaleri F, Scholer HR (2003) Nanog: A new recruit to the embryonic stem cell orchestra. Cell 113:551–552

Chaffee RR, Mazia D (1963) Echinochrome synthesis in hybrid sea urchin embryos. Dev Biol 7:502–512

Chiba S (2006) Notch signaling in stem cell systems. Stem Cells 24:2437–2447

Crawford B, Martin C (1998) Ultrastructure and differentiation of the larval esophageal muscle cells of the starfish Pisaster ochraceus. J Morphol 237:1–18
Dan K, Okazaki K (1956) Cyto-embryological studies of sea urchins. III. Role of the secondary mesenchyme cells in the formation of the primitive gut in sea urchin larvae. Biol Bull 110: 29–42
Davidson EH, Cameron RA, Ransick A (1998) Specification of cell fate in the sea urchin embryo: Summary and some proposed mechanisms. Development 125:3269–3290
Davidson EH, Rast JP, Oliveri P, Ransick A, Calestani C, Yuh CH, Minokawa T, Amore G, Hinman V, Arenas-Mena C et al (2002a) A provisional regulatory gene network for specification of endomesoderm in the sea urchin embryo. Dev Biol 246:162–190
Davidson EH, Rast JP, Oliveri P, Ransick A, Calestani C, Yuh CH, Minokawa T, Amore G, Hinman V, Arenas-Mena C, Otim O, Brown CT, Livi CB, Lee PY, Revilla R, Rust AG, Pan Z, Schilstra MJ, Clarke PJ, Arnone MI, Rowen L, Cameron RA, McClay DR, Hood L, Bolouri H (2002b) A genomic regulatory network for development. Science 295:1669–1678
Decker GL, Lennarz WJ (1988) Skeletogenesis in the sea urchin embryo. Development 103: 231–247
Delfino-Machín M, Chipperfield TR, Rodrigues FSLM, Kelsh RN (2007) The proliferating field of neural crest stem cells. Dev Dyn 236:3242–3254
Denham M, Conley B, Olsson F, Cole TJ, Mollard R (2005) Stem cells: An overview. Curr Protoc Cell Biol 23:23.1
Driesch H (1892) The potency of the first two cleavage cells in echinoderm development. Experimental production of partial and double formations. In: BH Willier, JM Oppenheimer (eds) Foundations of Experimental Embryology pp 38–55 (1974). Hafner, New York
Duloquin L, Lhomond G, Gache C (2007) Localized VEGF signaling from ectoderm to mesen-chyme cells controls morphogenesis of the sea urchin embryo skeleton. Development 134:2293–2302
Egana AL, Ernst SG (2004) Sphedgehog is expressed by pigment cell precursors during early gastrulation in *Strongylocentrotus purpuratus*. Dev Dyn 231:370–378
Ettensohn CA (1990) The regulation of primary mesenchyme cell patterning. Dev Biol 140: 261–271
Ettensohn CA (1992) Cell interactions and mesodermal cell fates in the sea urchin embryo. Development Suppl 43–51
Ettensohn CA, Malinda KM (1993) Size regulation and morphogenesis: A cellular analysis of skeletogenesis in the sea urchin embryo. Development 119:155–167
Ettensohn CA, McClay DR (1988) Cell lineage conversion in the sea urchin embryo. Dev Biol 125:396–409
Ettensohn CA, Ruffins SW (1993) Mesodermal cell interactions in the sea urchin embryo: Properties of skeletogenic secondary mesenchyme cells. Development 117:1275–1285
Ettensohn CA, Sweet HC (2000) Patterning the early sea urchin embryo. Curr Top Dev Biol 50:1–44
Ettensohn CA, Kitazawa C, Cheers MS, Leonard JD, Sharma T (2007) Gene regulatory networks and developmental plasticity in the early sea urchin embryo: Alternative deployment of the skeletogenic gene regulatory network. Development 134:3077–3087
Evans T, Rosenthal ET, Youngblom J, Distel D, Hunt T (1983) Cyclin: A protein specified by maternal mRNA in sea urchin eggs that is destroyed at each cleavage division. Cell 33:389–396
Fernandez-Guerra A, Aze A, Morales J, Mulner-Lorillon O, Cosson B, Cormier P, Bradham C, Adams N, Robertson AJ, Marzluff WF, Coffman JA, Geneviere AM (2006) The genomic repertoire for cell cycle control and DNA metabolism in *S. purpuratus*. Dev Biol 300: 238–251
Figeac N, Daczewska M, Marcelle C, Jagla K (2007) Muscle stem cells and model systems for their investigation. Dev Dyn 236:3332–3342
Fink RD, McClay DR (1985) Three cell recognition changes accompany the ingression of sea urchin primary mesenchyme cells. Dev Biol 107:66–74

Fuchs S, Sommer L (2007) The neural crest: Understanding stem cell function in development and disease. Neurodegener Dis 4:6–12

Geneviere-Garrigues AM, Barakat A, Doree M, Moreau JL, Picard A (1995) Active cyclin B-cdc2 kinase does not inhibit DNA replication and cannot drive prematurely fertilized sea urchin eggs into mitosis. J Cell Sci 108:2693–2703

Gibson AW, Burke RD (1985) The origin of pigment cells in embryos of the sea urchin *Strongylocentrotus purpuratus*. Dev Biol 107:414–419

Gibson AW, Burke RD (1987) Migratory and invasive behavior of pigment cells in normal and animalized sea urchin embryos. Exp Cell Res 173:546–557

Griffiths M (1965) A study of the synthesis of naphthaquinone pigments by the larvae of two species of sea urchins and their reciprocal hybrids. Dev Biol 11:433–447

Gustafson T, Kinnander H (1956) Microaquaria for time-lapse cinematographic studies of morphogenesis in swimming larvae and observations on sea urchin gastrulation. Exp Cell Res 21:36–51

Gustafson T, Wolpert L (1963) Studies on the cellular basis of morphogenesis in the sea urchin embryo. Formation of the coelom, the mouth, and the primary pore-canal. Exp Cell Res 29: 561–582

Hara Y, Katow H (2005) Exclusive expression of hedgehog in small micromere descendants during early embryogenesis in the sea urchin, *Hemicentrotus pulcherrimus*. Gene Exp Patterns 5:503–510

Hardin J (1996) The cellular basis of sea urchin gastrulation. Curr Top Dev Biol 33: 159–262

Hardin J, Illingworth CA (2006) A homologue of snail is expressed transiently in subsets of mesenchyme cells in the sea urchin embryo and is down-regulated in axis-deficient embryos. Dev Dyn 235:3121–3131

Harkey MA, Whiteley HR, Whiteley AH (1992) Differential expressionof the msp 130 gene among skeletal lineage cells in the sea urchin embryo: A three dimensional in situ hybridization analysis. Mech Dev 37:173–184

Hibino T, Loza-Coll M, Messier C, Majeske AJ, Cohen AH, Terwilliger DP, Buckley KM, Brockton V, Nair SV, Berney K, Fugmann SD, Anderson MK, Pancer Z, Cameron RA, Smith LC, Rast JP (2006) The immune gene repertoire encoded in the purple sea urchin genome. Dev Biol 300:349–365

Hirokawa T, Komatsu M, Nakajima Y (2008) Development of the nervous system in the brittle star Amphipholis kochii. Dev Gens Evol 218:15–21

Hodor PG, Ettensohn CA (1998) The dynamics and regulation of mesenchymal cell fusion in the sea urchin embryo. Dev Biol 199:111–124

Hörstadius S (1939) The mechanics of sea urchin development, studied by operative methods. Biol Rev 14:132–179

Howard-Ashby M, Materna SC, Brown CT, Chen L, Cameron RA, Davidson EH (2006) Gene families encoding transcription factors expressed in early development of *Strongylocentrotus purpuratus*. Dev Biol 300:90–107

Ishimoda-Takagi T, Chino I, Sato H (1984) Evidence for the involvement of muscle tropomyosin in the contractile elements of the coelom-esophagus complex in sea urchin embryos. Dev Biol 105:365–376

Juliano CE, Voronina E, Stack C, Aldrich M, Cameron AR, Wessel GM (2006) Germ line determinants are not localized early in sea urchin development, but do accumulate in the small micromere lineage. Dev Biol 300:406–415

Katow H (2005) Multipotential cells in echinoderm: Origin and role in embryogenesis. Zool Sci 22:1387

Katow H, Yaguchi S, Kiyomoto M, Washio M (2004) The 5-HT receptor cell is a new member of secondary mesenchyme cell descendants and forms a major blastocoelar network in sea urchin larvae. Mech Dev 121:325–337

Katow H, Yaguchi S, Kyozuka K (2007) Serotonin stimulates [Ca2+]i elevation in ciliary ectodermal cells of echinoplutei through a serotonin receptor cell network in the blastocoel. Exp Biol 210:403–412
Khaner O, Wilt F (1991) Interactions of different vegetal cells with mesomeres during early stages of sea urchin development. Development 112:881–890
Kiyomoto M, Zito F, Sciarrino S, Matranga V (2004) Commitment and response to inductive signals of primary mesenchyme cells of the sea urchin embryo. Dev Growth Diff 46:107–113
Kiyomoto M, Zito F, Costa C, Poma V, Sciarrino S, Matranga V (2007) Skeletogenesis by transfated secondary mesenchyme cells is dependent on extracellular matrix–ectoderm interactions in *Paracentrotus lividus* sea urchin embryos. Dev Growth Diff 49:731–741
Kleber M, Sommer L (2004) Wnt signaling and the regulation of stem cell function. Curr Opin Cell Biol 16:681–687
Le Douarin NM, Dupin E (2003) Multipotentiality of the neural crest. Curr Opin Genet Dev 13:529–536
Lee HY, Kleber M, Hari L, Brault V, Suter U, Taketo MM, Kemler R, Sommer L (2004) Instructive role of Wnt/beta-catenin in sensory fate specification in neural crest stem cells. Science 303:1020–1023
Leptin M (1991) Twist and snail as positive and negative regulators during *Drosophila* mesoderm development. Genes Dev 5:1568–1576
Livingston BT, Wilt FH (1989) Lithium evokes expression of vegetal-specific molecules in the animal blastomeres of sea urchin embryos. Proc Natl Acad Sci USA 86:3669–3673
Livingston BT, Killian CE, Wilt F, Cameron A, Landrum MJ, Ermolaeva O, Sapojnikov V, Maglott DR, Buchanan AM, Ettensohn CA (2006) A genome-wide analysis of biomineralization-related proteins in the sea urchin *Strongylocentrotus purpuratus*. Dev Biol 300:335–348
Malinda KM, Fisher GW, Ettensohn CA (1995) Four-dimensional microscopic analysis of the filopodial behaviour of primary mesenchyme cells during gastrulation in the sea urchin embryo. Dev Biol 172:552–566
Manzanares M, Locascio A, Nieto MA (2001) The increasing complexity of the Snail gene superfamily in metazoan evolution. Trends Genet 17:178–181
Materna SC, Davidson EH (2007) Logic of gene regulatory networks. Curr Opin Biotech 18: 351–354
Matsuno T, Tsushima M (2001) Carotenoids in sea urchins. In: Lawrence JM (ed) Edible sea urchins: Biology and ecology. Elsevier Science, Amsterdam
McClay DR, Logan CY (1996) Regulative capacity of the archenteron during gastrulation in the sea urchin. Development 122:607–616
McClay DR, Peterson RE, Range RC, Winter-Vann AM, Ferkowicz MJ (2000) A micromere induction signal is activated by beta-catenin and acts through notch to initiate specification of secondary mesenchyme cells in the sea urchin embryo. Development 127:5113–5122
McCoon PE, Blackstone E, Angerer RC, Angerer LM (1998) Sea Urchin FGFR muscle-specific expression: Posttranscriptional regulation in embryos and adults. Dev Biol 200:171–181
Meulemans D, Bronner-Fraser M (2004) Gene-regulatory interactions in neural crest evolution and development. Dev Cell 7:291–299
Miller J, Fraser SE, McClay D (1995) Dynamics of thin filopodia during sea urchin gastrulation. Development 121:2501–2511
Miller BA, Emlet RB (1999) Development of newly metamorphosed juvenile sea urchins (*Strongylocentrotus franciscanus*and *S. purpuratus*): Morphology, the effects of temperature and larval food ration, and a method for determining age. J Exp Mar Biol Ecol 235:67–90
Minokawa T, Hamaguchi Y, Amemiya S (1997) Skeletogenic potential of induced secondary mesenchyme cells derived from the presumptive ectoderm in echinoid embryos. Dev Genes Evol 206:472–476
Minokawa T, Wikramanayake AH, Davidson EH (2005) cis-Regulatory inputs of the wnt8 gene in the sea urchin endomesoderm network. Dev Biol 288:545–558
Molofsky AV, Pardal R, Morrison SJ (2004) Diverse mechanisms regulate stem cell self-renewal. Curr Opin Cell Biol 16:700–707

Monroy A, Oddo AM, Denicola M (1951) The carotenoid pigments during early development of the egg of the sea urchin *Paracentrotus lividus*. Exp Cell Res 2:700–702
Morales AV, Nieto MA (2004) The Snail gene family during gastrulation. In: Stern CD (ed) Gastrulation: From cells to embryos. Cold Spring Harbor Lab Press, New York, pp 631–641
Moreau JL, Marques F, Barakat A, Schatt P, Lozano JC, Peaucellier G, Picard A, Geneviere AM (1998) Cdk2 activity is dispensable for the onset of DNA replication during the first mitotic cycles of the sea urchin early embryo. Dev Biol 200:182–197
Morrison SJ, Kimble J (2006) Asymmetric and symmetric stem-cell divisions in development and cancer. Nature 441:1068–1074
Nakajima Y, Kaneko H, Murray G, Burke R (2004) Divergent patterns of neural development in larval echinoderm and asteroids. Evol Dev 6:95–104
Niwa H (2007) How is pluripotency determined and maintained? Dev 134:635–646
Niwa H, Miyazaki J, Smith AG (2000) Quantitative expression of Oct-3/4 defines differentiation, dedifferentiation or self-renewal of ES cells. Nat Genet 24:372–376
Obaya AJ, Sedivy JM (2002) Regulation of cyclin-Cdk activity in mammalian cells. Cell Mol Life Sci 59:126–142
Ohlstein B, Kai T, Decotto E, Spradling A (2004) The stem cell niche: Theme and variations. Curr Opin Cell Biol 16:693–699
Okazaki K (1975) Spicule formation by isolated micromeres of the sea urchin embryo. Am Zool 15:567–581
Okazaki K, Dan K (1954) The metamorphosis of partial larvae of *Peronella japonica*Mortensen, a sand dollar. Biol Bull 106:83–99
Pearse JS, Cameron RA (1991) Echinodermata: Echinoidea. In: Giese AC, Pearse JS, Pearse VB (eds) Reproduction of Marine Invertebrates, vol 6. Boxwood Press, Pacific Grove, pp 514–662
Pehrson JR, Cohen LH (1986) The fate of the small micromeres in sea urchin development. Dev Biol 113:522–526
Peterson RE, McClay DR (2003) Primary mesenchyme cell patterning during the early stages following ingression. Dev Biol 254:68–78
Raible DW (2006) Development of the neural crest: Achieving specificity in regulatory pathways. Curr Opin Cell Biol 18:698–703
Ransick A, Davidson EH (2006) cis-regulatory processing of Notch signaling input to the sea urchin glial cells missing gene during mesoderm specification. Dev Biol 297:587–602
Ransick A, Rast JP, Minokawa T, Calestani C, Davidson EH (2002) New early zygotic regulators expressed in endomesoderm of sea urchin embryos discovered by differential array hybridization. Dev Biol 246:132–147
Röttinger E, Saudemont A, Duboc V, Besnardeau L, McClay D, Lepage T (2008) FGF signals guide migration of mesenchymal cells, control skeletal morphogenesis of the skeleton and regulate gastrulation during sea urchin development. Development 135:352–365
Ruffins SW, Ettensohn CA (1993) A clonal analysis of secondary mesenchyme cell fates in the sea urchin embryo. Dev Biol 160:285–288
Ruffins SW, Ettensohn CA (1996) A fate map of the vegetal plate of the sea urchin (*Lytechinus variegatus*) mesenchyme blastula. Development 122:253–263
Ryberg E, Lundgren B (1979) Some aspects on pigment cell distribution and function in developing echinopluteus of *Psamechinus miliaris*. Dev Growth Diff 21:129–140
Sea urchin genome sequencing consortium et al (2006) The genome of the sea urchin *Strongylocentrotus purpuratus*. Science 314:941–952
Service M, Wardlaw AC (1984) Echinochrome-a as a bactericidal substance in the coelomic fluid of *Echinus esculentus*. Comp Biochem Physiol 79B:161–165
Sherwood DR, McClay DR (1999) LvNotch signaling mediates secondary mesenchyme specification in the sea urchin embryo. Development 126:1703–1713
Sherwood DR, McClay DR (2001) LvNotch signaling plays a dual role in regulating the position of the ectoderm-endoderm boundary in the sea urchin embryo. Development 128: 2221–2232

Shoguchi E, Tokuoka M, Kominami T (2002) In situ screening for genes expressed preferentially in secondary mesenchyme cells of sea urchin embryos. Dev Genes Evol 212:407–418

Silva JR (2000) The onset of phagocytosis and identity in the embryo of *Lytechinus variegatus*. Dev Comp Immunol 24:733–739

Smith AB (2004) Echinoderm roots. Nature 430:411–412

Smith LC, Rast JP, Brockton V, Terwilliger DP, Nair SV, Buckley KM, Majeske AJ (2006) The sea urchin immune system. Invert Surviv J 3:25–39

Sommer L (2005) Checkpoints of melanocyte stem cell development. Sci STKE 298:pe42

Stephens L, Hardin J, Keller R, Wilt F (1986) The effects of aphidicolin on morphogenesis and differentiation in the sea urchin embryo. Dev Biol 118:64–69

Sweet HC, Hodor PG, Ettensohn CA (1999) The role of micromere signaling in Notch activation and mesoderm specification during sea urchin embryogenesis. Development 126:5255–5265

Sweet HC, Gehring M, Ettensohn CA (2002) LvDelta is a mesoderm-inducing signal in the sea urchin embryo and can endow blastomeres with organizer-like properties. Development 129:1945–1955

Tamboline CR, Burke RD (1992) Secondary mesenchyme of the sea urchin embryo: Ontogeny of blastocoelar cells. J Exp Zool 262:51–60

Tanaka S, Dan K (1990) Study of the lineage and cell cycle of small micromeres in embryos of the sea urchin, *Hemicentrotus pulcherrimus*. Dev Growth Differ 32:145–156

Tokuoka M, Setoguchi C, Kominami T (2002) Specification and differentiation processes of secondary mesenchyme-derived cells in embryos of the sea urchin *Hemicentrotus pulcherrimus*. Dev Growth Diff 44:239–250

Vance KW, Goding CR (2004) The transcription network regulating melanocyte development and melanoma. Pigment Cell Res 17:318–325

Voronina E, Lopez M, Juliano CE, Gustafson E, Song JL, Extavour C, George S, Oliveri P, McClay D, Wessel G (2008) Vasa protein expression is restricted to the small micromeres of the sea urchin, but is inducible in other lineages early in development. Dev Biol 314:276–286

Walton KD, Croce JC, Glenn TD, Wu SY, McClay RD (2006) Genomics and expression profiles of the Hedgehog and Notch signaling pathways in sea urchin development. Dev Biol 300:153–164

Weston JA (1991) Sequential segregation and fate of developmentally restricted intermediate cell populations in the neural crest lineage. Curr Top Dev Biol 25:133–153

Wikramanayake AH, Peterson R, Chen J, Huang L, McClay DR, Klein WH (2004) Nuclear b-catenin-dependent Wnt8 signaling in vegetal cells of the early sea urchin embryo regulates gastrulation and differentiation of endoderm and mesodermal cell lineages. Genesis 39: 194–205

Wilt FH (1999) Matrix and mineral in the sea urchin larval skeleton. J Struct Biol 126:216–226

Wilt FH (2002) Biomineralization of the spicules of sea urchin embryos. Zool Sci 19:253–261

Wilt FH (2005) Developmental biology meets materials science: Morphogenesis of biomineralized structures. Dev Biol 280:15–25

Wilt FH, Killian CE, Livingston BT (2003) Development of calcareous skeletal elements in invertebrates. Diff 71:237–250

Wray GA (1996) Parallel evolution of nonfeeding larvae in echinoids. Syst Biol 45:308–322

Wray GA, McClay DR (1988) The origin of spicule forming cells in a "primitive" sea urchin (*Eucidaris tribuloides*) which appears to lack primary mesenchyme cell. Development 103:305–315

Wray GA, Raff RA (1989) Evolutionary modification of cell lineage in the direct-developing sea urchin *Heliocidaris erythrogramma*. Dev Biol 132:458–470

Wu J, Saint-Jeannet JP, Klein PS (2003) Wnt-frizzled signaling in neural crest formation. Trends Neurosc 26:40–45

Wu SY, McClay DR (2007) The Snail repressor is required for PMC ingression in the sea urchin embryo. Development 134:1061–1070

Wu SH, Yang YP, McClay DR (2008) Twist is an essential regulator of the skeletogenic gene regulatory network in the sea urchin embryo. Dev Biol 319:406–415

Yaguchi S, Kanoh K, Amemiya S, Katow H (2000) Initial analysis of immunocytochemical surface properties, location and formation of the serotonergic apical organ in sea urchin embryos. Dev Growth Diff 42:479–488

Yajima M (2007a) Evolutionary modification of mesenchyme cells in sand dollars in the transition from indirect to direct development. Evol Dev 9:258–267

Yajima M (2007b) A switch in the cellular basis of skeletogenesis in late-stage sea urchin larvae. Dev Biol 307:272–281

Yajima M, Kiyomoto M (2006) Study of larval and adult skeletogenic cells in the developing sea urchin larvae. Biol Bull 211:183–192

Yokota Y, Matranga V, Smolenicka Z (eds) (2002) Sea urchin: From basic biology to aquaculture. Balkema, The Netherlands

Young RS (1958) Development of pigment in the larvae of the sea urchin, *Lytechinus variegatus*. Biol Bull 114:394–403

Zandonella C (2005) Stem-cell therapies: The first wave. Nature 435:877–878

Zhan M (2008) Genomic studies to explore self-renewal and differentiation properties of embryonic stem cells. Frontiers Biosc 13:276–283

Zhou Q, Chipperfield H, Melton DA, Wong WH (2007) A gene regulatory network in mouse embryonic stem cells. Proc Natl Acad Sci 104:16438–16443

Zito F, Costa C, Sciarrino S, Poma V, Russo R, Angerer LM, Matranga V (2003) Expression of univin, a TGF-β growth factor, requires ectoderm-ECM interaction and promotes skeletal growth in the sea urchin embryo. Dev Biol 264:217–227

Chapter 9
Cell Dynamics in Early Embryogenesis and Pluripotent Embryonic Cell Lines: From Sea Urchin to Mammals

Anne-Marie Genevière, Antoine Aze, Yasmine Even, Maria Imschenetzky, Clara Nervi, and Luigi Vitelli

Abstract From oogenesis through fertilization and gastrulation, embryos use various mechanisms to regulate cell expansion, keeping a strict balance between cell proliferation, cell differentiation and cell death. While rapid divisions are necessary at the initial stage to ensure early embryo survival, further developmental transitions are marked by changes in cell cycle and transcriptional regulation. Pluripotency and capability of self-renewal are maintained in a low percentage of cells, the embryonic stem cells (ESCs), which will be later used as a cellular source for tissue replacement. Clearly, some essential characteristics of cell cycle and transcriptional regulation of the early embryo will be conserved in ESC lines. We addressed here the peculiarities of these developmental programs in early embryos and pluripotent embryonic cell lines, considering examples from marine invertebrates to mammals. Finally, we discussed the importance of transcription regulation and chromatin remodelling and their peculiar features in embryonic cells from these species.

Keywords Pluripotent cell lines · early development · cell cycle · chromatin remodelling · transcription factors

9.1 Introduction

The fusion of a male and a female gamete forms a first unique totipotent cell, which gives rise to a progeny of cells becoming increasingly restricted in their differentiation potential as the embryo develops. Totipotency, the self-organizing ability to generate all somatic, germ-line and extraembryonic tissues, is conserved for only few divisions, the number of which depends on the species. The persistence of the capacity to generate all adult tissues, defines pluripotent embryonic cells (Solter 2006). As pluripotent cells from blastocyst are committed toward one of the

A.-M. Genevière (✉)
Université Pierre et Marie Curie-Paris6, CNRS, UMR 7628, Banyuls-sur-mer, BP44, Avenue du Fontaulé, F-66651, France
e-mail: anne-marie.geneviere@obs-banyuls.fr

B. Rinkevich, V. Matranga (eds.), *Stem Cells in Marine Organisms*,
DOI 10.1007/978-90-481-2767-2_9, © Springer Science+Business Media B.V. 2009

germ layers that constitute the early embryo, they lose pluripotency but transiently conserving multipotency, the ability to generate multiple cell types within a particular lineage. A pool of stem cells persists in several adult tissues able to generate the type of cells present in the tissue where they reside. These different transitions are regulated in a species specific manner.

It is to note that pioneering experiments on pluripotency have been conducted in sea urchin embryos, an invertebrate model for mammalian development. Indeed in 1891, Hans Driesch (Driesch 1891) demonstrated that a single blastomere separated from a sea-urchin egg at the 2-cell stage was able to develop into a smaller but normal pluteus larva and not into a partial larva, as was the case for the ascidian egg. Similar results were obtained with blastomeres from the 4-cell stage. The analysis of "equipotentiality" was extended by Jacques Loeb (Loeb 1894). This regulative phenomenon aroused wide interest and laid foundation of the concepts of pluripotency of embryonic stem cells (ESCs). However, ESCs were isolated and established *in vitro* about one hundred years later, initially from the mouse (Evans and Kaufman 1981; Martin 1981) and then from primates: monkey and human (Thomson et al. 1998; Thomson et al. 1995).

9.2 Cell Fate Determination in Embryo, ESC *In Situ*

9.2.1 Stem Cell as an Evolutionary Unit

Despite the above mentioned pioneering experiments of Driesch and Loeb, it has been thought for years that ESCs were restricted to mammals until those cells were also identified and characterized in birds (Pain et al. 1996), in fish (Ma and Collodi 1999) and also in planarians (Sanchez Alvarado 2007). It is now believed that the existence of stem cells is a conserved feature in evolution (Weissman 2000). The limited number of species by which stem cells could be established reveals the difficulty to empirically define ESC culture conditions and to determine for each species the embryonic stage from which successful isolation of pluripotent cells can be obtained. This strengthens the necessity of further investigation of the mechanisms controlling the maintenance of pluripotency *in vivo* and *in vitro*.

The passage from a primary totipotent cell to an embryo made of spatially organized differentiated cells is a complex process. While bilaterian body plans are amazingly diverse, in most of the species a common set of mechanisms are used to direct development and cell fate determination. In vertebrates, but also in several invertebrates, a proper embryo patterning initially depends on asymmetrical distribution of maternal mRNAs and proteins in the cytoplasm of the egg. Later, a complex gene regulation program, dependent on inductive interactions between early blastomeres, controls the establishment of the embryonic axes and of the three primary germ layers, ectoderm, mesoderm and endoderm. In many organisms, transcription is activated at early stage, while sometimes at low level, and often many cycles before the transition from synchronous to asynchronous cell divisions. However, the progressive switch from a maternal to a zygotic control of patterning takes place at different stages according to the species, during the 10th division in sea urchin and zebrafish embryos (Davidson 1986; Kane and Kimmel 1993), the 12th division (4000 cell stage) in xenopus (Newport and Kirschner 1982) or the two-cell stage in mouse (Schultz 1993).

If we only consider deuterostomes, which include the chordates, the hemichordates and the echinoderms, the control of embryonic development has been investigated in several model species. In sea urchin, a non-chordate deuterostome, despite unfertilized eggs are polarized along the animal-vegetal axis, individualized blastomeres are totipotent until the four-cell stage. Then, embryos show an early split into lineage-specific cell programs. Unequal cleavages in the fourth and fifth divisions induce the formation of mesomeres, macromeres and small and large micromeres. The 60-cell embryo contains five different polyclonal lineage elements, which at this stage have apparently already undergone specification.

In chordates, germ layers are specified at the blastula stage and are then spatially positioned within the embryo by morphogenetic processes occurring during gastrulation. Embryonic cells prior to gastrulation have the capacity to respond to various signalling pathways and are therefore uncommitted (Gardner and Beddigton 1988; Snape et al. 1987). In xenopus, since zygotic transcription is not apparent until the mid-blastula transition (Newport and Kirschner 1982), the asymmetric localization of specific maternal mRNAs and proteins underlies the patterning of the early embryo. Although the xenopus oocyte is radially symmetrical, it contains distinct RNA populations distributed along an animal-vegetal polarity. A cortical rotation during the first cell cycle produces a distinctive pro-dorsal/ventral asymmetry that will establish the dorsal-ventral axis (Chung and Malacinski 1980; Danilchik and Denegre 1991; Scharf and Gerhart 1980; Vincent et al. 1987). The cytoplasmic patterns are spatially set by embryonic cleavage, but their effects on gene expression are not apparent until the onset of zygotic transcription.

In mammals the pre-implantation development involves six cleavage divisions to form a blastocyst. Blastocyst consists of approximately 60 cells, with i) an inner cell mass (ICM) containing pluripotent primitive ectoderm (PEct) cells, and ii) specialized outer trophoectoderm cells required for implantation and development of the placenta. All the somatic tissues and germ cells in adult derive from the ICM. Following implantation, the ICM develops to form the epiblast cells of the early egg cylinder (Nichols et al. 1998). These cells respond to signals from the surrounding extra-embryonic tissues that direct differentiation and initiate gastrulation.

9.2.2 Molecules Regulating Germ Layer Specification and Axial Patterning in Embryos

Recent years have seen a rapid advancement in the elucidation of the molecular players implicated in the patterning of invertebrate and vertebrate embryos. In spite of apparent time and spatial heterogeneity in the organization of axial patterning or germ layer specification, molecular mechanisms remain frequently conserved. Several groups of polypeptide growth factors such as transforming growth factor-βs (TGF-βs) and fibroblast growth factors (FGFs) have been shown to function as the extracellular signals responsible for the induction and patterning of germ layers

(Harland and Gerhart 1997; Munoz-Sanjuan and Brivanlou 2001; Whitman 2001). The canonical Wnt signalling pathway for example seems to be used in bilaterian and cnidarian to establish axial patterning (Croce and McClay 2006; Stern 2006). Another example is the crucial role of the Bone Morphogenetic Proteins (BMP) and their antagonist chordin in establishment of the dorso-ventral axis (for review see Gerhart 2006). Finally, a key player in mesoderm and endoderm formation and establishment of left-right asymmetry is the Nodal signalling pathway, nodal being expressed in a signalling centre, the Speman organizer in vertebrates (Duboc and Lepage 2008; Schier 2003). The central role in axis specification is conserved in sea urchin. Surprisingly, however the Nodal signalling pathway functions in ectoderm patterning and not in specification of endoderm and mesoderm (Duboc et al. 2005).

9.3 ESCs in Culture

While sea urchin or xenopus ESC lines have never been reported, pluripotent cells lines have been established from several mammalian species including mouse (mESC), monkey (moESC) and human (hESC), focusing on the pre-implantation stage of development. Recently, isolation of mouse pluripotent cells from epiblast, at post-implantation stage, has also been successfully achieved (Brons et al. 2007; Tesar et al. 2007). Pluripotent stem cell (PSC) lines with properties similar to ESC lines have been established by reprogramming mouse and primate adult cells with ectopic expression of transcription factors (TFs) (for review see Rossant 2008). Two major technologies are currently capable of reprogramming somatic cell nuclei back into a pluripotent state: a virus-mediated transduction to transfer specific reprogramming factors (Oct-4, Sox-2, c-Myc and Klf4) directly into the chromatin of a somatic cell giving rise to the so-called induced pluripotent stem (iPS) cells (Nakagawa et al. 2008; Okita et al. 2007; Park et al. 2008; Takahashi et al. 2007b; Takahashi and Yamanaka 2006; Wernig et al. 2007; Yu et al. 2007), and the transfer of a differentiated nuclei (SCNT, for somatic cell nuclear transfer) into a metaphase II enucleated oocyte (Barberi et al. 2003; Byrne et al. 2007; Rideout et al. 2002). These technologies are very promising to produce isogenic PSCs genetically identical to the donor patient, as reported in recent exhaustive reviews (Byrne 2008; Jaenisch and Young 2008; Liu 2008).

ESCs have been maintained by using various empirical combinations of feeder cells or conditioned media supplemented either by serum or by factors involved in the signalling pathways that direct embryo patterning (Vallier and Pedersen 2005). Despite the apparent common origin and the similar pluripotency of mouse and human ESCs, recent studies have revealed the involvement of different signalling pathways to maintain their pluripotency. As detailed in § 9.4.1. mESCs depend on leukaemia inhibitory factor (LIF) and BMP, whereas their human counterparts rely on activin (INHBA)/nodal (NODAL) and FGF (Brons et al. 2007). However, very recent studies have shown that extrinsic stimuli are dispensable for the derivation, propagation and pluripotency of ESCs. Indeed, mESC self-renewal has been achieved by the elimination of inductive differentiation signals from mitogen-activated protein kinase (MEK/ERK) pathway, including autocrine FGF-4. Briefly, according to A. Smith's group, ESC would be more similar to unicellular organisms with an intrinsic program for self-replication that does not require extrinsic stimuli (Ying et al. 2008).

According to Silva and Smith (Silva and Smith 2008), three properties define ESC: unlimited self-renewal, *in vitro* primary chimeras formation and germline transmission. Under appropriate conditions, ESCs can be propagated in culture for extended periods of time as a stable self-renewing population. ESCs retain their pluripotency and can differentiate into the same range of cell types as those formed in the embryo from the ICM. In mouse, ESCs can be tested for their developmental potential following injection into recipient blastocysts. In this case, the injected mESCs contribute to the germ line and all tissues of an adult animal. An alternative assay, especially for hESC, is to inject ESCs into immunocompromised mice where they form teratomas, a tumor comprising mesoderm, ectoderm and endoderm cells.

While functional assays of ESC developmental potential are a standard for ESC analysis, molecular markers are of practical issues. In mammalian species, some of these markers are "critical" TFs expressed in the ICM and ESCs. The POU domain TF Oct3/4, the homeodomain TF nanog and the high mobility group protein Sox-2 are the most used markers (Boiani and Scholer 2005; Niwa 2007) of a TF network detailed in § 9.4.4.

9.3.1 Isolation of Pluripotent ESC Lines

Murine ESCs were first isolated in 1981 from blastocyst stage embryos (Evans and Kaufman 1981; Martin 1981). Their pluripotency is maintained *in vitro* if they are cultured under conditions that promote proliferative and undifferentiated state, for example in the presence of LIF and BMP4 (Suda et al. 1987). When released from the influence of these cytokines suppress, *in vitro* or following re-introduction into the blastocyst, ESCs differentiate in a way similar to the PEct cells present in ICM. However, it is unclear if the *in vitro* pluripotent state of ESCs is identical to the transient pluripotent state of PEct cells *in vivo*. Optimization of mESC feeder-free culture methods may be helpful to clarify this aspect.

Monkey ESCs were first derived by Thompson's group in 1995 from *in vivo*-produced blastocyst (Thomson et al. 1995). To date, about fifty ESC lines have been derived from the Old Word macaques, cynomolgus and rhesus monkey. Primate and mouse ESCs are similar in their ability to self-renew and differentiate into cells representing all three embryonic germ layers but differ in their growth requirements and molecular signatures (Fluckiger et al. 2006). MoESC lines established in non-human primates (Rhesus, cynomolgus and marmoset monkey) similarly to hESC, are LIF independent (Pau and Wolf 2004; Suemori et al. 2001; Thomson et al. 1995). Moreover, they are able to differentiate in many cell types, including hematopoietic, neural and insulin-producing cells, which is relevant for clinical studies (Wianny et al. 2008).

The importance of moESCs is due to the higher genetical, anatomical and physiological similarities between monkey and human compared to rodent. MoESCs are thus accurate models in regenerative medicine, as in the case of neurodegenerative diseases, in strategies to prevent immune rejection and to test the feasibility, safety, and efficacy of ESC based treatments (Byrne et al. 2006). However, until recently, relatively little was known about the genetic programming of moESCs. The recent attempts to describe the transcription profiles of monkey ESCs are thus critical to provide information about monkey stemness genes (Byrne et al. 2006). Strategy to avoid host-graft rejection after transplantation of ESCs into a patient could relay on the availability of thousands of ESC lines with a MHC matching between ESC lines and patients. Histocompatible ESC lines have been derived in rhesus monkey following somatic SCNT (Byrne et al. 2007). The originated monkey SCNT-ESCs were in term of morphology, stemness markers and transcriptional profile, indistinguishable from control ESC lines.

Human cell lines. Initial reports on the derivation of pluripotent stem cells from the human blastocyst (Thomson et al. 1998) have been abundantly confirmed. Since the first derivations, technologies for hESC maintenance and manipulation have been successfully improved including set up of defined media (Laslett et al. 2007), as reported in § 9.4.1 of this review.

However to date, we have a limited understanding of the basic phenotype of hESC lines. The assumption that hESCs are positive for specific surface antigens (SSEA-3, SSEA-4, TRA-1-60, CD9), express specific genes of pluripotent cells (e.g., Oct-4, Nanog) and are capable of indefinite renewal and differentiation into the derivatives of all three embryonic germ layers, indicates that they can be regarded as a unique cellular entity. Actually, the hESCs phenotype is closer to a heterogeneous population of cells, including undifferentiated cells and cells that have already started their differentiation program. Unfortunately, many studies have considered hESC in culture as a homogeneous cell population giving rise to a plethora of data that should be carefully interpreted due to the complexity of the cellular system. More detailed information regarding recent studies on TFs in hESCs is provided in § 9.4.4 this review.

9.3.2 *ESC Differentiation*

In vitro, ESCs can exhibit a perpetual pluripotent state while *in vivo* pluripotency is only transient. In ESC lines, two models have been proposed for the structure of stem cell differentiation hierarchies (Laslett et al. 2007): (i) a series of binary choices between alternate and discrete cell states, driven by a serial cascade of expression of specific TFs; (ii) a progression through a differentiation hierarchy in a continuum that may be reversibly traversed.

Emerging concepts support that lineage commitment and formation of the three specific lineages of the mammalian pre-implantation embryo (ICM, trophoectoderm and extraembryonic endoderm) would be the result of a dynamic interplay of a network of key regulatory TFs (Rossant 2008). This implies that in early development the process of lineage choice begins before the disappearance of stem cell maintaining factors. This process may occur through a set of opposing reciprocal interactions between key TFs. A similar model has been proposed for the lineage priming model in hematopoiesis, in which the expression of genes characteristic of multiple differentiation lineages is observed in stem or progenitor cells that have not yet undertaken overt commitment (Orkin 2003).

9.4 Molecular Mechanisms Regulating Pluripotency

With their embryonic equivalent, pluripotent ESCs share two remarkable features: first, their cell division rate is very high and second they have extraordinary developmental plasticity. Specific molecular mechanisms are required to support self-renewal and to avoid cell commitment. Among the specific cell features that have been associated to stemness are usually underlined: (i) inhibition of certain signalling pathways, (ii) cell cycle characteristics, (iii) a peculiar organisation of DNA replication, (iv) activation of particular networks of TFs and (v) changes in chromatin state.

9.4.1 Signalling Pathways Required to ESC Self-Renewal

Different culture conditions used by different laboratories as well as variability in batches of fetal calf serum have complicated the studies of self-renewing signaling pathways in ESCs. Moreover, the signaling requirements for maintenance of human, monkey and murine ES cells differ significantly.

The best characterized effector of mESC self-renewal is LIF. LIF directs ES cell renewal through activation of the latent TF STAT3 (Matsuda et al. 1999; Niwa et al. 1998). The LIF/gp130 heterodimeric receptors recruits and activate STAT3 that translocates into the nucleus and stimulates the expression of specific genes (Burdon et al. 2002). According to results from different groups, c-myc is a direct transcriptional target of STAT3 (Matsuda et al. 1999; Niwa et al. 1998). Moreover, GSK3β dependent phosphorylation pathway controls c-myc levels. When LIF signal decreases, GSK3β is rapidly activated and c-myc is phosphorylated on threonine 58, ubiquitinylated and proteasome-dependent degraded. Probably, PI3K is also involved in the suppression of GSK3β activity (Jirmanova et al. 2002). PI3K signalling is involved in many aspects of cell behavior and plays a crucial role in mESC self-renewal. A major target of PI3K pathway is protein kinase B (PKB/AKT) but many downstream effectors have not been defined (Jirmanova et al. 2002). Another pathway, the BMP signalling pathway, acts in combination with LIF to sustain self-renewal and preserve multilineage differentiation. In serum-free cultures LIF is insufficient to block neural differentiation and maintain pluripotency. BMP, which induces expression of the inhibitor of differentiation/DNA binding (Id) genes via the SMAD (for signalling mediators and antagonists of the TGF-β superfamily) transducing pathway, is necessary to block entry into neural lineages (Ying et al. 2003). Recent results have shown that BMP signalling inhibits differentiation by

P38 MAP kinase suppression (Qi et al. 2004). These findings support a model by which ESC self-renewal results from a coordinated series of events involved both in maintenance of the pluripotent state and the blockade of differentiation pathways.

Very few studies have investigated the signalling pathways in moESCs. However, moESCs do not respond to LIF in culture and they require feeder cells to proliferate in an undifferentiated state. Sumi et al. clearly show that the LIF/gp130/STAT3 pathway, while active in moESCs and supporting self-renewal in murine ES cells, is dispensable for the maintenance of undifferentiated state in moESCs. An unidentified LIF-independent signalling pathway is probably responsible for moESC self-renewal (Sumi et al. 2004). Additional studies need to be performed to clarify the role of other self-renewal sustaining factors in primates ESCs.

Culture conditions for hESC self-renewal are also quite different to the ones for mESC maintenance. More precisely, as in monkey, LIF is not required for hESC self-renewal. The complexity of mouse embryonic fibroblast (MEF)-conditioned medium used for hESC culture is unfortunately a too complicated mixture to be able to define the critical factors. However, FGF2 appears to be a key component of defined media and Activin A, a member of TGF-beta family plays an important role in mantaining hESC self-renewal. Finally, PI3K signaling activation is critical for hESC self-renewal and this pathway has to be inactivated to promote hESC differentiation. Recent optimization of hESC culture conditions has moved from cultures on MEFs in fetal bovine serum-containing medium towards feeder-free culture methods using more defined animal substance-free culture media (Skottman and Hovatta 2006). In the defined medium now used to propagate hESCs (Brons et al. 2007), Activin A, FGF2 and insulin/IGF appear to be the consensus factors that play a key role in regulation of PI3K, SMAD and ERK signalling.

9.4.2 Cell Cycle

ESCs have the capacity to maintain a permanent proliferative state. However, the mechanisms linking this cell cycle control mode and pluripotency are unclear. Moreover, a still unresolved question is whether the molecular mechanisms regulating ES cell division *in vitro* are closely reminiscent of those controlling cell cycle in the embryo.

9.4.2.1 Cell Cycle Features During Early Embryogenesis and in ES Cells in Culture

Early embryogenesis is marked by shorter cell cycles with altered gap phases and an uncoupling of cell growth and cell cycle progression.

In sea urchin, eggs are stored in a haploid state, thus fertilization directly triggers entry into mitotic cycles, rather than meiosis resumption as in most animal eggs. After one cycle in which short G1-and G2 phases precede DNA replication and mitosis respectively, a rapid succession of S- and M phases mark the cleavage stages. These early divisions rely on translation of maternally stored mRNA or post translational modifications of existing proteins stored in the eggs until the cell cycle becomes dependent on zygotic transcription and develops into a more conventional somatic cycle with gap phases.

In Xenopus the first cell cycle after fertilization, which takes place in M-phase of meiosis II, displays a very long G2 phase while the subsequent cleavage cycles consist of rapid oscillations between S and M phase without intervening gap phases until cycle 13. After this midblastula transition (MBT) when zygotic transcription is activated and cells become motile, the cell cycle lengthens and cleavages become asynchronous (Graham and Morgan 1966).

In mammals, the early cycles after fertilization of the egg are similar to canonical cell cycles.

In mouse, the 2 cell stage is reached at 1.5 days d.p.c. The next four cell divisions have an average time of approximately 12 h reaching the 32 cells early blastocyst at 3.5 d.p.c. (Hogan et al. 1994). Interestingly, the formation of polar outer cells (OCs) and apolar inner cells (ICs) is associated to differences in cell cycle length in these two populations (MacQueen and Johnson 1983). However, during the egg cylinder stage at 6.5 days d.p.c, cell cycle length decreases dramatically reaching in the proliferative zone an average as little as 2–3 h (Snow 1977). When the embryonic germ layer starts to differentiate, the cell division rate slows down like a somatic cell cycle (about 24 h).

In vitro, the rapid division of ESCs, 11–16 h in mESC as opposed to 24 h in somatic cells, is associated with unusual cell cycle properties (Savatier et al., 1994; Solter et al. 1971; Stead et al. 2002). In human, mouse and rhesus monkey the shorter cell cycle of ESCs is due to an abbreviated G1 phase (Becker et al. 2006; Fluckiger et al. 2006). Furthermore ES cell proliferation is not restricted by serum starvation or contact inhibition (Smith 2001).

Changes in cell cycle dynamics occur during ESC differentiation while it is still unclear if these cell cycle changes are intimately associated with loss of pluripotency. Until now no direct links have been demonstrated between the striking features of cell cycle control in the embryo or in ES cell lines and the pluripotent state (Ohtsuka and Dalton 2008).

9.4.2.2 Cell Cycle Regulators

Cell cycles are driven by a molecular clock in which the cyclin-dependent kinases (Cdks) play a key role. In the somatic cell cycles, while cyclin B/Cdk1 is essential to trigger mitosis, Cdk2 associated to cyclin E and cyclin A controls S phase progression and cyclin D/Cdk4-6 allows crossing the restriction R point.

Cyclin Dependent Kinases

In sea urchin the core cell cycle genes Cdk1 and Cdk2 and their respective regulatory subunits cyclin A, cyclin B and cyclin E have been successively identified (Evans et al. 1983; Geneviere-Garrigues et al. 1995; Kurokawa et al. 1997; Moreau et al. 1998; Sumerel et al. 2001). While both kinases are constitutively expressed during early development, the cyclin B/Cdk1 activity peaks at each M-phase of the cleavage stages (Geneviere-Garrigues et al. 1995; Meijer et al. 1991). In contrast, Cdk2 kinase activity is stable during early embryogenesis while dispensable for the activation of the first DNA replication after fertilization (Moreau et al. 1998; Sumerel et al. 2001). The cyclin E protein levels dropped rapidly between the two cell stage and early blastula stage even though the mRNA declined later (Sumerel et al. 2001). The Cdk2 kinase activity decreased as development proceeds after the 5th-6th cleavage, with a significant drop occurring at the morula stage, despite the fact that, during this time, there is at least a 50-fold higher rate of DNA synthesis/embryo than during the first cell cycle. Cdk2 kinase activity declined to barely detectable levels by the mesenchyme blastula stage.

In Xenopus, cyclin A1, B1 and B2 concentrations decline just after fertilization and then fluctuate during each division, the activity of Cdk1 closely paralleling the cyclin B oscillations (Hartley et al. 1996). At MBT the level of cyclin B1 showed a marked increase corresponding to zygotic expression of the protein and Cdk1 becomes clearly phosphorylated on Tyr 15, a regulatory phosphorylation absent in cycles 2–12 (Ferrell et al. 1991; Hartley et al. 1996). Cyclin A1 disappears rapidly after the MBT while cyclin A2 protein concentration, very low in early embryos, increases rapidly, reaching a constant level in adult tissues (Finkielstein et al. 2002). During embryonic cycles, cyclin A is not required for DNA synthesis (Fang and Newport 1991), and the cyclin A/Cdk2 complexes do not form until the onset of zygotic transcription (Rempel et al. 1995). In contrast to cyclins A and B, the cyclin E level remains constant throughout the first 12 cleavage stages though cyclin E/Cdk2 activity slightly oscillates twice per cell cycle, independently on protein synthesis, transcription and nucleo-cytoplasmic ratio (Hartley et al. 1996; Hartley et al. 1997; Howe and Newport 1996). It has been proposed that oscillatory activity of cyclin E/Cdk2 is governed by changes in Cdk2 phosphorylation state involving a concerted action of maternal forms of the wee1 and CDC25 regulatory kinase and phosphatase, respectively (Ciliberto et al. 2003; Wroble et al. 2007). Smooth oscillations in cyclin E/Cdk2 activity constitute a maternal timer that regulates the timing of the events of MBT (Hartley et al. 1997). One of these events is the rapid degradation of the maternally supplied cyclin E itself accompanied by a decrease in cyclin E/Cdk2 activity (Hartley et al. 1996; Howe et al. 1995; Rempel et al. 1995).

As shown by knockout studies, Cdks and most cyclins are functionally redundant in mammals with the exception of cyclin A2 and cyclin B1 (Ciemerych et al. 2002; Parisi et al. 2003). The homozygous null mutants of cyclin A2 survive until 5.5 d.p.c (Murphy 1999) and the homozygous null mutants of cyclin B1 can survive before 10 d.p.c. (Brandeis et al. 1998). No specific role for the regulation of cell cycle in pluripotent cells has been revealed by these knockout experiments during mouse development. On the other hand, the small amount of pluripotent cells in the mammalian embryo make very difficult to undertake biochemical analysis of cyclin-Cdk complexes.

However, the control of the dramatic changes in cell cycle characteristics occurring during development of mammalian embryos has been clarified by recent studies on the cell cycle regulators of ESCs in culture, particularly in mouse (Stead et al.

2002; White et al. 2005). The protein levels of cyclin E1, A2 and B1 have been described and compared to other cell types (Faast et al. 2004; Savatier et al. 1996; Stead et al. 2002). All cyclins are highly expressed in mESC and most notably, Cdk activities are robust throughout the cell cycle. Whereas in somatic cells cyclin E-Cdk2 activity is periodic and peaks at the G1 to S transition, mESC have constitutive cyclin E-Cdk2 activity independent of the cell cycle phase. The only Cdk showing periodic activity is the mitotic kinase Cdk1-cyclin B (Stead et al. 2002).

Some differences arouse from the comparison of ES cell cycle in mouse and primate. For example, cyclin A is not expressed in moESCs, suggesting a slightly different cell cycle regulation (Fluckiger et al. 2006).

In conclusion, it appears that cyclin B-Cdk1 is the only Cdk activity fluctuating upon cell cycle progression during embryonic development and in the established ESC lines and would thus be sufficient to drive cell division (Santamaria et al. 2007). Moreover, cells are committed to S phase in spite of the continual cyclin E kinase activity.

p16INK4/Cdk4-Cyclin D/pRb Pathway

In mammalian somatic cells, cyclin D-Cdk4/6 is specifically involved in the G0 to G1 phase transition. Mitogenic signaling induces the synthesis of D-type cyclins and possibly the proper folding and transport of Cdk4 and/or Cdk6 to the nucleus. Active complexes of Cdk4 or Cdk6 and D-type cyclins phosphorylate members of the retinoblastoma (Rb) protein family, which includes pRb, p107 and p130. Rb proteins function to repress transcription by binding and modulating the activity of TFs, such as E2F family members (Cobrinik 2005).

Recent data from *D. melanogaster*, *C. elegans*, *S. purpuratus* and mice suggest that cyclin D/Cdk4 complexes are not essential for cell proliferation or early development in these species and may be involved more in the regulation of cell growth and/or developmental patterning than in cell cycle progression (Datar et al. 2000; Malumbres et al. 2004; Meyer et al. 2000; Moore et al. 2002; Park and Krause 1999).

In the sea urchin *S. purpuratus* genome only one homologue has been identified for the Cdk4/6 family. Similarly, a single homologue of Cdk4/6 is found in *D. melanogaster*, *C. elegans* and *C. intestinalis*, supporting the view that the gene duplication is specific to vertebrates (Fernandez-Guerra et al. 2006). Cdk4 mRNA and protein are constitutively expressed in sea urchin eggs and throughout embryonic development. In contrast, cyclin D mRNA is barely detectable in eggs and early embryos, increasing dramatically in embryos at the early blastula stage and remains at a constant level throughout embryogenesis. An increase in Cdk4 kinase activity occurs concomitantly with the increase in cyclin D mRNA. It was hypothesized from knockdown experiments that cyclin D and Cdk4 are required for normal development and perhaps the patterning of the developing embryo, but may not be directly involved in regulating entry into the cell cycle (Moore et al. 2002). On the other hand, the sea urchin genome contains two Rb-like and two E2F/DP like genes, which mirror the situation in flies, suggesting that this is the primitive bilaterian repertoire of these genes (Fernandez-Guerra et al. 2006). According to the transcriptome analysis these transcripts would be all expressed during early development (Samanta et al. 2006), a result which require confirmation by *in situ* hybridization and Northern blot or real-time PCR analyses. Notably, INK4 Cdk-inhibitors are absent of the sea urchin genome.

In Xenopus neither cyclin D1 nor cyclin D2 are detectable prior to MBT (Schnackenberg et al. 2007; Vernon and Philpott 2003a). Cdk4 begins to be detected throughout the animal pole at stage 8. Maternal transcripts of pRb are highly represented in oocytes, while the protein level increases tenfold between unfertilized egg and stage 7.5, remaining then constant until stage 11.5. Zygotic transcription of pRb is initiated at stage 13, and persists throughout early development and in all examined tissues of the adult frog, while with different expression levels (Destree et al. 1992). However, either loss or overexpression of pRb has no effect on either cell cycling or tissue differentiation (Cosgrove and Philpott 2007). pRb is maintained in a hyperphosphorylated, inactive form during early development. This maintenance of pRb in its inactive state may facilitate the high level of cell proliferation required for the rapid growth of young embryos.

Similarly, pRb pathway is dispensable during early mouse development (Ciemerych and Sicinski 2005; Wikenheiser-Brokamp 2006). Rb–/– mice die during gestation with anemia, placental defects, and widespread hyperproliferation and apoptosis, whereas p130–/– and p107–/– mice can develop normally (Cobrinik et al. 1996; LeCouter et al. 1998a,b; Lee et al. 1996; Mulligan et al. 1998). However and intriguingly, most of the abnormalities seen in Rb null embryos are secondary to the effect of Rb loss in placental tissue (Wu et al. 2003). Rb null embryos rescued by wild-type placenta survive until birth. Thus Rb would not be essential for the early stages of embryonic development in mice. To explain these results a study of Iwamory et col. shows a low expression level of Rb at mRNA and protein level, before the late gastrula stage (Iwamori et al. 2002). Interestingly, forced Rb expression using a plasmid vector injected into zygotes, produces a growth arrest before morulae stage. In addition, Xie et col. (Xie et al. 2005) found phosphorylated Rb protein during pre-implantation development that could be responsible for shortened G1 phase.

In accordance with data obtained by gene targeting experiments in mouse embryo, there is compelling evidence that cell cycle control in mouse as well as primate ESCs is not dependent of a functional p16INK4/Cdk4/6-Cyclin D/pRb:E2F pathway (Burdon et al. 2002; Dannenberg et al. 2000; Fluckiger et al. 2006; Sage et al. 2000; Savatier et al. 1996; White et al. 2005). While most Cdk activities are active through the cell cycle, in mESC the Cdk4-cyclin D1 complexes exhibit low activity (Faast et al. 2004). In hESCs the mRNA levels for cyclin D2 and Cdk4 are upregulated when hESCs enter into the abbreviated G1 phase. This data suggest that cyclin D2-Cdk4 could contribute to G1 cell cycle progression in hESCs. Unfortunately, protein expression or activities were not investigated in this study and the cyclin E level was not explored during the proliferation of hESCs (Becker et al. 2006). A substantial Cdk6 activity associated to cyclin D3 but resistant to p16INK4A inhibition was identified in mESCs (Faast et al. 2004). This unusual

Cdk6 behaviour lead the authors to suggest than Cdk6-cyclin D3 either was not wired into the cell signalling machinery or would have a role beside cell cycle control in mESCs. Murine ESCs express the pocket proteins pRb and p107, but they are hyperphosphorylated and consequently held in a biochemically inactive state (Savatier et al. 1994; Stead et al. 2002). E2F target genes are thus actively transcribed throughout the cell cycle (Stead et al. 2002). On the other hand, mESCs that lack all three Rb gene family members (Rb, p107 and p130) are incapable of differentiation and cell cycle withdrawal (Dannenberg et al. 2000) while single or double knockout show no defect in differentiation probably due to functional compensation in the Rb family members.

The presence of an inactivated pRB in ES cells is again reminiscent of the situation encountered during early development. The constitutive Cdk2-cyclin E activity would determine Rb hyperphosphorylation and this could be a simple way to explain the lack of R point in ES cells (Orford and Scadden 2008). Probably, in this way the mESC pass over the early G1 point by bypassing the restriction point. This suggests that the acquisition of R point when the Cdk2-cyclin E activity comes under the control of active Rb is a hallmark of the capability of ESCs to undergo multilineage differentiation. This model is consistent with the idea that early G1 is a key point that can affect cell fate decision to enter the cell cycle or to be committed to differentiate (Massague 2004).

9.4.2.3 Checkpoint Responses

Upon activation of the DNA-damage pathway somatic cells will respond either by a short cell cycle pause allowing to repair their DNA content or by a prolonged arrest if the damage is more severe. Usually the parallel activation of apoptotic mechanisms ensures that abnormal cells are eliminated if no repair occurs (Norbury and Zhivotovsky 2004). The ATM/ATR kinases play a central role in the transduction of the DNA-damage signal by triggering a phosphorylation cascade involving chk1/2 kinases. While the cell cycle delay is mediated by the inhibition of CDC25 phosphatases activity, an arrest of cell division involves transcriptional activation of the p53/p21 pathway (for review see Bartek et al. 2004).

Another striking feature of the mitotic cycle of ES cells is the lack of cell cycle arrest at the G1 checkpoint in response to DNA damage or nucleotide depletion (Aladjem et al. 1998; Hong and Stambrook 2004). Similarly, the cell cycles during early cleavage stage in embryos usually do not engage checkpoints in response to damaged or unreplicated DNA (Anderson et al. 1997; Hensey and Gautier 1997; Newport and Dasso 1989). Moreover, mESCs do not express Cdk inhibitory proteins of the INK family (p15, p16, p18, p19), neither p21Cip, p27kip1 and p57 (Faast et al. 2004; Savatier et al. 1996; Stead et al. 2002). This peculiar absence of tumor suppressor function is an intriguing aspect of mESC cell cycle which also raises the question of checkpoint control in ES cells and embryos.

It has been previously assumed that rapidly developing sea urchin early embryos lack DNA-damage checkpoint mechanisms. However, recent results demonstrate the presence of a fully functional checkpoint mechanism that can be activated by genotoxic agents in these embryos (Le Bouffant et al. 2007). At low concentration, embryos were able to repair the DNA damage and recover from checkpoint arrest, whereas at high doses they underwent morphological and biochemical changes characteristic of apoptosis. Furthermore, the full repertoire of proteins involved in the checkpoint pathway involved in cell cycle pause, including ATM, ATR, Chk1,Chk2, and a single CDC25 homologue, has been identified in the sea urchin genome. With the exception of ATM, all these transcripts have been found expressed during early embryogenesis in the transcriptome analysis (Fernandez-Guerra et al., 2006). Other interesting features of the sea urchin genome are the presence of a single p53 homologue, and only one homologue of the CDK inhibitor p21/p27 family. No information is yet available on the function of these proteins in sea urchin embryo neither in the cell differentiation process nor in the DNA-damage response. This simplified checkpoint pathway would be an interesting background for a comparative analysis of the response to DNA damage.

During the first 12 cell divisions that precede MBT, Xenopus embryos lack checkpoints that halt the cell cycle in response to DNA damage. In fact, DNA damaging agents like ionizing radiations failed to activate a checkpoint response and to arrest cell cycle of early embryos, whereas the same treatment did activates a response efficiently in late embryos (Anderson et al. 1997; Finkielstein et al. 2001). One possibility would be that one or more of the biochemical components required for DNA damage checkpoint signal transduction are absent in pre-MBT embryos and, thus, zygotic transcription could be necessary to produce elements of the checkpoint pathway. However, the addition of exogenous DNA with free ends mimicking DNA double-strand breaks (DSB) induces checkpoint activation in pre-MBT embryos or egg extracts, suggesting all components of a functional checkpoint pathway exist before the MBT (Anderson et al. 1997; Conn et al. 2004; Kumagai et al. 1998). Moreover, the presence of damage-free DNA, either as uncut plasmid DNA or Xenopus sperm chromatin, sensitize the DSB-induced checkpoint producing ATM autophosphorylation, ATM-dependent Chk1 and Chk2 phosphorylation by recruitment of activated ATM from the soluble fraction to the undamaged threshold DNA. It is thus hypothesized that the capability of checkpoint activation in response to DNA damage, acquired at MBT, is dependent on DNA-to-cytoplasmic ratio (Conn et al. 2004; Peng et al. 2007; Peng et al. 2008). The checkpoint signal generated at early time being insufficient to prevent establishment of the robust maternally provided cyclin B-Cdk1 activity. Interestingly, in Xenopus embryos a dramatic change in the response to DNA damage occurs at the MBT. When ionizing radiation is administered any time before the MBT, Xenopus embryos initiate apoptosis after the MBT. However, if ionizing radiation is given after the MBT, apoptosis is prevented and cell cycle progression arrest in G1 (Finkielstein et al. 2001). This arrest is a direct consequence of an increased amount of the Cdk inhibitor p27Xic1, which binds to and inhibits both cyclin D1-Cdk4 and cyclin A2-Cdk2 complexes. This promotes a delay in the G1/S transition, allowing more time for DNA repair, and blocks apoptosis, which might occur if S phase were initiated with damaged DNA (Finkielstein et al. 2001). In xenopus the principal cyclin-dependent kinase inhibitor Xic1, which shows structural and functional characteristics of p21Cip1, p27Kip1 and p57Kip2 (Shou and Dunphy 1996; Su et al. 1995), is expressed early during development but strongly accumulates after stage 10 (Finkielstein et al. 2001) particularly in the developing myotome (Vernon and Philpott 2003b) and in cells destined to become primary neurons (Vernon et al. 2003).

In mES the two known signalling pathways that mediate the checkpoint are compromised. The checkpoint kinase, Chk2, which participates in both pathways is sequestered at centrosomes in ES cells and does not phosphorylate its substrates (i.e. p53 and Cdc25A) that must be modified to produce a G1 arrest. Ectopic expression of Chk2 does not rescue the p53-mediated pathway, but does restore the pathway mediated by Cdc25A. Wild type ES cells exposed to ionizing radiation do not accumulate in G1 but do so in S-phase and in G2. ES cells that ectopically express Chk2 undergo cell cycle arrest in G1 as well as G2, and appear to be protected from apoptosis (Hong et al. 2007).

Thus, in embryos as well as in ES cells in culture, responses to DNA-damage are disturbed. Even if molecular mechanisms are fully or partially present, their request does not induce cell cycle arrest. Indeed, ES cells cannot tolerate DNA damage since mutations may be passed on to future generations. ES cells, therefore, must have robust mechanisms to protect the integrity of their genomes including protective mechanisms which eliminate those ES cells that have acquired deleterious mutations.

9.4.3 Regulation of Replication in ES Cells and Embryos

To further explore the molecular mechanisms underpinning the proliferative features of undifferentiated ES cells, control of DNA-replication process was investigated. In somatic cells prior to S phase, replication origins are "licensed" by loading prereplicative complexes (pre-RC) onto them. Assembly of pre-RC is marked by the sequential binding of ORC (for origin recognition complex), Cdc6, Cdt1 and finally MCMs (for mini chromosome maintenance) to the origin of replication (reviewed in Bell 2002; Spradling 1999). In metazoan, geminin regulates pre-RC assembly, preventing over-replication of DNA in proliferating cells. The expression of the replicative factors was compared between mouse ES and differentiated somatic cells (Fujii-Yamamoto et al. 2005). All the pre-RC factors investigated are highly transcribed in ESCs while at protein level only Cdc6 is highly and rather constitutively expressed. While proteasome-dependent degradation of CDC6 did not seem to take place, cell-cycle dependent phosphorylation and chromatin association/dissociation of the different replication factors indeed occurred in a manner similar to that of somatic differentiated cells, as were observed with CDC6 or MCM. While the geminin protein expression is only slightly increased in ESCs, its genetic ablation in the mouse prevents formation of ICM, causes premature endoreduplication and cell commitment to the trophoblast cell lineage, suggesting an essential role of this protein to form pluripotent cells (Gonzalez et al. 2006). The high level expression of some replication factors found in ESCs is also a characteristic of early xenopus or drosophila embryonic cells and could thus be related to their high proliferative capacity.

Once the MCMs are loaded, the origins are ready for activation leading to the onset of DNA synthesis. Phosphorylations by Cdks and Ddk (Dbf4-dependent kinase) activate pre-RC for initiation of replication. Ddk consists of a catalytic subunit Cdc7 and a regulatory subunit that confers substrate specificity: Dbf4/ASK (for Activator of S phase kinase). Inactivation of Cdc7 or a loss of the ASK gene in ESCs result in growth arrest with rapid cessation of DNA synthesis (Kim et al. 2003; Yamashita et al. 2005), confirming the key role of Cdc7/ASK in the progression of the ESC cycle. Interestingly, Fuji-Yamamoto and col. showed a particularly high level of ASK expression in ESCs, correlated with the abundance of MCM2 and MCM4 phosphorylation (Fujii-Yamamoto et al. 2005).

The patterns of ES cell-DNA replication in the few loci where they have been studied, as the beta-globin, the X-inactivation center (Xic) or the HoxB loci, evidence a replication executed from numerous origins of replication (oris). Similarly, a high frequency of DNA ori with no detectable dependence on specific DNA sequences is observed during early xenopus embryogenesis (Hyrien et al. 1995). Moreover, at that stage chromatin organises at S phase entry into short loops and replicons, allowing recruitment of a large amount of ORC proteins (Lemaitre et al. 2005). These similarities lead Lemaitre et al. to hypothesize that stemness could be in part accounted by a specific organisation of DNA replication reflecting the ones encountered in developing embryos. Later on, when cells differentiate and lose totipotency this flexible organisation would be replaced by a more specific pattern of replication dependent of the transcriptional program assigned to the differentiating cells. Furthermore, these authors demonstrated that a mitotic remodelling of the replicon and chromosome structure is a prerequisite for differentiated nuclei reprogramming when transplanted into eggs in a cloning perspective.

9.4.4 Transcription Factors Regulatory Networks in ESCs and Early Embryo Development

ESCs are highly regulated at the transcriptional level (Kim et al. 2008). Different approaches, including genetic studies, were able to identify critical pluripotency factors in ESC (Mitsui et al. 2003; Nichols et al. 1998). In summary, a TF regulatory network including Oct-4, Nanog and Sox-2, is conserved between human and mouse ESCs and is responsible for maintenance of pluripotency. Oct-4 is a POU family TF which binds the octamer sequence 5′-ATGCAAAT-3′; Sox2 is a high mobility group (HMG) DNA binding domain-containing TF that binds to the consensus motifs 5′-CATTGTT-3′; Nanog is a homeobox-containing TF recently found dispensable for expression of somatic cell pluripotency but specifically required for germ cells formation (Chambers et al. 2007). The function of Oct-4, Nanog and Sox-2 during embryogenesis has been extensively reviewed by different authors (Niwa et al. 1998; Silva and Smith 2008).

An important breakthrough was recently made by Takahashi and Yamanaka who, by transducing the transcription factors Oct4, Sox2, c-Myc and Kruppel-like factor 4 (Klf4), successfully reprogrammed mouse/human embryonic/adult somatic cells to pluripotent ES-like stem cells (Takahashi et al. 2007a; Takahashi and Yamanaka 2006). The ectopic expression of these reprogramming factors in somatic cells has been suggested to initiate a sequence of epigenetic events, including changes in DNA methylation and chromatin modifications in domains encompassing genes that play essential roles in the maintenance of ES pluripotency and lineage specification, thereby triggering the pluripotent state of iPS cells (Jaenisch and Young 2008; Niwa 2007; Takahashi and Yamanaka 2006). More recently, Jiang et al. showed that the Kruppel-like factors are required for ESC self-renewal, since the simultaneous depletion of Klf2, Klf4, and Klf5 leads to ESC differentiation (Jiang et al. 2008). Chromatin immunoprecipitation coupled to microarray assay showed that Klfs and Nanog TFs share many common target genes. Moreover, the data indicate that the Klf and Nanog circuitries form a highly interconnected transcriptional regulatory network required for maintenance of ESC pluripotency *in vitro* (Jiang et al. 2008). To gain a more complete understanding of the regulatory network that maintains the pluripotent state, Kim et col. (Kim et al. 2008), investigated target promoters of Oct4, Sox2, Klf4 and c-Myc and of additional TF proteins-interacting partners of Nanog and Oct4 (Dax1, Nac1, Zpf281 and Rex1). The results showed two different classes of target genes: one class of promoters bound by few factors tend to be inactive or repressed and a second class of promoters bound by at least four or more TFs are very active in the pluripotent state but severely repressed during ESC differentiation.

Endogenous Nanog and Oct4, both can associate with the nucleosome remodelling and co-repressor complex NuRD, transcriptional repressor Sin3a and TF PML to form multiple transcriptional repressive complexes also including histone deacetylase 1/2 activities and metastasis-associated proteins 1/2 (MTA 1/2), an emerging family of novel transcriptional co-regulators (Liang et al. 2008). This particular complex has been defined NODE (for Nanog and Oct4 associated deacetylase). Building of the NODE complex may represent the initial block that the Nanog and Oct4 repressor complexes exert on target genes to control ES cell fate. The assembly of additional regulatory complexes on these sites acts as a triggering step for ES cell differentiation. Studies on histone 3 (H3) bivalent modification identified large region of H3 lysine 27 methylation harboring smaller region of H3 lysine 4 methylation, thus suggesting a complex pattern of signals and markers characterizing the differentiation potential of ESCs (Bernstein et al. 2006).

A small number of key TF including Oct3/4 and Cdx2 play a critical role during mammalian embryonic development. The blastomeres are totipotent and practically identical until the eight cell stage. At the eight cell-stage, the blastomeres divide either symmetrically to generate the two polar OCs or asymmetrically to generate one apolar IC and one polar OC. As a consequence, ICs and OCs are respectively the precursors of the PEct cells in the ICM and the trophectoderm cells are established between the eight and 16-cell stage (Yamanaka et al. 2006). Until early morula stage, these two factors are expressed in all blastomere. Interestingly, when ICs and OCs are formed, Cdx2 is found in the OCs and Oct3/4 is present in the ICs (Niwa et al. 2005). Moreover at the morula stage, the ICs express the Nanog homeodomain TF (Mitsui et al. 2003). Recently it has been shown that histone arginine methyltransferase Carm1 regulates Nanog expression levels in early blastomeres (Torres-Padilla et al. 2006). Apparently the role of Nanog could be to promote the PEct development. On the other hand, the TF GATA6 appears essential for the development of primitive endoderm (Pend) cells. In summary, common and complex transcriptional networks have been shown to establish and preserve pluripotency in ESCs despite differences in self-renewal signals.

9.4.5 Chromatin Characteristics in Early Embryogenesis and ESC Lines

Chromatin decondensation and remodelling is concomitant to the activation of paternal genome after fertilization. These chromatin rearrangements are crucial for the formation of the totipotent zygotic nucleus. Nuclear swelling and massive chromatin decondensation also characterizes the epigenetic reprogramming of somatic nuclei injected into unfertilized eggs as was described in Xenopus laevis cloning (Gurdon 1976). Similar nuclear changes were also observed in rodent cloning (Mullins et al. 2004). The correlation with a decondensed chromatin architecture is further reinforced by a recent report concerning hESCs. As mentioned earlier, it was shown that hESCs harbor highly decondensed territories corresponding to the location of the Oct3/4 which is responsible for pluripotency and of c- myc that is involved in cell cycle control. This high level of chromatin decondensation was correlated with the nuclear distribution of heterochromatin protein 1 variants (HP1), in particular HP1 alpha (Bartova et al. 2008). From all these information it emerges that a decondensed chromatin architecture is related to ESC pluripotency.

9.4.5.1 From Gametes to the Totipotent Zygote

Compelling evidence indicates that chromatin architecture is influenced by a set of epigenetic mechanisms: histone post-translational modifications, the association to DNA of particular histone variants, the binding of specific non histone chromosomal proteins, the activity of proteins associated to chromatin remodelling and also the level of DNA methylation (reviewed by Imhof 2006).

In the newly formed embryo resident maternal genome and paternal genome are very different in term of architecture. In the sperm, the genome is haploid and arranged in most species in a highly condensed and silent chromatin. This spatial arrangement is disrupted after fertilization. During sperm chromatin remodelling in most species, the complete set of basic chromosomal proteins that are packing sperm chromatin are lost (for reviews see Sutovsky and Schatten 2000; Wright 1999).

In sea urchin, remodelling of sperm chromatin leads to a complete replacement of the sperm specific histones (SpH) by cleavage stage (CS) histones that are recruited from maternally inherited pools (reviewed by Imschenetzky et al. 2003). It was further demonstrated that the release of the SpH from sperm DNA is a prerequisite for their degradation and is facilitated by a histone chaperone present in an inactive state in unfertilized eggs and activated after fertilization (Iribarren et al. 2008).

In most vertebrates sperm chromatin is packed by two distinct types of arginine and cysteine rich protamine molecules P1 and P2, which interact with DNA forming an extremely dense nucleo-protamine complex characterized by large toroïdal-coiled subunits containing approximately 50 kb of DNA per subunit (Balhorn et al. 1999). Both protamines appear to be fundamental for the organization of sperm chromatin, since transgenic knockout of either P1 or P2 renders the males infertile (Cho et al. 2001). The reduction of sperm disulfide bonds in these protamines mediated by glutathion is a prerequisite for sperm chromatin decondensation (Sutovsky and Schatten 1997). This step is followed by the loss of protamines from sperm chromatin which is completed at the time of the second polar body extrusion, marking the end of the second meiosis. Proteins N1 and N2 as well as members of the nucleoplasmin/nucleophosmin family of proteins (NPMs) function as histone chaperones during transitions that occur throughout male chromatin remodelling in batracians (reviewed by Philpott et al. 2000). At present the NPM family of proteins is well known in molecular, functional, structural and phylogenetic terms (Frehlick et al. 2006). The histone chaperones NPMs have also been involved in the remodelling of somatic nuclei when transplanted into Xenopus eggs cytoplasm. This remodelling by NPMs results in the replacement of the somatic H1 variants (H1 and H1o) by the oocyte-specific B4 histone variant and the chromosomal protein HMG1 (Dimitrov and Wolffe 1996).

Interestingly, in early mouse embryogenesis a particular histone variant H3.3 is incorporated into the parental genome, probably through the chaperone HIRA. Although specific functions for zygotic histone variants are not well understood, the H3.3 could play a role in regulating epigenetic mechanisms *in vivo* (Torres-Padilla et al. 2006). Further epigenetic asymmetries are present in the parental genome of mice due to difference in histone modifications and localization of epigenetic modifiers (Erhardt et al. 2003). Specific patterns of histone modification characterize the paternal pronucleus (Groth et al. 2007), and a global DNA demethylation has been described in the paternal genome but not in the maternal genome (Mayer et al. 2000). Recently, Nakamura T et al., showed that PGC7/Stella, a maternal factor essential for early development, protects the maternal genome from demethylation and maintains the DNA methylation state of several imprinted loci and epigenetic asymmetry (Nakamura et al. 2007).

9.4.5.2 From the Totipotent Zygote to the Early Embryo

After the fusion of both pronuclei, the chromatin of the totipotent zygote of sea urchin is formed by five cleavage stage histone variants (CS variants). This chromatin organization persists during the two initial cleavage divisions characterizing the chromatin of the totipotent initial blastomeres described initially by the pioneering work of Hans Driesch in 1981. These CS variants are very distantly related to histones that are expressed afterwards during sea urchin development in terms of their coding sequence. In addition, the genes encoding these histone variants contain introns and polyA addition signals. The mRNA encoding these histones are maternally inherited and possesses long non-translated sequences that may be related to the translational control of the expression of these proteins (Mandl et al. 1997). Interestingly, the CS histones were found to be preserved and segregated into specific embryonic territories in late larval stages of development (Oliver et al. 2003). In addition to the CS histone variants, in sea urchin two sets of histone variants are sequentially expressed in a developmentally regulated manner: the early α variants that are assembled into chromatin from middle cleavage stages (16 blastomers) until hatching and the late variants that are predominant from blastula stage onward (Giudice 1999).

In vertebrates, unfertilized egg or zygotic chromatin has been less investigated than sperm chromatin. In general, it is believed that egg chromatin is organized into nucleosomes by somatic histones. However, the histone residing in the linker region appears to be oocyte-specific and closely related to the CS1 variant from sea urchin. As reported, the H1M (B4) gene in batracians displays a significant sequence homology with CS1 from sea urchin (Dimitrov and Wolffe 1996). The replacement of linker histone B4 by its somatic H1 counterpart *in vitro*, leads to a chromatin reorganization that occludes the binding site for the TF TFIII A (Crane-Robinson 1999). In mouse a gene encoding a 34 kDa histone variant homologous to the oocyte–specific variants B4 from the frog and CS1 from sea urchin was reported. It was postulated that this mammalian histone variant, named $H1^{oo}$, may play a critical role in the control of gene expression presumably through the perturbation of chromatin structure (Tanaka et al. 2001). Thus far it is unknown if the specific histone variants present in very early embryos, CS1 in sea urchin, H1M in batracians and $H1^{oo}$ in mammalians, may be involved in the determination of particular chromatin organization and specific gene expression of the ESCs.

Indeed, at this stage of development, a key requirement for the zygote is to establish a transcriptional active genome from a quiescent one using maternally inherited factors. These include TF Oct3/4 and Sox 2, epigenetic factors such as Ezh2 and Eed belonging to the Polycomb group of proteins and the chromatin remodelling factor BRG-1/brm-1 (Schuettengruber et al. 2007). BRG-1 loss-of-function results in downregulation of zygotic genome activation and consequently arrest at the two-cell stage (Bultman et al. 2006).

In mouse, when the embryo genome starts to be activated from late zygote to the two cell stage, the epigenetic status of the parental genome is still distinct by the DNA methylation status. Differences in DNA methylation are maintained until two cleavage stage and decline during pre-implantation development (Mayer et al. 2000). The analysis of X inactivation in the ICM and trophoectoderm cells reveals that the "imprinted" paternal X chromosome is inactivated during pre-implantation. At early blastocyst stage, the paternal X chromosome expresses the non-coding Xist RNA, loses active chromatin marks H3K4me2 and H3K4me3, acquires inactive chromatin marks H3K9me2 and H3K9me3 and finally H2A is ubiquitinated (Heard 2004). The inactive epigenetic marks associated with the paternal X chromosome inactivation are erased at the late blastocyst stage solely in PEct cells. This event could play a major role for the establishment of PEct cells pluripotent state. Notably, the paternal X chromosome remains inactivated in the extra-embryonic trophectoderm and PEnd cells (Heard 2004). Interestingly, the maintenance of genomic imprints is different in the embryo where DNA methylation is established probably due to an evolutionary adaptation for a long duration stage as the adult. Finally, in the mouse, ICs and OCs do not differ in their epigenetic patterns during morula stage. However, after commitment, epigenetic modification at appropriate gene loci might reveal the identities of PEct, trophoectoderm and PEnd cells in the blastocyst.

9.4.5.3 ESC Lines

Chromosomal protein dynamics in ESCs cells is considerably higher than those corresponding to differentiated cells. As reported by the group of Misteli using fluorescent recovery after photobleaching (FRAP) technology at a single–cell level the dissociation constants of histone H3 on genomic DNA in mouse ESCs is on the order of seconds, while in differentiated cells in the order of minutes. These authors had further demonstrated that the hyperdynamic binding of chromosomal proteins to DNA is required for the pluripotent potential of ESCs (Meshorer et al. 2006; Phair et al. 2004). Histones post-translational modifications have a major role in the regulation of chromatin structural changes related with epigenetic mechanisms that determine the transition from pluripotency to cell linage determination of the ESCs. Among these histones posttranslational modifications, H3 Lys-4 methylation and H3 and H4 acetylation are signals promoting transcription, in contrast H3 Lys-9 and H3 Lys-27 methylations are signals that induce compact chromatin domains associated to genetic silencing (reviewed by Gan et al. 2007).

9.5 Concluding Remarks

Evidence is emerging on the complex regulatory networks establishing the peculiar features of pluripotent cells at early stages of deuterostome embryo development. Whereas, pioneering studies on sea urchin embryos were fundamental to establish the concepts of cell "equipotentiality" and therefore that of "pluripotency", ESC lines from marine species are not yet available. To date, ESC lines have been only established from several mammalian species at pre- and post-implantation stages of development. Common mechanisms underlie the coordination of the cell cycle progression at early embryonic stages and in self-renewing ESCs in culture. Particularly, mESC, hESC and moESC lines are all characterized by a short G1 phase in which the pRb pathway is dispensable. Moreover Cdk1-cyclin B seems to be the only Cdk activity necessary to drive cell cycle in these pluripotent cells. Finally, ESCs are characterized by peculiar regulation of DNA replication and a disturbed response to DNA-damage. However, all the regulators of these unusual cell cycle properties and the relationship of the cell cycle dynamics resulting in the pluripotent cell state have not been fully identified. Moreover, transcriptional and epigenetic networks initiating the early mammalian developmental stages and regulating cell pluripotency and self-renewal appear to be evolutionary conserved in ESC lines. Thus, one challenge for the future will be to unravel the contribution of individual signal cascades in the complex and highly controlled regulatory networks and are preserved during the evolution and dictate cell phenotypic changes in pluripotent cells and early embryos.

Acknowledgments We thank Dr. H. Escriva and C. Iribarren for helpful discussions and suggestions. This work was partially supported by grants from: Ligue contre le cancer (France), Italian Association for Cancer Research (AIRC); Italian Health Ministry and Ministry for the Instruction, University and Research (PRIN).

References

Aladjem MI, Spike BT, Rodewald LW, Hope TJ, Klemm M, Jaenisch R, Wahl GM (1998) ES cells do not activate p53-dependent stress responses and undergo p53-independent apoptosis in response to DNA damage. Curr Biol 8(3):145–155

Anderson JA, Lewellyn AL, Maller JL (1997) Ionizing radiation induces apoptosis and elevates cyclin A1-Cdk2 activity before but not after the midblastula transition in Xenopus. Mol Biol Cell 8(7):1195–1206

Balhorn R, Cosman M, Thornton K, Krishnan VV, Corzett M, Bench G, Kramer C, Lee JI, Hud NV, Allen M, Prieto M, Meyer-Ilse W, Brown JT, Kirz J, Zhang X, Bradbury EM, Maki G, Braun RE, Breed W (1999) Protamine mediated condensation of DNA in mammalian sperm. In: Gagnon C (ed) Stem cells in Marine Organisms. Cache Rivers Press, Vienna, pp 55–70

Barberi T, Klivenyi P, Calingasan NY, Lee H, Kawamata H, Loonam K, Perrier AL, Bruses J, Rubio ME, Topf N, Tabar V, Harrison NL, Beal MF, Moore MA, Studer L (2003) Neural subtype specification of fertilization and nuclear transfer embryonic stem cells and application in parkinsonian mice. Nat Biotechnol 21(10):1200–1207

Bartek J, Lukas C, Lukas J (2004) Checking on DNA damage in S phase. Nat Rev Mol Cell Biol 5(10):792–804

Bartova E, Krejci J, Harnicarova A, Kozubek S (2008) Differentiation of human embryonic stem cells induces condensation of chromosome territories and formation of heterochromatin protein 1 foci. Differentiation 76(1):24–32

Becker KA, Ghule PN, Therrien JA, Lian JB, Stein JL, van Wijnen AJ, Stein GS (2006) Self-renewal of human embryonic stem cells is supported by a shortened G1 cell cycle phase. J Cell Physiol 209(3):883–893

Bell SP (2002) The origin recognition complex: From simple origins to complex functions. Genes Dev 16(6):659–672

Bernstein BE, Mikkelsen TS, Xie X, Kamal M, Huebert DJ, Cuff J, Fry B, Meissner A, Wernig M, Plath K, Jaenisch R, Wagschal A, Feil R, Schreiber SL, Lander ES (2006) A bivalent chromatin structure marks key developmental genes in embryonic stem cells. Cell 125(2): 315–326

Boiani M, Scholer HR (2005) Regulatory networks in embryo-derived pluripotent stem cells. Nat Rev Mol Cell Biol 6(11):872–884

Brandeis M, Rosewell I, Carrington M, Crompton T, Jacobs MA, Kirk J, Gannon J, Hunt T (1998) Cyclin B2-null mice develop normally and are fertile whereas cyclin B1-null mice die in utero. Proc Natl Acad Sci USA 95(8):4344–4349

Brons IG, Smithers LE, Trotter MW, Rugg-Gunn P, Sun B, Chuva de Sousa Lopes SM, Howlett SK, Clarkson A, Ahrlund-Richter L, Pedersen RA, Vallier L (2007) Derivation of pluripotent epiblast stem cells from mammalian embryos. Nature 448(7150):191–195

Bultman SJ, Gebuhr TC, Pan H, Svoboda P, Schultz RM, Magnuson T (2006) Maternal BRG1 regulates zygotic genome activation in the mouse. Genes Dev 20(13):1744–1754

Burdon T, Smith A, Savatier P (2002) Signalling, cell cycle and pluripotency in embryonic stem cells. Trends Cell Biol 12(9):432–438

Byrne JA (2008) Generation of isogenic pluripotent stem cells. Hum Mol Genet 17(R1):R37–R41

Byrne JA, Mitalipov SM, Clepper L, Wolf DP (2006) Transcriptional profiling of rhesus monkey embryonic stem cells. Biol Reprod 75(6):908–915

Byrne JA, Pedersen DA, Clepper LL, Nelson M, Sanger WG, Gokhale S, Wolf DP, Mitalipov SM (2007) Producing primate embryonic stem cells by somatic cell nuclear transfer. Nature 450(7169):497–502

Chambers I, Silva J, Colby D, Nichols J, Nijmeijer B, Robertson M, Vrana J, Jones K, Grotewold L, Smith A (2007) Nanog safeguards pluripotency and mediates germline development. Nature 450(7173):1230–1234

Cho C, Willis WD, Goulding EH, Jung-Ha H, Choi YC, Hecht NB, Eddy EM (2001) Haploinsufficiency of protamine-1 or -2 causes infertility in mice. Nat Genet 28(1):82–86

Chung HM, Malacinski GM (1980) Establishment of the dorsal/ventral polarity of the amphibian embryo: Use of ultraviolet irradiation and egg rotation as probes. Dev Biol 80(1): 120–133

Ciemerych MA, Kenney AM, Sicinska E, Kalaszczynska I, Bronson RT, Rowitch DH, Gardner H, Sicinski P (2002) Development of mice expressing a single D-type cyclin. Genes Dev 16(24):3277–3289

Ciemerych MA, Sicinski P (2005) Cell cycle in mouse development. Oncogene 24(17):2877–2898

Ciliberto A, Petrus MJ, Tyson JJ, Sible JC (2003) A kinetic model of the cyclin E/Cdk2 developmental timer in Xenopus laevis embryos. Biophys Chem 104(3):573–589

Cobrinik D (2005) Pocket proteins and cell cycle control. Oncogene 24(17):2796–2809

Cobrinik D, Lee MH, Hannon G, Mulligan G, Bronson RT, Dyson N, Harlow E, Beach D, Weinberg RA, Jacks T (1996) Shared role of the pRB-related p130 and p107 proteins in limb development. Genes Dev 10(13):1633–1644

Conn CW, Lewellyn AL, Maller JL (2004) The DNA damage checkpoint in embryonic cell cycles is dependent on the DNA-to-cytoplasmic ratio. Dev Cell 7(2):275–281

Cosgrove RA, Philpott A (2007) Cell cycling and differentiation do not require the retinoblastoma protein during early Xenopus development. Dev Biol 303(1):311–324

Crane-Robinson C (1999) How do linker histones mediate differential gene expression? Bioessays 21(5):367–371

Croce JC, McClay DR (2006) The canonical Wnt pathway in embryonic axis polarity. Semin Cell Dev Biol 17(2):168–174

Danilchik MV, Denegre JM (1991) Deep cytoplasmic rearrangements during early development in Xenopus laevis. Development 111(4):845–856

Dannenberg JH, van Rossum A, Schuijff L, te Riele H (2000) Ablation of the retinoblastoma gene family deregulates G(1) control causing immortalization and increased cell turnover under growth-restricting conditions. Genes Dev 14(23):3051–3064

Datar SA, Jacobs HW, de la Cruz AF, Lehner CF, Edgar BA (2000) The Drosophila cyclin D-Cdk4 complex promotes cellular growth. Embo J 19(17):4543–4554

Davidson E (1986) Gene activity in early development, 3rd edn. Academic Press, New York

Destree OH, Lam KT, Peterson-Maduro LJ, Eizema K, Diller L, Gryka MA, Frebourg T, Shibuya E, Friend SH (1992) Structure and expression of the Xenopus retinoblastoma gene. Dev Biol 153(1):141–149

Dimitrov S, Wolffe AP (1996) Remodeling somatic nuclei in Xenopus laevis egg extracts: Molecular mechanisms for the selective release of histones H1 and H1(0) from chromatin and the acquisition of transcriptional competence. Embo J 15(21):5897–5906

Driesch H (1891) Entwicklungsmechanische studien. I. Der Werth der beiden ersten Furchungszellen in der Echinodermenentwicklung. Experimentelle Erzeugung von Tell und Doppelbildungen. Z Wiss Zool 53:160–178 and 183–184

Duboc V, Lepage T (2008) A conserved role for the nodal signaling pathway in the establishment of dorso-ventral and left-right axes in deuterostomes. J Exp Zool B Mol Dev Evol 310(1): 41–53

Duboc V, Rottinger E, Lapraz F, Besnardeau L, Lepage T (2005) Left-right asymmetry in the sea urchin embryo is regulated by nodal signaling on the right side. Dev Cell 9(1):147–158

Erhardt S, Su IH, Schneider R, Barton S, Bannister AJ, Perez-Burgos L, Jenuwein T, Kouzarides T, Tarakhovsky A, Surani MA (2003) Consequences of the depletion of zygotic and embryonic enhancer of zeste 2 during preimplantation mouse development. Development 130(18): 4235–4248

Evans MJ, Kaufman MH (1981) Establishment in culture of pluripotential cells from mouse embryos. Nature 292(5819):154–156

Evans T, Rosenthal ET, Youngblom J, Distel D, Hunt T (1983) Cyclin: A protein specified by maternal mRNA in sea urchin eggs that is destroyed at each cleavage division. Cell 33(2): 389–396

Faast R, White J, Cartwright P, Crocker L, Sarcevic B, Dalton S (2004) Cdk6-cyclin D3 activity in murine ES cells is resistant to inhibition by p16(INK4a). Oncogene 23(2):491–502

Fang F, Newport JW (1991) Evidence that the G1-S and G2-M transitions are controlled by different cdc2 proteins in higher eukaryotes. Cell 66(4):731–742

Fernandez-Guerra A, Aze A, Morales J, Mulner-Lorillon O, Cosson B, Cormier P, Bradham C, Adams N, Robertson AJ, Marzluff WF, Coffman JA, Geneviere AM (2006) The genomic repertoire for cell cycle control and DNA metabolism in S. purpuratus. Dev Biol 300(1): 238–251

Ferrell JE Jr, Wu M, Gerhart JC, Martin GS (1991) Cell cycle tyrosine phosphorylation of p34cdc2 and a microtubule-associated protein kinase homolog in Xenopus oocytes and eggs. Mol Cell Biol 11(4):1965–1971

Finkielstein CV, Chen LG, Maller JL (2002) A role for G1/S cyclin-dependent protein kinases in the apoptotic response to ionizing radiation. J Biol Chem 277(41):38476–38485

Finkielstein CV, Lewellyn AL, Maller JL (2001) The midblastula transition in Xenopus embryos activates multiple pathways to prevent apoptosis in response to DNA damage. Proc Natl Acad Sci USA 98(3):1006–1011

Fluckiger AC, Marcy G, Marchand M, Negre D, Cosset FL, Mitalipov S, Wolf D, Savatier P, Dehay C (2006) Cell cycle features of primate embryonic stem cells. Stem Cells 24(3): 547–556

Frehlick LJ, Eirin-Lopez JM, Jeffery ED, Hunt DF, Ausio J (2006) The characterization of amphibian nucleoplasmins yields new insight into their role in sperm chromatin remodeling. BMC Genomics 7:99

Fujii-Yamamoto H, Kim JM, Arai K, Masai H (2005) Cell cycle and developmental regulations of replication factors in mouse embryonic stem cells. J Biol Chem 280(13): 12976–12987

Gan Q, Yoshida T, McDonald OG, Owens GK (2007) Concise review: Epigenetic mechanisms contribute to pluripotency and cell lineage determination of embryonic stem cells. Stem Cells 25(1):2–9

Gardner RL, Beddington RS (1988) Multi-lineage 'stem' cells in the mammalian embryo. J Cell Sci Suppl 10:11–27

Geneviere-Garrigues AM, Barakat A, Doree M, Moreau JL, Picard A (1995) Active cyclin B-cdc2 kinase does not inhibit DNA replication and cannot drive prematurely fertilized sea urchin eggs into mitosis. J Cell Sci 108 (Pt 7):2693–2703

Gerhart J (2006) The deuterostome ancestor. J Cell Physiol 209(3):677–685

Giudice G (1999) Genes and their products in sea urchin development. Curr Top Dev Biol 45: 41–116

Gonzalez MA, Tachibana KE, Adams DJ, van der Weyden L, Hemberger M, Coleman N, Bradley A, Laskey RA (2006) Geminin is essential to prevent endoreduplication and to form pluripotent cells during mammalian development. Genes Dev 20(14):1880–1884

Graham CF, Morgan RW (1966) Changes in cell cycle during early amphibian development. Dev Biol 14:439–460
Groth A, Rocha W, Verreault A, Almouzni G (2007) Chromatin challenges during DNA replication and repair. Cell 128(4):721–733
Gurdon JB (1976) Injected nuclei in frog oocytes: Fate, enlargement, and chromatin dispersal. J Embryol Exp Morphol 36(3):523–540
Harland R, Gerhart J (1997) Formation and function of Spemann's organizer. Annu Rev Cell Dev Biol 13:611–667
Hartley RS, Rempel RE, Maller JL (1996) In vivo regulation of the early embryonic cell cycle in Xenopus. Dev Biol 173(2):408–419
Hartley RS, Sible JC, Lewellyn AL, Maller JL (1997) A role for cyclin E/Cdk2 in the timing of the midblastula transition in Xenopus embryos. Dev Biol 188(2):312–321
Heard E (2004) Recent advances in X-chromosome inactivation. Curr Opin Cell Biol 16(3): 247–255
Hensey C, Gautier J (1997) A developmental timer that regulates apoptosis at the onset of gastrulation. Mech Dev 69(1–2):183–195
Hogan B, Beddington R, Costantini F, Lacy E (1994) Manipulating the Mouse Embryo: Cold Spring Harbor Laboratory Press
Hong Y, Cervantes RB, Tichy E, Tischfield JA, Stambrook PJ (2007) Protecting genomic integrity in somatic cells and embryonic stem cells. Mutat Res 614(1–2):48–55
Hong Y, Stambrook PJ (2004) Restoration of an absent G1 arrest and protection from apoptosis in embryonic stem cells after ionizing radiation. Proc Natl Acad Sci USA 101(40): 14443–14448
Howe JA, Howell M, Hunt T, Newport JW (1995) Identification of a developmental timer regulating the stability of embryonic cyclin A and a new somatic A-type cyclin at gastrulation. Genes Dev 9(10):1164–1176
Howe JA, Newport JW (1996) A developmental timer regulates degradation of cyclin E1 at the midblastula transition during Xenopus embryogenesis. Proc Natl Acad Sci USA 93(5): 2060–2064
Hyrien O, Maric C, Mechali M (1995) Transition in specification of embryonic metazoan DNA replication origins. Science 270(5238):994–997
Imhof A (2006) Epigenetic regulators and histone modification. Brief Funct Genomic Proteomic 5(3):222–227
Imschenetzky M, Puchi M, Morin V, Medina R, Montecino M (2003) Chromatin remodeling during sea urchin early development: Molecular determinants for pronuclei formation and transcriptional activation. Gene 322:33–46
Iribarren C, Morin V, Puchi M, Imschenetzky M (2008) Sperm nucleosomes disassembly is a requirement for histones proteolysis during male pronucleus formation. J Cell Biochem 103(2):447–455
Iwamori N, Naito K, Sugiura K, Tojo H (2002) Preimplantation-embryo-specific cell cycle regulation is attributed to the low expression level of retinoblastoma protein. FEBS Lett 526(1 3):119–123
Jaenisch R, Young R (2008) Stem cells, the molecular circuitry of pluripotency and nuclear reprogramming. Cell 132(4):567–582
Jiang J, Chan YS, Loh YH, Cai J, Tong GQ, Lim CA, Robson P, Zhong S, Ng HH (2008) A core Klf circuitry regulates self-renewal of embryonic stem cells. Nat Cell Biol 10(3):353–360
Jirmanova L, Afanassieff M, Gobert-Gosse S, Markossian S, Savatier P (2002) Differential contributions of ERK and PI3-kinase to the regulation of cyclin D1 expression and to the control of the G1/S transition in mouse embryonic stem cells. Oncogene 21(36):5515–5528
Kane DA, Kimmel CB (1993) The zebrafish midblastula transition. Development 119(2):447–456
Kim J, Chu J, Shen X, Wang J, Orkin SH (2008) An extended transcriptional network for pluripotency of embryonic stem cells. Cell 132(6):1049–1061

Kim JM, Yamada M, Masai H (2003) Functions of mammalian Cdc7 kinase in initiation/monitoring of DNA replication and development. Mutat Res 532(1–2):29–40
Kumagai A, Guo Z, Emami KH, Wang SX, Dunphy WG (1998) The Xenopus Chk1 protein kinase mediates a caffeine-sensitive pathway of checkpoint control in cell-free extracts. J Cell Biol 142(6):1559–1569
Kurokawa D, Akasaka K, Mitsunaga-Nakatsubo K, Shimada H (1997) Cloning of cyclin E cDNA of the sea urchin, Hemicentrotus pulcherrimus. Zool Sci 14(5):791–794
Laslett AL, Grimmond S, Gardiner B, Stamp L, Lin A, Hawes SM, Wormald S, Nikolic-Paterson D, Haylock D, Pera MF (2007) Transcriptional analysis of early lineage commitment in human embryonic stem cells. BMC Dev Biol 7:12
Le Bouffant R, Cormier P, Cueff A, Belle R, Mulner-Lorillon O (2007) Sea urchin embryo as a model for analysis of the signaling pathways linking DNA damage checkpoint, DNA repair and apoptosis. Cell Mol Life Sci 64(13):1723–1734
LeCouter JE, Kablar B, Hardy WR, Ying C, Megeney LA, May LL, Rudnicki MA (1998a) Strain-dependent myeloid hyperplasia, growth deficiency, and accelerated cell cycle in mice lacking the Rb-related p107 gene. Mol Cell Biol 18(12):7455–7465
LeCouter JE, Kablar B, Whyte PF, Ying C, Rudnicki MA (1998b) Strain-dependent embryonic lethality in mice lacking the retinoblastoma-related p130 gene. Development 125(23): 4669–4679
Lee SY, Jeon DG, Lee JS, Hwang CS, Huh K, Lee TW, Hong SI (1996) Deletion of Rb1 gene in late osteosarcoma from survivor of unilateral retinoblastoma – a case report. J Korean Med Sci 11(1):94–98
Lemaitre JM, Danis E, Pasero P, Vassetzky Y, Mechali M (2005) Mitotic remodeling of the replicon and chromosome structure. Cell 123(5):787–801
Liang J, Wan M, Zhang Y, Gu P, Xin H, Jung SY, Qin J, Wong J, Cooney AJ, Liu D, Songyang Z (2008) Nanog and Oct4 associate with unique transcriptional repression complexes in embryonic stem cells. Nat Cell Biol 10(6):731–739
Liu SV (2008) iPS cells: A more critical review. Stem Cells Dev 17(3):391–397
Loeb J (1894) Ueber eine einfache Methode, zwei oder mehr zusammengewachsene Embryonen aus einem Ei hervorzubringen. Pflügers Arch 55:525–530
Ma C, Collodi P (1999) Preparation of primary cell cultures from lamprey. Methods Cell Sci 21(1):39–46
MacQueen HA, Johnson MH (1983) The fifth cell cycle of the mouse embryo is longer for smaller cells than for larger cells. J Embryol Exp Morphol 77:297–308
Malumbres M, Sotillo R, Santamaria D, Galan J, Cerezo A, Ortega S, Dubus P, Barbacid M (2004) Mammalian cells cycle without the D-type cyclin-dependent kinases Cdk4 and Cdk6. Cell 118(4):493–504
Mandl B, Brandt WF, Superti-Furga G, Graninger PG, Birnstiel ML, Busslinger M (1997) The five cleavage-stage (CS) histones of the sea urchin are encoded by a maternally expressed family of replacement histone genes: Functional equivalence of the CS H1 and frog H1M (B4) proteins. Mol Cell Biol 17(3):1189–1200
Martin GR (1981) Isolation of a pluripotent cell line from early mouse embryos cultured in medium conditioned by teratocarcinoma stem cells. Proc Natl Acad Sci USA 78(12):7634–7638
Massague J (2004) G1 cell-cycle control and cancer. Nature 432(7015):298–306
Matsuda T, Nakamura T, Nakao K, Arai T, Katsuki M, Heike T, Yokota T (1999) STAT3 activation is sufficient to maintain an undifferentiated state of mouse embryonic stem cells. Embo J 18(15):4261–4269
Mayer W, Niveleau A, Walter J, Fundele R, Haaf T (2000) Demethylation of the zygotic paternal genome. Nature 403(6769):501–502
Meijer L, Azzi L, Wang JY (1991) Cyclin B targets p34cdc2 for tyrosine phosphorylation. Embo J 10(6):1545–1554
Meshorer E, Yellajoshula D, George E, Scambler PJ, Brown DT, Misteli T (2006) Hyperdynamic plasticity of chromatin proteins in pluripotent embryonic stem cells. Dev Cell 10(1):105–116

Meyer CA, Jacobs HW, Datar SA, Du W, Edgar BA, Lehner CF (2000) Drosophila Cdk4 is required for normal growth and is dispensable for cell cycle progression. Embo J 19(17):4533–4542

Mitsui K, Tokuzawa Y, Itoh H, Segawa K, Murakami M, Takahashi K, Maruyama M, Maeda M, Yamanaka S (2003) The homeoprotein Nanog is required for maintenance of pluripotency in mouse epiblast and ES cells. Cell 113(5):631–642

Moore JC, Sumerel JL, Schnackenberg BJ, Nichols JA, Wikramanayake A, Wessel GM, Marzluff WF (2002) Cyclin D and cdk4 are required for normal development beyond the blastula stage in sea urchin embryos. Mol Cell Biol 22(13):4863–4875

Moreau JL, Marques F, Barakat A, Schatt P, Lozano JC, Peaucellier G, Picard A, Geneviere AM (1998) Cdk2 activity is dispensable for the onset of DNA replication during the first mitotic cycles of the sea urchin early embryo. Dev Biol 200(2):182–197

Mulligan GJ, Wong J, Jacks T (1998) p130 is dispensable in peripheral T lymphocytes: Evidence for functional compensation by p107 and pRB. Mol Cell Biol 18(1):206–220

Mullins LJ, Wilmut I, Mullins JJ (2004) Nuclear transfer in rodents. J Physiol 554(Pt 1):4–12

Munoz-Sanjuan I, Brivanlou AH (2001) Early posterior/ventral fate specification in the vertebrate embryo. Dev Biol 237(1):1–17

Murphy M (1999) Delayed early embryonic lethality following disruption of the murine cyclin A2 gene. Nat Genet 23(4):481

Nakagawa M, Koyanagi M, Tanabe K, Takahashi K, Ichisaka T, Aoi T, Okita K, Mochiduki Y, Takizawa N, Yamanaka S (2008) Generation of induced pluripotent stem cells without Myc from mouse and human fibroblasts. Nat Biotechnol 26(1):101–106

Nakamura T, Arai Y, Umehara H, Masuhara M, Kimura T, Taniguchi H, Sekimoto T, Ikawa M, Yoneda Y, Okabe M, Tanaka S, Shiota K, Nakano T (2007) PGC7/Stella protects against DNA demethylation in early embryogenesis. Nat Cell Biol 9(1):64–71

Newport J, Dasso M (1989) On the coupling between DNA replication and mitosis. J Cell Sci Suppl 12:149–160

Newport J, Kirschner M (1982) A major developmental transition in early Xenopus embryos: II. Control of the onset of transcription. Cell 30(3):687–696

Nichols J, Zevnik B, Anastassiadis K, Niwa H, Klewe-Nebenius D, Chambers I, Scholer H, Smith A (1998) Formation of pluripotent stem cells in the mammalian embryo depends on the POU transcription factor Oct4. Cell 95(3):379–391

Niwa H (2007) How is pluripotency determined and maintained? Development 134(4):635–646

Niwa H, Burdon T, Chambers I, Smith A (1998) Self-renewal of pluripotent embryonic stem cells is mediated via activation of STAT3. Genes Dev 12(13):2048–2060

Niwa H, Toyooka Y, Shimosato D, Strumpf D, Takahashi K, Yagi R, Rossant J (2005) Interaction between Oct3/4 and Cdx2 determines trophectoderm differentiation. Cell 123(5): 917–929

Norbury CJ, Zhivotovsky B (2004) DNA damage-induced apoptosis. Oncogene 23(16):2797–2808

Ohtsuka S, Dalton S (2008) Molecular and biological properties of pluripotent embryonic stem cells. Gene Ther 15(2):74–81

Okita K, Ichisaka T, Yamanaka S (2007) Generation of germline-competent induced pluripotent stem cells. Nature 448(7151):313–317

Oliver MI, Rodriguez C, Bustos P, Morin V, Gutierrez S, Montecino M, Geneviere AM, Puchi M, Imschenetzky M (2003) Conservative segregation of maternally inherited CS histone variants in larval stages of sea urchin development. J Cell Biochem 88(4):643–649

Orford KW, Scadden DT (2008) Deconstructing stem cell self-renewal: Genetic insights into cell-cycle regulation. Nat Rev Genet 9(2):115–128

Orkin SH (2003) Priming the hematopoietic pump. Immunity 19(5):633–634

Pain B, Clark ME, Shen M, Nakazawa H, Sakurai M, Samarut J, Etches RJ (1996) Long-term in vitro culture and characterisation of avian embryonic stem cells with multiple morphogenetic potentialities. Development 122(8):2339–2348

Parisi T, Beck AR, Rougier N, McNeil T, Lucian L, Werb Z, Amati B (2003) Cyclins E1 and E2 are required for endoreplication in placental trophoblast giant cells. Embo J 22(18): 4794–4803

Park IH, Zhao R, West JA, Yabuuchi A, Huo H, Ince TA, Lerou PH, Lensch MW, Daley GQ (2008) Reprogramming of human somatic cells to pluripotency with defined factors. Nature 451(7175):141–146
Park M, Krause MW (1999) Regulation of postembryonic G(1) cell cycle progression in Caenorhabditis elegans by a cyclin D/CDK-like complex. Development 126(21):4849–4860
Pau KY, Wolf DP (2004) Derivation and characterization of monkey embryonic stem cells. Reprod Biol Endocrinol 2:41
Peng A, Lewellyn AL, Maller JL (2007) Undamaged DNA transmits and enhances DNA damage checkpoint signals in early embryos. Mol Cell Biol 27(19):6852–6862
Peng A, Lewellyn AL, Maller JL (2008) DNA damage signaling in early Xenopus embryos. Cell Cycle 7(1):3–6
Phair RD, Scaffidi P, Elbi C, Vecerova J, Dey A, Ozato K, Brown DT, Hager G, Bustin M, Misteli T (2004) Global nature of dynamic protein-chromatin interactions in vivo: Three-dimensional genome scanning and dynamic interaction networks of chromatin proteins. Mol Cell Biol 24(14):6393–6402
Philpott A, Krude T, Laskey RA (2000) Nuclear chaperones. Semin Cell Dev Biol 11(1):7–14
Qi X, Li TG, Hao J, Hu J, Wang J, Simmons H, Miura S, Mishina Y, Zhao GQ (2004) BMP4 supports self-renewal of embryonic stem cells by inhibiting mitogen-activated protein kinase pathways. Proc Natl Acad Sci USA 101(16):6027–6032
Rempel RE, Sleight SB, Maller JL (1995) Maternal Xenopus Cdk2-cyclin E complexes function during meiotic and early embryonic cell cycles that lack a G1 phase. J Biol Chem 270(12):6843–6855
Rideout WM, 3rd, Hochedlinger K, Kyba M, Daley GQ, Jaenisch R (2002) Correction of a genetic defect by nuclear transplantation and combined cell and gene therapy. Cell 109(1):17–27
Rossant J (2008) Stem cells and early lineage development. Cell 132(4):527–531
Sage J, Mulligan GJ, Attardi LD, Miller A, Chen S, Williams B, Theodorou E, Jacks T (2000) Targeted disruption of the three Rb-related genes leads to loss of G(1) control and immortalization. Genes Dev 14(23):3037–3050
Samanta MP, Tongprasit W, Istrail S, Cameron RA, Tu Q, Davidson EH, Stolc V (2006) The transcriptome of the sea urchin embryo. Science 314(5801):960–962
Sanchez Alvarado A (2007) Stem cells and the Planarian Schmidtea mediterranea. C R Biol 330(6–7):498–503
Santamaria D, Barriere C, Cerqueira A, Hunt S, Tardy C, Newton K, Caceres JF, Dubus P, Malumbres M, Barbacid M (2007) Cdk1 is sufficient to drive the mammalian cell cycle. Nature 448(7155):811–815
Savatier P, Huang S, Szekely L, Wiman KG, Samarut J (1994) Contrasting patterns of retinoblastoma protein expression in mouse embryonic stem cells and embryonic fibroblasts. Oncogene 9(3):809–818
Savatier P, Lapillonne H, van Grunsven LA, Rudkin BB, Samarut J (1996) Withdrawal of differentiation inhibitory activity/leukemia inhibitory factor up-regulates D-type cyclins and cyclin-dependent kinase inhibitors in mouse embryonic stem cells. Oncogene 12(2): 309–322
Scharf SR, Gerhart JC (1980) Determination of the dorsal-ventral axis in eggs of Xenopus laevis: Complete rescue of uv-impaired eggs by oblique orientation before first cleavage. Dev Biol 79(1):181–198
Schier AF (2003) Nodal signaling in vertebrate development. Annu Rev Cell Dev Biol 19:589–621
Schnackenberg BJ, Palazzo RE, Marzluff WF (2007) Cyclin E/Cdk2 is required for sperm maturation, but not DNA replication, in early sea urchin embryos. Genesis 45(5):282–291
Schuettengruber B, Chourrout D, Vervoort M, Leblanc B, Cavalli G (2007) Genome regulation by polycomb and trithorax proteins. Cell 128(4):735–745
Schultz RM (1993) Regulation of zygotic gene activation in the mouse. Bioessays 15(8):531–538
Shou W, Dunphy WG (1996) Cell cycle control by Xenopus p28Kix1, a developmentally regulated inhibitor of cyclin-dependent kinases. Mol Biol Cell 7(3):457–469

Silva J, Smith A (2008) Capturing pluripotency. Cell 132(4):532–536
Skottman H, Hovatta O (2006) Culture conditions for human embryonic stem cells. Reproduction 132(5):691–698
Smith AG (2001) Embryo-derived stem cells: Of mice and men. Annu Rev Cell Dev Biol 17: 435–462
Snape A, Wylie CC, Smith JC, Heasman J (1987) Changes in states of commitment of single animal pole blastomeres of Xenopus laevis. Dev Biol 119(2):503–510
Snow MHL (1977) Gastrulation in the mouse: Growth and regionalization of the epiblast. J Embryol ExpMorphol 42:293–303
Solter D (2006) From teratocarcinomas to embryonic stem cells and beyond: A history of embryonic stem cell research. Nat Rev Genet 7(4):319–327
Solter D, Skreb N, Damjanov I (1971) Cell cycle analysis in the mouse EGG-cylinder. Exp Cell Res 64(2):331–334
Spradling AC (1999) ORC binding, gene amplification, and the nature of metazoan replication origins. Genes Dev 13(20):2619–2623
Stead E, White J, Faast R, Conn S, Goldstone S, Rathjen J, Dhingra U, Rathjen P, Walker D, Dalton S (2002) Pluripotent cell division cycles are driven by ectopic Cdk2, cyclin A/E and E2F activities. Oncogene 21(54):8320–8333
Stern CD (2006) Evolution of the mechanisms that establish the embryonic axes. Curr Opin Genet Dev 16(4):413–418
Su JY, Rempel RE, Erikson E, Maller JL (1995) Cloning and characterization of the Xenopus cyclin-dependent kinase inhibitor p27XIC1. Proc Natl Acad Sci USA 92(22):10187–10191
Suda Y, Suzuki M, Ikawa Y, Aizawa S (1987) Mouse embryonic stem cells exhibit indefinite proliferative potential. J Cell Physiol 133(1):197–201
Suemori H, Tada T, Torii R, Hosoi Y, Kobayashi K, Imahie H, Kondo Y, Iritani A, Nakatsuji N (2001) Establishment of embryonic stem cell lines from cynomolgus monkey blastocysts produced by IVF or ICSI. Dev Dyn 222(2):273–279
Sumerel JL, Moore JC, Schnackenberg BJ, Nichols JA, Canman JC, Wessel GM, Marzluff WF (2001) Cyclin E and its associated cdk activity do not cycle during early embryogenesis of the sea Urchin. Dev Biol 234(2):425–440
Sumi T, Fujimoto Y, Nakatsuji N, Suemori H (2004) STAT3 is dispensable for maintenance of self-renewal in nonhuman primate embryonic stem cells. Stem Cells 22(5):861–872
Sutovsky P, Schatten G (1997) Depletion of glutathione during bovine oocyte maturation reversibly blocks the decondensation of the male pronucleus and pronuclear apposition during fertilization. Biol Reprod 56(6):1503–1512
Sutovsky P, Schatten G (2000) Paternal contributions to the mammalian zygote: Fertilization after sperm-egg fusion. Int Rev Cytol 195:1–65
Takahashi K, Okita K, Nakagawa M, Yamanaka S (2007a) Induction of pluripotent stem cells from fibroblast cultures. Nat Protoc 2(12):3081–3089
Takahashi K, Tanabe K, Ohnuki M, Narita M, Ichisaka T, Tomoda K, Yamanaka S (2007b) Induction of pluripotent stem cells from adult human fibroblasts by defined factors. Cell 131(5):861–872
Takahashi K, Yamanaka S (2006) Induction of pluripotent stem cells from mouse embryonic and adult fibroblast cultures by defined factors. Cell 126(4):663–676
Tanaka M, Hennebold JD, Macfarlane J, Adashi EY (2001) A mammalian oocyte-specific linker histone gene H1oo: Homology with the genes for the oocyte-specific cleavage stage histone (cs-H1) of sea urchin and the B4/H1M histone of the frog. Development 128(5):655–664
Tesar PJ, Chenoweth JG, Brook FA, Davies TJ, Evans EP, Mack DL, Gardner RL, McKay RD (2007) New cell lines from mouse epiblast share defining features with human embryonic stem cells. Nature 448(7150):196–199
Thomson JA, Itskovitz-Eldor J, Shapiro SS, Waknitz MA, Swiergiel JJ, Marshall VS, Jones JM (1998) Embryonic stem cell lines derived from human blastocysts. Science 282(5391): 1145–1147

Thomson JA, Kalishman J, Golos TG, Durning M, Harris CP, Becker RA, Hearn JP (1995) Isolation of a primate embryonic stem cell line. Proc Natl Acad Sci USA 92(17):7844–7848
Torres-Padilla ME, Bannister AJ, Hurd PJ, Kouzarides T, Zernicka-Goetz M (2006) Dynamic distribution of the replacement histone variant H3.3 in the mouse oocyte and preimplantation embryos. Int J Dev Biol 50(5):455–461
Vallier L, Pedersen RA (2005) Human embryonic stem cells: An in vitro model to study mechanisms controlling pluripotency in early mammalian development. Stem Cell Rev 1(2):119–130
Vernon AE, Devine C, Philpott A (2003) The cdk inhibitor p27Xic1 is required for differentiation of primary neurones in Xenopus. Development 130(1):85–92
Vernon AE, Philpott A (2003a) The developmental expression of cell cycle regulators in Xenopus laevis. Gene Exp Patterns 3(2):179–192
Vernon AE, Philpott A (2003b) A single cdk inhibitor, p27Xic1, functions beyond cell cycle regulation to promote muscle differentiation in Xenopus. Development 130(1):71–83
Vincent JP, Scharf SR, Gerhart JC (1987) Subcortical rotation in Xenopus eggs: A preliminary study of its mechanochemical basis. Cell Motil Cytoskeleton 8(2):143–154
Weissman IL (2000) Stem cells: Units of development, units of regeneration, and units in evolution. Cell 100(1):157–168
Wernig M, Meissner A, Foreman R, Brambrink T, Ku M, Hochedlinger K, Bernstein BE, Jaenisch R (2007) In vitro reprogramming of fibroblasts into a pluripotent ES-cell-like state. Nature 448(7151):318–324
White J, Stead E, Faast R, Conn S, Cartwright P, Dalton S (2005) Developmental activation of the Rb-E2F pathway and establishment of cell cycle-regulated cyclin-dependent kinase activity during embryonic stem cell differentiation. Mol Biol Cell 16(4):2018–2027
Whitman M (2001) Nodal signaling in early vertebrate embryos: Themes and variations. Dev Cell 1(5):605–617
Wianny F, Bernat A, Huissoud C, Marcy G, Markossian S, Cortay V, Giroud P, Leviel V, Kennedy H, Savatier P, Dehay C (2008) Derivation and cloning of a novel rhesus embryonic stem cell line stably expressing tau-green fluorescent protein. Stem Cells 26(6):1444–1453
Wikenheiser-Brokamp KA (2006) Retinoblastoma family proteins: Insights gained through genetic manipulation of mice. Cell Mol Life Sci 63(7–8):767–780
Wright SJ (1999) Sperm nuclear activation during fertilization. Curr Top Dev Biol 46:133–178
Wroble BN, Finkielstein CV, Sible JC (2007) Wee1 kinase alters cyclin E/Cdk2 and promotes apoptosis during the early embryonic development of Xenopus laevis. BMC Dev Biol 7:119
Wu L, de Bruin A, Saavedra HI, Starovic M, Trimboli A, Yang Y, Opavska J, Wilson P, Thompson JC, Ostrowski MC, Rosol TJ, Woollett LA, Weinstein M, Cross JC, Robinson ML, Leone G (2003) Extra-embryonic function of Rb is essential for embryonic development and viability. Nature 421(6926):942–947
Xie Y, Sun T, Wang QT, Wang Y, Wang F, Puscheck E, Rappolee DA (2005) Acquisition of essential somatic cell cycle regulatory protein expression and implied activity occurs at the second to third cell division in mouse preimplantation embryos. FEBS Lett 579(2):398–408
Yamanaka Y, Ralston A, Stephenson RO, Rossant J (2006) Cell and molecular regulation of the mouse blastocyst. Dev Dyn 235(9):2301–2314
Yamashita N, Kim JM, Koiwai O, Arai K, Masai H (2005) Functional analyses of mouse ASK, an activation subunit for Cdc7 kinase, using conditional ASK knockout ES cells. Genes Cells 10(6):551–563
Ying QL, Nichols J, Chambers I, Smith A (2003) BMP induction of Id proteins suppresses differentiation and sustains embryonic stem cell self-renewal in collaboration with STAT3. Cell 115(3):281–292
Ying QL, Wray J, Nichols J, Batlle-Morera L, Doble B, Woodgett J, Cohen P, Smith A (2008) The ground state of embryonic stem cell self-renewal. Nature 453(7194):519–523
Yu J, Vodyanik MA, Smuga-Otto K, Antosiewicz-Bourget J, Frane JL, Tian S, Nie J, Jonsdottir GA, Ruotti V, Stewart R, Slukvin II, Thomson JA (2007) Induced pluripotent stem cell lines derived from human somatic cells. Science 318(5858):1917–1920

Chapter 10
Regeneration in Hemichordates and Echinoderms

Amanda L. Rychel and Billie J. Swalla

Abstract Hemichordates are closely related to echinoderms, which are remarkable for their powers of regeneration. Among hemichordates, some enteropneust worms show dramatic regeneration, though this process has not been as well studied as in echinoderms. Unlike echinoderms, which are pentameral in adult form, hemichordates exhibit bilateral symmetry throughout the life cycle. Adult body regeneration in hemichordates may therefore show similar molecular patterning to chordate regeneration. In this chapter, we review the original literature about regeneration in hemichordates. We present our results from *Glossobalanus berkeleyi* and *Ptychodera flava*, the latter of which reliably regenerates anterior structures in the laboratory. When *P. flava* is bisected, the wound at the anterior end of the posterior half heals, followed by outgrowth of a blastema that becomes the new proboscis. After the proboscis develops, the mouth opens, the collar folds up, and gill slits are formed by tissue remodeling in the posterior worm fragment. Renewed interest in hemichordate regeneration, combined with the fact that it can be elicited in the lab, sets the stage for using molecular markers for stem cells and differentiated tissue in order to characterize the cellular and molecular events occurring during hemichordate regeneration. Hemichordate data presented here suggests that the ability to regenerate using stem cells may be common in ambulacrarian deuterostomes, the sister group to chordates. Understanding the molecular basis of regeneration in hemichordates may lead the way to methods for stimulating regeneration in vertebrates, including in humans for therapeutic purposes.

Keywords Hemichordates · Deuterostomes · Regeneration · Stem cells

---

B.J. Swalla (✉)
Department of Biology, Center for Developmental Biology, and Institute for Stem Cell and Regenerative Medicine, University of Washington, Seattle, WA 98195, USA;
Friday Harbor Laboratories, University of Washington, Friday Harbor,WA 98250-9299, USA;
Kewalo Marine Laboratory, University of Hawaii, Honolulu, HI 96813, USA
e-mail: bjswalla@u.washington.edu

B. Rinkevich, V. Matranga (eds.), *Stem Cells in Marine Organisms*,
DOI 10.1007/978-90-481-2767-2_10,

10.1 Introduction

Regeneration is a phenomenon widespread among metazoans. Most animal groups, with nematodes and cephalochordates as possible exceptions, include some species that regenerate certain tissues (Sánchez Alvarado 2000). Morgan classified regeneration into two different types: regeneration that requires cell proliferation (epimorphosis), and regeneration that does not require cell proliferation (morphallaxis) (Morgan 1898; Morgan 1901; Sánchez Alvarado 2000). Epimorphosis is frequently seen in the form of a blastema, a mass of proliferating undifferentiated cells that precedes regeneration. Morphallaxis, on the other hand, involves remodeling of existing tissues without extensive cell proliferation. Epimorphic blastema regeneration is the characteristic mode of regeneration found in vertebrates. Limbs and tails of anuran tadpoles (Dent 1962; Shimizu-Nishikawa et al. 2003; Mochii et al. 2007), urodele amphibians (Iten and Bryant 1976; Brockes 1997; Echeverri and Tanaka 2005) and fins of zebrafish (Stoick-Cooper et al. 2007; Yokoyama et al. 2007) form blastemas when replacing missing structures. In tunicates, it appears that regeneration occurs most frequently through epimorphosis. In colonial tunicates, whole body regeneration is accomplished via epimorphosis without a blastema, a process similar to asexual vascular budding (Goldin 1948; Rinkevich et al. 2007; Ballarin and Manni 2009; Sköld et al. 2009). Neural regeneration in the solitary adult tunicate, *Ciona,* probably involves non-blastemal epimorphosis and morphallaxis as well (Schultze 1900; Bollner et al. 1993; Bollner et al. 1995; Bollner et al.1997).

Because regeneration in a given animal may employ both epimorphosis and morphallaxis, the terms distalization and intercalation have been proposed by Agata et al. (2007) to describe regeneration processes. Frequently, the first step during regeneration is wound healing. At this point, whether or not a blastema is formed, there must be a new patterning axis established in order for the new tissues to be replaced correctly. This process is called distalization, and occurs whether regeneration is accomplished via epimorphosis or morphallaxis. Next, tissues that were lost must be replaced, and this process is called intercalation. Intercalation also can involve epimorphosis or morphallaxis.

A third distinction that can be made between regeneration modes is whether or not regeneration results in simply repairing a damaged individual or results in two complete individuals – resulting in asexual reproduction. Regeneration is considered bi-directional when bisection of an animal will result in two fully functional animals, and unidirectional if only one half regenerates. Within deuterostomes, bi-directional regeneration is characteristic of some echinoderms (Emson and Wilkie 1980; Mladenov and Burke 1984; Vickery et al. 2001b; Vickery et al. 2002; Eaves and Palmer 2003; Knott et al. 2003; McGovern 2003; Rubilar et al. 2005; Candia Carnevali 2006) and possibly of hemichordates (Gilchrist 1923; Packard 1968; Petersen and Ditadi 1971) but not of solitary ascidians or vertebrates (chordates).

Here, we begin by reviewing regeneration mechanisms and asexual reproduction in echinoderms, since they are closely related to hemichordates, and the processes of regeneration in them are much better understood. Next, we will review original literature concerning asexual reproduction and regeneration in hemichordates, present new data from studies on regeneration in enteropneust worms that is ongoing in our laboratory, and suggest where we expect future studies to be most fruitful.

10.2 Regeneration and Asexual Reproduction in Echinoderms

Within ambulacraria (echinoderms and hemichordates) (Swalla and Smith 2007), echinoderms have been documented to have extensive regeneration (Candia Carnevali 2006). There are five extant classes of echinoderms: crinoids (sea lilies), asteroids (sea stars), ophiuroids (brittle stars), holothuroids (sea cucumbers) and echinoids (sea urchins and sand dollars). Some species in each echinoderm class have been reported to regenerate (Candia Carnevali et al. 2009; D'Ancona Lunetta 2009). Crinoids are the most basal class of echinoderms and their ability to regenerate new arms has been well documented (Candia Carnevali et al. 1997; Candia Carnevali et al. 1998; Thorndyke et al. 2001a; Thorndyke et al. 2001b; Patruno et al. 2003), even in the fossil record (Oji 2001), suggesting that regeneration is an ancestral trait of echinoderms.

Both bi- and unidirectional regeneration modes are found in echinoderms. Examples of unidirectional regeneration are found in sea star arms and radial central nerve cords (Cuenot 1948; Thorndyke et al. 2001a) and sea cucumber gut and muscle regeneration after spontaneous evisceration (Dolmatov and Ginanova 2001; Garcia-Arraras and Greenberg 2001). Of all classes of echinoderms, adult sea urchins have the most limited regenerative capacity, yet they are still able to regenerate several of their body parts, including spines and pedicellariae (Heatfield and Travis 1975a,b; Drager et al. 1989; Dubois and Ameye 2001).

Some adult asteroids, ophiuroids, and holothuroids are able to regenerate bi-directionally when arms are cut or autonomized (Emson and Wilkie 1980; Mladenov and Burke 1984; Mazzone and Byrne 2001; Mazzone et al. 2003). This regeneration mechanism furthermore allows some species to reproduce asexually. Spontaneous autonomous fission in adults is the primary means of reproduction in some species (McGovern 2003; Rubilar et al. 2005), which probably leads to clonal populations. Echinoderm larvae of nearly every class (echinoids, asteroid, ophiuroids, and holothuroids) commonly clone themselves both in culture and nature through asexual budding (Bosch 1988; Bosch et al. 1989; Jaeckle 1994; Balser 1998; Vickery and McClintock 2000; Eaves and Palmer 2003; Knott et al. 2003; Sköld et al. 2009). Larval cloning in echinoids and ophiuroids appears to occur by a recapitulation of usual developmental processes based on visual observation of the process (Eaves and Palmer 2003; Knott et al. 2003). It will be especially informative to examine how closely echinoderm adult fission and larval cloning mimic normal development on a cellular and molecular basis.

In addition to the natural process of larval cloning in planktotrophic larvae, in some cases, surgically bisected echinoderm larvae can regenerate bi-directionally (Vickery et al. 2001b, 2002). Bi-directional regeneration in bisected larva is seen in sea urchins (echinoids) and sea stars (asteroids). Sand dollars (echinoids) were observed to regenerate completely the posterior half, but the anterior half was reported to only partially regenerate over two weeks (Vickery et al. 1999). Mesenchymal cells were seen migrating into the cut site, and were similar in appearance to mesenchymal cells forming the blastema in adult echinoderm arm regeneration (Bonasoro et al. 1998; Thorndyke et al. 1999; Candia Carnevali and Bonasoro 2001). Larval halves lacking coeloms and hence, lacking coelomocytes, were still able to regenerate. This indicates that coelomocytes are not required for this process, at least in echinoderm larvae (Vickery et al. 2002). Subtractive hybridization was done to isolate genes involved in the process of larval regeneration after bisection and revealed nine differentially expressed genes (Vickery et al. 2001a). At the time of publication in 2001, eight of these genes had no known homology with any gene in GenBank, indicating the potential power of this technique in marine invertebrates for discovering new genes important in regeneration. Whether regeneration occurs in injured holothuroid, crinoid, and ophiuroid larvae remains to be seen. Holothuroid (Hörstadius 1925, 1928, 1973; Dolmatov 1991) and non-feeding crinoid larvae (Runnström 1925) that have been bisected have had limited success in achieving full regeneration, but they have not been studied as intensively as other echinoderm classes for their regenerative capacity. Ophiuroid larvae, though known to clone from the cast off larval arms (Balser 1998), have not yet been shown to regenerate following surgical bisection. It is also unknown whether or not crinoid larvae or hemichordate tornaria larvae (which are morphologically similar to echinoderm larvae) are able to regenerate or clone.

10.3 Developmental Genes Implicated in Echinoderm Regeneration

Molecular understanding of regeneration in echinoderms has only recently been explored. So far, only a few genes have been implicated in being differentially regulated during regeneration in echinoderms, including *Hox1*, and *BMP2/4* and *univin*, from the TGF-ß superfamily of signaling molecules (Table 10.1). *Hox1* is expressed at low levels in normal sea star radial nerves, and is upregulated during regeneration of the nerve (Thorndyke et al. 2001a,b). *Hox* gene expression has also been seen in other regeneration systems. For instance, *Hox* genes are expressed during planarian (Bayascas et al. 1997; Bayascas et al. 1998; Saló et al. 2001) and urodele amphibian limb regeneration (Simon and Tabin 1993; Stocum 1996; Brockes 1997; Torok et al. 1998; Carlson et al. 2001; Nicolas et al. 2003). It is thought that *Hox* expression is important for repatterning the body or limb axis during regeneration, as during embryonic development. Experiments with urodele amphibians show that there are *Hox* genes expressed during regeneration that are regeneration specific (Brown and Brockes 1991; Stocum 1996; Torok et al. 1998; Carlson et al. 2001; Christen et al. 2003).

Bone morphogenetic proteins (BMPs) are part of the TGF-β (transforming growth factor- β) superfamily of signaling molecules (Hogan 1996; Newfeld et al. 1999). BMPs play important roles in animal embryonic development as well as in wound healing (O'Kane and Ferguson 1997) and regeneration in the tail and limb

Table 10.1 Developmental genes expressed during echinoderm arm regeneration

Taxon	Gene	Developmental expression	Nonregenerating adult expression	Regeneration expression pattern	References
Asterias rubens (sea star)	*Hox1*	?	Lower levels (rt-PCR) in arm radial nerve	Upregulated (rt-PCR) expression arm radial nerve	Thorndyke et al. 2001a
Amphiura filiformis (brittle star)	*univin*	?	Uniform expression in mesenchyme of arm radial water canal	Separate proximal and distal expression regions in radial water canal mesenchyme at 3 weeks after amputation	Thorndyke et al. 2001b; Bannister et al. 2005
Antedon Bifida (crinoid)	*BMP2/4*	?	Detectable by rt-PCR but not by *in situ* in normal adult arm	Early in blastema, then in coelomic canal epithelium, later in most proximal newly formed tissue	Patruno et al. 2003

of frog tadpoles (Beck et al. 2006). *BMP2* and *BMP4* are normally expressed during frog tail and limb development, and if either are antagonized after tail and limb amputation, these structures fail to regenerate (Beck et al. 2006). Echinoderms have two TGF- β superfamily members that have been implicated in echinoderm arm regeneration: *BMP2/4* (Patruno et al. 2003) and a TGF- β gene related to *Xenopus Vg1* (Range et al. 2007), *univin* (Bannister et al. 2005), although the functional role of these genes in normal post embryonic echinoderm development is unknown. In a crinoid, *BMP2/4* is expressed early in arm regeneration in the blastema, later in the coelomic canal epithelium, and in advanced stages, expression is found only proximally where new tissue is being differentiated (Thorndyke et al. 2001a,b,c; Patruno et al. 2003). A TGF- β similar to sea urchin *univin* is expressed during several stages of arm regeneration in brittle stars (ophiuroids), and expression is seen in coelomocytes in both normal and regenerating adults (Bannister et al. 2005).

Both *BMP2/4* and *univin* are known to be important in early embryonic development in echinoids. *Univin* promotes embryonic skeletal growth in sea urchin embryos and larvae (Zito et al. 2003). In early development *univin* is upstream of *nodal* signaling and is important in the dorsoventral (oral-aboral) axis formation of the sea urchin embryo (Range et al. 2007). *BMP2/4* is also implicated in patterning the dorsoventral axis of sea urchins (Duboc et al. 2004). While it is not yet known if *BMP2/4* or a gene similar to *univin* is implicated in hemichordate regeneration, some evidence exists for similar embryonic function for *BMP2/4* in hemichordates and echinoderms. Expression of *BMP2/4* in morphologically similar echinoderm holothuroid and hemichordate ptychoderid larvae was seen specifically in the hydropore region (Harada et al. 2002), suggesting that they are homologous structures. *BMP2/4* expression has not been reported in indirect developing hemichordate *Ptychodera flava* embryos (Harada et al. 2002) but is important for body patterning in the direct developing hemichordate, *Saccoglossus kowalevskii*. *BMP2/4* expression in *S. kowalevskii* begins during gastrulation, where it is expressed on one side of the embryo. Expression continues throughout development in the same place, but it narrows to a stripe on the ectoderm at the dorsal midline by the juvenile stage (Lowe et al. 2006). Knock down of *BMP2/4* expression in *S. kowalevskii* results in a ventralized embryo since the mouth forms circumferentially and gill slits that normally form on the dorsal side are lacking, whereas overexpression of *BMP2/4* results in a dorsalized embryo lacking a mouth (Lowe et al. 2006). Therefore, *BMP2/4* plays an important role in dorsal-ventral axis formation in sea urchin and direct developing hemichordate embryos.

Hemichordates are less well characterized than echinoderms in their regenerative abilities, but some species of enteropneust worms have remarkable regeneration (Hadfield 1975). However, this subject has received little research attention in the intervening decades. Our lab is currently studying regeneration in several different enteropneust hemichordate species.

10.4 Sexual and Asexual Reproduction in Hemichordates

Regeneration ability often correlates with reproductive mode. Animals that are able to reproduce asexually are often also able to overcome injury via regeneration. Therefore, asexual reproduction and regeneration ability have been considered related phenomena (Sánchez Alvarado 2000; Bely and Wray 2001; van Bekkum 2004; Saló 2006). For instance, in groups such as tunicates where some forms reproduce only sexually, and others both asexually and sexually, those that are able to asexually reproduce are highly regenerative compared to obligate sexual reproducers (Kawamura et al. 2008; Tiozzo et al. 2008).

Hemichordates follow one of two distinct life histories, the solitary enteropneust worms (Fig. 10.1A,B), and the colonial pterobranchs (Fig. 10.1C,D). Solitary enteropneusts comprise the indirect developing families Ptychoderidae (Fig. 10.1A), Spengelidae (not shown), and a direct developing family Harrimaniidae (Fig. 10.1B) (Cameron et al. 2000). Despite the fact that all enteropneust hemichordates are obligate sexual reproducers, and besides developmental mode differences (indirect vs. direct), they are distinguished by differences in morphology. In the adult enteropneusts, the anterior proboscis of ptychoderids is much shorter and wider than the long thin proboscis of harrimaniids (Fig. 10.1A,B) (Kowalevsky 1866; Spengel 1893; Hyman 1959). Also in ptychoderids, the gill bar skeleton has adjacent gill bars connected by small synapticulae, their gonads are contained within prominent ridges or wings in the branchial region of the trunk, and in the posterior trunk hepatic saccules or outpockets of the gut are found (Fig. 10.1A). However, no synapticulae, genital ridges, or hepatic saccules are present in harrimaniids (Hyman 1959; Aronowicz and Lowe 2006).

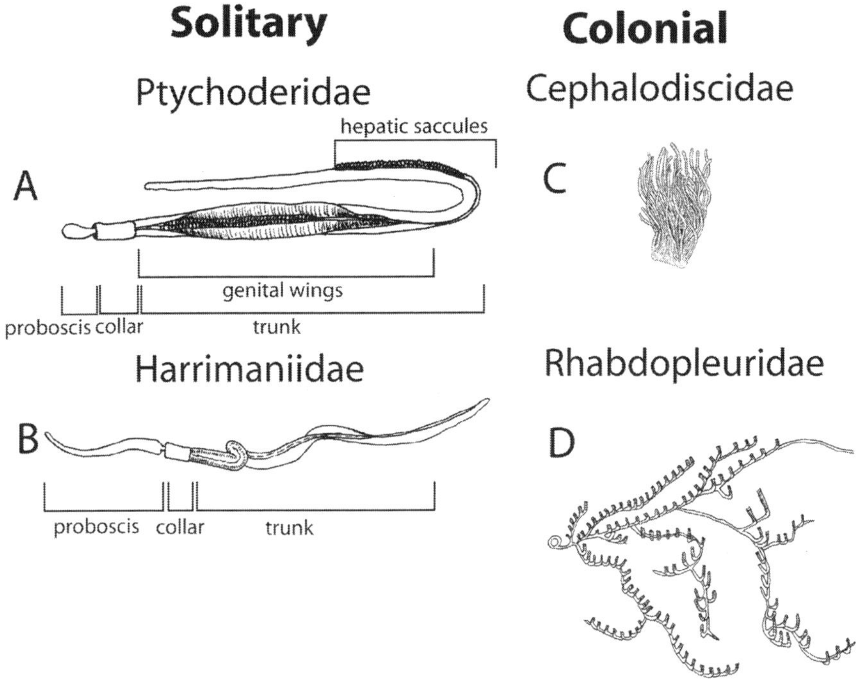

Fig. 10.1 Comparison of the body plans between two solitary enteropneust worm families, Ptychoderidae (**A**) and Harrimaniidae (**B**), and two colonial pterobranch families, Cephalodiscidae (**C**) and Rhabdopleuridae (**D**). (**A**) modified from (Hyman 1959), (**B**) modified from (Marion 1886), (**C**) modified from (Andersson 1907), and (**D**) modified from (Schepotieff 1907a). Ptychoderid worms (**A**) have a shorter, stouter proboscis compared to harrimaniid worms (**B**), and possess hepatic sacs and genital ridges or wings (**A**) that are lacking in harrimaniids (**B**). Colonial hemichordates form colonies by budding asexually. *Cephalodiscus* species (**C**) can be either sessile or pelagic, but all form small colonies. In contrast, *Rhabdopleura* species (**D**) grow within a characteristic cartilaginous coenecium, and individuals will not be viable if buds are separated from the colony

In contrast, the colonial pterobranchs are extremely morphologically divergent from solitary enteropneusts and include the direct developing families Cephalodiscidae (Fig. 10.1C) and Rhabdopleuridae (Fig. 10.1D) (Cameron et al. 2000). Pterobranch hemichordates reproduce both asexually and sexually (Fig. 10.2A), but regeneration ability has not yet been documented. Cephalodiscid pterobranch zooids are usually seen with 1–14 buds growing from the most posterior end of the stalk, the number of buds depending on the species (Figs. 10.1C and 10.2A). Adult asexually reproduced cephalodiscid zooids secrete their own coenecium (external collagenous tube) and thereby remain attached to their parent zooid although the living tissues of individual parent and offspring may or may not remain physically linked (Masterman 1900; Schepotieff 1908; Lester 1985; Schiaparelli et al. 2004) (Fig. 10.2A). By contrast, budding in rhabdopleurid pterobranchs occurs along the stolons that connect individuals to one another, so that

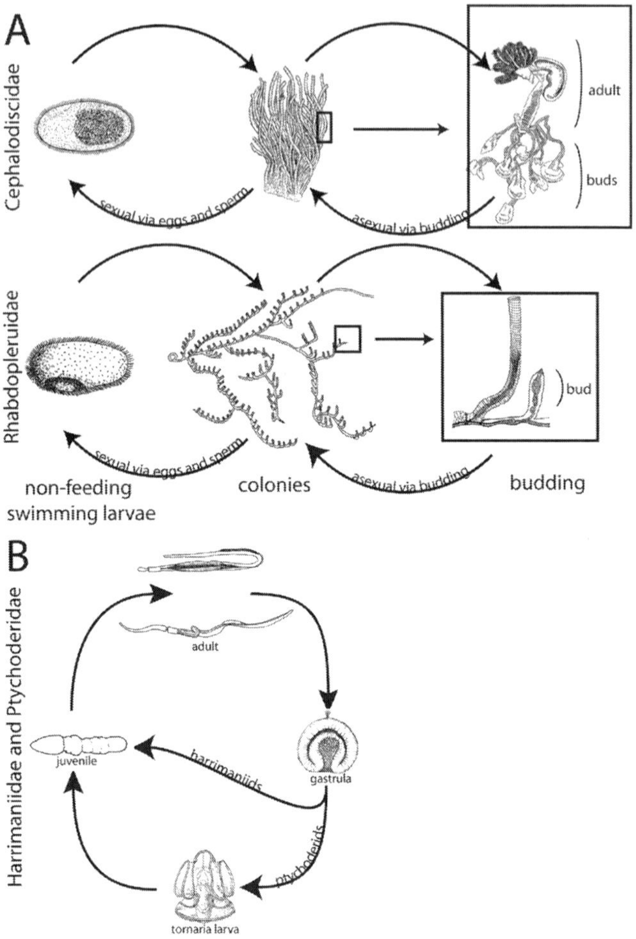

Fig. 10.2 Life cycles of colonial and solitary hemichordates. A. Colonial hemichordate life cycle. *Cephalodiscus* larva modified from (Schepotieff 1909), *Cephalodiscus* colony modified from (Andersson 1907), *Cephalodiscus* individual modified from (John 1931); *Rhabdopleura* larva modified from (Lester 1988); *Rhabdopleura* colony modified from (Schepotieff 1907b), budding *Rhabdopleura* modified from (Schepotieff 1907b). B. Solitary hemichordate life cycle (adult hemichordates modified from (Hyman 1959)). Both cephalodiscids and rhabdopleurids can reproduce asexually via budding (individual shown on the *right* magnified in *box*) and sexually via eggs and sperm (*left side*) (**A**). When reproducing sexually, both cephalodiscids and rhabdopleurids develop into swimming, non-feeding larvae and then directly develop into settled zooids. The larva of *Rhabdopleura* is characterized by a ventral groove on the anterior end, and the *Cephalodiscus* larva has a prominent yolky mass localized to the posterior end. Asexual budding in cephalodiscids occurs with new individuals budding off of the parent stalk (**A**, *top right*), while in rhabdopleurids, budding occurs from the tip or lateral edges of the growing stolon that connects individual zooids in a colony (**A**, *bottom right*). Solitary hemichordates reproduce sexually (**B**). Harrimaniids go through direct development, whereas ptychoderids develop indirectly first into a feeding planktonic tornaria larva (*bottom*) before settling and developing into a juvenile

the individuals are distinct (Figs. 10.1D and 10.2A). Both terminal and lateral buds develop along a given stretch of stolon (Schepotieff 1907b; Stebbing 1970; Dilly 1975) (Fig. 10.2A). This type of clonal reproduction and attachment of adult zooids is comparable to social ascidians which form asexual buds from stolons that connect individual zooids (Zeng et al. 2006; Tiozzo et al. 2008). The differences in budding between groups of pterobranchs results in clumped grouping of zooids in cephalodiscids (Fig. 10.1C) compared with more dispersed individuals in colonies of rhabdopleurids (Fig. 10.1D).

Solitary hemichordates, on the other hand, are obligate sexual reproducers (Fig. 10.2B). It is possible, however, that fragile enteropneusts that are broken by wave action or predation are, in some species, able to recover via regeneration of resulting pieces, which may lead to two or more new individuals from a single severed worm. Although both harrimaniids and ptychoderid solitary enteropneusts are obligate sexual reproducers, they are not equally able to regenerate.

10.5 Regeneration in Solitary Enteropneust Worms

We focus here on regeneration in solitary enteropneust hemichordates, since very little is documented for colonial hemichordate regeneration. First, we describe the normal developmental mode in hemichordates, and then their regeneration potential. The direct developing harrimaniids progress from embryogenesis directly into a juvenile worm (Fig. 10.2B) (Bateson 1884, 1885; Hyman 1959; Cameron et al. 2000), although they do spend a short period of time as a non-feeding lecithotropic larva. Ptychoderids are indirect developers, and following embryogenesis, pass through a planktonic feeding stage (tornaria) prior to settling and developing adult structures (Fig. 10.2B) (Metschnikoff 1869; Hadfield 1975; Urata and Yamaguchi 2004).

Regeneration has been reported just once in a direct developing harrimaniid (Tweedell 1961). In *Saccoglossus kowalevskii*, regeneration is fairly limited in lab conditions as they are only able to regenerate the most anterior structures after amputation, that is, within the proboscis and collar (Fig. 10.1B). In contrast, several ptychoderid species (*Ptychodera flava*, *Glossobalanus minutus*, *Balanoglossus australiensis*, and *Glossobalanus crozieri*) have been reported to have remarkable regeneration (Willey 1899; Dawydoff 1909, 1948; Rao 1955; Packard 1968; Petersen and Ditadi 1971). Most of the published observations have been from posterior trunk pieces containing only gonad, gut, and hepatic sacculations that are able to regenerate all new anterior structures, including the proboscis and collar, which contain the heart, kidney, collar nerve cord, stomochord, and proboscis skeleton, and the branchial region containing the pharyngeal slits and skeleton (Willey 1899; Dawydoff 1909, 1948; Rao 1955; Packard 1968; Petersen and Ditadi 1971).

All of the detailed reports of ptychoderid regeneration have several features in common. Anterior regeneration from an amputated trunk appears to incorporate both epimorphosis and morphallaxis (Dawydoff 1909; Rao 1955). We have repeated many of the observations of Dawydoff (1909) and Rao (1955) in studies of regeneration in *Ptychodera flava* over eight days in running seawater tables at Kewalo Marine Laboratory in Honolulu, HI. The first step, wound closure, is accomplished by the endoderm growing together with ectoderm, and we see this occurring in all amputated worms within two days (Figs. 10.3A,B,D,E and 10.4D) (Rao 1955). Based on histological data, Rao (1955) observed mesenchyme cells or coelomic cells from the coelomic fluid of trunk region migrate into the wound site. Epimorphosis is characteristic of the beginning of the regenerative process: within two days the wound has closed, then at four days, cells have moved into the space between the endoderm and ectoderm (Figs. 10.3F and 10.4E). A small blastemal structure was seen at day four in six of seven animals bisected. This blastema precedes the formation of the proboscis and appears to be the distalization step in hemichordate regeneration. Next, the proboscis begins to grow out from the proliferated tissue, which at day six, was present in six out of eight amputated animals along with a mouth opening (Fig. 10.3G–I). Rao (1955) concluded that the mouth only breaks through once the ectoderm differentiation is complete. We have observed endoderm on lateral sides of the proboscis evaginated by day six (Fig. 10.4G), although no collar rudiments were visible externally. Then, by day eight, collar buds were visible externally in five out of seven amputated worms. During this time, around day six to eight, we have also observed the stomochord forming via endoderm evagination as in normal development (Figs. 10.4I and 10.5B) (Rao 1955). These collar halves later fuse first on the ventral side (Fig. 10.3H), and secondarily on the dorsal side (Fig. 10.3I,J). The dorsal fusion of the two collar halves creates a new dorsal collar nerve tube. Around this time, mesenchyme cells with ectoderm and endoderm form a rudiment that eventually gives rise to the proboscis and heart/kidney complex (Figs. 10.3G–I and 10.4A,I).We have noticed that the proboscis and collar tissue are much lighter in color in regenerated Ptychodera than in non-regenerated adults (Fig. 10.3G–J), indicating that this is newly proliferated tissue that is not yet pigmented. Further evidence for the role of cell proliferation in this process is the presence of PCNA positive nuclei in the proboscis ectoderm, mesenchyme (Fig. 10.5A,C,E), and evaginating endoderm (Fig. 10.5A,B,D) in a day six regenerating animal. It remains to be determined whether or not the mesenchyme are an undifferentiated population of stem cells or whether they dedifferentiate into stem cells once wounding has occurred. Evidence for tissue remodeling to make way for new gill slits to form in the regenerating trunk comes from the paucity of cells in the endoderm that are posterior to evaginating collar endoderm in a day six regenerate (Fig. 10.4F). We propose that these missing cells may have undergone apoptosis and stem cells will later proliferate and remodel the endoderm to form gill endoderm.

Dawydoff (1909, 1948), who examined regeneration in *Glossobalanus minutus,* noted that while any region of an amputated trunk would regenerate anterior structures, those amputated more anteriorly regenerate more readily. Also, we have seen that the anterior two body portions of *P. flava*, the proboscis and collar, regenerate more rapidly (within two weeks), than structures of the third body portion, the

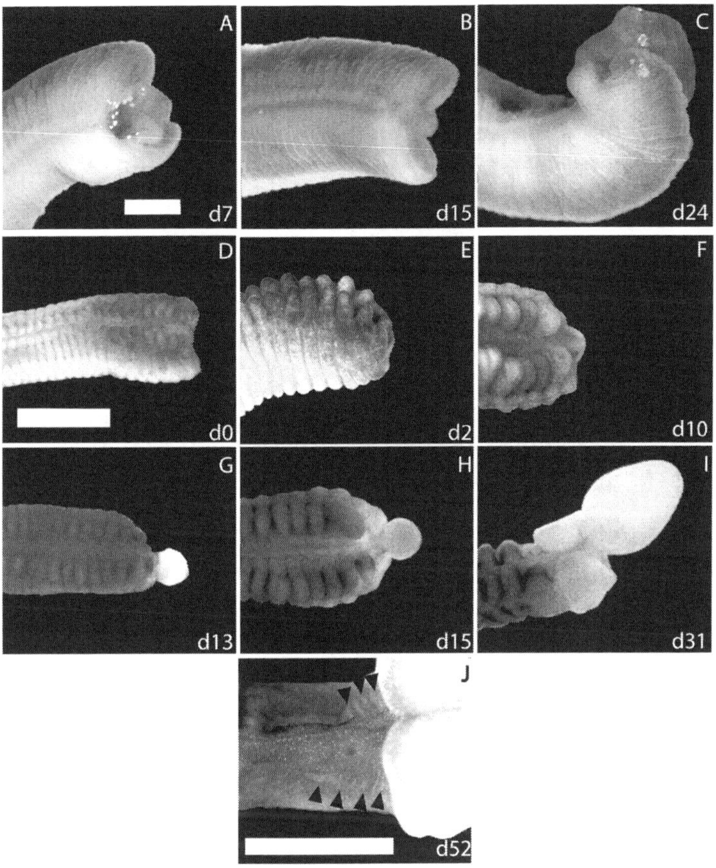

Fig. 10.3 Regeneration in the Ptychoderidae. Anterior regeneration is compared in *Glossobalanus berkeleyi*. (**A–C**) and *Ptychodera flava* (**D–J**). Regeneration in a single *Glossobalanus berkeleyi* individual over 24 days that begins, but does not progress (**A–C**). When *Glossobalanus berkeleyi* is bisected (**A**, day 7) the posterior half is able to heal the wound in the cut end at the anterior (**B**). Here, wound healing is complete at 15 days (**B**). Regeneration appears to be stalled at this step, since subsequently no regeneration blastema forms, and no regrowth occurs (**B,C**). In this case, the closed end reopened and became necrotic within 24 days (**C**). Stages of regeneration in a selection of *P. flava* individuals (**D–J**). The timing of regeneration is slightly variable in *Ptychodera flava*, but it always follows the same sequence. When *P. flava* is bisected (**D**; day 0), the open end of the posterior half heals within 2 days (**E**), and a regeneration blastema forms within 5–10 days (**F**). Development of anterior structures is accomplished rapidly thereafter with a proboscis/mouth forming first in 9–15 days (**G**), closely followed by the emergence of the two halves of the collar within 15 days (**H**). By 31 days the collar is completely joined (**I**). Once the proboscis and collar are formed, gill slits are added in an anterior to posterior fashion by remodeling existing trunk tissue (**J**). Arrowheads indicate developing gill slits in a day 52 *P. flava* regenerate. *Scale bar* equals 2 mm in **A**, **D**, and **J**

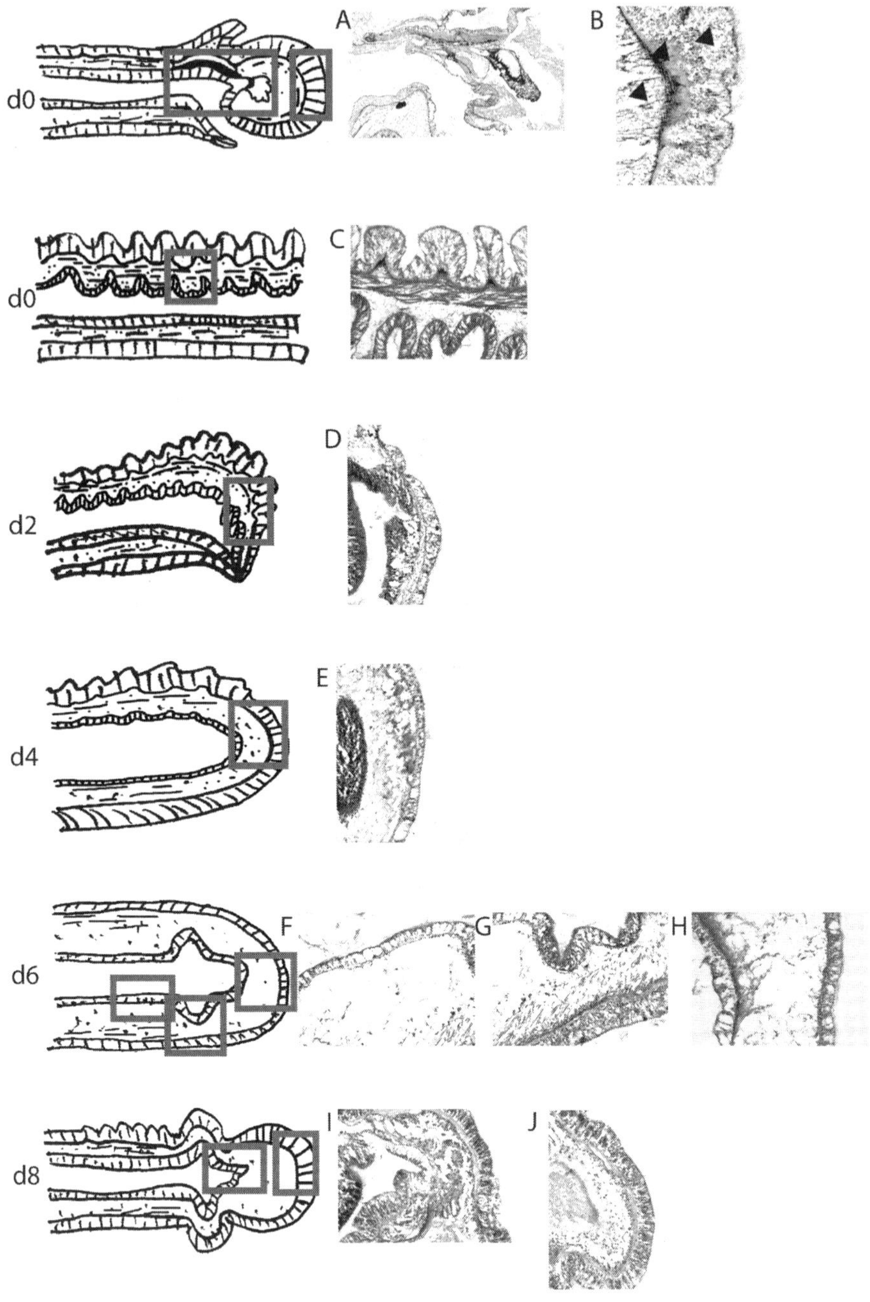

Fig. 10.4

trunk. In one instance, we observed a worm that had been regenerating for nearly two months with nine to ten gill slits in the trunk (Fig. 10.3 J). While ptychoderids in general may have greater powers of regeneration than harrimaniids, it is possible that not all ptychoderid species regenerate. We have, on several occasions, cut individuals of a Pacific Northwest ptychoderid hemichordate, *Glossobalanus berkeleyi*, and have not yet seen complete regeneration (Table 10.2; Fig. 10.3A–C). In *Ptychodera flava*, regeneration proceeded very reliably in the lab (Table 10.3), and in each case, regeneration began by forming the most anterior structures first, with more posterior structures following. The proboscis formed first, followed by the collar, then gill slits began to develop in an anterior to posterior manner, similar to how they form in normal development (Fig. 10.3 J) (Aronowicz and Lowe 2006; Rychel et al. 2006; Rychel and Swalla 2007). In conclusion, anterior regeneration in *P. flava* is a complex process likely involving both cell death of old tissues followed by cell proliferation and tissue remodeling. The source of the stem cells for the new tissues generated in the proboscis and collar is still unknown, but our lab is actively pursuing their identity. We are beginning with a cellular and molecular characterization of hemichordate regeneration and expect this to be an excellent model system once the *Ptychodera flava* genome is sequenced.

10.6 Future Experiments and Prospectus

Some hemichordate species show extensive regeneration of the anterior structures of the worm and represent a new model system to study stem cells and regeneration. In *Ptychodera flava*, the entire proboscis, mouth, and collar regenerate from the posterior half within two weeks after bisection of the worm, and gill slits are

Fig. 10.4 Trichrome stained longitudinal sections in regenerating *Ptychodera flava*. *Boxes* on the line drawing on the *left* indicate the area shown in the magnified sections on the *right*. The anterior of a normal adult worm (d0), the proboscis, contains the heart/kidney/stomochord (**A**). In the proboscis tip differentiated muscle, nervous layer, and ectoderm are seen, and are indicated by *arrowheads* (**B**). In the posterior hepatic region of a normal worm (d0), differentiated ectoderm, muscle and gut endoderm are present along with mesenchymal cells between ectoderm and endoderm (**C**). At day two of regeneration (d2), ectoderm and endoderm have grown together, and the ectoderm is very thin relative to normal proboscis ectoderm. Mesenchymal cells are sparse, but they all have long filopodia (**D**). At day four of regeneration (d4), the ectoderm has thickened and mesenchymal cells with very abundant extensions proliferate in the space between the ectoderm and endoderm (**E**). At day six (d6) in sections lateral to where the proboscis has formed, spaces between cells are visible in more posterior endoderm (**F**). More anteriorly, evaginating collar endoderm is present (**G**). Mesenchymal cells with long extensions are still present in the space between endoderm and ectoderm at the anterior end of the section (**H**). At day eight (d8), evaginating stomochord endoderm is present surrounded by mesenchyme cells that will form the heart/kidney rudiment (**I**), and the mesenchyme cells in the proboscis are smaller, with fewer extensions than mesenchymal cells seen in earlier stages (**J**). At eight days, the regenerated proboscis ectoderm is thicker than six days (**J**), but not as thick as a normal adult (**B**)

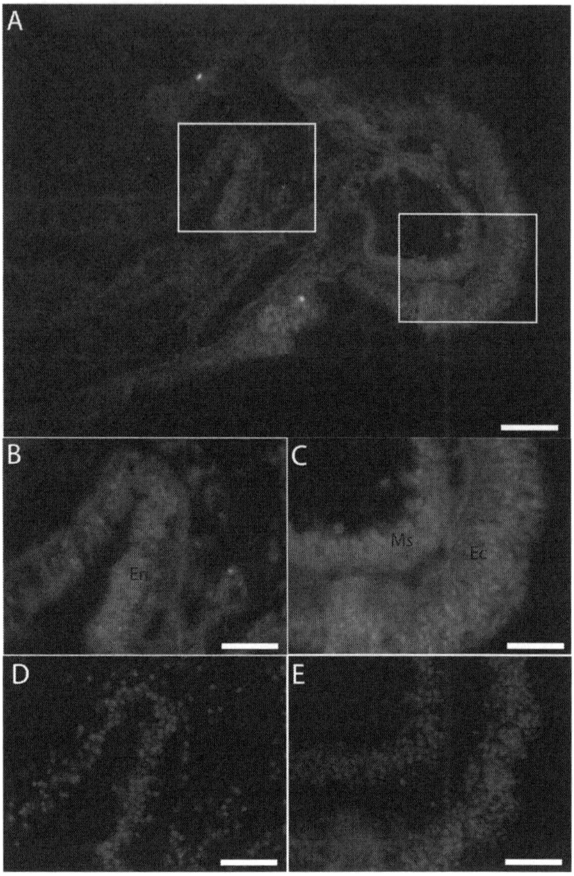

Fig. 10.5 Six day regenerating *Ptychodera flava* cell proliferation. In a longitudinal section stained with an anti-PCNA antibody, cells with nuclei in S phase (PCNA and DAPI positive) are seen concentrated at the anterior of the animal (**A**). *Boxes* in (**A**) indicate where higher power magnifications are seen in (**B–E**). Cell nuclei that are positive for PCNA (**B,C**) and DAPI (**D,E**). PCNA labels cells in S phase, suggesting that cell division is high in the evaginating endoderm (**B,D**), as well as proboscis ectoderm and mesenchyme (**C,E**). En: endoderm, Ms: mesenchyme, Ec: ectoderm. *Scale bar* in (**A**) is 100 μm, *scale bar* in (**B–E**) is 50 μm

slowly added over a longer time scale. In our observations, the anterior structures are regenerated more quickly than more posterior ones, and this would be selectively advantageous to allow eating quickly. We have on several occasions bisected the harrimaniid, *Saccoglossus bromophenolosus*, and have not observed any regeneration. This combined with evidence from *Saccoglossus kowalevskii* (Tweedell 1961) suggest that the direct developing harrimaniids show much less regeneration than the ptychoderid worms that have planktonic tornaria larvae. Since these families are also separated phylogenetically (Cameron et al. 2000), it is difficult to know whether regeneration differences are due to phylogenetic history, or if reduced regeneration potential is a consequence of direct development. We know that echinoderms vary in regenerative ability, since among the classes of echinoderms, echinoids have the most limited regenerative ability as adults. It would be interesting to see if echinoderms varied in their regenerative abilities in a phylogenetic way within classes and/or if the differences seen are also linked to direct development. It is unknown

Table 10.2 *Glossobalanus berkeleyi* amputation trials

Trial	Number amputated	Percent of survival 7 days	Percent of survival 14 days	Percent of survival 21 days	Longest number of days followed
1	4	100	100	100	52
2	3	100	100	67	27
3	10	50	50	20	25

Table 10.3 *Ptychodera flava* regeneration concurrent trials

Trial	Number amputated	Trial length in days	Percent of survival at end of trial	Percent of survivors regenerating	Typical regeneration stage at end of trial	Percent of animals at typical regeneration stage
1	10	2	80	100	wound closure	100
2	10	4	40	100	blastema	86
3	10	6	80	100	small proboscis + mouth	75
4	10	8	70	100	larger proboscis + 2 collar buds	71

whether hemichordate tornaria larvae are capable of cloning, a common ability in brittle star, sea star, sea urchin, and sea cucumber larvae. If they can, then it suggests this is an ancestral feature of dipleurula ambulacrarian larvae; while if not, then this would be a phenomenon that would be restricted to echinoderm larvae.

In light of the current excitement about stem cells and regeneration for medical purposes, the study of the cellular behaviors and molecular mechanisms of hemichordate stem cells and regeneration is relevant since they have a body plan with many features in common with vertebrates (Aronowicz and Lowe 2006; Rychel and Swalla 2007; Swalla 2007). The hemichordate body plan develops in an anterior to posterior fashion, similar to vertebrates, and expression of the *Hox* genes begins right at the first gill slit (Lowe et al. 2003; Aronowicz and Lowe 2006; Swalla 2006) just as *Hox* gene expression begins dorsal to the first gill slit or arch in the midbrain of vertebrates. These results show that an A-P axis based on *Hox* expression is conserved between hemichordates and vertebrates. The coelomic cells that are necessary for regeneration in echinoderms and hemichordates are likely made up of mesenchymal or stem cells, that are capable of differentiating into a variety of cell types. The in-depth study of these cells will be fascinating. We hope to understand how they are mobilized, multiply and then subsequently differentiate after injury in hemichordates.

Acknowledgments We would like to extend a special "thank you" to Dr. Michael Hadfield of Kewalo Marine Laboratory for generously sharing his knowledge and space over the course of these studies on *Ptychodera flava*. We would also like to thank members of the Swalla lab, Dr. Elena Levine Keeling (California Polytechnic State University) and Federico Brown, along with Dr. Maria Byrne (University of Sydney), Dr. Joel Smith (California Institute of Technology), Dr. Alexa Bely (University of Maryland) and Dr. Barbara Wakimoto (University of Washington) for critically reading the manuscript during its preparation. The research was supported by a Royalty Research Grant from the University of Washington to BJS. ALR was supported by NRSA Grant Number T32 HD007183-26A1 from NIH-NICHD.

References

Agata K, Saito Y, Nakajima E (2007) Unifying principles of regeneration I: Epimorphosis versus morphallaxis. Dev Growth Diff 49:73–78

Andersson KA (1907) Die Pterobranchier der schwedischen Südpolarexpedition 1901–1903. Wiss Ergebn Schwed Südpolar-Exp 5:1–122

Aronowicz J, Lowe CJ (2006) Hox gene expression in the hemichordate Saccoglossus kowalevskii and the evolution of deuterostome nervous systems. Integr Comp Biol 46: 890–901

Ballarin L, Manni L (2009) Stem cells in sexual and asexual reproduction of *Botryllus schlosseri* (Ascidiacea, Tunicata): An overview. In: Matranga V, Rinkevich B (eds) Marine Stem Cells. Springer, New York, in press

Balser EJ (1998) Cloning by ophiuroid echinoderm larvae. Biol Bull 194:187–193

Bannister R, McGonnell IM, Graham A, Thorndyke MC, Beesley PW (2005) Afuni, a novel transforming growth factor-β gene is involved in arm regeneration by the brittle star *Amphiura filiformis*. Dev Genes Evol 215:393–401

Bateson W (1884) The early stages in the development of *Balanoglossus*. Q J Microsc Sci 24: 208–236

Bateson W (1885) The later stages in the development of *Balanoglossus kowalevskii*. Q J Microsc Sci 25:81–122

Bayascas J, Castillo E, Muñoz-Mármol AM, Baguñà J, Saló E (1998) Synchronous and early activation of planarian *Hox* genes and the re-specification of body axes during regeneration in *Dugesia* (*G.*) *tigrina* (Turbellaria; Tricladida). Hydrobiologia 383:125–130

Bayascas JR, Castillo E, Munoz-Marmol AM, Salo E (1997) Planarian *Hox* genes: Novel patterns of expression during regeneration. Development 124:141–148

Beck CW, Christen B, Barker D, Slack JMW (2006) Temporal requirement for bone morphogenetic proteins in regeneration of the tail and limb of *Xenopus* tadpoles. Mech Dev 123: 674–688

Bely AE, Wray GA (2001) Evolution of regeneration and fission in annelids: Insights from engrailed- and orthodenticle-class gene expression. Development 128:2781–2791

Bollner T, Beesley PW, Thorndyke MC (1993) Substance P- and cholecytokinin-like immunoreactivity during post-metamorphic development of the central nervous system in the ascidian *Ciona intestinalis*. Cell Tissue Res 272:545–552

Bollner T, Beesley PW, Thorndyke MC (1997) Investigation of the contribution from peripheral GnRH-like immunoreactive 'neuroblasts' to the regenerating central nervous system in the protochordate *Ciona intestinalis*. Proc Biol Sci 264:1117–1123

Bollner T, Howalt S, Thorndyke MC, Beesley PW (1995) Regeneration and post-metamorphic development of the central nervous system in the protochordate *Ciona intestinalis*: A study with monoclonal antibodies. Cell Tissue Res 279:421–432

Bonasoro F, Candia Carnevali MD, Moss C, Thorndyke M (1998) Epimorphic versus morphalactic mechanisms in arm regeneration of crinoids and asteroids: Pattern of cell proliferation and lineage. In: Mooi R, Telford M (eds) Echinoderms. Balkema, Rotterdam

Bosch I (1988) Reproduction by budding in natural populations of bipinaria larvae from the sea star genus Luidia. In: Burke RD, Mladinov PV, Lambert P, Parsley RL (eds) Echinoderm Biology. Balkema, Rotterdam, p 728
Bosch I, Rivkin RB, Alexander SP (1989) Asexual reproduction by oceanic planktotrophic echinoderm larvae. Nature 337:169–170
Brockes JP (1997) Amphibian limb regeneration: Rebuilding a complex structure. Science 276: 81–87
Brown R, Brockes JP (1991) Identification and expression of a regeneration-specific homeobox gene in the newt limb blastema. Development 111:489–496
Cameron CB, Garey JR, Swalla BJ (2000) Evolution of the chordate body plan: New insights from phylogenetic analyses of deuterostome phyla. Proc Natl Acad Sci USA 97:4469–4474
Candia Carnevali MD (2006) Regeneration in echinoderms: Repair, regrowth, cloning. Invert Surv J 3:64–76
Candia Carnevali MD, Bonasoro F (2001) Microscopic overview of crinoid regeneration. Microsc Res Techn 55:403–426
Candia Carnevali MD, Bonasoro F, Biale A (1997) Pattern of bromodeoxyuridine incorporation in the advanced stages of arm regeneration in the feather star *Antedon mediterranea*. Cell Tissue Res 289:363–374
Candia Carnevali MD, Bonasoro F, Patruno M, Thorndyke MC (1998) Cellular and molecular mechanisms of arm regeneration in crinoid echinoderms: The potential of arm explants. Dev Genes Evol 208:421–430
Candia Carnevali MD, Thorndyke MR, Matranga V (2009) Regenerating in echinoderms: A promise to understand stem cells potential. In: Matranga V, Rinkevich R (eds) Marine Stem Cells. Springer, New York, in press
Carlson MR, Komine Y, Bryant SV, Gardiner DM (2001) Expression of Hoxb13 and Hoxc10 in developing and regenerating Axolotl limbs and tails. Dev Biol 229:396–406
Christen B, Beck CW, Lombardo A, Slack JM (2003) Regeneration-specific expression pattern of three posterior Hox genes. Dev Dyn 226:349–355
Cuenot L (1948) Anatomie, ethologie et systematique des echinodermes. In: Grassé PP (ed) Traité de zoologie. Masson et Cie, Paris
D'Ancona Lunetta G (2009) Stem cells in *Holothuria polii* and *Sipunculus nudus*. In: Matranga V, Rinkevich B (eds) Marine Stem Cells. Springer, New York, in press
Dawydoff C (1909) Beobachtungen über den regenerationsprozess bei den enteropneusten. Z Wiss Zool 93:237–305
Dawydoff C (1948) Embranchement des stomocordés. In: Grassé PP (ed) Traité de Zoologie. Echinodermes, Stomocordés, Protochordés, vol 11. Masson et Cie, Paris, 367–449
Dent JN (1962) Limb regeneration in larvae and metamorphosing individuals of the South African clawed toad. J Morphol 110:61–77
Dilly PN (1975) The dormant buds of *Rhabdopleura compacta* (Hemichordata). Cell Tissue Res 159:387–397
Dolmatov IY (1991) Regeneration of tentacles in early pentactulae of the holothurian *Eupentacta fraudatrix*. Biologiya Morya Vladivostok 5:99–101
Dolmatov IY, Ginanova TT (2001) Muscle regeneration in holothurians. Microsc Res Tech 55: 452–463
Drager BJ, Harkey MA, Iwata M, Whiteley AH (1989) The expression of embryonic primary mesenchyme genes of the sea urchin, *Strongylocentrotus purpuratus*, in the adult skeletogenic tissues of this and other species of echinoderms. Dev Biol 133:14–23
Duboc V, Röttinger E, Besnardeau L, Lepage T (2004) *Nodal* and *BMP2/4* signaling organizes the oral-aboral axis of the sea urchin embryo. Dev Cell 6:397–410
Dubois P, Ameye L (2001) Regeneration of spines and pedicellariae in echinoderms: A review. Microsc Res Tech 55:427–437
Eaves AA, Palmer AR (2003) Reproduction: Widespread cloning in echinoderm larvae. Nature 425:146–146

Echeverri K, Tanaka EM (2005) Proximodistal patterning during limb regeneration. Dev Biol 279:391–401

Emson RH, Wilkie IC (1980) Fission and autotomy in echinoderms. Oceanogr Mar Biol Ann Rev 18:155–250

Garcia-Arraras JE, Greenberg MJ (2001) Visceral regeneration in holothurians. Microsc Res Tech 55:438–451

Gilchrist JDF (1923) A form of dimorphism and asexual reproduction in *Ptychodera capensis* (Hemichordata). J Linn Soc Lond Zool 35:393–398

Goldin A (1948) Regeneration in *Perophora viridis*. Biol Bull 94:184–193

Hadfield MG (1975) Hemichordata. In: Giese AC, Pearce JS (eds) Reproduction of Marine Invertebrates (vol 2 Entoprocts and Lesser Coelomates). Academic Press, London, pp 185–240

Harada Y, Shoguchi E, Taguchi S, Okai N, Humphreys T, Tagawa K, Satoh N (2002) Conserved expression pattern of *BMP-2/4* in hemichordate acorn worm and echinoderm sea cucumber embryos. Zool Sci 19:1113–1121

Heatfield BM, Travis DF (1975a) Ultrastructural studies of regenerating spines of the sea urchin *Strongylocentrotus purpuratus*. I. Cell types without spherules. J Morphol 145:13–49

Heatfield BM, Travis DF (1975b) Ultrastructural studies of regenerating spines of the sea urchin *Strongylocentrotus purpuratus*. II. Cell types with spherules. J Morphol 145:51–71

Hogan BL (1996) Bone morphogenetic proteins: Multifunctional regulators of vertebrate development. Genes Dev 10:1580–1594

Hörstadius S (1925) Entwicklungsmechanische studien an Holothuria poli Delle Chiaje. Ark Zool B17 8:1–6

Hörstadius S (1928) Über die determination des keimes bei echinodermen. Acta Zool Stockh 9:1–191

Hörstadius S (1973) Experimental embryology of echinoderms. Clarendon Press, Oxford

Hyman LH (1959) The Enterocoelous Coelomates: Phylum Hemichordata. In: Hyman LH (ed) The invertebrates. Hemichordata. Smaller coelomate groups Chaetognatha, Hemichordata, Pogonophora, Phoronida, Ectoprocta, Brachiopoda, Sipuncuhda. The coelomate bilateria, vol 5. McGraw-Hill, New York, pp 72–207

Iten LE, Bryant SV (1976) Stages of tail regeneration in the adult newt, *Notophtalmus viridescens*. J Exp Zool 196:283–292

Jaeckle WB (1994) Multiple modes of asexual reproduction by tropical and subtropical sea star larvae: An unusual adaptation for genet dispersal and survival. Biol Bull 186:62–71

John CC (1931) Cephalodiscus. Discovery Reports 3:223–260

Kawamura K, Sugino Y, Sunanaga T, Fujiwara S (2008) Multipotent epithelial cells in the process of regeneration and asexual reproduction in colonial tunicates. Dev Growth Diff 50:1–11

Knott KE, Balser EJ, Jaeckle WB, Wray GA (2003) Identification of asteroid genera with species capable of larval cloning. Biol Bull 204:246–255

Kowalevsky A (1866) Anatomie des Balanoglossus Delle Chiaje. Mém l'Acad Imp Sci St Petersburg 7:1–18

Lester SM (1985) *Cephalodiscus* sp. (Hemichordata: Pterobranchia): Observations of functional morphology, behavior and occurrence in shallow water around Bermuda. Mar Biol 85: 263–268

Lester SM (1988) Ultrastructure of adult gonads and development and structure of the larva of *Rhabdopleura normani* (Hemichordata: Pterobranchia). Acta Zool Stockh 69:95–109

Lowe CJ, Terasaki M, Wu M, Freeman RM, Runft L, Kwan K, Haigo S, Aronowicz J, Lander E, Gruber C, Smith M, Kirschner M, Gerhart J (2006) Dorsoventral patterning in hemichordates: Insights into early chordate evolution. PLoS Biol 4:e291

Lowe CJ, Wu M, Salic A, Evans L, Lander E, Stange-Thomann N, Gruber CE, Gerhart J, Kirschner M (2003) Anteroposterior patterning in hemichordates and the origin of the chordate nervous system. Cell 113:853–865

Marion A (1886) Études zoologique sur deux espèces d'entéropneustes. Arch Zool Exp Gén Ser 2 4:305–326

Masterman A (1900) On the further anatomy and the budding processes of Cephalodiscus. Trans Roy Soc Edinb 39:507–527
Mazzone F, Byrne M (2001) The haemal sinus – a possible conduit for migratory cells involved in repair and regeneration of the radial nerve cord in *Coscinasterias muricata* following autonomy. In: Barker MF (ed) Echinoderms 2000. Balkema, Lisse, pp 167–191
Mazzone F, Byrne M, Thorndyke M (2003) Regeneration in the sea star Coscinasterias muricata following autonomy. In: Feral JP, David B (eds) Echinoderm research 2001. Balkema, Lisse, pp 209–213
McGovern TM (2003) Plastic reproductive strategies in a clonal marine invertebrate. Proc Royal Soc B: Biol Sci 270:2517–2522
Metschnikoff E (1869) Ueber die systematische stellung von *Balanoglossus*. Zool Anz 4:139–143, 153–157
Mladenov PV, Burke RD (1984) Echinodermata: Asexual propagation. In: Adiyodi KG, Adiyodi RG (eds) Reproductive biology of invertebrates, vol VI part B. Asexual propagation and reproductive strategies. Oxford and Hill, New Delhi, pp 339–383
Mochii M, Taniguchi Y, Shikata I (2007) Tail regeneration in the Xenopus tadpole. Dev Growth Diff 49:155–161
Morgan TH (1898) Experimental studies of the regeneration of *Planaria maculata*. Arch Entw Mech Org 7:364–397
Morgan TH (1901) Regeneration. The Macmillan Company, New York
Newfeld SJ, Wisotzkey RG, Kumar S (1999) Molecular evolution of a developmental pathway: Phylogenetic analyses of transforming growth factor-β family ligands, receptors and smad signal transducers. Genetics 152:783–795
Nicolas S, Papillon D, Perez Y, Caubit X, Le Parco Y (2003) The spatial restrictions of 5' HoxC genes expression are maintained in adult newt spinal cord. Biol Cell 95:589–594
O'Kane S, Ferguson MWJ (1997) Transforming growth factor βs and wound healing. Int J Biochem Cell Biol 29:63–78
Oji T (2001) Fossil record of echinoderm regeneration with special regard to crinoids. Microsc Res Techn 55:397–402
Packard A (1968) Asexual reproduction in *Balanoglossus* (Stomochordata). Proc R Soc Lond B Biol Sci 171:261–272
Patruno M, McGonnell I, Graham A, Beesley P, Candia Carnevali MD, Thorndyke M (2003) Anbmp2/4 is a new member of the transforming growth factor-β superfamily isolated from a crinoid and involved in regeneration. Proc R Soc Lond B 27:1341–1347
Petersen JA, Ditadi ASF (1971) Asexual reproduction in *Glossobalanus crozieri* (Ptychoderidae, Enteropneusta, Hemichordata). Mar Biol 9:78–85
Range R, Lapraz F, Quirin M, Marro S, Besnardeau L, Lepage T (2007) Cis-regulatory analysis of *nodal* and maternal control of dorsal-ventral axis formation by Univin, a TGF-β related to Vg1. Development 134:3649–3664
Rao KP (1955) Morphogenesis during regeneration in an enteropneust. J Animal Morphol Physiol 1:1–7
Rinkevich Y, Paz G, Rinkevich B, Reshef R (2007) Systemic bud induction and retinoic acid signaling underlie whole body regeneration in the urochordate *Botrylloides leachi*. PLoS Biology 5:e71
Rubilar T, Pastor de Ward CT, Díaz de Vivar ME (2005) Sexual and asexual reproduction of *Allostichaster capensis* (Echinodermata: Asteroidea) in Golfo Nuevo. Mar Biol 146: 1083–1090
Runnström J (1925) Regulatorische bildung von cölomanlagen bei seeigelkeimen mit gehemmter urdarmbildung. W Roux Arch Entw Mech Org 105:114–119
Rychel AL, Smith SE, Shimamoto HT, Swalla BJ (2006) Evolution and development of the chordates: Collagen and pharyngeal cartilage. Mol Biol Evol 23:541–549
Rychel AL, Swalla BJ (2007) Development and evolution of chordate cartilage. J Exp Zool B Mol Dev Evol 308B:325–335

Saló E (2006) The power of regeneration and the stem-cell kingdom: Freshwater planarians (Platyhelminthes). BioEssays 28:546–559

Saló E, Tauler J, Jimenez E, Ramón Bayascas J, Gonzalez-Linares J, Garcia-Fernàndez J, Baguñà J (2001) Hox and ParaHox genes in flatworms: Characterization and expression. Am Zool 41:652–663

Sánchez Alvarado A (2000) Regeneration in the metazoans: Why does it happen? Bioessays 22:578–590

Schepotieff A (1907a) Die Anatomie von Rhabdopleura. Zool Jahrb Abt Anat 23:463–534

Schepotieff A (1907b) Knospungsprozess und gehäuse von Rhabdopleura. Zool Jahrb Abt Anat 24:193–238

Schepotieff A (1908) Knospungsprozess von Cephalodiscus. Zool Jahrb Abt Anat 25:405–494

Schepotieff A (1909) Die Pterobranchier des Indischen Ozeans. Zool Jahrb Abt Anat 24:429–448

Schiaparelli S, Cattaneo-Vietti R, Mierzejewski P (2004) A "protective shell" around the larval cocoon of *Cephalodiscus densus* Andersson, 1907 (Graptolithoidea, Hemichordata). Polar Biol 27:813–817

Schultze LS (1900) Die regeneration des ganglions von Ciona intestinalis L. und über das verhältnis der regeneration und knospung zur keimblätterlehre. Jena Z Natwiss 33:263–344

Shimizu-Nishikawa K, Takahashi J, Nishikawa A (2003) Intercalary and supernumerary regeneration in the limbs of the frog, *Xenopus laevis*. Dev Dyn 227:563–572

Simon HG, Tabin CJ (1993) Analysis of *Hox-4.5* and *Hox-3.6* expression during newt limb regeneration: Differential regulation of paralogous *Hox* genes suggest different roles for members of different *Hox* clusters. Development 117:1397–1407

Sköld H, Obst M, Sköld M, Åkesson B (2009) Asexual reproduction in marine metazoans. In: Matranga V, Rinkevich B (eds) Marine Stem Cells. Springer, New York, in press

Spengel JW (1893) Die enteropneusten des golfes von Neapel. Fauna und Flora des Golfes von Neapel Monogr 18:1–755

Stebbing ARD (1970) Aspects of the reproduction and life cycle of *Rhabdopleura compacta* (Hemichordata). Mar Biol 5:205–212

Stocum DL (1996) A conceptual framework for analyzing axial patterning in regenerating urodele limbs. Int J Dev Biol 40:773–783

Stoick-Cooper CL, Weidinger G, Riehle KJ, Hubbert C, Major MB, Fausto N, Moon RT (2007) Distinct Wnt signaling pathways have opposing roles in appendage regeneration. Development 134:479–489

Swalla BJ (2006) Building divergent body plans with similar genetic pathways. Heredity 97: 235–243

Swalla BJ (2007) New insights into vertebrate origins. In: Moody S (ed) Principles of Developmental Genetics. Elsevier Science/Academic Press, San Diego, pp 114–128

Swalla BJ, Smith AB (2008) Deciphering deuterostome phylogeny: Molecular, morphological and palaeontological perspectives. Proc R Soc Lond, 363:1557–1568

Thorndyke M, Chen WC, Moss C, Candia Carnevali MD (1999) Regeneration in echinoderms: Cellular and molecular aspects. In: Candia Carnevali MD, Bonasoro F (eds) Echinoderm Research. Balkema, Rotterdam, pp 159–164

Thorndyke MC, Chen WC, Beesley PW, Patruno M (2001a) Molecular approach to echinoderm regeneration. Microsc Res Tech 55:474–485

Thorndyke MC, Patruno M, Chen WC, Beesley PW (2001b) Stem cells and regeneration in invertebrate deuterostomes. Symp Soc Exp Biol 53:107–120

Thorndyke MC, Patruno M, Moss C, Beesley PW, Mallefet J (2001c) Cellular and molecular bases of arm regeneration in brittlestars. In: Barker M (ed) Echinoderms 2000. Swets & Zeitlinger, Lisse, pp 323–326

Tiozzo S, Brown FD, De Tomaso AW (2008) Regeneration and stem cells in ascidians. In: Bosch TCG (ed) Stem Cells from Hydra to Man. Springer, New York, pp 95–112

Torok MA, Gardiner DM, Shubin NH, Bryant SV (1998) Expression of HoxD genes in developing and regenerating Axolotl limbs. Dev Biol 200:225–233

Tweedell KS (1961) Regeneration of the enteropneust *Saccoglossus kowalevskii*. Biol Bull 120:118–127

Urata M, Yamaguchi M (2004) The development of the enteropneust hemichordate *Balanoglossus misakiensis* Kuwano. Zool Sci 21:533–540

van Bekkum DW (2004) Phylogenetic aspects of tissue regeneration: Role of stem cells: A concise overview. Blood Cells Mol Dis 32:11–16

Vickery MC, Vickery MS, McClintock JB, Amsler CD (2001a) Utilization of a novel deuterostome model for the study of regeneration genetics: Molecular cloning of genes that are differentially expressed during early stages of larval sea star regeneration. Gene 262:73–80

Vickery MCL, Vickery MS, Amsler CD, McClintock JB (2001b) Regeneration in echinoderm larvae. Microsc Res Techn 55:464–473

Vickery MS, McClintock JB (2000) Effects of food concentration and availability on the incidence of cloning in planktotrophic larvae of the sea star *Pisaster ochraceus*. Biol Bull 199:298–304

Vickery MS, Vickery MC, McClintock JB (2002) Morphogenesis and organogenesis in the regenerating planktotrophic larvae of asteroids and echinoids. Biol Bull 203:121–133

Vickery MS, Vickery MCL, McClintock JB (1999) Regeneration in echinoid larvae. Am Zool 39:51A

Willey A (1899) Enteropneusta from the South Pacific with notes on the West Indian species. Willey's Zoological Results 3:223–334

Yokoyama H, Ogino H, Stoick-Cooper CL, Grainger RM, Moon RT (2007) Wnt/beta-catenin signaling has an essential role in the initiation of limb regeneration. Dev Biol 306:170–178

Zeng L, Jacobs MW, Swalla BJ (2006) Coloniality has evolved once in stolidobranch ascidians. Integr Comp Biol 46:255–268

Zito F, Costa C, Sciarrino S, Poma V, Russo R, Angerer LM, Matranga V (2003) Expression of univin, a TGF-β growth factor, requires ectoderm-ECM interaction and promotes skeletal growth in the sea urchin embryo. Dev Biol 264:217–227

Chapter 11
Stem Cells in Sexual and Asexual Reproduction of *Botryllus schlosseri* (Ascidiacea, Tunicata): An Overview

Loriano Ballarin and Lucia Manni

Abstract *Botryllus schlosseri*, a colonial ascidian reproducing both sexually and asexually, is an excellent model for the study of adult stem cells, since budding blastozooids are derived from a disc of stem cells at the parental atrial chamber and the overlying epidermis. At the colony level, adult zooids undergo cyclical generation changes during which the adults die, are resorbed and then replaced by their buds which reach functional maturity and start filtering. At the same time, blood cell renewal takes place thanks to a new haemopoietic wave. Several experimental manipulations were used to evaluate the behaviour of stem cells during bud morphogenesis under different developmental constraints. When all zooids and buds of a colony are extirpated, new developed zooids are derived from totipotent blood cells which aggregate on the walls of the blood vessels of the tunic. Stem cells can also be transferred from one colony to another through vasculature during fusion of allogeneic compatible colonies. Germ stem cells persist in a host colony so that heterochtonous offspring can be collected after many blastogenetic generations. The persistence of somatic stem cells alters the fusibility pattern of the colony. Pluripotent stem cells are also involved in the embryonic development of larval and juvenile organs; these embryonic cells were analysed and compared with bud stem cells for their potentials and gene expression.

Keywords Palleal budding · Vascular budding · Embryogenesis · Somatic stem cells · Germ stem cells

11.1 Introduction

Tunicates are marine, filter-feeding chordate invertebrates, both sessile and pelagic, featuring the presence of a peculiar embedding tissue: the test or tunic. The majority of tunicates, and the most frequently studied, are ascidians, sessile animals living in

---

L. Ballarin (✉)
Department of Biology, University of Padova, via U. Bassi 58/B, 35131 Padova, Italy
e-mail: loriano.ballarin@unipd.it

B. Rinkevich, V. Matranga (eds.), *Stem Cells in Marine Organisms*,
DOI 10.1007/978-90-481-2767-2_11,

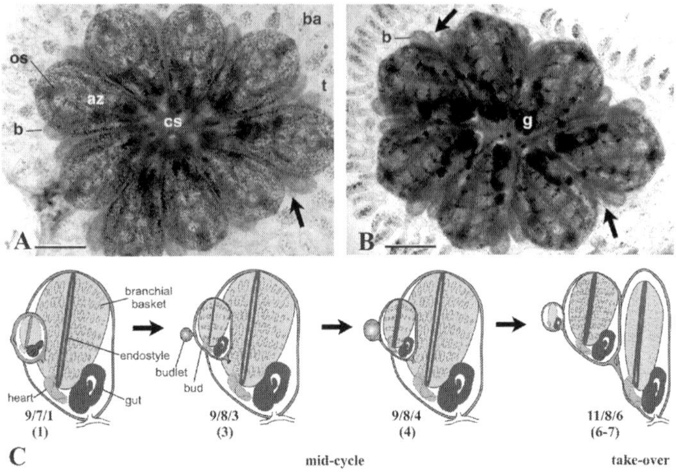

Fig. 11.1 A colony of *Botryllus schlosseri* in dorsal (**A**) and ventral (**B**) views. *Arrow*, budlets; az, adult zooid; b, bud; ba, blood ampulla; cs, cloacal siphon; g, gut; os, oral siphon; t, tunic. *Scale bars*: 650 μm. (**C**) Sketch of colonial blastogenic cycle. Days from the beginning of the cycle are in *brackets*

all seas as solitary or colonial species, with tadpole-like larvae sharing the main chordate features, i.e., a branchial basket, a muscular tail, a rod-like notochord in the tail, and a dorsal nerve cord in the form of a hollow tube expanded anteriorly in a brain.

In the last decade, solitary ascidian species (*Ciona intestinalis*, *Ciona savignyi*, *Halocynthia roretzi*) have been proposed as model organisms for the study of the genetic control of embryogenesis and differentiation, and the combined efforts of some laboratories have led to partial or full sequencing of their genome. This has aroused the interest of many researchers in these animals. Colonial ascidians are less well-known at the molecular level, but offer the possibility of comparisons, in the same organism, of different developmental pathways (embryogenesis and blastogenesis) leading to the same end-product, i.e., the adult filter-feeding zooid.

Botryllus schlosseri (Fig. 11.1A), a cosmopolitan colonial ascidian common in shallow temperate waters, was introduced in the laboratory more than 50 years ago (Sabbadin 1955) as a model organism for the study of diverse biological phenomena, such as sexual and asexual reproduction, regeneration, immunobiology and allorecognition (Manni et al. 2007). This species is easily found in the field and can grow and reproduce successfully in laboratory aquaria. In addition, colonies are embedded in a soft, transparent tunic which allows direct and detailed observation of their constituents under the microscope.

New colonies arise from tadpole-like larvae, 1.2 mm in length, which, after a brief swimming period, settle on a suitable substrate and metamorphose into fully functional oozooids, approximately 0.5 mm in length. Each oozooid is the founder of a new colony and bears a single palleal bud on its right side. After a week, at 19°C, the oozooid is resorbed in the course of a process called take-over (see below) and replaced by its bud. The latter develops into a mature, filter-feeding zooid (blastozooid) which, in turn, produces two or more palleal buds, on each side of the body, capable of originating budlets and replacing the adult generation when it dies. Since adult zooids are cyclically resorbed and replaced by growing buds, a healthy colony can grow in size having up to thousands of zooids and buds, kept synchronized in their development by the common vascular system. Therefore, a colony is a clone, which progressively grows as new zooid generations are produced.

11.2 The *B. schlosseri* Colony and the Colonial Blastogenetic Cycle

In a typical colony, zooids originating by blastogenesis, 1.5 mm in length, are grouped in star-shaped systems embedded in the common gelatinous tunic. Oral siphons open in the anterior part of each zooid, and atrial siphons converge into a central cloacal chamber, connected with the exterior by a common cloacal siphon (Fig. 11.1A). Three blastogenetic generations are usually present in a colony: (i) adult zooids, (ii) palleal buds (primary buds) on zooids, capable of replacing them, and (iii) budlets (secondary buds) on buds, which will grow to buds and, finally, to zooids (Fig. 11.1B). Zooids, buds, and budlets are connected by a circulatory system composed of a network of vessels of epidermal origin, joined to a marginal vessel which runs along the contour of the colony. Sausage-like blind ends, known as ampullae, depart from these vessels towards the tunic surface and store blood cells.

According to Berrill (1941) and Sabbadin (1955), eleven developmental stages are recognised in blastogenesis: stages 1–6 refer to secondary buds and stages 7–8 to primary buds; stage 9 identifies adult zooids which turn into stage 10 when sexually mature; and stage 11 indicates regressing zooids at take-over (Table 11.1). At 19°C, the life span of a zooid, from its appearance as a bud primordium to its death at take-over, takes place during three weekly blastogenetic cycles, defined as the period of time between the opening of siphons in the zooids of a new blastogenetic generation and take-over, when the adults regress, a new blastogenetic generation reaches functional maturity, the secondary buds grow to primary buds, and new budlet primordia appear.

As zooid, bud and budlet development is closely coordinated, the stage of a colony during its blastogenetic cycle is univocally defined and expressed by a formula of three numbers separated by slashes (e.g., 9/8/4), each referring to the developmental stages of zooids, buds and budlets, respectively, as defined by Sabbadin (1955, 1958) (Fig. 11.1C).

The cycle starts with stage 9/6/0, a brief interval without observable budlet generation immediately following take-over, proceeds to stages 9/7/1 and 9/8/2-5, and ends with take-over, at stage 11/8/6, lasting about 24–36 h at 19°C, during which adult zooids are progressively resorbed and replaced by growing buds. Stages 9/8/2-5 correspond to the mid-cycle, as defined by Lauzon et al. (1992): during this period, the tissues of filtering zooids are not involved in important morphogenetic

Table 11.1 Stages of zooid development of *Botryllus schlosseri* (after Sabbadin 1955)

Secondary bud	
1	Thickening disc on parent atrial wall
1+	Initial arching of bud primordium.
2	Disc expanded and arched in hemisphere.
2+	Skewing of hemisphere toward anterior end of parent.
3	Double vesicle stage: inner vesicle originating from closure of disc, outer vesicle from parent epidermis
3+	Bud elongation according to anteroposterior axis, which diverges anteriorly from that of parent.
4	Appearance of atrial folds (peribranchial chamber rudiments) and gut primordium.
5	Branchial and peribranchial chambers and gut recognisable.
6	Heart recognisable.
Primary bud	
7	Appearance of atrial wall thickening, representing bud rudiments of following generation.
8	Heart beating (8^1, heart beating slowly; 8^2, heart beating at normal rhythm).
Adult	
9	Functional maturity. (9^1: oral siphon aperture; 9^2: atrial siphon aperture; 9^3: common cloacal siphon aperture).
10	Gonad maturation
11	Resorption. 11^1: siphons retraction and closure; 11^2: general shrinkage of zooids; 11^3: further contraction of zooids and branchial dissolution; 11^4: heart beat stops.

or renewal processes, as indicated by the low frequency of dividing or dying cells (Tiozzo et al. 2006).

11.3 Stem Cells in Asexual Development

11.3.1 Palleal Budding

B. schlosseri is an excellent model organism for the study of the involvement of stem cells in blastogenesis. New bud primordia form as thickenings in morphogenetic fields, well localised in space and time, in the parental mantle, i.e., the peribranchial wall, the overlying epidermis and the mesenchymal tissues between them. Analogously to *Polyandrocarpa misakiensis* (Kawamura et al. 2008), buds inherit the parental epidermis directly, whereas their internal organs (e.g., branchial basket, gut, nervous system, gonads, heart) form from the multipotent somatic stem cells located in the peribranchial epithelium and mantle mesenchymal cells. The budding of *B. schlosseri* requires cell proliferation from the bud primordium. From this point of view it resembles epimorphosis, but a certain degree of morphallaxis cannot be excluded, as reported for *P. misakiensis* (Kawamura and Nakauchi

Table 11.2 Comparison of contralateral blastogenetic potentials in *B. schlosseri* buds from various laboratory colonies (from Sabbadin 1994)

Bud location	Number of budlets Produced	Resorbed
right, anterior	339	57 (16.8%)
right, posterior	81	74 (91.3%)
left, anterior	339	227 (67.0%)
left, posterior	14	14 (100%)

1986, 1991; Kawamura et al. 1995), and considering that, in *Botryllus*, the budding fragments can give rise to adult zooids (Sabbadin 1958; Majone 1977).

Once formed, a bud primordium arches progressively to form the characteristic stage of double vesicle (stage 3 in Table 11.1). The inner vesicle folds to form the branchial and atrial chambers, as well as gut and nervous system; the heart and gonads derive from the aggregation of stem cells in the mesenchyme.

Buds show bilateral but asymmetric blastogenetic and gonadogenetic potentials: the former is higher on the right side than on the left (Table 11.2), whereas the latter, in terms of number and size of testes and ovaries, is higher on the left side (Sabbadin 1955, 1994; Sabbadin and Zaniolo 1979) (Table 11.3). The final size of buds is influenced by competition with other buds of the same zooid (Sabbadin 1955, 1958). Early budlets display a high differentiative potential, especially on the right side, being able to split into two; the posterior budlet undergoes slower morphogenesis or apoptotic death.

In experimental manipulation, such as the ablation of right buds from zooids close to take-over, quiescent pluripotent cells of the atrophic left buds can resume growth and develop reverse bilateral asymmetry (*situs inversus viscerum et cordis*): in this case, the digestive tract of the zooids is located on the right and the heart on the left. Once developed, this epigenetic modification is transferred to the following blastogenetic generations; this reversion also influences the blastogenetic and gonadogenetic powers, which are now higher on the left and right sides, respectively (Sabbadin 1956; Sabbadin et al. 1975, 1991). Reversed bilateral asymmetry, not genetically inheritable by the zooid offspring, highlights the great degree of developmental plasticity of the somatic cells of the atrial epithelium, the fate of which is influenced by both genetic and epigenetic factors.

Table 11.3 Comparison of contralateral testes and ovaries in zooids and buds from 6 colonies of *B. schlosseri* (from Sabbadin and Zaniolo 1979)

Stage	Testes left	right	Ovaries left	right
Adults	150	106	–	–
Primary buds	–	–	28	11
Secondary buds	–	–	32	22

11.3.2 Vascular Budding

The high potential of circulatory stem cells can be experimentally demonstrated in colonial ascidians of the genus *Botrylloides*, in which vascular budding is used as a common method of reconstituting colonies from the colonial matrix after the hibernation (or aestivation) period (Oka and Watanabe 1959; Burighel et al. 1976; Rinkevich et al. 1995; Rinkevich et al. 2007). In *B. schlosseri*, vascular budding can be induced in colonies deprived of all their zooids (adults and buds) and reduced to the peripheral tunic with colonial circulatory vessels containing blood kept in motion by the contractile ampullae. In these conditions, new budlets appear and mature into functional zooids, thus restoring the colony to its original phenotype (Sabbadin et al. 1975; Voskoboynik et al. 2007). Like the palleal buds, vascular buds pass through a double vesicle stage, in which the outer layer derives from the tunic vascular epithelium, continuous with the zooid epidermis, and the inner one from totipotent blood cells, presumably haemoblasts or lymphocytes, which aggregate on the walls of the tunic blood vessels (Voskoboynik et al. 2007), as reported by Freeman (1964) in *Perophora viridis*.

Zooid anterior-posterior polarity depends on vascularisation, as shown in the case of isolated palleal buds, since the vessel supplying them always causes the entrance point to become the posterior end of the zooid (Sabbadin et al. 1975). Conversely, bilateral asymmetry seems to be predetermined in the bud primordium, since isolated palleal buds develop the same type of asymmetry as their parents. Vascular buds also show the same asymmetry as the parental matrix, either normal or reversed (Sabbadin and Zaniolo 1979). This suggests that the information for bilateral asymmetry, which is precociously transmitted to internal vesicle, comes from the outer, epidermal layer of buds represented, in vascular budding, by the vessel epithelium.

11.4 Palleal Budding, Vascular Budding and the "Double-Vesicle Stage"

As already noted, palleal buds arise from specialised staminal areas of the peribranchial epithelium as small thickenings, which arch and then close to form a vesicle (the inner bud vesicle) enveloped by the epidermis (the outer bud vesicle); the intermediate layer is composed of mesenchyme. This stage is defined "double vesicle" stage (Fig. 11.2A,B). Most organogenesis stems from the inner vesicle. Therefore, the thickened, button-like area of the peribranchial epithelium is constituted of pluripotent cells, which are able to differentiate into many cell types, such as the ciliated cells of the branchial basket, the motor and sensory neurons of the cerebral ganglion, and the digestive and endocrine cells of the gut (Manni and Burighel 2006). In the case of vascular budding, the bud primordium forms by aggregation of blood cells, which are received into an evagination of the vessel epithelium and organised early into the "double vesicle". This bud grows and produces budlets by normal palleal budding, which finally become filtering adults (Sabbadin et al. 1975; Voskoboynik et al. 2007).

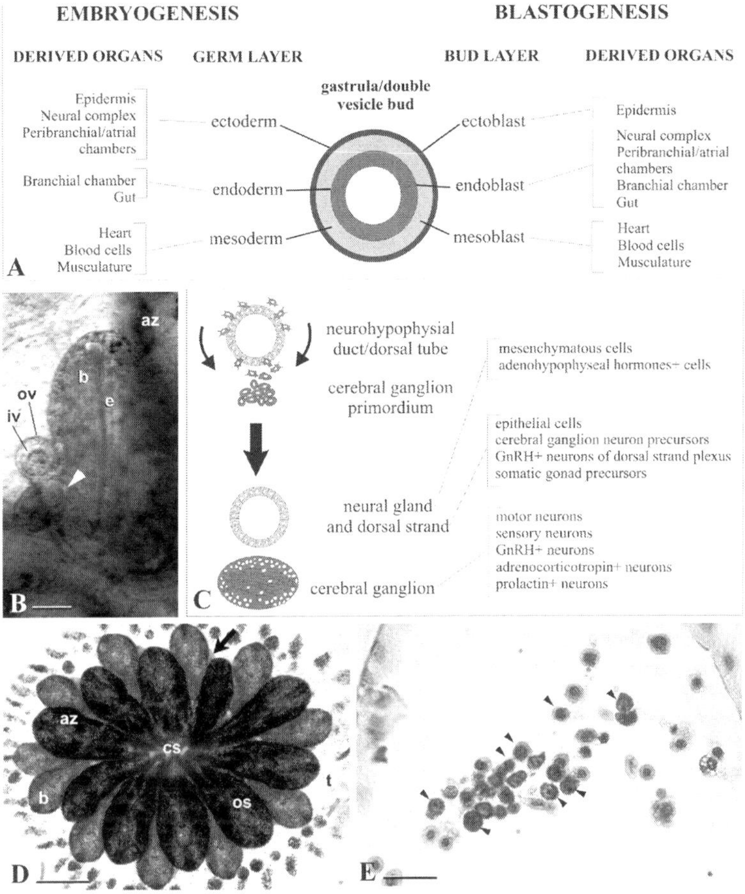

Fig. 11.2 (**A**) Comparative sketch to show organ derivation from germ layers (*left*) and bud layers (*right*) during embryogenesis (gastrula stage) and blastogenesis (double vesicle stage), respectively. (**B**) *In vivo* view of budlet at double vesicle stage. *Arrowhead*, posterior right budlet of bud (b); *Scale bars*, 100 μm. (**C**) Sketch of cell types derived from rudiment of neural complex (neural gland, dorsal strand, cerebral ganglion). See text for references. (**D**) Dorsal view of a colony of *Botryllus schlosseri* at take-over. Arrow: bullet. *Scale bar*, 1 mm. (**E**) Section of marginal ampulla at takeover, showing abundance of circulating haemoblasts(arrowheads). *Scale bar*, 20 μm. az: adult zooid; b: bud; ba: blood ampulla; e: bud endostyle; iv: inner vesicle of anterior right budlet; ov: outer vesicle of anterior right budlet

According to Brien (1948), the double vesicle bud may be considered "a triploblastic vesicle of the gastrula type", taking into consideration the two epithelial vesicles and the mesenchyme between them. The three bud layers were called ectoblast, endoblast and mesoblast by the above author, irrespective of the real germ layer from which they originated (Fig. 11.2A). As regards their fate, the ectoblast represents the epidermis; the endoblast undergoes invagination/evagination processes, epithelial fusion and cellular delamination for the formation of several organs; and the mesoblast differentiates connective tissue, blood cells, heart, gonads and musculature (Sabbadin 1955; Burighel 1970; Burighel and Brunetti 1971; Nunzi et al. 1979; Casagrande et al. 1993; Burighel et al. 1998).

Since most organogenesis stems from the endoblast (deriving from the peribranchial epithelium in the palleal bud and haemoblasts in the vascular budding), this bud layer is characterised by multipotential regardless of its embryonic derivation. In fact, in *B. schlosseri*, although the outer vesicle always forms from the parent's epidermis (being ectodermic in origin) and the intervening layer comes from parental mantle mesenchyme (and thus mesodermic in origin), the inner vesicle varies in origin, according to the type of budding (it is ectodermic in palleal budding, and mesodermic in vascular budding). Thus, considering budding as a short pathway towards forming adults, with its own organogenetic program which does not involve traditional germ layers and the phylotypic stage, it appears that inner palleal/vascular bud vesicles are *de facto* equivalent from a functional point of view, being composed of multipotent cells capable of differentiating the same structures.

11.5 Pluripotential of Nervous System Precursors in Embryo and Bud

In *Botryllus*, both sexual and asexual reproduction produce similar zooids, although development starts from a fertilised egg in the first case and from somatic stem cells in the second. This has aroused the interest of many authors, thanks to the possibility of studying and comparing the differentiation of pluripotent cells and their behaviour both during determinative development of zygotes and regulative development of buds deriving from cells of different origin (Manni and Burighel 2006).

In the colony, a clonal relationship exists between oozooid and blastozooid, and they display the same body organization; moreover, where the cytological differentiation has been documented (e.g., the neural complex and branchial fissures), organs show the same morphology, tissue components and cell types (Manni et al. 1999, 2002).

In a series of studies, Manni and Burighel (2006) compared embryogenesis and blastogenesis in detail, demonstrating that there are several similarities in the formation of the neural complex, which in ascidians is composed of the cerebral ganglion, the neural gland and the dorsal strand. Data evidence the pluripotential of the neural complex rudiment, which is able to produce the following cell types: motor neurons, concentrated in the ganglion; sensory neurons associated with the secondary sensory cells of a mechanoreceptor organ, the coronal organ; ganglionic neurons immunoreactive to GnRH, adrenocorticotropin and prolactin; mesenchymatous cells of the neural gland, some of them immunoreactive to adenohypophyseal hormones, and epithelial cells of the dorsal strand, with neurogenic potential (reviewed in Manni et al. 1999; Kawamura et al. 2002). Moreover, the epithelial cells of the dorsal strand are probably involved in the formation of the somatic gonad rudiment (Takamura et al. 2002) and give rise to GnRH-positive neurons located in the dorsal strand plexus (Mackie 1995) (Fig. 11.2C).

This pluripotential is manifested during morphogenetic processes which display several similarities. In both embryogenesis and blastogenesis (Burighel et al. 1998; Manni et al. 1999), the neural complex forms from a blind tubular structures: the neurohypophysial duct and the dorsal tube, in embryo and in bud, respectively. The tubular structures derive from localised, thickened epithelia – the anterior neural plate in embryo and the dorsal region of inner vesicle in bud (Mackie and Burighel 2005). The two tubes have the same anatomical relationship with adjacent organs, being dorsal, and grow anteriorly to reach and open into the prospective oral siphon. In both cases, the cerebral ganglion is composed of neuroblasts, which delaminate from the wall of the tubular structures and migrate for a short distance to reach their final position under the neural gland. Moreover, the gene *Bs-Pitx*, a homologue of the panhypophyseal *Pitx* gene of vertebrates, is expressed in comparable prospective regions of embryos and buds and labels the same territories in the two developmental pathways (the rudiment of the neural gland aperture, some anterior-ventral neurons of the forming cerebral ganglion, a ring of tissue round the mouth) (Tiozzo et al. 2004).

As a result, starting from different points (a mosaic embryo and a cluster of pluripotent somatic cells, respectively), embryonic and blastogenetic development leads to the formation of comparable pluripotent territories, in which the same genes are activated and similar final products are reached.

11.6 Blood Cell Renewal at Take-Over

The presence and the role of blood stem cells become evident during take-over (Fig. 11.2D), in which blastozooids undergo massive cell death, involving 20–30% of the total haemocyte population, as evidenced by both DNA fragmentation (TUNEL and COMET) assays and labelling of phosphatidylserine on the outer leaflet of the plasma membrane by fluorescent annexin-V (Cima et al. 2003). All these events must be ascribed to apoptosis, as shown by the characteristic morphology of dying cells and the significantly increased caspase-3 activity of haemocyte lysates during take-over (Ballarin et al. 2008). During this period, there is a significant increase in circulating globular, macrophage-like phagocytes, with engulfed matter inside their vacuoles, and a parallel decrease in hyaline amoebocytes, representing the phagocytes which actively sense and ingest foreign particles. The number of circulating globular phagocytes drops after take-over, whereas the frequency of ameboid phagocytes increases: this suggests that renewal of circulating haemocytes must occur to replace senescent cells engulfed by phagocytes; the new amoebocytes likely derive from the differentiation of multipotent stem cells entering the circulation at the generation change (Fig. 11.2E). This is confirmed by both cytofluorimetric and histochemical observations, demonstrating that a new population of small, undifferentiated cells, ascribed to haemoblasts according to their morphology (Ballarin and Cima 2005), appears at take-over (Ballarin et al. 2008).

These observations are consistent with the results of Sala (1973) showing that *in vitro* cultures of mesenchymal cells from *Botryllus* larvae can develop most blood cell types, starting from haemoblast-like cells.

11.7 Circulating Somatic and Germ Stem Cells

An interesting and useful feature of *B. schlosseri* is the possibility of subcloning the colony: fragments of a limited number of systems can detach naturally or be separated experimentally from the parent colony. When small subclones (1-2 systems) from different colonies are juxtaposed and left to adhere to a supporting glass slide, their leading ends facing each other, they can fuse their tunics and anastomose their vascular vessels (Sabbadin 1962, 1982) if they share at least one allele at the fusibility-histocompatibility locus. Fusion confers benefits on the chimaeric colony such as increase in size, with consequent selective advantage in substrate competition; earlier achievement of sexual maturity (which occurs only over a minimum size); and increased genetic diversity (Rinkevich and Shapira 1999; Stoner et al. 1999; Laird et al. 2005; De Tomaso 2006; Nyholm et al. 2006). However, the histocompatibility gene limits fusion to kin colonies, in order to avoid somatic or germ cell parasitism (Weissman 2000; Laird et al. 2005). In fact, fusion of colonies involves exchange of circulating stem cells, with the formation of chimaeric colonies having cells with two different genotypes which compete with each other to create their phenotype in the new colonial environment. Evidence of "winner" and "loser" genotypes in somatic and germ cells was first demonstrated by Sabbadin and Zaniolo (1979), who analysed the phenotypes and collected the descendants of chimaeric colonies. In a series of elegant experiments, the authors demonstrated that, as in other clonal organisms (Buss 1987), *B. schlosseri* colonies can recycle undifferentiated germ cells through successive generations. When parabiotic pairs of colonies of the opposite genotypes AAbb and aaBB, relative to two Mendelian loci controlling pigmentation, were separated after a few days of fusion and double-crossed, as males and females, to colonies of the double recessive genotype aabb, offspring phenotypically Ab from aaBB x aabb crosses and phenotypically aB from AAbb x aabb crosses were obtained. This heterochthonous offspring derived from primordial germ cells exchanged during the period of fusion, able to differentiate to gametes of either sex. It was possible to obtain heterochthonous offspring even at the 14th and 15th blastogenetic generations after the interruption of fusion.

The persistence of cells from a previously fused alien colony was also demonstrated in terms of fusibility alteration, still evident after more than 200 days from the separation of fused colonies, and by the chimaeric electrophoretic patterns of the enzyme phosphoglucoisomerase (Sabbadin and Astorri 1988) or AFLP markers (Rinkevich et al. 1998). These results strongly suggested that toti- or pluripotent stem cells could be transferred into compatible colonies and differentiate in their new environmental conditions.

This possibility has recently been studied with modern techniques by labelling isolated cells and transplanting them into the circulation of compatible host colonies

(Laird et al. 2005). The peripheral vascular ampullae can easily be reached by a capillary mounted on a microinjector, and haemocytes can be injected into other colonies, either isogenic or allogenic, in order to study their effects on the host colony (Sabbadin and Ballarin 1990; Laird et al. 2005). The approach allows the formation of chimaeras and the identification of heterologous components, avoiding the need to fuse colonies and collect descendants. It is thus very suitable for studying the basis of stem cell differentiation, following cell commitment before transplantation and the influence of the various landing niches, and the several environmental factors, in directing final cell differentiation (see also the chapter "Stem cells and chimerism in colonial urochordates" by Voskoboynik, Rinkevich and Weissman). Using this technique, Laird et al. (2005) identified a population of multipotent stem cells from *B. schlosseri* colonies, which were able to give rise to both somatic and germ cells. This may be interpreted in the light of the presence of either separate stem cells for somatic and germ lines, or pluripotent stem cells which differentiate according to micro-environmental signals received from the niche in which they land after transplantation.

11.8 Concluding Remarks

Today, stem cells have aroused great interest among scientists, especially for the ultimate possibility of using them to treat genetic diseases. However, basic knowledge of stem cell biology is still poor, and simple animal models are required, offering the possibility of studying the molecular basis of stem cell regulation, their origin, the nature of their niches and their differentiation pathways.

B. schlosseri represents an interesting model for stem cell studies as it has the following potentials:

11.8.1 Somatic Stem Cells can Give Rise to New Zooids

This is commonly observed in palleal and vascular budding. However, the bud-originating tissues are different in the two situations and this may lead to diverse developmental pathways, worthy to study. The recent report of differences in the morphology of zooids originating from vascular budding as compared with palleal budding (Voskoboynik et al. 2007) argues in favour of this hypothesis.

The ultimate fate of stem cells is under genetic and epigenetic control, as indicated by left buds which, although having the potential for normal development (which always occurs when right buds are extirpated), frequently undergo apoptosis and are resorbed. In this case, the stem cells interrupt their differentiation and enter a death pathway; this suggests the presence of "checkpoints" in development, the nature of which is not known.

In addition, the epigenetic control of somatic stem cell differentiation offers the possibility of studying the cell interactions underlying axis specification during development. In *Botryllus*, a reverse phenotype, transmittable asexually, can be induced when the normal bud developmental programme is altered. The new positional information is also maintained by vascular buds which have the same (normal or reversed) morphology as the parental matrix, suggesting the interesting role of colonial epithelium in the specification of body axes.

11.8.2 Germ Stem Cells are Subpopulations of Circulating Cells

This means that they can re-circulate within a colony and pass from one generation to another. In addition, they can enter an alien colony, if it fuses with the donor colony, where they can colonise the gonads and compete with autologous germ stem cells to produce gametes (Weissman 2000; Laird et al. 2005). Homologues of *vasa*, a marker of undifferentiated germ cells, have recently been identified in two botryllid ascidians, *Botryllus primigenus* and *Botrylloides leachi*, expressed in developing gonads and in a subset of circulating hemocytes (Sunanaga et al. 2006; Brown and Swalla 2007). The expression is inducible *de novo* in the case of vascular budding, indicating epigenetic control also in the case of germline determination, the molecular basis of which is still unknown.

11.8.3 Pluripotent Stem Cells Gradually Evolve into Lineage-Specific Stem Cells

This is clearly evident in angiogenesis, blood cell renewal, and the development of the central nervous system. In addition, stem cells can re-circulate through the colonial vasculature and enter a fusible alien colony, as indicated by the alteration of fusibility of the ex-parabionts. This offers the possibility of fine study of stem cell differentiation, focussing on the development of single organs, and following the colonisation of new niches represented by colonies which have received alien cells through fusion or microinjection.

The combined efforts of the various laboratories now involved in researches on *Botryllus* stem cell will certainly offer new clues, in the near future, to understand the molecular control of stem cell differentiation in a simple chordate closely related to vertebrates.

Acknowledgments Grant information: grants from the Ministero della Università e Ricerca Scientifica e Tecnologica to L.B. and L.M.

References

Ballarin L, Cima F (2005) Cytochemical properties of *Botryllus schlosseri* haemocytes: Indications for morpho-functional characterisation. Eur J Histochem 49:255–264

Ballarin L, Menin A, Tallandini L, Matozzo V, Burighel P, Basso G, Fortunato E, Cima F (2008) Haemocytes and blastogenetic cycle in the colonial ascidian *Botryllus schlosseri*: A matter of life and death. Cell Tissue Res 331:555–564

Berrill NJ (1941) The development of the bud in *Botryllus*. Biol Bull 80:169–184
Brien P (1948) Morphologie et reproduction des tuniciers. In: Grassé PP (ed) Traité de Zoologie, vol 11. Masson, Paris, pp 553–1040
Brown FD, Swalla BJ (2007) *Vasa* expression in a colonial ascidian, *Botrylloides violaceus*. Evol Dev 9:165–177
Burighel P (1970) Sviluppo e differenziamento del tubo digerente nel blastozooide dell'ascidia coloniale *Botryllus schlosseri* (Pallas). Boll Zool 37:177–192
Burighel P, Brunetti R (1971) The circulatory system in the blastozooid of the colonial ascidian *Botryllus schlosseri* (Pallas). Boll Zool 38:273–289
Burighel P, Brunetti R, Zaniolo G (1976) Hibernation of the colonial ascidian *Botrylloides leachi* (Savigny): Histological observations. Boll Zool 43:293–301
Burighel P, Lane NJ, Zaniolo G, Manni L (1998) Neurogenic role of the neural gland in the development of the ascidian, *Botryllus schlosseri* (Tunicata, Urochordata). J Comp Neurol 394:230–241
Buss L (1987) The evolution of individuality. Princeton University Press, Princeton
Casagrande L, Martinucci GB, Burighel P (1993) Origin and differentiation of branchial stigmata in the compound ascidian *Botryllus schlosseri* (Tunicata). Animal Biol 2:111–121
Cima F, Basso G, Ballarin L (2003) Apoptosis and phosphatidylserine-mediated recognition during the take-over phase of the colonial life-cycle in the ascidian *Botryllus schlosseri*. Cell Tissue Res 312:369–376
De Tomaso AW (2006) Allorecognition polymorphism versus parasitic stem cells. Trends Gen 22:485–490
Freeman G (1964) The role of blood cells in the process of asexual reproduction in the tunicate *Perophora viridis*. J Exp Zool 156:157–184
Kawamura K, Nakauchi M (1986) Mitosis and body patterning during morphallactic development of palleal buds in ascidians. Dev Biol 116:39–50
Kawamura K, Nakauchi M (1991) Homeostatic integration of stem cell dynamics during palleal budding of ascidians. Zool Sci 8:11–22
Kawamura K, Hashimoto K, Nakauchi M (1995) Development of irradiated tunicate buds: Is cell division required for morphallaxis? Dev Growth Diff 37:487–496
Kawamura K, Kouki T, Kawahara G, Kikuyama S (2002) Hypophyseal development in vertebrates from amphibians to mammals. Gen Comp Endocrinol 126:130–135
Kawamura K, Sugino Y, Sunanaga T, Fujiwara S (2008) Multipotent epithelial cells in the process of regeneration and asexual reproduction in colonial tunicates. Dev Growth Diff 50:1–10
Laird DJ, De Tomaso AW, Weissman IL (2005) Stem cells are units of natural selection in a colonial ascidian. Cell 123:1351–1360
Lauzon RJ, Ishizuka KJ, Weissman IL (1992) A cyclical, developmentally-regulated death phenomenon in a colonial urochordate. Dev Dyn 194:71–83
Mackie GO (1995) On the 'visceral nervous system' of *Ciona*. J Mar Biol Ass UK 75:141–151
Mackie GO, Burighel P (2005) The nervous system in adult tunicates: Current research directions. Canad J Zool 83:151–183
Majone F (1977) Regeneration of isolated bud fragments of *Botryllus schlosseri*. Acta Embryol Exp 1:11–19
Manni L, Burighel P (2006) Common and divergent pathways in alternative developmental processes of ascidians. BioEssays 28:902–912
Manni L, Lane NJ, Sorrentino M, Zaniolo G, Burighel P (1999) Mechanism of neurogenesis during the embryonic development of a tunicate. J Comp Neurol 412:527–541
Manni L, Lane NJ, Zaniolo G, Burighel P (2002) Cell reorganisation during epithelial fusion and perforation: The case of ascidian branchial fissures. Dev Dyn 224:303–313
Manni L, Zaniolo G, Cima F, Burighel P, Ballarin L (2007) *Botryllus schlosseri*: A model ascidian for the study of asexual reproduction. Dev Dyn 236:335–352
Nyholm SV, Passegue E, Ludington WB, Voskoboynik A, Mitchel K, Weissman IL, De Tomaso AW (2006) *Fester*, a candidate allorecognition receptor from a primitive chordate. Immunity 25:163–173

Nunzi MG, Burighel P, Schiaffino S (1979) Muscle cell differentiation in the ascidian heart. Devel Biol 68:371–380

Oka H, Watanabe H (1959) Vascular budding in *Botrylloides*. Biol Bull 117:340–346

Rinkevich B, Shapira M (1999) Multi-partner urochordate chimeras outperform two-partner chimerical entities. Oikos 87:315–320

Rinkevich B, Shlemberg Z, Fishelson L (1995) Whole-body protochordate regeneration from totipotent blood cells. Proc Natl Acad Sci USA 92:7695–7699

Rinkevich B, Weissman IL, De Tomaso AW (1998) Transplantation of Fu/HC incompatible zooids in *Botryllus schlosseri* results in chimerism. Biol Bull 195:98–106

Rinkevich Y, Paz G, Rinkevich B, Reshef R (2007) Systemic bud induction and retinoic acid signaling underlie whole body regeneration in the urochordate *Botrylloides leachi*. PLoS Biol 5:e71. doi:10.1371/journal.pbio.0050071

Takamura K, Fujimura M, Yamaguchi Y (2002) Primordial germ cells originate from the endodermal strand cells in the ascidian *Ciona intestinalis*. Dev Genes Evol 212:11–18

Tiozzo S, Christiaen L, Deyts C, Manni L, Joly JS, Burighel P (2004) Embryonic versus blastogenetic development in the compound ascidian *Botryllus schlosseri*: Insights from Pitx expression patterns. Dev Dyn 232:468–478

Tiozzo S, Ballarin L, Burighel P, Zaniolo G (2006) Programmed cell death in vegetative development: Apoptosis during the colonial life-cycle of the ascidian *Botryllus schlosseri*. Tissue Cell 38:193–201

Sabbadin A (1955) Osservazioni sullo sviluppo, l'accrescimento e la riproduzione di *Botryllus schlosseri* (Pallas), in condizioni di laboratorio. Boll Zool 22:243–263

Sabbadin A (1956) *Situs inversus viscerum* provocato sperimentalmente in *Botryllus schlosseri* (Pallas) Ascidiacea. Rendiconti Accademia Nazionale Lincei 20:659–666

Sabbadin A (1958) Analisi sperimentale dello sviluppo delle colonie di *Botryllus schlosseri* (Pallas) [Ascidiacea]. Arch Ital Anat Embr 63:178–221

Sabbadin A (1962) Le basi genetiche della capacità di fusione fra colonie in *Botryllus schlosseri* (Ascidiacea). Rendiconti Accademia Nazionale Lincei, 32:1031–1035

Sabbadin A (1982) Formal genetics of ascidians. Am Zool 22:765–773

Sabbadin A (1994) Coloniality in ascidians and its adaptive value, with special reference to *Botryllus schlosseri*. Animal Biol 3:157–163

Sabbadin A, Astorri C (1988) Chimeras and histocompatibility in the colonial ascidian *Botryllus schlosseri*. Dev Comp Immunol 12:737–747

Sabbadin A, Ballarin L (1990) The reaction of the colonial ascidian *Botryllus schlosseri* to allogeneic blood. Preliminary results. Acta Embryol Morphol Exp ns 11:211–212

Sabbadin A, Zaniolo G (1979) Sexual differentiation and germ cell transfer in the colonial ascidian *Botryllus schlosseri*. J Exp Zool 207:289–304

Sabbadin A, Zaniolo G, Majone F (1975) Determination of polarity and bilateral asymmetry in palleal and vascular buds of the ascidian *Botryllus schlosseri*. Dev Biol 46:79–87

Sabbadin A, Burighel P, Zaniolo G (1991) Studies on the colonial ascidian *Botryllus schlosseri*. A review of some lines of research. Acta Embryol Morphol Exp ns 12:259–279

Sala M (1973) Differenziamento *in vitro* di cellule della linea sanguigna da elementi mesenchimali di larva di *Botryllus schlosseri*. Atti Memorie Accademia Patavina Scienze Lettere Arti 85:275–293

Stoner DS, Rinkevich B, Weissman IL (1999) Heritable germ and somatic cell lineage competitions in chimeric colonial protochordates. Proc Natl Acad Sci USA 96:9148–9153

Sunanaga T, Saito Y, Kawamura K (2006) Postembryonic epigenesis of Vasa-positive germ cells from aggregated hemoblasts in the colonial ascidian, *Botryllus primigenus*. Dev Growth Diff 48:87–100

Voskoboynik A, Simon-Blecher N, Soen Y, Rinkevich B, De Tomaso AW, Ishizuka KJ, Weissman IL (2007) Striving for normality: Whole body regeneration through a series of abnormal generations. FASEB J 21:1335–1344

Weissman IL (2000) Stem cells: Units of development, units of regeneration, and units in evolution. Cell 100:157–168

Chapter 12
Stem Cells, Chimerism and Tolerance: Lessons from Mammals and Ascidians

Ayelet Voskoboynik, Baruch Rinkevich, and Irving L. Weissman

Abstract Chimerism is the presence of cells derived from more than one individual in a given individual. This phenomenon has been detected in a wide variety of multicellular organisms, including vertebrates. In mammals natural chimerism can develop from bidirectional trafficking of cells between multiple fetuses or between the fetus and its mother during pregnancy. Because stem cells are the only self-renewing cells in a tissue, it is likely that chimerism is sustained by stem cells. Colonial marine ascidians, like *Botryllus schlosseri*, may serve as an evolutionary model of chimerism. Colonies are initially formed by asexual reproduction of a founder individual, and the progeny clone members are united under a single gelatinous tunic by a network of anastomosed extracorporeal blood vessels. In these organisms, pairs of genetically distinct colonies can establish a natural chimerism upon physical contact by anastomosis of extracorporeal blood vessels between colonies. The ability to create a chimeric entity between colonies is determined by a single, highly polymorphic, fusion/histocompatibility locus (Fu/HC). Colonies that share at least one Fu/HC allele (mainly kin under *in situ* conditions) fuse vessels upon contact. A pair that does not share any Fu/HC allele does not, but instead has an inflammatory immune response that forms a permanent scar between colonies. Following fusion, cells transmigrate between colonies and, in some cases, replace the germline and/or the somatic tissues of the host. The replacement of host tissues by a donor genotype is pre-determined genetically and follows hierarchies of "winner strains" replacing "loser strains" tissues.

In both mammals and ascidians, natural creation of a chimera entity is mainly restricted to kin; long-term chimerism can be established by stem cells; and tolerance of donor tissues can be mediated by chimerism. While several studies and observations across different species, tissues and systems link chimerism to tolerance, its actual role in tolerance induction or maintenance is only understood in

---

A. Voskoboynik (✉)
Institute of Stem Cell Biology and Regenerative Medicine, Stanford University School of Medicine, Stanford, CA 94305, USA; Department of Developmental Biology, Stanford University Hopkins Marine Station, Pacific Grove, CA 93950, USA
e-mail: ayeletv@stanford.edu

B. Rinkevich, V. Matranga (eds.), *Stem Cells in Marine Organisms*,
DOI 10.1007/978-90-481-2767-2_12,

experimentally induced mammalian models. Here we review the chimerism phenomenon in mammals and ascidians, discuss the possible role of stem cells as mediators of chimerism and the possible role of cellular chimerism as mediators of tolerance.

Keywords Ascidian · Mammals · Stem cells · Chimerism · Tolerance

12.1 Chimerism in Mammals

12.1.1 Fetal/Maternal Micro Chimerism

During mammalian pregnancy, fetal cells are detected in the maternal circulation and organs and maternal cells are detected in the umbilical cord and in fetal blood samples and organs (human and mice; Tuffrey et al. 1969; Walknowska et al. 1969; Barnes and Holliday 1970; Gaillard et al. 1978; Herzenberg et al. 1979; Liegeois et al. 1981; Lo et al. 1989, 1996, 1998; Hamada et al. 1993; Bianchi et al. 1996, 1997, 2002; Piotrowski and Croy 1996; Oosterwijk et al. 1998; Troeger et al. 1999; Zhou et al. 2000; Marleau et al. 2003; O'Donoghue et al. 2004; Kaplan and Land 2005; Khosrotehrani et al. 2005; Loubiere et al. 2006). Using fluorescent-activated cell sorting and polymerase chain reaction amplification of a Y-chromosome sequence, Bianchi et al. (1996) detected male CD34+ cells in the blood of women who carried male fetuses up to 27 years earlier. It been suggested that these women with sons became chimeras (termed as microchimerism- Mc; Bianchi et al. 1996). Further studies reported that in humans, microchimerism is a common phenomenon recorded in a high percentage of studied women up to five decades after giving birth (Bianchi et al. 1996; O'Donoghue et al. 2004; Loubiere et al. 2006). Microchimerism is affiliated with very low frequencies of circulating allogeneic cells in either mother/progeny hosts (estimated frequencies are 1 allogeneic cell in 10^5–10^7 host cells; Bianchi et al. 1996; O'Donoghue et al. 2004; Loubiere et al. 2006). These studies detected allogeneic fetal hematologic progenitor cells and mesenchymal stem cells in the maternal soma and circulation (Bianchi et al. 1996; O'Donoghue et al. 2004). Considering half-live values of circulating peripheral blood cells from few hours to several months only, it has been suggested that chimeric humans may harbor long-surviving populations of maternal/fetal stem cells origin. Indeed, studies demonstrated that blood chimerism can be initiated by a single hematopoietic stem cell (engraftment irradiated mice (Smith et al. 1991; Cao et al. 2004)). O'Donoghue et al. (2004) cultured *in vitro* fetal mesenchymal cells from maternal bone marrow and ribs that differentiated *in vitro* into adipocytes and other cells with markers shared with muscle, nerve and bone cells, although marker expression *in vitro* does not always correlate with functional cells of the marker lineage *in vivo* (Anderson et al. 2001; Wagers and Weissman 2004). Long- term maternal and fetal Mc was found in cells mediating the host's adaptive and innate immunity, including cells of the lymphoid (T and B lymphocytes, NK cells) and the myeloid lineages (monocyte/macrophages; Loubiere et al. 2006). Allogeneic cells have been identified in many organs including kidneys, livers and hearts of healthy women with or without prior pregnancies (chimerism in different organs were detected in 23 out of 75 tested women, age at death 29–93; Koopmans et al. 2005). Moreover, several studies demonstrated in post-mortem analyses the replacement of significant tissues of clinically affected maternal organs by fetal cells. For example, pathologic thyroid samples from a woman with a goiter revealed a complete replacement of a thyroid section by male cells (Srivatsa et al. 2001). A liver biopsy from a woman with hepatitis C revealed thousands of male cells, originating from a fetus that had been terminated 17 years earlier (Johnson et al. 2002). Other studies in human and rodents also suggest that the frequency of fetal cell microchimerism increases in diseased or injured organs (Nelson et al. 1998; Artlett et al. 1998; Klintschar et al. 2006; Christner and Jimenez 2004; Imaizumi et al. 2002; Wang et al. 2004; Tan et al. 2005). Children with severe combined immunodeficiency (SCID) syndromes, often are engrafted with maternal cells, and they engraft maternal hematopoietic grafts more easily than paternal grafts (Pollack et al. 1982; Cowan et al. 2008; Neven et al. 2008) however, sometimes the maternally engrafted patients have a T cell mediated maternal vs offspring graft vs host immune reaction (Flomenberg et al., 1983; Conley et al. 1984; Suda et al. 1984; Le Deist et al. 1987; Vaidya et al. 1991).

Kaplan and Land (2005) investigated the effect of maternal-fetal histocompatibility levels on the presence and numbers of maternal Mc in mice. While detecting maternal chimerism in the brain and lymphoid tissue (spleen and thymus) of all tested mice (age 0–45 days), regardless of maternal-fetal histocompability level, MHC homozygous progeny had higher numbers of maternal cells compared to MHC heterozygous progeny, suggesting maternal fetal histocompatibility influence on microchimerism. In humans, histocompatibility between mother and fetus also affects the frequency and extent of fetal cell microchimerism, and the presence of specific tissue type antigens such as maternal HLA-DQA1$*$0501 influences levels of microchimerism (Lambert et al. 2000; Nelson 2001). Data from humans suggest low-grade transfer and long term persistence of fetal cells in mothers and mother cells in their progeny. However, the rarity of allogeneic cells in maternal/progeny tissues, the use of indirect methods which include enhancement techniques and the absence of a classic negative control (in humans every tested sample can potentially carry microchimerism) raises the possibilities of artifact generation. Detection of longterm persistence of microchimerism in model organisms by direct visualization is needed to further eliminate the possibilities of artifacts and strengthen the impact of these important set of results.

12.1.2 Natural Chimerism in Twins

Cells movement between sibling twins during human pregnancy is an additional type of natural chimerism.

Human twin chimerism was discovered on 1953, when blood typing tests revealed that some people have blood cells of more than a single blood group (Dunsford et al. 1953). Based on routine blood grouping protocol, a procedure detecting allogeneic mixture of blood cells at levels exceeding 5%, twin chimerism in humans was considered as a rare phenomenon, illuminated by 40 reported cases (Tippett 1983). Using a more sensitive method, van Dijk et al. (1996) found that 8% of non-identical human twins and 21% of triplets are chimeric, revealing also higher ratios of allogeneic cells than in fetal-maternal Mc (about 1:1000 vs $1{:}10^5$, respectively). Given that most of multiple-embryo conceptions involve the loss of one embryo or more in early pregnancy (Boklage 1990), twin blood cells may be a significant source of natural chimerism, even among single births. Thus, the frequency of natural human chimerism represents an under-estimated figure (Boklage 2006; Wolinsky 2007). Twin chimerism with high levels of allogeneic blood cells yields tolerance to sibling's antigens. All twin chimeras, which were identified by a routine blood grouping (frequency >5% of allogeneic cell population), were able to accept blood and mutual skin grafts from their chimeric-partner (Tippett 1983). Human chimeras with low frequencies of the other twin cell population, representing chimeric frequencies of ~0.1% (those identified by van Dijk et al. 1996), are not tolerant to the foreign antigens of their twin and could not accept blood grafts.

Liegeois (1983) hypothesized that fetal and maternal Mc play a role in tolerance induction and maintenance during pregnancy. This hypothesis, although never tested directly, may reveal an evolutionary explanation (positive selective force) for the existence of Mc phenomenon, so common in mammals.

12.1.3 Natural Tetragametic Chimerism

Tetragametic chimerism is an uncommon case of congenital chimerism, developed from equal contributions of two distinct origins, recognizing both as self. It occurs naturally through fertilization of two ova by two spermatozoa, followed by fusion of the zygotes and development of an organism with intermingled cells of origin (Tippett 1983). Examples were found in various mammalian species (Mintz and Palm 1969; Tucker et al. 1978; Bordenave and Babinet 1984; Sumantri et al. 1997), including humans (Watkins et al. 1981; Bromilow and Duguid 1989; Verp et al. 1992; Green et al. 1994; Sybert 1994; Strain et al. 1995, 1998; Uehara et al. 1995; Repas-Humpe et al. 1999). Tetragametic chimerism is identified by the presence of two populations of red cells (Watkins et al. 1981), ambiguous genitalia and hermaphroditism (Green et al. 1994; Strain et al. 1998; Repas-Humpe et al. 1999), and/or by patchy skin or eye pigmentation (Sybert 1994). Tetragametic chimerism may also be localized to the germ line of phenotypically normal fertile women (Mayr et al. 1979; Yu et al. 2002). Yu et al. (2002) reported a woman with various tissues demonstrating the existence of up to four genetic components. This woman had only one cell line in her blood but her germ-line revealed two types of allogeneic germ cells. This example of tetragametic state has important implications for organ or stem-cell transplantation events, because, even with only one cell line in her blood, she was tolerant to cells from family members with any combination of the four familial human leukocyte antigens (HLA) haplotypes (Yu et al. 2002). These results are consistent with studies of tetragametic mice (Mintz and Palm 1969).

In a mouse model of tetragametic chimerism, blastomeres from two embryos were co-cultured to form a chimera. Twelve of 34 mice had only one red cell population in their blood even though they had two cell lines in other tissues (Mintz and Palm 1969). This finding can be explained by the development of a single cell line from a clone, or because of stem cell competition. The latter option is supported by the findings of another study of tetragametic rams, where one of two red-cell lines has completely disappeared in the tetragametic animals within five years (Tucker et al. 1978).

12.1.4 Callitrichid Primates, Living with Chimerism

Callitrichid primates (marmosets and tamarins reviewed in Haig 1999) represent an interesting evolutionary example for natural chimerism where a state of germ cell chimerism is revealed by about 50% of the population (Hampton 1973). Callitirichid females typically give birth to two chimeric offspring with allogeneic mixed cells in blood, bone marrow, lymph nodes and spleen (Gengozian et al. 1964; Haig 1999). In marmosets, major histocompatibility complex (MHC) class I antigens are significantly less diversified than those of other primates. The lack of MHC class I diversity limits the variety of peptides that can be presented to T cells for immune surveillance. As a result, marmosets are particularly susceptible to viral infections (Watkins et al. 1990). The less diversified immune system in marmosets can be either the result of high level of chimerism, or the key machinery for the observed somatic and germ cells chimerism in these organisms.

12.1.5 Bovine Fraternal Twins, the Discovery of Tolerance

In 1945, Owen found that most bovine fraternal twins are born with red blood cells genetically belonging to the other twin zygotic lineage, and retain this chimerism for life. Based on earlier observations, indicating that bovine twins share a single placenta and blood circulation during fetal phase (Lillie 1916), Owen conjectured that blood cells and their precursors (which he called "embryonal cells ancestral to the erythrocytes") move from one twin to the other in the uterus, allowing mixing of blood cells from both genotypes (Owen 1945). Bovine fraternal twins frequently exhibit a complete lifelong mutual tolerance to each other's leukocytes and a temporary (up to 15 months) tolerance to each other's skin grafts (Anderson et al. 1951; Billingham et al. 1952; Stone et al. 1965, 1971; Emery and McCullagh 1980a,b). Burnet and Fenner (1949) highlighted Owen's discovery in their prominent monograph "The Production of Antibodies", where they predicted the existence of tolerance as a ubiquitous phenomenon, and developed their notion of "self markers" to explain why the body does not react against self. In 1954, Owen et al. further detected another remarkable aquiered tolerance phenomenon connected to embryonic exposure. They reported that a rhesus D negative pregnant mother is less likely to produce antibodies against rhesus D positive child, in cases where the grandmother is rhesus D positive. This observation led to the discovery that a high percent of individuals are more tolerate of non-inherited maternal HLA antigens (NIMA) than non-inherited paternal HLA antigens NIPA (Claas et al. 1988, 1989). Furthermore, it was hypothesized that exposure of a child to NIMA during pregnancy may lead to NIMA-specific tolerance later in life. Chimerism probably has an important role in this acquired tolerance effect (Owen et al. 1954; van den Boogaardt et al. 2006).

12.1.6 The Association Between Chimerism and Transplantation Tolerance

Tolerance in transplantation is defined as a state of unresponsiveness to donor antigens while maintaining immune reactivity to other foreign antigens. Tolerance is necessary for success of transplantation and prevention of immunological rejection. The high degree of HLA (human leukocyte antigens; humans MHC) polymorphism makes it very difficult to find a well-matched donor (Tiercy 2002). Enhancing tolerance by partial MHC matching and induction of chimerism proved to be promising immunologic approaches for successful transplantation (Opelz et al. 1973; Persijn et al. 1979; Colson et al. 1995a; Prabhune et al. 2000; Monaco 2003; Starzl 2004; Ichinohe et al. 2005; Taieb et al. 2007). Billingham et al. (1953) were the first to induce specific tolerance to solid organs by engrafting bone marrow cells or splenic leukocytes. Tolerance was achieved by exposing gestational or newborn mice to the donor cellular transplant antigenic stimuli before the immune systems' maturation. Engrafted mice did not reject allogeneic material, accepting skin from mice of the leukocyte donor strain, but not from any other strain (6–8 weeks after birth; Billingham et al. 1953; Billingham and Brent 1956). Tolerance to skin alograft was permanent or transient depending on the mouse strain (Billingham and Brent 1956). This acquired tolerance to donor tissues was associated with leukocyte chimerism, which was detected in the animals' lymphoid organs (Billingham et al. 1953; Billingham and Brent 1956). Similarly, chickens hatched from eggs that had been joined in parabiosis or exposed to allogeneic adult blood, exhibited varying degrees of tolerance to solid tissues (depending on the strain and the method applied; Hasek 1953a,b). A few years later, tolerance was produced by transplanting hematolymphoid cells into lethally irradiated adult mice (Main and Prehn 1955; Mannick et al. 1959; Hume et al. 1960; Murray et al. 1960). However, where donor and recipient did not "share" a good histocompatibility match, leukocytes engraftment in newborn/irradiated chimeric mice resulted in acute or chronic graft versus host disease (GVHD; Billingham and Brent 1956; Trentin 1956; Uphoff 1957; Billingham et al. 1959; Barnes et al. 1962; Mathe et al. 1963; Bach et al. 1968; Gatti et al. 1968). Histocompatibility differences also affect the ability to induce tolerance by chimerism; it was easier to induce tolerance across partial MHC barriers than across full MHC barriers (Weissman 1963, 2002; Brent 1997). Further studies on GVHD development revealed involvement of T, B and NK cells within the donor's bone marrow (Tyan1973; Korngold and Sprent 1978; Gale and Reisner 1986; Vallera and Blazar 1989; Shizuru et al. 1996). The incidence and severity of GVHD were reduced or prevented by transplantation of either donor T cell depleted (TCD) bone marrow, or purified hematopoietic stem cells (HSC; O'Reilly 1983; Mitsuyasu et al. 1986; Maraninchi et al. 1987; Martin et al. 1987; Anasetti et al. 1989; Ash et al. 1990; Armitage 1994; Shizuru et al. 1996). However, the incidence of graft failure was significantly increased (from 3% without TCD, to 8–20% in HLA-identical and 20–50% in HLA-non identical recipients; O'Reilly 1983; Mitsuyasu et al. 1986; Maraninchi et al. 1987; Martin et al. 1987; Anasetti et al. 1989; Ash et al. 1990; Armitage 1994; Shizuru et al. 1996). Kaufman et al. (1994) further discovered that in a fully allogeneic environment, a donor derived accessory cell population "facilitating cell" is required for HSC engraftment, (Kaufman et al. 1994; Gandy et al. 1999; Colson et al. 2007).

During the 1960s and 1970s, much thought was given to blood chimerism and its relation towards tolerance induction in solid organ transplantation. Studies demonstrated that maintenance of specific tolerance in the host requires the presence and the replication of donor cells (Billingham and Brent 1956; Simonsen 1956; Mitchison 1959, 1961; Lubaroff and Silvers 1973; Sharabi et al. 1992). Hasek and Haskova (1958) observed that when a larger variety of allogeneic donors had contributed to the pool, the spectrum of induced tolerance was broader. Other studies described cases where recipients became tolerant to only part of the antigenic spectrum of injected cells (termed split tolerance; Billingham and Silvers 1959, 1960; Lustgraaf et al. 1960; Heslop and Nisbet 1962). By injecting different numbers of male splenic cell suspensions (MSA) into newborn female hosts, Weissman (1973) induced specific tolerance to donor skin allografts. He found a positive correlation between the number of injected cells and the incidence of tolerance induction. When male splenic cell suspensions were injected into adult hosts, an accelerated rejection of the donor skin allograft was observed. Additionally, the incidence of skin allograft rejection in the adult hosts had a positive correlation to the number of injected cells (Fig. 12.1; Weissman 1973). Weissman further injected splenic cell suspensions from the tolerant females to both newborn and adult isologous females. Cell doses, which developed second-set rejection in adults, induced tolerance in the newborn, showing the transferability of the tolerance to the newly born mice (Weissman 1973). Minimal doses for second-set rejection, permit the estimation for male antigen concentrations in tolerant female spleens. The percent of donor chimerism was always very low (Weissman 1973). Opelz et al. (1973) demonstrated that patients treated with multiple pretransplant blood transfusions had a significantly higher graft survival compared with nontransfused patients, whereas leukocyte depleted transfusions were not associated with this effect, indicating that leukocytes are essential for tolerance (Persijn et al. 1979). In 1984, Ildstad and Sachs succeeded to induce temporary tolerance to skin grafts in adult mice by mixed chimerism. Tolerance was induced through transplanting of T-cell depleted bone marrow components from host and donor (mixed chimerism) into irradiated hosts (Ildstad and Sachs 1984). This group further demonstrated that a mixed chimerism state enables donor specific organ transplantations of skin, vascularized cardiac and islet allografts across major and minor histocompatibility barriers, without employing a long-term immunosuppression treatment (Ildstad and Sachs 1984; Colson et al. 1995a,b; Li et al. 1995). When multiple histocompatibility barriers were present, levels of donor chimerism were significantly lower (37.5

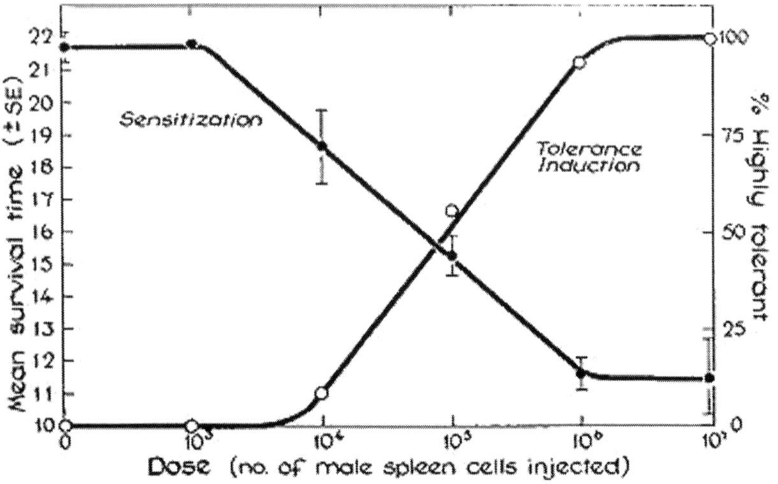

Fig. 12.1 Enhancement of rejection and tolerance induction in mature and newly born mice respectively, via chimerism. Dose response curves for sensitization and tolerance induction to male-specific antigen (MSA). The data for sensitization represent the mean survival time±SE of male tail skin grafts on presensitized syngeneic female hosts. In the 10^4–10^6 cell dose range, each succeeding data point is significantly different (p<0.02 to p<0.001) from the data point preceding it (t test). The data for tolerance induction represent the percentage of hosts highly tolerant (>60 days) of male tail skin grafts when grafted 2 months following neonatal inoculation of the indicated number of male spleen cells. An analysis of the distribution of acceptances and rejections revealed at successive data points from 10^3 to 10^6 cells injected (p<0.01 to P<0.001). (From Weissman IL (1973) Transplantation 15:265–269)

± 12.5% vs. 94.6 ± 3.8%; Colson et al. 1995a). Foster et al. (1998, 2001) showed that mixed chimeric recipients with a level of donor chimerism above 60% did not develop a rejection reaction to transplanted limb for more than 100 days, whereas animals with a chimerism level below 20% developed clinical and histological signs of moderate rejection (Foster et al. 1998, 2001). Gandy and Weissman (1998) and Shizuru et al. (2000), demonstrated that purified allogeneic HSC grafts induce tolerance to donor solid organ. Donor chimerism resulting from the engraftment of transplanted HSC permitted a long-term survival of donor-matched neonatal heart grafts while third party heart grafts were rejected. A comprehensive review of data collected from successful long-term organ transplants in human (liver, intestine, kidney; Starzl 2004) demonstrated that relatively high levels of donor leukocyte chimerism is detected in the recipient circulation during the first two months following transplantation. After that period, the frequency of donor cells in the host circulation was reduced (undetectable by flow cytometry). Long-term persistence of low levels of donor leukocytes in blood and tissue of transplant recipients was detected in any successful long-term transplant as long as 29 years after transplantation (30 studied patients; Starzl et al. 1992, 1993a,b). The implanted organ allografts become a chimeric mixture of donor and recipient cells (Kashiwagi et al. 1969; Iwaki et al. 1991; Murase et al. 1991; Starzl et al. 1993b; Randhawa et al. 1994).

The long-term persistence of multilineage microchimerism implies that hematolymphopoietic precursors and stem cells are part of the passenger leukocyte population of organ grafts (Starzl et al. 1996). Based on these observations, a hypothesis for "chimerism induced tolerance" was generated, suggesting an applicable approach for transplantation medicine (Starzl 2004). In 1961, Weissman and Lustgraaf suggested that tolerance induction is connected to routing, timing frequency, and level of chimerism. However, more than 40 years later, very little information has been accumulated regarding the dynamic of chimeric cells in mammalian hosts, patterns of host/alien cells competition, possible cellular parasitism, niche occupation and immunoregulatory mechanisms for routing, timing and frequencies of chimeric cells.

During the last 50 years, an evolutionary similar model phenomenon, the natural chimerism of Botryllid ascidians, has been the focus of many studies. Some of the major questions that were addressed included the genetic control of self-nonself recognition, the nature of cells that are involved with chimerism, the dynamics of chimeric cells and their ability to induce tolerance or rejection in the host.

12.2 Chimerism in Ascidians

12.2.1 Natural Long-Term Chimerism in Ascidians

Natural chimerism in marine colonial organisms has been observed for over a century (Giard 1872; Bancroft 1903), and has been demonstrated experimentally in many sessile modular animals including sponges, bryzoans, cnidarians and ascidians (Buss 1982; Grosberg 1988; Hughes 1989; Ilan and Loya 1990; Chadwick-Furman and Rinkevich 1994; Frank et al. 1997; Barki et al. 2002).

About 3–10% of the colonial ascidian *Botryllus schlosseri* populations and one-third of *Diplosoma listerianum* colonies in the field are chimeras, revealing mixtures of several different genetic entities (Stoner and Weissman 1996; Sommerfeldt and Bishop 1999; Ben-Shlomo et al. 2001, 2006, 2008; Paz et al. 2003).

12.2.2 Genetically Controlled Allorecognition System in Colonial Organisms Restrict the Development of Chimeras to Kin

Under natural conditions, genetically controlled allorecognition systems in colonial marine organisms restrict the creation of chimeras to self and kin (either from day 1 as detected in botryllus and hydractinia; Oka and Watanabe 1975; Sabbadin 1962; Muller 1964; Scofield et al. 1982; Buss et al. 1984; Grosberg 1988; Buss and Grosberg 1990; Cadavid et al. 2004, or following maturation of the immune system as detected in hard and soft corals; Frank et al. 1997; Barki et al. 2002). Botryllid ascidians are the primary group of marine organisms studied in the context of natural chimerism. When conspecific botryllid colonies contact each other, several

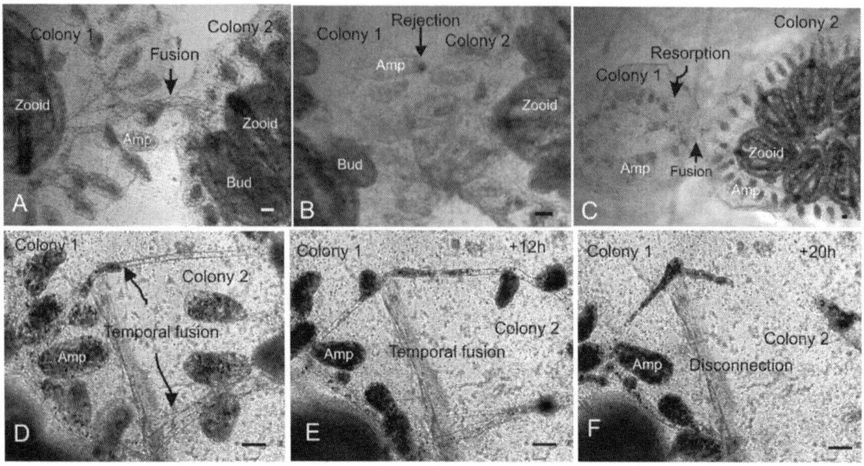

Fig. 12.2 Fusion and rejection reactions in *B. schlosseri*. Several morphological outcomes of fusion and non-fusion reactions as observed in *B. schlosseri* (**A**) Vasculature fusion between two compatible colonies; (**B**) Rejection, and development of dark points of rejection at the contact area between two non compatible colonies; (**C**) Resorption of one of the partner zooids and buds 360 h following fusion; (**D–F**) Temporary fusion, followed by vasculature disconnection 20 h following fusion. Amp-ampullae; h-hour; *scale bar*, 100 μm

morphological and cellular allogeneic processes and reactions develop, including: fusion, rejection, indifference (no reaction) and a temporary fusion followed by disconnection (Fig. 12.2; Bancroft 1903; Oka and Watanabe 1957; Sabbadin 1962, 1982; Mukai 1967; Tanaka and Watanabe 1973; Scofield et al. 1982; Saito and Watanabe 1982, 1984; Scofield et al. 1983; Hirose et al. 1988; 1990; Boyd et al. 1990; Sabbadin et al. 1992: Rinkevich and Weissman 1992a,b; Rinkevich et al. 1994a,b, 1998; Saito et al. 1994; Chadwick-Furman and Weissman 1995; Ballarin et al. 1995, 1998, 2002; Cima et al. 2004, 2006; Rinkevich 2005).

Oka and Watanabe investigated fusibility between *Botryllus primigenus* colonies and showed that fusion and non fusion in *Botryllus primigenus* are genetically controlled (Oka and Watanabe 1957, 1960, 1967; Oka 1970; Watanabe and Taneda 1982). A series of fusibility experiments between different generations in defined groups (Fig. 12.3) led Oka and Watanabe to hypothesize that fusibility (colony specificity) in *Botryllus primigenous* is controlled by a single polymorphic gene locus with multiple, codominantly expressed alleles, whereas sharing one of these alleles is required for fusion. They also concluded that under natural conditions, colonies are heterozygous for the gene controlling fusibility (Mukai and Watanabe 1957). Weissman et al. (1990) named this gene Fu/HC for fusibility/histocompatibility.

Crosses between compatible *Botryllus primigenus* colonies yielded two interesting outcomes: (1) a cross between compatible colonies that share one allele in their fusibility locus resulted in colonies homozygous for the fusibility gene whereas heterozygotes identical to the maternal colony genotypes were missing;

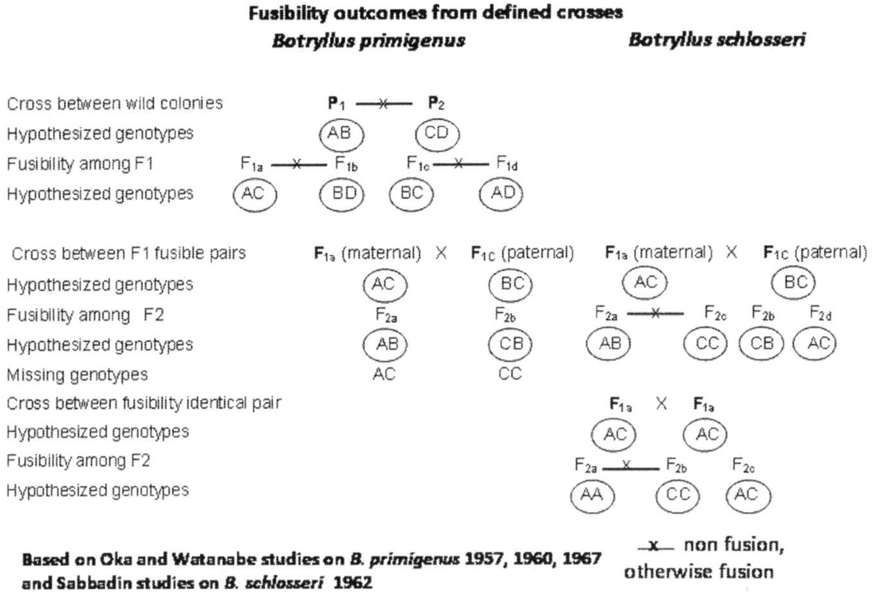

Fig. 12.3 A few examples of fusibility experiments between defined crosses that led Oka and Watanabe to propose their hypothesis of the genetic control of fusibility in *B. primigenus*, and supporting evidences that came from crosses that Sabbadin (1962) conducted in *B. schlosseri*. In *B. primigenus* wild colonies usually do not fuse. Progeny always fuse with their parental or maternal colony. Crossing two mutually fusible colonies yield fusible progeny. In progeny from a cross between compatible colonies that share one allele in their fusibility locus, the homozygous for the fusibility gene and heterozygote identical to the maternal colony genotypes are missing. Homologous heterozygote colonies, which share both alleles in the gene controlling fusibility are sterile. In *B. schlosseri*, the progeny from a cross between compatible colonies segregate into four groups. *B. schlosseri* homologous heterozygote colonies, sharing both alleles in the gene controlling fusibility are producible and reveal two mutually incompatible groups of homozygotes and one group of heterozygotes compatible with all the progeny. —X— non fusion, otherwise fusion

(2) Homologous heterozygote colonies, sharing both FuHC alleles were sterile (Fig. 12.3). Based on these results, Oka and Watanabe suggested that the fusibility gene is involved also in self-sterility, i.e., fertilization occurs only when the alleles controlling spermatozoa/ova fusions, are different (Oka and Watanabe 1957, 1960, 1967; Oka 1970). The sexual incompatibility phenomenon in *Botryllus primigenus* prevented Oka and Watanabe from directly testing and proving their fusibility genetic control hypothesis.

Sabbadin (1962) examined fusibility frequencies of pairs of *Botryllus schlosseri* and further confirmed Oka and Watanabe's genetic control hypothesis on fusibility. He found that in *Botryllus schlosseri*, the progeny from a cross between compatible colonies segregate, as predicted by Oka and Watanabe, into three groups, two mutually incompatible groups of homozygotes and one group of heterozygotes compatible with all the progeny (Sabbadin 1962; Fig. 12.3). Further support came from a study by Scofield et al. (1982) who showed that genetic control of fusibility in Monterey *Botryllus* newly hatched oozooids is similar to the one described by Sabbadin for older colonies of *Botryllus schlosseri* and is due to a high degree of polymorphism at a single locus.

Inspired by the discovery of a genetic basis for allograft acceptance or rejection in the *Botryllus* system, Burnet (1971) suggested studying this organism as a model system for the evolution of self-recognition. He claimed that: "although self recognition in ascidians is not analogous to the immunological processes of vertebrates, it presents a primitive type of "self and not self" recognition from which adaptive immunity may have evolved" (Burnet 1971).

An extensive study of the botryllid ascidians self-nonself recognition system in the last 25 years revealed that while the colonial ascidian Fu/HC and the mammalian MHC share phenomenological features, including polymorphism and specificity, their molecular structure is different. Similar to the MHC, the Fu/HC is highly polymorphic (Karakashian and Milkman 1967; Scofield et al. 1982; 1983; Grosberg and Quinn 1986; Grosberg 1987; Rinkevich et al. 1995; De Tomaso et al. 2005; Ben-Shlomo et al. 2008). However, structural homologies were not found (Rinkevich and Weissman 1992b; Fagan and Weissman 1997; Pancer et al. 1993, 1996a,b,c, 1997; Muller et al. 1994; Khalturin et al. 2003; De Tomaso et al. 2005; Nyholm et al. 2006). Isolation and characterization of a candidate Fu/HC gene (cFu/HC) in *Botryllus schlosseri* revealed a gene encoding a type I transmembrane protein with multiple extracellular immunoglobulin domains that include extracellular epidermal growth factor (EGF) repeats. The cFu/HC is expressed in tissues associated with the natural fusion/non-fusion reaction; whereas its polymorphism predicts the outcome of these reactions (De Tomaso et al. 2005). Another gene, *Fester*, a putative polymorphic receptor involved in histocompatibility and tightly linked to the Fu/HC, was also isolated and characterized (Nyholm et al. 2006). The products of *Fester* are type I transmembrane proteins with a unique structure, consisting of three terminal domains and a short consensus repeat (SCR) domain, which is found also in complement receptor proteins, but exhibits no meaningful homology to known proteins (Nyholm et al. 2006). An analysis of Fester mRNA species shows a massive polymorphism due to differential splicing of the ectodomain exons (Nyholm et al. 2006). siRNA treatment targeting Fester blocked fusion of compatible pairs, and also blocked rejection reactions in incompatible colonies. In one case, incompatible colonies fused when blocking Fester with monoclonal antibodies. These observations suggest that products of *Fester* encode the Fu/HC-encoded-ligand receptors determining histocompatibility (Nyholm et al. 2006). The *Botryllus* cFu/HC is not homologous to any molecules of the vertebrate MHC-based histocompatibility system. Neither the Fu/HC nor Fester appears in the draft genomes of the solitary ascidians *Ciona intestinalis* and *Ciona savignyi* (De Tomaso et al. 2005; Nyholm et al. 2006). Moreover, the whole *Botryllus* Fu/HC locus does not have a syntenic region in the *Ciona* genome or in the genomes of vertebrates (De Tomaso et al. 2005).

Urochordates share many components of innate immunity with vertebrates (Lauzon et al. 1992, 1993; Ballarin et al. 1998; Davidson and Swalla 2002; Dehal et al. 2002; Nonaka and Miyazawa 2002; Azumi et al. 2003; Khalturin et al. 2003;

Litman et al. 2007; Oren et al. 2007; Rinkevich et al. 2007). Allogeneic colonies can tolerate each other and create a chimera when sharing at least 50% of Fu/HC genes. This is consistent with an immune system, like the NK system in vertebrates, wherein recognition of self prevents an immune reaction. In mice and humans, the adaptive immune T cell system rejects grafts even if one of two alleles are shared, as T cells make immune reaction against non-self allele gene products. However, genes involved in adaptive immunity, which include the polymorphic MHC class I and II glycoproteins that present internal peptides to T cells, the clonally expressed T-cell receptors (TCRs), immunoglobulins (Igs) and the recombination activating genes (RAG1, RAG2), have not been identified in urochordates (Klein 1989; Laird et al. 2000; Dehal et al. 2002; Kaufman. 2002; Azumi et al. 2003; Khalturin et al. 2003; De Tomaso et al. 2005; Litman et al. 2005).

12.2.3 Stem Cell Parasitism in the Chimera Entity

When discussing chimerism in *Botryllus*, Burnet (1971) hypothesized the emergence of intraspecific parasitism: "When free swimming marine protovertebrates ancestral first appeared, it is conceivable that one of the early results was the emergence of new ecological niche, survival by parasitism on ones own kind" (Burnet 1971). Eight years later, in a remarkable set of experiments, Sabbadin and Zaniolo (1979) demonstrated this kind of parasitism in *Botryllus schlosseri*. They created short term chimeras between defined genotypes (characterized by different pigment markers), separated the chimeras partners 2–7 days following fusion, crossed them with recessive genotypes on the pigment allele which could not affect the offspring pigmentation, and typed the offspring genotypes. In 31/53 tested chimeras, offspring populations matched both the chimeric host and donor genotypes. In few cases, the entire offspring populations matched solely the donor genotype, demonstrating potential dominancy at the germ cell parasitism level. Donor germ cells remained in the hosts for long periods, and chimeric colonies continued to produce the donor offspring for, at least the first 105 days following separation. By using molecular markers (microsatellites), Pancer et al. (1995), Stoner and Weissman (1996) and Stoner et al. (1999) confirmed these results and further showed that in a chimera, the blood, soma and germ cells, demonstrated the combine genotypes of both chimeric partners. Moreover, in most cases the circulating pluripotent cells of one partner parasitized either the soma or the germ line of the other partner and replaced the whole mass of gonads or the soma (bud/zooid) of several individuals in the host colony (termed germline or somatic cell parasitism; G/SCP). In a few cases, a complete takeover of donor genotype occurred and the whole mass of gonads in the chimeric colony expressed solely the donor's genotype (Sabbadin and Zaniolo 1979; Pancer et al. 1995; Stoner and Weissman 1996; Stoner et al. 1999). G/SCP was expressed and observed in all of the three types of chimerism (stable, resorption and separation; Sabbadin and Zaniolo 1979; Pancer et al. 1995; Stoner and Weissman 1996; Stoner et al. 1999). Under invariant environmental conditions, both germline and somatic cell parasitism followed repeatable hierarchies of "winner strains" and "loser strains" (Stoner et al. 1999). However, breeding experiments proved that only the hierarchical position of germ cell parasitism is sexually inherited (Stoner et al. 1999). The hierarchy of somatic parasitism in *Botryllus* chimeras is a plastic trait, as variations in the environmental conditions (such as seawater temperature) can be reversed; the winner – loser hierarchy at the somatic parasitism level (Rinkevich and Yankelevich 2004). This supported observations in chimeric fungi made by Buss (1982) that suggested a greater store of genetic variability and a wider range of physiological qualities in a chimera compared to individual organisms. As a result, chimerism may provide the chimeric entity with a wider range of responses than each individual, and the chimera can take advantage of its joint genotypes by adjusting its somatic constituent to variant condition in a changeable environment.

The long-term ability of cells from one genotype to replace the germline and somatic cells of the host, led to hypothesize that cell parasitism in the chimeras is mediated by stem cells (Sabbadin and Zaniolo 1979; Rinkevich and Weissman 1987; Pancer et al. 1995; Stoner and Weissman 1996; Stoner et al. 1999). By transplanting a single cell, which expresses high enzymatic activity of aldehyde dehydrogenase and a set of serial engraftment assays, Laird et al. (2005) revealed that adult stem cells are responsible for a stable long-term chimerism in *Botryllus schlosseri*. Yet, the location of these cells remained unknown (Laird et al. 2005). Recently, by using *in vivo* cell labeling, cell engraftment and time lapse imaging Voskoboynik et al. 2008 have demonstrated that the anterior ventral region of the sub-endostyle sinus (termed endostyle niche) harbors and exports somatic stem cells in *Botryllus* colonies. As few as 5–20 engrafted cells transplanted from the donor endostyle niche sufficed to generate a somatic chimerism in compatible hosts; however, no germline chimerism was demonstrated. The induction of somatic chimerism demonstrates a remarkable stemness capacity of the cells in the endostyle niche (Voskoboynik et al. 2008).

12.2.4 Induction of Tolerance and Intolerance by Chimerism

Chimerism may alter the fusibility trait in botryllid ascidians. Studies showed that chimeras might fuse with colonies, which they rejected before or reject colonies, with which they fused in former assays, assuming that the chimera entities harbor the challenged genotype's FuHC alleles (Mukai 1967; Taneda 1985; Sabbadin and Astorri 1988). Sabbadin and Astori performed experiments that demonstrated successive rejections, fusions or simultaneous fusions and rejections of tested genotypes by chimeric colonies, over a period of several months. They hypothesized that variations recorded in chimeric entities reflect dynamics and interactions within chimera's cellular components and competitive interactions of the different genomes within the chimera (Sabbadin and Astorri 1988). A few years later, Stoner and Weissman analyzed chimeric colonies' constituents and found that chimeras exhibited either a uniform pattern in which both genotypes were distributed throughout the entire colony or a sectorial pattern in which different genotypes were detected in some areas but not in others (Stoner and Weissman 1996). Colonies, which showed both rejection and fusion concurrently, probably have a sectorial distribution pattern of chimeric cells. The above studies suggest that temporal and spatial dynamics of chimeric cells have an important role in the induction of tolerance or intolerance state towards other colonies and demonstrate that chimerism can alter genetically inherited fusibility patterns in *Botryllus*.

12.3 Chimerism Insights from Mammals and Ascidians

Chimerism is a widespread phenomenon (Table 12.1). Although the cell type of origin which mediates chimerism is unknown, it has hypothesized that chimerism is mediated by stem cells (Bianchi et al. 1996; Cao et al. 2004; O'Donoghue and Fisk 2004; Laird et al. 2005; Bianchi 2007). Stem cells are undifferentiated cells defined by their ability at the single cell level to both self-renew and differentiate to produce non-self renewing oligolineage progenitors which in turn give rise to progeny that are more restricted in their differentiating potential (Weissman 2000).

Long-term persistence of few allogeneic cells in hosts were demonstrated in murine and human systems. Following stress allogeneic cells may also home to sites of damaged tissues and increase in numbers in diseased or injured organs (reviewed in Bianchi 2007). The retained allogeneic cells appear to have a multilineage capacity and express differentiation markers (Bianchi 2007). The types of stem

Table 12.1 Chimeras features across different systems

Feature Chimeric system	Stem cells mediated phenomena	Histocompatibility Compatibility	Tolerance state between chimera's partners	Chimerism is expressed in:
Colonial ascidians	Have been demonstrated	$\geq$50%	+	Germ and soma
Pregnancy (mammals)	Have been suggested	50%	Usually +	Soma
Fraternal twins (mammals)	Have been Suggested	75% possibility $\geq$50%	Depends on chimerism level	Soma
Tetragametic (mammals)	fertilization of 2 ova by 2 spermatozoa	100%	+	Germ and soma
Callitrichid primates (mammals)	Have been suggested	High chances $\geq$50% Defect MHC sys	+	Germ and soma
Transplantation (mammals)	Have been demonstrated	Easier to induce and maintain across full or partial MHC matches	Can be initiated and maintained via chimerism	Soma

cells (e.g. HSCs, MSCs) which mediate this phenomenon are unknown. As the medical implications of natural recruitment of stem cells for organ repair are important for future stem cell therapy, it is crucial to characterize and study the phenomenon of chimerism in depth. Another common feature between chimerism in mammals and ascidian is that it occurs naturally in individuals that share at least part of their histocompatibility genes (mainly kin; Table 12.1). Levels of allogeneic cells within hosts in mammals are usually restricted to a very small amount (Mc). In marine colonial organisms, on the other hand, chimerism associated with high frequencies of allogeneic cells and with germline parasitism. Burnet (1971, 1973) hypothesized that vertebrate adaptive immunity developed to limit or prevent uncontrolled germline parasitism. He further hypothesized that compared to the innate immune system, adaptive immunity has better capabilities to limit and control chimerism. Indeed, immunodeficient mammalian individuals show higher levels of chimerism (systemic *Lupus erythematosus*, systemic sclerosis and SCID syndrom; Pollack et al. 1982; Nelson 1996; Artlett et al. 1998; Nelson et al. 1998; Evans et al. 1999; Adams and Nelson 2004; Cowan et al. 2008; Neven et al. 2008), and it is easier to induce chimerism when the immune system is suppressed (e.g. Shizuru et al. 1996, 2000; Cao et al. 2004). Moreover, except in a tetragametic state which represents a special chimerism status, the only other mammalian system expressing germline chimerism carries a defective MHC system (marmosets; Hampton 1973). Germ cell parasitism holds a major threat to survival, where "super competitive" stem cell genotypes could homogenize an entire population (Burnet 1971; Buss 1982). Buss (1982) proposed that colonial organisms might develop a self/non-self histocompatibility system to limit inter-individual germ cell parasitism to histocompatible kin only (Buss 1982). Despite the severe cost of germ cell parasitism, chimerism between kin was developed, conserved and maintained along phyla. Several potential ecological benefits of the chimeric state in invertebrates were proposed (e.g. genetic variability, heterosis phenomenon, developmental synergism; size increase e.g. Buss 1982; Rinkevich 1993; Rinkevich and Shapira 1999: Rinkevich and Yankelevich 2004). Mammalian chimerism is strongly connected with induction and maintenance of tolerance (Owen 1945; Owen et al. 1954; Billingham et al. 1953; Weissman 1973; Ildstad and Sachs 1984; Colson et al. 1995a; Gandy and Weissman 1998; Shizuru et al. 2000; Andrassy et al. 2003; Starzl 2004).

Liegeois (1983) and Starzl (2004) hypothesized that chimerism has a role in tolerance induction and maintenance during pregnancy and following organ transplantation. This hypothesis, although never been tested directly, is consistent with recent findings which reveal how chimerism induces antigen-specific fetal tolerance to maternal antigens during human pregnancy (Mold et al. 2008). This study highlights the potential important role of cell chimerism for induction of tolerance to allogeneic tissues and provide and evolutionary explanation for this widespread phenomenon. It can also explain why chimera formation is restricted to closely related individuals, as in mammalian fetal/maternal microchimerism, where about 50% of histocompatibility genes are shared (Table 12.1). Moreover, in ascidians, when somatic cell parasitism induces development of a foreign entity within the host colony, chimerism serves as a state that enables "virtual pregnancy" (development of a semiallogeneic individual within the host colony). This suggests a possible role of chimerism in the evolution of viviparity.

Acknowledgments This study was supported by a grant from US-Israel Binational Science Foundation (2003-010). The authors thank Amir Voskobynik for critical reviewing the manuscript Karla J. Palmeri for editing and Stefano Tiozzo for translating Sabbadins' paper.

References

Adams KM, Nelson JL (2004) Microchimerism: An investigative frontier in autoimmunity and transplantation. JAMA 291:1127–1131

Anasetti C, Rybka W, Sullivan KM, Banaji M, Slichter SJ (1989) Graft-v-host disease is associated with autoimmune-like thrombocytopenia. Blood 73:1054–1058

Anderson D, Billingham RE, Lampkin GH, Medawar PB (1951) The use of skin grafting to distinguish between monozygotic and dizygotic twins in cattle. Heredity 5:379

Anderson DJ, Gage FH, Weissman IL (2001) Can stem cells cross lineage boundaries? Nature Medicine 7:393–395

Andrassy J, Kusaka S, Jankowska-Gan E, Torrealba JR, Haynes LD, Marthaler BR, Tam RC, Illigens BM, Anosova N, Benichou G, Burlingham WJ (2003) Tolerance to noninherited maternal MHC antigens in mice. J Immunol 171:5554–5561

Armitage JO (1994) Bone marrow transplantation. N Engl J Med 330:827–838

Artlett CM, Smith JB, Jimenez SA (1998) Identification of fetal DNA and cells in skin lesions from women with systemic sclerosis. N Engl J Med 338:1186–1191

Ash RC, Casper JT, Chitambar CR, Hansen R, Bunin N, Truitt RL, Lawton C, Murray K, Hunter J, Baxter-Lowe LA (1990) Successful allogeneic transplantation of T-cell-depleted bone marrow from closely HLA-matched unrelated donors. N Engl J Med 322:485–494

Azumi K, Takahashi H, Miki Y, Fujie M, Usami T, Ishikawa H, Kitayama A, Satou Y, Ueno N, Satoh N (2003) Construction of a cDNA microarray derived from the ascidian ciona intestinalis. Zoolog Sci 20:1223–1229

Bach FH, Albertini RJ, Joo P, Anderson JL, Bortin MM (1968) Bone-marrow transplantation in a patient with the wiskott-aldrich syndrome. Lancet 2:1364–1366

Ballarin L, Cima F, Floreani M, Sabbadin A (2002) Oxidative stress induces cytotoxicity during rejection reaction in the compound ascidian *Botryllus schlosseri*. Comp Biochem Physiol C Toxicol Pharmacol 133:411–418

Ballarin L, Cima F, Sabbadin A (1995) Morula cells and histocompatibility in the colonial ascidian *Botryllus schlosseri*. Zool Sci 12:757–764

Ballarin L, Cima F, Sabbadin A (1998) Phenoloxidase and cytotoxicity in the compound ascidian *Botryllus schlosseri*. Dev Comp Immunol 22:479–492

Bancroft FW (1903) Variation and fusion of colonies in compound ascidians. Proc Calif Acad Sci 3:137–186

Barki Y, Gateño D, Graur D, Rinkevich B (2002) Soft coral natural chimerism: A window in ontogeny allows the creation of entities, compounding incongruous parts. Marine Ecology Progress Series 231:91–99

Barnes DW, Loutit JF, Micklem HS (1962) "Secondary disease" of radiation chimeras: A syndrome due to lymphoid aplasia. Ann NY Acad Sci 99:374–385

Barnes RD, Holliday J (1970) The morphological identity of maternal cells in newborn mice. Blood 36:480–489

Ben-Shlomo R, Douek J, Rinkevich B (2001) Heterozygote deficiency and chimerism in remote populations of a colonial ascidian from New Zealand. Mar Ecol Prog Series 209:109–117

Ben-Shlomo R, Motro U, Paz G, Rinkevich B (2008) Pattern of sett lement and natural chimerism in the colonial urochordate *Botryllus schlosseri*. Genetica 132:51–58

Ben-Shlomo R, Paz G, Rinkevich B (2006) Postglacial-period and recent invasions shape the population genetics of botryllid ascidians along European Atlantic coasts. Ecosystems 9:1118–1127

Bianchi DW (2007) Fetomaternal cell trafficking: a story that begins with prenatal diagnosis and may end with stem cell therapy. J Pediatr Surg 42:12–18

Bianchi DW, Simpson JL, Jackson LG, Elias S, Holzgreve W, Evans MI, Dukes KA, Sullivan LM, Klinger KW, Bischoff FZ, Hahn S, Johnson KL, Lewis D, Wapner RJ, de la Cruz F (2002) Fetal gender and aneuploidy detection using fetal cells in maternal blood: Analysis of NIFTY I data. National institute of child health and development fetal cell isolation study. Prenat Diagn 22:609–615

Bianchi DW, Williams JM, Sullivan LM, Hanson FW, Klinger KW, Shuber AP (1997) PCR quantitation of fetal cells in maternal blood in normal and aneuploid pregnancies. Am J Hum Genet 61:822–829

Bianchi DW, Zickwolf GK, Weil GJ, Sylvester S, DeMaria MA (1996) Male fetal progenitor cells persist in maternal blood for as long as 27 years postpartum. Proc Natl Acad Sci USA 93: 705–708

Billingham RE, Brent L (1956) Acquired tolerance of foreign cells in newborn animals. Proc R Soc Lond B Biol Sci 146:78–90

Billingham RE, Brent L, Brown JB, Medawar PB (1959) Time of onset and duration of transplantation immunity. Transplant Bull 6:410–414

Billingham RE, Brent L, Medawar PB (1953) Actively acquired tolerance of foreign cells. Nature 172:603–606

Billingham RE, Lampkin GH, Medawar PB, Williams HL (1952) Tolerance to homografts, twin diagnosis and the freemartin condition in cattle. Heredity 6:200–212

Billingham RE, Silvers WK (1959) Inbred animals and tissue transplantation immunity, with an index of some inbred strains other than mice. Transplant Bull 6:399–406

Billingham RE, Silvers WK (1960) Studies on tolerance of the Y chromosome antigen in mice. J Immunol 85:14–26

Boklage CE (1990) Survival probability of human conceptions from fertilization to term. Int J Fertil 35:75, 79–80, 81–94

Boklage CE (2006) Embryogenesis of chimeras, twins and anterior midline asymmetries. Hum Reprod 21:579–591

Bordenave GR, Babinet C (1984) Tetraparental rabbits chimeric for their lymphoid system–I. allotype expression. Mol Immunol 21:353–361

Boyd HC, Weissman IL, Saito Y (1990) Morphologic and genetic verification that Monterey Botryllus and Woods Hole Botryllus are the same species. Biol Bull 178:239–250

Brent L (1997) The discovery of immunologic tolerance. Hum Immunol 52:75–81

Bromilow IM, Duguid J (1989) The liverpool chimaera. Vox Sang 57:147–149

Burnet FM (1971) “Self-recognition” in colonial marine forms and flowering plants in relation to the evolution of immunity. Nature 232:230–235

Burnet FM (1973) Multiple polymorphism in relation to histocompatibility antigens. Nature 245:359–361

Burnet FM, Fenner F (1949) The production of antibodies, 2nd edn. Macmillan, Melbourne, London

Buss LW (1982) Somatic cell parasitism and the evolution of somatic tissue compatibility. Proc Natl Acad Sci USA 79:5337–5341

Buss LW, Grosberg RK (1990) Morphogenetic basis for phenotypic differences in Hydroid competitive behavior. Nature 343:63–66

Buss LW, Mcfadden CS, Keene DR (1984) Biology of Hydractiniid hydroids. 2. histocompatibility effector system competitive mechanisms mediated by nematocyst discharge. Biol Bull 167:139–158

Cadavid LF, Powell AE, Nicotra ML, Moreno M, Buss LW (2004) An invertebrate histocompatibility complex. Genetics 167:357–365

Cao YA, Wagers AJ, Beilhack A, Dusich J, Bachmann MH, Negrin RS, Weissman IL, Contag CH (2004) Shifting foci of hematopoiesis during reconstitution from single stem cells. Proc Natl Acad Sci USA 101:221–226

Chadwick-Furman NE, Rinkevich B (1994) A complex allorecognition system in a reef-building coral: Delayed response, reversals and nontransitive hierarchies. Coral Reefs 13: 57–63

Chadwick-Furman NE, Weissman IL (1995) Life history plasticity in chimaeras of the colonial ascidian *Botryllus schlosseri*. P Roy Soc B-Biol Sci 262:157–162

Cima F, Sabbadin A, Ballarin L (2004) Cellular aspects of allorecognition in the compound ascidian *Botryllus schlosseri*. Dev Comp Immunol 28:881–889

Cima F, Sabbadin A, Zaniolo G, Ballarin L (2006) Colony specificity and chemotaxis in the compound ascidian *Botryllus schlosseri*. Comp Biochem Physiol A Mol Integr Physiol 145:376–382

Claas FH, Gijbels Y, van der Velden-de Munck J, van Rood JJ (1988) Induction of B cell unresponsiveness to noninherited maternal HLA antigens during fetal life. Science 241:1815–1817

Claas FH, Gijbels Y, von Veen A, de Waal LP, D'Amaro J, Persijn GG, van Rood JJ (1989) Selection of cross-match negative HLA-A and/or -B mismatched donors for highly sensitized patients. Transplant Proc 21:665–666

Colson YL, Shinde Patil VR, Ildstad ST (2007) Facilitating cells: Novel promoters of stem cell alloengraftment and donor-specific transplantation tolerance in the absence of GVHD. Crit Rev Oncol Hematol 61:26–43

Colson YL, Wren SM, Schuchert MJ, Patrene KD, Johnson PC, Boggs SS, Ildstad ST (1995a) A nonlethal conditioning approach to achieve durable multilineage mixed chimerism and tolerance across major, minor, and hematopoietic histocompatibility barriers. J Immunol 155: 4179–188

Colson YL, Zadach K, Nalesnik M, Ildstad ST (1995b) Mixed allogeneic chimerism in the rat. donor-specific transplantation tolerance without chronic rejection for primarily vascularized cardiac allografts. Transplantation 60:971–980

Conley ME, Nowell PC, Henle G, Douglas SD (1984) XX T cells and XY B cells in two patients with severe combined immune deficiency. Clin Immunol Immunopathol 31:87–95

Cowan MJ, Neven B, Cavazanna-Calvo M, Fischer A, Puck J (2008) Hematopoietic Stem Cell Transplantation for Severe Combined Immunodeficiency Diseases. Biol Blood Marrow TR 14:73–76

Christner PJ, Jimenez SA (2004) Animal models of systemic sclerosis: Insights into systemic sclerosis pathogenesis and potential therapeutic approaches. Curr Opin Rheumatol 16:746–752

Davidson B, Swalla BJ (2002) A molecular analysis of ascidian metamorphosis reveals activation of an innate immune response. Development 129:4739–4751

De Tomaso AW, Nyholm SV, Palmeri KJ, Ishizuka KJ, Ludington WB, Mitchel K, Weissman IL (2005) Isolation and characterization of a protochordate histocompatibility locus. Nature 438:454–459

Dehal P, Satou Y, Campbell RK, Chapman J, Degnan B, De Tomaso A, Davidson B, Di Gregorio A, Gelpke M, Goodstein DM et al. (2002) The draft genome of ciona intestinalis: Insights into chordate and vertebrate origins. Science 298:2157–2167

Dunsford I, Bowley CC, Hutchison AM, Thompson JS, Sanger R, Race RR (1953) A human blood-group chimera. Br Med J 2:80–81

Emery D, McCullagh P (1980a) Immunological reactivity between chimeric cattle twins. I. homograft reaction. Transplantation 29:4–9

Emery D, McCullagh P (1980b) Immunological reactivity between chimeric cattle twins. II. normal lymphocyte transfer. Transplantation 29:10–16

Evans PC, Lambert N, Maloney S, Furst DE, Moore JM, Nelson JL (1999) Long-term fetal microchimerism in peripheral blood mononuclear cell subsets in healthy women and women with scleroderma. Blood 93:2033–2037

Fagan MB, Weissman IL (1997) HSP70 genes and historecognition in botryllus schlosseri: Implications for MHC evolution. Hereditas 127:25–35

Foster RD, Ascher NL, McCalmont TH, Neipp M, Anthony JP, Mathes SJ (2001) Mixed allogeneic chimerism as a reliable model for composite tissue allograft tolerance induction across major and minor histocompatibility barriers. Transplantation 72:791–797

Foster RD, Fan L, Neipp M, Kaufman C, McCalmont T, Ascher N, Ildstad S, Anthony JP, Niepp M (1998) Donor-specific tolerance induction in composite tissue allografts. Am J Surg 176: 418–421

Frank U, Oren U, Loya Y, Rinkevich B (1997) Alloimmune maturation in the coral Stylophora pistillata is achieved through three distinctive stages, four months post metamorphosis. Proc Royal Soc Lond B 264:99–104

Flomenberg N, Dupont B, O'Reilly RJ, Hayward A, Pollack MS (1983) The use of T cell culture techniques to establish the presence of an intrauterine-derived maternal T cell graft in a patient with severe combined immunodeficiency (SCID). Transplantation 36:733–735

Gaillard MC, Ouvre E, Liegeois A, Lewin D (1978) The concentration of fetal cells in maternal haematopoietic organs during pregnancy. An experimental study in mice (author's transl). J Gynecol Obstet Biol Reprod (Paris) 7:1043–1050

Gale RP, Reisner Y (1986) Graft rejection and graft-versus-host disease: Mirror images. Lancet 1:1468–1470

Gandy KL, Domen J, Aguila H, Weissman IL (1999) CD8+TCR+ and CD8+TCR- cells in whole bone marrow facilitate the engraftment of hematopoietic stem cells across allogeneic barriers. Immunity 11:579–590

Gandy KL, Weissman IL (1998) Tolerance of allogeneic heart grafts in mice simultaneously reconstituted with purified allogeneic hematopoietic stem cells. Transplantation 65:295–304

Gatti RA, Meuwissen HJ, Allen HD, Hong R, Good RA (1968) Immunological reconstitution of sex-linked lymphopenic immunological deficiency. Lancet 2:1366–1369

Gengozian N, Batson JS, Eide P (1964) Hematologic and cytogenetic evidence for hematopoietic chimerism in the marmoset, *Tamarinus nigricollis*. Cytogenetics 10:384–393

Giard A (1872) Recherches sur les ascidies composees ou synascidies. Arch Zool Exp Gen 1: 501–704

Green AJ, Barton DE, Jeniks P, Pearson J, Yates JRW (1994) Chimerism shown by cytogenetics and DNA polymorphism analysis. J Med Genet 31:816–817

Grosberg RK (1987) Limited dispersal and proximity dependent mating success in *Botryllus schlosseri*. Evolution 41:372–384

Grosberg RK (1988) The evolution of allorecognition specificity in colonial invertebrates. Quarterly Rev Biol 63:377–412

Grosberg RK, Quinn JF (1986) The genetic control and consequences of kin recognition by the larvae of a colonial marine invertebrate. Nature 322:456–459

Haig D (1999) What is a marmoset? Am J Primatol 49:285–296

Hamada H, Arinami T, Kubo T, Hamaguchi H, Iwasaki H (1993) Fetal nucleated cells in maternal peripheral blood: Frequency and relationship to gestational age. Hum Genet 91:427–432

Hampton SH (1973) Germ cell chimerism in male marmosets. Am J Phys Anthropol 38: 265–268

Hasek M (1953a) Parabiosis of birds during their embryonic development. Chekhoslovatskaia Biol 2:29–31

Hasek M (1953b) Vegetative hybrization of animals by means of junction of the blood circulation during embryonic development. Chekhoslovatskaia Biol 2:267–282

Hasek M, Haskova V (1958) A contribution to the significance of individual antigenic specificity in homografting. Transplan Bull 5:69

Herzenberg LA, Bianchi DW, Schroder J, Cann HM, Iverson GM (1979) Fetal cells in the blood of pregnant women: Detection and enrichment by fluorescence-activated cell sorting. Proc Natl Acad Sci USA 76:1453–1455

Heslop B, Nisbet NW (1962) Split tolerance in inbred rats. Transplan Bull 29:471–473

Hirose E, Saito Y, Watanabe H (1988) Anew type of the manifestation of the colony specificity in the compound ascidian, *Botrylloides violaceus* Oka. Biol Bull 175:240–245

Hirose E, Saito Y, Watanabe H (1990) Allogeneic rejection induced by the cut surface contact in the compound ascidian, *Botrylloides simodensis*. Invertebr Dev Reprod 17:159–164

Hughes RN (1989) A functional biology of clonal animals. Chapman & Hall, London and New York

Hume DM, Jackson BT, Zukoski CF, Lee HM, Kauffman HM, Egdahl RH (1960) The homotransplantation of kidneys and of fetal liver and spleen after total body irradiation. Ann Surg 152:354–373

Ichinohe T, Teshima T, Matsuoka K, Maruya E, Saji H (2005) Fetal-maternal microchimerism: Impact on hematopoietic stem cell transplantation. Curr Opin Immunol 17:546–552

Ildstad ST, Sachs DH (1984) Reconstitution with syngeneic plus allogeneic or xenogeneic bone marrow leads to specific acceptance of allografts or xenografts. Nature 307: 168–170

Ilan M, Loya Y (1990) Ontogenetic variation in sponge histocompatibility responses. Biol Bull 179:279–286

Imaizumi M, Pritsker A, Unger P, Davies TF (2002) Intrthyroidal fetal microchimerism in pregnancy and postpartum. Endocrinology 143:247–253

Iwaki Y, Starzl TE, Yagihashi A, Taniwaki S, Abu-Elmagd K, Tzakis A, Fung J, Todo S (1991) Replacement of donor lymphoid tissue in small-bowel transplants. Lancet 337: 818–819

Johnson KL, Samura O, Nelson JL, McDonnell MdWM, Bianchi DW (2002) Significant fetal cell microchimerism in a nontransfused woman with hepatitis C: Evidence of long-term survival and expansion. Hepatology 36:1295–1297

Kaplan J, Land S (2005) Influence of maternal-fetal histocompatibility and MHC zygosity on maternal microchimerism. J Immunol 174:7123–7128

Karakashian S, Milkman R (1967) Colony fusion compatibility types in *Botryllus schlosseri*. Biol Bull 133:473

Kashiwagi N, Porter KA, Penn I, Brettschneider L, Starzl TE (1969) Studies of homograft sex and of gamma globulin phenotypes after orthotopic homotransplantation of the human liver. Surg Forum 20:374–376

Kaufman CL, Colson YL, Wren SM, Watkins S, Simmons RL, Ildstad ST (1994) Phenotypic characterization of a novel bone marrow-derived cell that facilitates engraftment of allogeneic bone marrow stem cells. Blood 84:2436–2446

Kaufman J (2002) The origins of the adaptive immune system: Whatever next? Nat Immunol 3:1124–1125

Khalturin K, Becker M, Rinkevich B, Bosch TC (2003) Urochordates and the origin of natural killer cells: Identification of a CD94/NKR-P1-related receptor in blood cells of botryllus. Proc Natl Acad Sci USA 100:622–627

Khosrotehrani K, Johnson KL, Guegan S, Stroh H, Bianchi DW (2005) Natural history of fetal cell microchimerism during and following murine pregnancy. J Reprod Immunol 66:1–12

Klein J (1989) Are invertebrates capable of anticipatory immune responses? Scand J Immunol 29:499–505

Klintschar M, Immel UD, Kehlen A, Schwaiger P, Mustafa T, Mannweiler S, Regauer S, Kleiber M, Hoang-Vu C (2006) Fetal microchimerism in Hashimito's thyroiditid: A quantitative approach. Eur J Endocrinol 154:237–241

Koopmans M, Kremer Hovinga IC, Baelde HJ, Fernandes RJ, de Heer E, Bruijn JA, Bajema IM (2005) Chimerism in kidneys, livers and hearts of normal women: Implications for transplantation studies. Am J Transplant 5:1495–1502

Korngold B, Sprent J (1978) Lethal graft-versus-host disease after bone marrow transplantation across minor histocompatibility barriers in mice. Prevention by removing mature T cells from marrow. J Exp Med 148:1687–1698

Laird DJ, De Tomaso AW, Cooper MD, Weissman IL (2000) 50 million years of chordate evolution: Seeking the origins of adaptive immunity. Proc Natl Acad Sci USA 97:6924–6926

Laird DJ, De Tomaso AW, Weissman IL (2005) Stem cells are units of natural selection in a colonial ascidian. Cell 123:1351–1360

Lambert NC, Evans PC, Hashizumi TL, Maloney S, Gooley T, Furst DE, Nelson JL (2000) Cutting edge: Persistent fetal microchimerism in T lymphocytes is associated with HLA-DQA10501: Implications in autoimmunity. J Immunol 164:5545–5548

Lauzon RJ, Ishizuka KJ, Weissman IL (1992) A cyclical, developmentally-regulated death phenomenon in a colonial urochordate. Dev Dyn 194:71–83

Lauzon RJ, Patton CW, Weissman IL (1993) A morphological and immunohistochemical study of programmed cell death in *Botryllus schlosseri* (tunicata, ascidiacea). Cell Tissue Res 272: 115–127

Le Deist F, Raffoux C, Griscelli C, Fischer A (1987) Graft vs graft reaction resulting in the elimination of maternal cells in a SCID patient with maternofetal GVHd after an HLA identical bone marrow transplantation. J Immunol 138:423–427

Li H, Colson YL, Ildstad ST (1995) Mixed allogeneic chimerism achieved by lethal and nonlethal conditioning approaches induces donor-specific tolerance to simultaneous islet allografts. Transplantation 60:523–529

Liegeois A (1983) In Edelman P, Sureau C (eds) Immunologie de le reproduction humanie, Edition Boz. Paris, pp 99–100

Liegeois A, Gaillard MC, Ouvre E (1981) Micro-chimerism in pregnant mice. Transplant Proceed 13:1250–1252

Lillie FR (1916) The theory of the free-martin. Science 43:611–613

Litman GW, Cannon JP, Dishaw LJ (2005) Reconstructing immune phylogeny: New perspectives. Nat Rev Immunol 5:866–879

Litman GW, Dishaw LJ, Cannon JP, Haire RN, Rast JP (2007) Alternative mechanisms of immune receptor diversity. Curr Opin Immunol 19:526–534

Lo ES, Lo YM, Hjelm NM, Thilaganathan B (1998) Transfer of nucleated maternal cells into fetal circulation during the second trimester of pregnancy. Br J Haematol 100:605–606

Lo YM, Lo ES, Watson N, Noakes L, Sargent IL, Thilaganathan B, Wainscoat JS (1996) Two-way cell traffic between mother and fetus: Biologic and clinical implications. Blood 88:4390–4395

Lo YM, Patel P, Wainscoat JS, Sampietro M, Gillmer MD, Fleming KA (1989) Prenatal sex determination by DNA amplification from maternal peripheral blood. Lancet 2:1363–1365

Loubiere LS, Lambert NC, Flinn LJ, Erickson TD, Yan Z, Guthrie KA, Vickers KT, Nelson JL (2006) Maternal microchimerism in healthy adults in lymphocytes, monocyte/macrophages and NK cells. Lab Invest 86:1185–1192

Lubaroff DM, Silvers WK (1973) Importance of chimerism in maintaining tolerance of skin allografts in mice. J Immunol 111:65–71

Lustgraaf EC, Fuson RB, Eichwald EJ (1960) Sex tolerance and split tolerance. Transplant Bull 26:145–150

Main JM, Prehn RT (1955) Successful skin homografts after the administration of high dosage X radiation and homologous bone marrow. J Natl Cancer Inst 15:1023–1029

Mannick JA, Lochte LH Jr, Ashley CA, Thomas ED, Ferrebee JW (1959) A functioning kidney homotransplant in the dog. Surgery 46:821–828

Maraninchi D, Gluckman E, Blaise D, Guyotat D, Rio B, Pico JL, Leblond V, Michallet M, Dreyfus F, Ifrah N (1987) Impact of T-cell depletion on outcome of allogeneic bone-marrow transplantation for standard-risk leukaemias. Lancet 2:175–178

Marleau AM, Greenwood JD, Wei Q, Singh B, Croy BA (2003) Chimerism of murine fetal bone marrow by maternal cells occurs in late gestation and persists into adulthood. Lab Invest 83:673–681

Martin PJ, Hansen JA, Storb R, Thomas ED (1987) Human marrow transplantation: An immunological perspective. Adv Immunol 40:379–438

Mathe G, Amiel JL, Schwarzenberg L, Cattan A, Schneider M (1963) Haematopoietic chimera in man after allogenic (homologous) bone-marrow transplantation. (control of the secondary syndrome. specific tolerance due to the chimerism). Br Med J 2:1633–1635

Mayr WR, Pausch V, Schnedl W (1979) Human chimaera detectable only by investigation of her progeny. Nature 277:210–211

Mintz B, Palm J (1969) Gene control of hematopoiesis. I. erythrocyte mosaicism and permanent immunological tolerance in allophenic mice. J Exp Med 129:1013–1027

Mitchison NA (1959) Blood transfusion in fowl: An example of immunological tolerance requiring the persistence of antigen. Bull Soc Int Chir 18:257–273

Mitchison NA (1961) Immunological tolerance and immunological paralysis. Br Med Bull 17: 102–106

Mitsuyasu RT, Champlin RE, Gale RP, Ho WG, Lenarsky C, Winston D, Selch M, Elashoff R, Giorgi JV, Wells J (1986) Treatment of donor bone marrow with monoclonal anti-T-cell antibody and complement for the prevention of graft-versus-host disease. A prospective, randomized, double-blind trial. Ann Intern Med 105:20–26

Mold JE, Michaelsson J, Burt TD, Muench MO, Beckerman KP, Busch MP, Lee TH, Nixon DF, McCune JM (2008) Maternal alloantigens promote the development of tolerogenic fetal regulatory T Cells in Utero. Science 322:1562–1565

Monaco AP (2003) Chimerism in organ transplantation: Conflicting experiments and clinical observations. Transplantation 75:13S–16S

Mukai H (1967) Experimental alteration of fusibility in compound ascidians. Sci Rep Tokyo Kyoiku Daigaku 13B:51–73

Mukai H, Watanabe H (1975) Distribution of fusion incompatibility types in natural populations of the compound ascidian *Botryllus primigenus*. Proc Jpn Acad 51:44–47

Muller AW (1964) Experimentelle Untersuchungen uber Stockentwicklung, pelypendifferenzierung und Sexualchimaren bei Hydractinia echinata. Wilhem Roux' Arch 155:182–268

Muller WE, Pancer Z, Rinkevich B (1994) Molecular cloning and localization of a novel serine protease from the colonial tunicate *Botryllus schlosseri*. Mol Mar Biol Biotechnol 3:70–77

Murase N, Demetris AJ, Woo J, Furuya T, Nalesnik M, Tanabe M, Todo S, Starzl TE (1991) Lymphocyte traffic and graft-versus-host disease after fully allogeneic small bowel transplantation. Transplant Proc 23:3246–3247

Murray JE, Merrill JP, Dammin GJ, B. DJ, Jr, Walter CW, Brooke MS, Wilson RE (1960) Study on transplantation immunity after total body irradiation: Clinical and experimental investigation. Surgery 48:272–284

Nelson J (2001) HLA relationships of pregnancy, microchimerism and autoimmune disease. J Reprod Immunol 52:77–84

Nelson JL (1996) Maternal-fetal immunology and autoimmune disease: Is some autoimmune disease auto-alloimmune or allo-autoimmune? Arthritis Rheum 39:191–194

Nelson JL, Furst DE, Maloney S, Gooley T, Evans PC, Smith A, Bean MA, Ober C, Bianchi DW (1998) Microchimerism and HLA-compatible relationships of pregnancy in scleroderma. Lancet 351:559–562

Neven B, Cavazanna-Calvo M, Fischer A (2008) Late Immunologic and clinical outcomes for children with SCID. Biol Blood Marrow TR 14:76–78

Nonaka M, Miyazawa S (2002) Evolution of the initiating enzymes of the complement system. Genome Biol 3:REVIEWS1001

Nyholm SV, Passegue E, Ludington WB, Voskoboynik A, Mitchel K, Weissman IL, De Tomaso AW (2006) Fester, A candidate allorecognition receptor from a primitive chordate. Immunity 25:163–173

O'Donoghue K, Chan J, de la Fuente J, Kennea N, Sandison A, Anderson JR, Roberts IA, Fisk NM (2004) Microchimerism in female bone marrow and bone decades after fetal mesenchymal stem-cell trafficking in pregnancy. Lancet 364:179–182

O'Donoghue K, Fisk NM (2004) Fetal stem cells. Best Pract Res Clin Obstet Gynaecol 18:853–875

Oka H (1970) Colony specificity in compound ascidians. The genetic control of fusibility. In: Yukawa H (ed) Profiles of Japanese Science and Scientists. Kodansha, Tokyo

Oka H, Watanabe H (1957) Colony specificity in compound ascidians as tested by fusion experiments (a preliminary report). Proc Jpn Acad 33:657–659

Oka H, Watanabe H (1960) Problems of colony specificity in compound ascidians. Bull Mar Biol Stat Asamushi 10:153–155

Oka H, Watanabe H (1967) Problems of colony specificity, with special reference to the fusibility of ascidians. Kagaku 37:307 313

Oosterwijk JC, Mesker WE, Ouwerkerk-van Velzen MC, Knepfle CF, Wiesmeijer KC, Beverstock GC, van Ommen GJ, Kanhai HH, Tanke HJ (1998) Fetal cell detection in maternal blood: A study in 236 samples using erythroblast morphology, DAB and HbF staining, and FISH analysis. Cytometry 32:178–185

Opelz G, Sengar DP, Mickey MR, Terasaki PI (1973) Effect of blood transfusions on subsequent kidney transplants. Transplant Proc 5:253–259

O'Reilly RJ (1983) Allogeneic bone marrow transplantation: Current status and future directions. Blood 62:941–946

Oren M, Douek J, Fishelson Z, Rinkevich B (2007) Identification of immune-relevant genes in histoincompatible rejecting colonies of the tunicate *Botryllus schlosseri*. Dev Comp Immunol 31:889–902

Owen RD (1945) Immunogenetic consequences of vascular anastomoses between bovin twins. Science 102:400–401

Owen RD, Wood HR, Foord AG, Sturgeon P, Baldwin LG (1954) Evidence for actively tolerance to Rh antigens. Proc Natl Acad Sci USA 40:420–424

Pancer Z, Diehl-Seifert B, Rinkevich B, Muller WE (1997) A novel tunicate (botryllus schlosseri) putative C-type lectin features an immunoglobulin domain. DNA Cell Biol 16:801–806

Pancer Z, Gershon H, Rinkevich B (1993) cDNA cloning of a putative protochordate FK506-binding protein. Biochem Biophys Res Commun 197:973–977

Pancer Z, Gershon H, Rinkevich B (1995) coexistence and possible parasitism of somatic and germ cell lines in chimeras of the colonial urochordate *Botryllus schlosseri*. Biol Bull 189:106–112

Pancer Z, Cooper EL, Muller WE (1996a) A tunicate (*Botryllus schlosseri*) cDNA reveals similarity to vertebrate antigen receptors. Immunogenetics 45:69–72

Pancer Z, Leuck J, Rinkevich B, Steffen R, Muller I, Muller WE (1996b) Molecular cloning and sequence analysis of two cDNAs coding for putative anionic trypsinogens from the colonial urochordate *Botryllus schlosseri* (ascidiacea). Mol Mar Biol Biotechnol 5:326–333

Pancer Z, Scheffer U, Muller I, Muller WE (1996c) Cloning of sponge (geodia cydonium) and tunicate (botryllus schlosseri) proteasome subunit epsilon (PRCE): Implications about the vertebrate MHC-encoded homologue LMP7 (PRCC). Biochem Biophys Res Commun 228:406–410

Paz G, Douek J, Mo CQ, Goren M, Rinkevich B (2003) Genetic structure of *Botryllus schlosseri* (Tunicata) populations from the Mediterranean coast of Israel. Mar Ecol Prog Ser 250:163–174

Persijn GG, Cohen B, Lansbergen Q, van Rood JJ (1979) Retrospective and prospective studies on the effect of blood transfusions in renal transplantation in the netherlands. Transplantation 28:396–401

Piotrowski P, Croy BA (1996) Maternal cells are widely distributed in murine fetuses in utero. Biol Reprod 54:1103–1110

Pollack SM, Kirkpatrick D, Kapoor N, Dupont B, O'Reilly RJ (1982) Identification by HLA typing of intrauterine-derived maternal T cells in four patients with severe combined immunodeficiency. N Eng J Med 307:662–666

Prabhune KA, Gorantla VS, Maldonado C, Perez-Abadia G, Barker JH, Ildstad ST (2000) Mixed allogeneic chimerism and tolerance to composite tissue allografts. Microsurgery 20: 441–447

Randhawa PS, Starzl T, Ramos HC, Nalesnik MA, Demetris J (1994) Allografts surviving for 26 to 29 years following living-related kidney transplantation: Analysis by light microscopy, in situ hybridization for the Y chromosome, and anti-HLA antibodies. Am J Kidney Dis 24:72–77

Repas-Humpe LM, Humpe A, Lynen R, Glock B, Dauber EM, Simson G, Mayr WR, Kohler M, Eber S (1999) A dispermic chimerism in a 2-year-old caucasian boy. Ann Hematol 78:431–434

Rinkevich B (1993) Immunological resorption in *Botryllus schlosseri* (Tunicata) chimeras is characterized by multilevel hierarchical organization of histocompatibility alleles. A speculative endeavor. Biol Bull 184:342–345

Rinkevich B (2005) Rejection patterns in botryllid ascidian immunity: The first tier of allorecognition. Canad J Zool 83:101–121

Rinkevich B, Lilker-Levav T, Goren M (1994b) Allorecognition/Xenorecognition responses in Botrylloides subpopulations from the Mediterranean coast of Israel. Jour Exp Zool 270: 302–313

Rinkevich B, Porat R, Goren M (1995) Allorecognition elements on Urochordate histocompatibility locus indicate unprecedented extensive polymorphism. Proc Biol Sci 259:319–324
Rinkevich B, Shapira M (1999) Multi-partner urochordate shimeras outperform two-partner chimerical entities. Oikos 87:315–320
Rinkevich B, Tartakover S, Gershon H (1998) Contribution of morula cells to allogeneic responses in the colonial urochordate *Botryllus schlosseri*. Mar Biol 131:227–236
Rinkevich B, Weissman IL (1987) Chimeras in colonial invertebrates- a synergidtic symbiosis or somatic and germ cell parasitism. Symbiosis 4:117–134
Rinkevich B, Weissman IL (1992a) Allogeneic resorption in colonial protochordates: Consequences of nonself recognition. Dev Comp Immunol 16:275–286
Rinkevich B, Weissman IL (1992b) Incidents of rejection and indifference in Fu/HC incompatible protochordate colonies. J Exp Zool 263:105–111
Rinkevich B, Weissman IL, Shapira M (1994a) Alloimmune hierarchies and stress-induced reversals in the resorption of chimeric protochordate colonies. Proc R Soc Lon Ser B Biol Sci 258: 215–220
Rinkevich B, Yankelevich I (2004) Environmental split between germ cell parasitism and somatic cell synergism in chimeras of a colonial urochordate. J Exp Biol 207:3531–3536
Rinkevich Y, Douek J, Haber O, Rinkevich B, Reshef R (2007) Urochordate whole body regeneration inaugurates a diverse innate immune signaling profile. Dev Biol 312:131–146
Sabbadin A (1962) Le basi geneticha della capacita di fusion fra colonies in *Botryllus schlosseri* (Ascidiacea). Rend Accad Naz Lincei Ser 32:1031–1035
Sabbadin A (1982) Formal genetics of ascidians. Am Zool 22:765
Sabbadin A, Astorri C (1988) Chimeras and histocompatibility in the colonial ascidian botryllus schlosseri. Dev Comp Immunol 12:737–747
Sabbadin A, Zaniolo G (1979) Sexual differentiation and germ cell transfer in the colonial ascidian *Botryllus schlosseri*. J Exp Zool 207:289–304
Sabbadin A, Zaniolo G, Ballarin L (1992) Genetic and cytological aspects of histocompatibility in ascidians. Boll Zool 59:167–173
Saito Y, Hirose E, Watanabe H (1994) Allorecognition in compound ascidians. Int J Dev Biol 38:237–247
Saito Y, Watanabe H (1982) Colony specificity in the compound ascidian *Botryllus scalaris*. Proc Jpn Acad 58B:105–108
Saito Y, Watanabe H (1984) Partial biochemical characterization of humoral factors involved in the nonfusion reaction of a botryllid ascidian, *Botrylloides simodensis*. Zool Sci 1:229–235
Scofield VL, Schlumpberger JM, Nagashima LS (1983) Morphology and genetics of rejection reactions between oozooids from the tunicate *Botryllus schlosseri*. Biol Bull 165:733–744
Scofield VL, Schlumpberger JM, West LA, Weissman IL (1982) Protochordate allorecognition is controlled by a MHC-like gene system. Nature 295:499–502
Sharabi Y, Abraham VS, Sykes M, Sachs DH (1992) Mixed allogeneic chimeras prepared by a non-myeloablative regimen: Requirement for chimerism to maintain tolerance. Bone Marrow Transplant 9:191–197
Shizuru JA, Jerabek L, Edwards CT, Weissman IL (1996) Transplantation of purified hematopoietic stem cells: Requirements for overcoming the barriers of allogeneic engraftment. Biol Blood Marrow Transplant 2:3–14
Shizuru JA, Weissman IL, Kernoff R, Masek M, Scheffold YC (2000) Purified hematopoietic stem cell grafts induce tolerance to alloantigens and can mediate positive and negative T cell selection. Proc Natl Acad Sci USA 97:9555–9560
Simonsen M (1956) Actively acquired tolerance to heterologous antigens; immunological consequences of transfusing chicken embryos with human blood. Acta Pathol Microbiol Scand 39:21–34
Smith LG, Weissman IL, Heimfeld S (1991) Clonal analysis of hematopoietic stem cell differentiation in vivo. Proc Natl Acad Sci USA 88:2788–2792

Sommerfeldt AD, Bishop JDD (1999) Random amplified polymorphic DNA (RAPD) analysis reveals extensive natural chimerism in a marine protochordate. Mol Ecol 8: 885–890

Srivatsa B, Srivatsa S, Johnson KL, Samura O, Lee SL, Bianchi DW (2001) Microchimerism of presumed fetal origin in thyroid specimens from women: A case-control study. Lancet 358:2034–2038

Starzl TE (2004) Chimerism and tolerance in transplantation. Proc Natl Acad Sci USA 101(Suppl 2):14607–14614

Starzl TE, Demetris AJ, Murase N, Ildstad S, Ricordi C, Trucco M (1992) Cell migration, chimerism, and graft acceptance. Lancet 339:1579–1582

Starzl TE, Demetris AJ, Trucco M, Murase N, Ricordi C, Ildstad S, Ramos H, Todo S, Tzakis A, Fung JJ (1993a) Cell migration and chimerism after whole-organ transplantation: The basis of graft acceptance. Hepatology 17:1127–1152

Starzl TE, Demetris AJ, Trucco M, Zeevi A, Ramos H, Terasaki P, Rudert WA, Kocova M, Ricordi C, Ildstad S (1993b) Chimerism and donor-specific nonreactivity 27 to 29 years after kidney allotransplantation. Transplantation 55:1272–1277

Starzl TE, Murase N, Thomson A, Demetris AJ (1996) Liver transplants contribute to their own success. Nat Med 2:163–165

Stone WH, Cragle RG, Johnson DF, Bacon JA, Bendel S, Korda N (1971) Long-term observations of skin grafts between chimeric cattle twins. Transplantation 12:421–428

Stone WH, Cragle RG, Swanson EW, Brown DG (1965) Skin grafts: Delayed rejection between pairs of cattle twins showing erythrocyte chimerism. Science 148:1335–1336

Stoner DS, Rinkevich B, Weissman IL (1999) Heritable germ and somatic cell lineage competitions in chimeric colonial protochordates. Proc Natl Acad Sci USA 96:9148–9153

Stoner DS, Weissman IL (1996) Somatic and germ cell parasitism in a colonial ascidian: Possible role for a highly polymorphic allorecognition system. Proc Natl Acad Sci USA 93: 15254–15259

Strain L, Dean JC, Hamilton MP, Bonthron DT (1998) A true hermaphrodite chimera resulting from embryo amalgamation after in vitro fertilization. N Engl J Med 338:166–169

Strain L, Warner JP, Johnston T, Bonthron DT (1995) A human parthenogenetic chimaera. Nat Genet 11:164–169

Suda T, Thompson LF, O'Connor RD, Bastian JF (1984) Phenotype and function of engrafted maternal T cells in patients with severe combined immunodeficiency. J Immunol 133: 2513–2517

Sumantri C, Boediono A, Ooe M, Saha S, Suzuki T (1997) Fertility of sperm from a tetraparental chimeric bull. Anim Reprod Sci 46:35–45

Sybert VP (1994) Hypomelanosis of ito: A description, not a diagnosis. J Invest Dermatol 103:141S–143S

Taieb A, Clavijo-Alvarez JA, Hamad GG, Lee WP (2007) Immunologic approaches to composite tissue allograft. J Hand Surg [Am] 32:1072–1085

Tan XW, Liao H, Sun L, Okabe M, Xiao ZC, Dawe GS (2005) Fetal microchimerism in the maternal mouse brain: A novel population of fetal progenitor or stem cells able to cross the blood-brain barrier? Stem Cells 23:1443–1452

Tanaka K, Watanabe H (1973) Allogeneic inhibition in a compound ascidian, *Botryllus primigenus* oka. I. Processes and features of nonfusion reaction. Cell Immunol 7:410–426

Taneda Y (1985) Simultaneous occurrence of fusion and nonfusion reaction in two colonies in contact of the compound ascidian *Botryllus priminegus*. Dev Comp Immunol 9:371

Tiercy JM (2002) Molecular basis of HLA polymorphism: Implications in clinical transplantation. Transpl Immunol 9:173–180

Tippett P (1983) Blood group chimeras. A review. Vox Sang 44:333–359

Trentin JJ (1956) Mortality and skin transplantability in x-irradiated mice receiving isologous, homologous or heterologous bone marrow. Proc Soc Exp Biol Med 92:688–693

Troeger C, Zhong XY, Burgemeister R, Minderer S, Tercanli S, Holzgreve W, Hahn S (1999) Approximately half of the erythroblasts in maternal blood are of fetal origin. Mol Hum Reprod 5:1162–1165
Tucker EM, Dain AR, Moor RM (1978) Sex chromosome chimaerism and the transmission of blood group genes by tetraparental rams. J Reprod Fertil 54:77–83
Tuffrey M, Bishun NP, Barnes RD (1969) Porosity of the mouse placenta to maternal cells. Nature 221:1029–1030
Tyan ML (1973) Modification of severe graft-versus-host disease with antisera to the (theta) antigen or to whole serum. Transplantation 15:601–604
Uehara S, Nata M, Nagae M, Sagisaka K, Okamura K, Yajima A (1995) Molecular biologic analyses of tetragametic chimerism in a true hermaphrodite with 46,XX/46,XY. Fertil Steril 63:189–192
Uphoff DE (1957) Genetic factors influencing irradiation protection by bone marrow. I. the F1 hybrid effect. J Natl Cancer Inst 19:123–130
Vaidya S, Mamiok R, Daeschner CW 3rd, Williams J, Ruth J, Goldblum RM (1991) Supression of graft versus host reaction in severe combined immunodeficiency with maternal-fetal T cell engraftment. Am J Pediatr Hematol Oncol 13:172–175
Vallera DA, Blazar BR (1989) T cell depletion for graft-versus-host-disease prophylaxis. A perspective on engraftment in mice and humans. Transplantation 47:751–60
van den Boogaardt DEM, van Rood JJ, Roelen DL, Claas FH (2006) The influence of inherited and noninherited parental antigens on outcome after transplantation. Transpl Int 19:360–371
van Dijk BA, Boomsma DI, de Man AJ (1996) Blood group chimerism in human multiple births is not rare. Am J Med Genet 61:264–268
Verp MS, Harrison HH, Ober C, Oliveri D, Amarose AP, Lindgren V, Talerman A (1992) Chimerism as the etiology of a 46,XX/46,XY fertile true hermaphrodite. Fertil Steril 57: 346–349
Voskoboynik A, Soen Y, Rinkevich Y, Ueno H, Rosner A, Reshef R, Ishizuka KJ, Palmeri KJ, Moiseeva E, Rinkevich B, Weissman IL (2008) Identification of the endostyle as a stem cell niche in a colonial chordate. Cell Stem Cell 3:456–464
Wagers AJ, Weissman IL (2004) Plasticity of adult stem cells. Cell 116:639–648
Walknowska J, Conte FA, Grumbach MM (1969) Practical and theoretical implications of fetal-maternal lymphocyte transfer. Lancet 1:1119–1122
Wang Y, Iwatani H, Ito T, Horimoto N, Yamato M, Matsui I, Imai E, Hori M (2004) Fetal cells in mother rats contribute to the remodeling of liver and kidney after injury. Biochem Biophys Res Commun 325:961–967
Watanabe H, Taneda Y (1982) Self or non-self recognition in compound ascidians. Am Zool 22:775–782
Watkins DI, Chen ZW, Hughes AL, Hodi FS, Letvin NL (1990) Genetically distinct cell populations in naturally occurring bone marrow-chimeric primates express similar MHC class I gene products. J Immunol 144:3726–3735
Watkins WM, Yates AD, Greenwell P, Bird GW, Gibson M, Roy TC, Wingham J, Loeb W (1981) A human dispermic chimaera first suspected from analyses of the blood group gene-specified glycosyltransferases. J Immunogenet 8:113–128
Weissman IL (1963) Tissue transplantation Specific "Conditioning" of the host. In: Eichwald EJ (ed) Advances in biological and medical physics, vol 9. Academic press, INC, New York
Weissman IL (1973) Transfer of tolerance. Transplantation 15:265–269
Weissman IL (2000) Stem Cells units of development, units of regeneration and units in evolution. Cell 1:157–168
Weissman IL (2002) The road ended up at stem cells. Immunol Rev 185:159–174
Weissman IL, Lustgraaf EC (1961) Antibody formation and repressor systems. Transplant Bull 28:134–135
Weissman IL, Saito Y, Rinkevich B (1990) Allorecognition histocompatibility in a protochordate species: Is the relationship to MHC somatic [error;semantic] or structural?. Immunol Rev 113:227–241

Wolinsky H (2007) A mythical beast. increased attention highlights the hidden wonders of chimeras. EMBO Rep 8:212–214

Yu N, Kruskall MS, Yunis JJ, Knoll JH, Uhl L, Alosco S, Ohashi M, Clavijo O, Husain Z, Yunis EJ (2002) Disputed maternity leading to identification of tetragametic chimerism. N Engl J Med 346:1545–1552

Zhou L, Yoshimura Y, Huang Y, Suzuki R, Yokoyama M, Okabe M, Shimamura M (2000) Two independent pathways of maternal cell transmission to offspring: Through placenta during pregnancy and by breast-feeding after birth. Immunology 101:570–580

Chapter 13
Effect of Bacterial Infection on Stem Cell Pattern in Porifera

Werner E.G. Müller, Márcio Reis Custódio, Matthias Wiens, Carla Zilberberg, Amélie Châtel, Isabel M. Müller, and Heinz C. Schröder

Abstract Multicellular organisms derived from one common ancestor, the Urmetazoa. The only living fossils, which can testify about the earliest evolutionary processes in Metazoa on the molecular level are sponges (phylum: Porifera). The present study outlines that stem cells may play essential roles in cellular specialization, embryogenesis and sponge Bauplan formation, using the demosponge *Suberites domuncula* as a model. Data indicate that the archaeocytes represent, besides the germ/embryonic cells, totipotent stem cells. First marker genes have been identified, which are expressed in totipotent stem cells and in cells from gemmules. Furthermore, genes have been described that are characteristic for the three main cell lineages in sponges; they all originate from archaeocytes and are involved in the differentiation of skeletal cells, epithelial cells and contractile cells. Finally it is shown that after exposure to the endotoxin LPS (lipopolysaccharide) a differential gene expression occurs, leading to an upregulation of the gene encoding perforin and a concomitant down-regulation of noggin, a stem cell marker. In parallel with this process an increased phosphorylation of the mitogen-activating protein kinase p38 occurs. This modification of the p38 kinase has been quantified with a novel ELISA assay. Our data suggest that in response to bacterial infection the number of stem cells in sponges decreases.

Keywords Sponges · Cell culture · Infection · Stem cells · Primmorph · Strees-response genes · Gene expression · Sponge genes.

13.1 Introduction

Based on molecular biological and cell biological data it is now established that the approximately 30 phyla [including also the Porifera (sponges)], integrated in

W.E.G. Müller (✉)
Institut für Physiologische Chemie, Abteilung Angewandte Molekularbiologie, Universität, Duesbergweg 6, D-55099 Mainz, Germany
e-mail: wmueller@uni-mainz.de

B. Rinkevich, V. Matranga (eds.), *Stem Cells in Marine Organisms*,
DOI 10.1007/978-90-481-2767-2_13, © Springer Science+Business Media B.V. 2009

the multicellular animal kingdom of Metazoa, share one common ancestor, the Urmetazoa (Müller et al. 1997; Müller 2001). Other than metazoan organisms, colonies of animal cells, such as choanoflagellates, display no division of labor. In Metazoa the cells have been differentiated to perform a series of functions, like digestion, sensation, contraction or secretion (Grunz 2004). The determination of metazoan cells to a distinct fate occurs via an alteration in the pattern of gene activity/expression, which thus allows a specification into distinct roles and functions. Specialized cells are the basis for pattern formation, a process during which a spatial and temporal pattern of cell activities is organized within the well-ordered organism (Nüsslein-Volhard and Wieschaus 1980). Metazoa are grouped into (i) the morphologically more diverse bilaterians, which themselves are further subdivided into Protostomia, including Ecdysozoa and Lophotrochozoa, and Deuterostomia and (ii) the non-bilaterian metazoans (phyla Porifera, Placozoa, Coelenterata and Ctenophora; Brusca and Brusca 1990).

During ontogeny a body plan is established, which also defines the main axes of the multicellular animal. From Porifera to Arthropoda (Protostomia) and Vertebrata (Deuterostomia) the level of complexity of the axes (antero-posterior ends [oral-aboral polarity], dorsal-ventral sides) increases (Müller 2005). The two phyla Porifera and Coelenterata possess only two epithelial layers, the ecto-epithelium surrounding the body and the endo-epithelium (Bergquist 1978; Garrone 1978), which encloses the digestive cavity(ies); they are termed diploblastic animals. The two groups are defined by only one apical-basal polarity (Wiens et al. 2003). The triploblastic, bilaterian animals contain in addition to these external layers a third, middle, mesodermal cell layer which originates usually from the endoderm. Body pattern formation can be studied during embryogenesis or during differentiation of embryonic cells, e.g. in three dimensional cell cultures. These morphogenetic processes are based on and controlled by differential spatial and temporal expression of genes that initiate or maintain a large number of signaling pathways (Galliot and Miller 2000).

Segmentation (a process during which very similar functional units along a body axis are formed almost simultaneously [in insects]) or somite formation (sequential formation of units along a body axis [in vertebrates]) are features of triploblastic organisms. The formation of a body axis is controlled by characteristic sets of genes; e.g. in insects the parasegments are delimited by the function of the pair-rule genes and the subsequent segments by the segment polarity genes (Wolpert 1998). Ancestors of these genes exist in diploblasts, Coelenterata and Porifera, as single molecules (Hoshiyama et al. 1998; Wiens et al. 2003).

During the past few years it was elaborated that the Porifera (sponges), as the phylogenetically oldest metazoan phylum, possess already the basic structural and functional elements required for the construction of a body plan (Perovic et al. 2003). They evolved approximately 800 million years ago (Müller and Schäcke 1996). Sponges are grouped according to the inorganic composition of their skeleton (spicules) into the classes of Demospongiae and Hexactinellida, which possess hydrated, amorphous, noncrystalline silica spicules, and the class of Calcarea, whose skeletal spicules are formed from calcium carbonate (Simpson 1984).

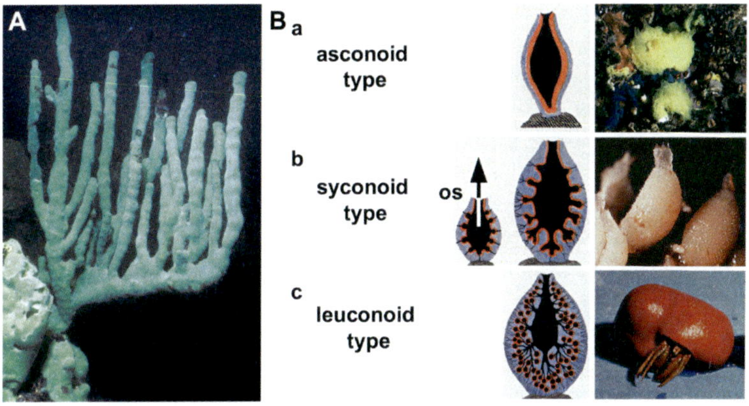

Fig. 13.1 Body plan organization in the phylum Porifera. (**A**) Sponges, here *Lubomirskia baicalensis*, conduct water from the external milieu through their inhalant openings at the surface to the exhalent apertures, the oscules. This highly complex aquiferous system is formed from differentiated somatic cells, which interact in a tuned manner through cell-cell/cell-matrix interaction. The water flow through a specimen can be visualized with fluorescein; the pulsatory extrusion of the water current can be seen (not shown). (**B**) The three different organization/body plan patterns of sponges are shown. The choanoderm layer is in *red* and the pinacoderm layer in *blue*. (a) In the asconoid type, the sack-like organization shows an unfolded choanoderm. (b) In the syconoid type the folding of the choanoderm layer results in the formation of choanocyte chambers, which are connected with the incurrent canals. After passing the chambers the excurrent canals lead the water current to the oscule (os). (c) The organization of the leuconoid type. This complex canal network is composed of discrete choanocyte chambers, connected with a network of incurrent and excurrent canals. In all three types of body organization a basal site (attachment to the substratum) and an oscular/apical pole (oscule; opening at which the excurrent canals extrude water from the central cavity) exists, which determines the axis (*arrow*). Representative species of the different organization levels are shown: (a) *Clathrina coriacea*, a calcareous sponge of the asconoid type. Magnification ×0.2. (b) A syconoid type of organization as seen in *Sycon raphanus* (Calcarea); ×3. (c) *Suberites domuncula* a siliceous sponge (Demospongiae) of the leuconoid type; ×0.5

In sponges the water flow is directed from the lateral surface of the specimens, through the porocytes and the canals to the lacunae and the choanoderm [choanocyte chambers]; Fig. 13.1. These chambers are composed of two kinds of epithelial cell layers, formed by choanocytes [flagellated cells] and by cone cells [cells with a double-conical shape, which hang into the chamber formed by choanocytes]. These layers surround the mesohyl; in this central body the sponge cells are loosely embedded in a matrix, composed primarily of collagen, galectin and glycoconjugates (Müller et al. 1997). This matrix surrounds motile archaeocytes, which are pluripotent, and other, differentiated cells, e.g. collencytes and lophocytes, that are involved in the formation of collagen as well as sclerocytes that form spicules. In addition, myocytes exist, which allow contraction of the sponge body upon mechanical irritation (Simpson 1984).

According to the folding pattern of the two epithelial layers of the pinacoderm, in correlation to the choanoderm, three types of poriferan body plan complexity are distinguished; the asconoid, the syconoid and the leuconoid type (Bergquist 1978). The folding of the choanoderm allows the formation of spherical chambers, the choanocyte chambers, of different structural degrees (Fig. 13.1). In the asconoid type, the lateral opening(s) direct the water current to the continuous layer of choanocytes which face a single atrium and then to the oscule (Fig. 13.1B-a). This organization is seen in some Calcarea, e.g. *Clathrina coriacea*. In the syconoid type the folding pattern forms choanocyte chambers and the water flow enters the sponge through the porocytes into choanocyte chambers through which the water is pressed into the atrium and finally via the oscule again to the external milieu (Fig. 13.1B-b). This type of organization is seen in Calcarea and Demospongiae; as an example the calcareous sponge *Sycon raphanus* is shown here. The leuconoid type derives from the syconoid type of organization. Here, the porocyte openings lead from the vestibule to the connecting incurrent canals and into the choanocyte chambers. Subsequently, the water current is pressed into excurrent canals which open into the atrium and then leaves the organism via the oscule (Fig. 13.1B-c). Most sponge species, like the marine (siliceous) demosponge *Suberites domuncula* (Fig. 13.1B-c) show the leuconoid type of organization. The Hexactinellida have an organization pattern which is reminiscent of the syconoid type and are composed of a choanodermal syncytium (see: Müller et al. 2004). Recently, the biochemical and molecular basis of the axis formation in sponges, here shown with the freshwater sponge *Lubomirskia baicalensis* as example, had been demonstrated (Fig. 13.1A; Wiens et al. 2006).

By molecular biological techniques – in the meantime more than 30,000 ESTs have been identified from the sponge *S. domuncula* – and subsequent functional classification it could be disclosed that the basic strategies of body plan formation and gene expression patterns found in Porifera are characteristic for Metazoa in general (Müller et al. 2001). Furthermore, an *in vitro* 3D-cell [three-dimensional] culture system, termed primmorphs, was established (Custodio et al. 1998; Müller et al. 1999a). Culturing primmorphs is a newly developed technique to grow sponge cells *in vitro*; these cells have the potency to proliferate and to differentiate. This system allows an understanding of the roles of the morphogenetic and pattern forming genes. Amazingly the cell-cell- and cell-matrix adhesion molecules found in sponges share high sequence and functional similarity to those of higher metazoan phyla (Müller et al. 2004; Fig. 13.2). The extracellular binding sites to the ligands, but also the intracellular domains of these cell membrane receptors remained conserved throughout the animal kingdom. As it is also functionally proven, the receptors are provided with the properties of outside-in signaling (Wimmer et al. 1999b). This system allowed to study the effects of solute morphogenic factors (e.g. myotrophin; Schröder et al. 2000), or secreted molecules (e.g. epidermal growth factor; Perović-Ottstadt et al. 2004a), as well as of their receptors, that are involved in axis formation (Frizzled receptor; Adell et al. 2003b), and of transcription factors that are required for polarity formation (e.g. the organizer-specific factor *LIM homeobox protein*, Wiens et al. 2003; or Forkhead, Adell et al. 2003a) were discovered (Müller et al. 2004; Fig. 13.2).

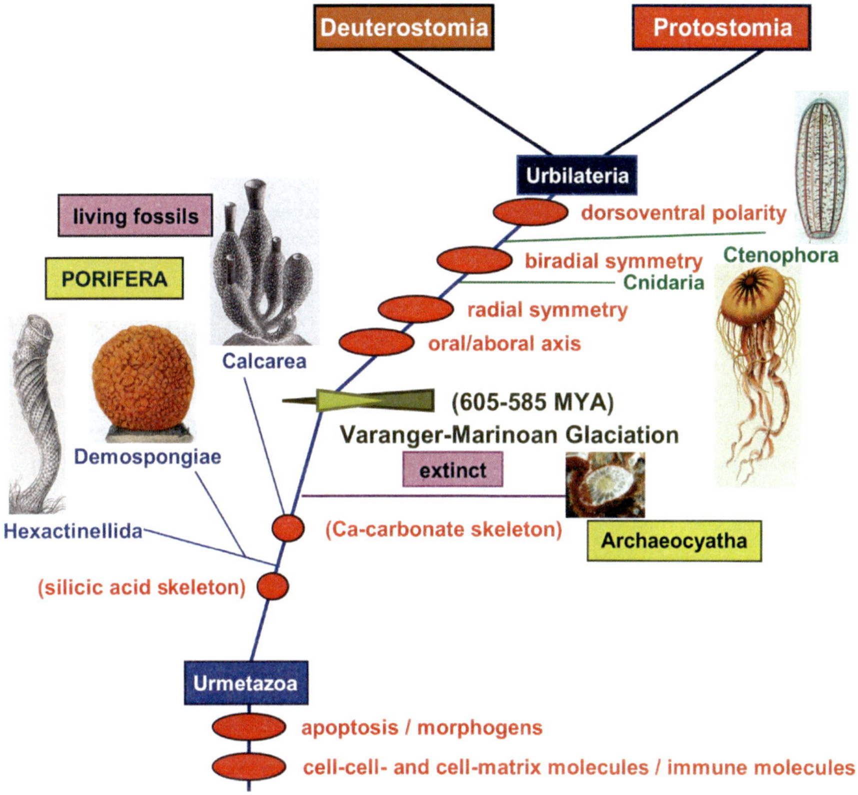

Fig. 13.2 Phylogenetic position of the Porifera within the metazoan kingdom. The three poriferan classes (Hexactinellida and Demospongiae emerged first, and later the Calcarea) evolved from the common ancestor of all metazoan phyla, the Urmetazoa. The major evolutionary novelties which have to be attributed to the Urmetazoa are those molecules which mediate apoptosis, control morphogenesis, the immune molecules and cell adhesion molecules. The three classes of Porifera are the model systems which comprise a genetic reservoir of molecules that direct including pattern formation characteristic for Metazoa; e.g. the transcription factors, paired box homeodomain molecules, LIM-class homeodomain, T-box [Brachyury] or winged helix [Forkhead]. As a sister group to the Calcarea, the Cnidaria evolved. Subsequently, the Ctenophora emerged which comprises not only an oral/aboral polarity but also a biradial symmetry. In the Cnidaria the *paired box* transcription factors have been identified. Finally the Urbilateria developed from the Diploblasts (two epithelial layers), which are built from three germ layers (Triploblasts). They diverged into the Protostomia, with the crown species *D. melanogaster* and *C. elegans*, and the Deuterostomia with *H. sapiens*. Within triploblastic animals the homeobox genes are arranged in clusters

13.2 Basis of Metazoan Pattern Formation: The Stem Cell System in Sponges

Multicellular animals are characterized by the existence of a series of differentiated somatic cells, in addition to the totipotent germ cells (Cory and Adams 1998).

During metazoan evolution the number of distinct cell types increased steadily (Müller et al. 2003a; Müller 2006). The different somatic cell types derive from the zygote through the respective stem cell stage; the differentiation proceeds in niches that form a microenvironment in which defined expression patterns of signaling molecules and local environmental factors direct the fate of the cells (Jones 2001). These niches modify their regulatory property in response to a changing environment to ensure that stem cell activities meet the needs of an organism for a given differentiated cell type. Two major types of stem cells derive from the zygote, the germline stem cells and the somatic stem cells. While the germline stem cells retain their total differentiation capacity, this properties is restricted in the somatic stem cells which gradually lose their stem cell propensity. However, recent studies indicate that this traditional view of an irreversible loss of the stem cell ability during maturation of somatic cells might not completely reflect the physiological situation (Wagner and Weissman 2004). It appears that such fixed "points of no return" do not always exist but that at all levels of differentiation from the pluripotent progenitor cells to the committed progenitors to the lineage progenitors and finally to the terminally differentiated cells, the propensity to act as stem cell is retained, even though with a decreasing intensity. In consequence, the differentiation lineages of somatic cells are dynamic and plastic and the strong distinction between embryonic stem cells and adult stem cells should be reconsidered.

It is generally agreed that the archaeocytes are the toti-/pluripotent cells in sponges from which the other cells originate; evidence was presented indicating a localization of archaeocytes not only within the mesohyl but also in the endopinacoderm layer (Borojevic 1966, 1970, 1971; Simpson 1984). Archaeocytes give rise to the major classes of differentiated somatic cells, (i) the epithelial cells, pinacocytes and choanocytes, (ii) the cells forming the skeleton, collencytes and sclerocytes, and (iii) the contractile cells, myocytes (Fig. 13.3). However, archaeocytes give not only rise to the different somatic cells but also to the germ cells from which the embryos originate (Diaz 1977 and 1979). Another line of differentiation of the archaeocytes is to the thesocytes, the totipotent cells of gemmules which are asexual propagative dissemination bodies. Pinacocytes, collencytes/sclerocytes and myocytes are cells with a low stem cell propensity, implying that these somatic cells are "terminally" differentiated (Simpson 1984).

At present, the study of embryonic stem cells in sponges is limited, since no technique to induce mass production of embryos under controlled conditions has yet been successful. As a substitution, the three-dimensional cell culture has been established for *S. domuncula* (Fig.13.4A); Custodio et al. 1998; Müller et al. 1999a. Under suitable conditions dissociated, single cells (Fig. 13.4D) form special types of cell aggregates, the primmorphs (Fig. 13.4E). They contain cells of high proliferation and differentiation capacity. Furthermore, *S. domuncula* has the capacity to propagate sexually [a free-floating egg is shown in Fig. 13.4B] and also asexually, via reproduction bodies termed gemmules (Fig. 13.4C).

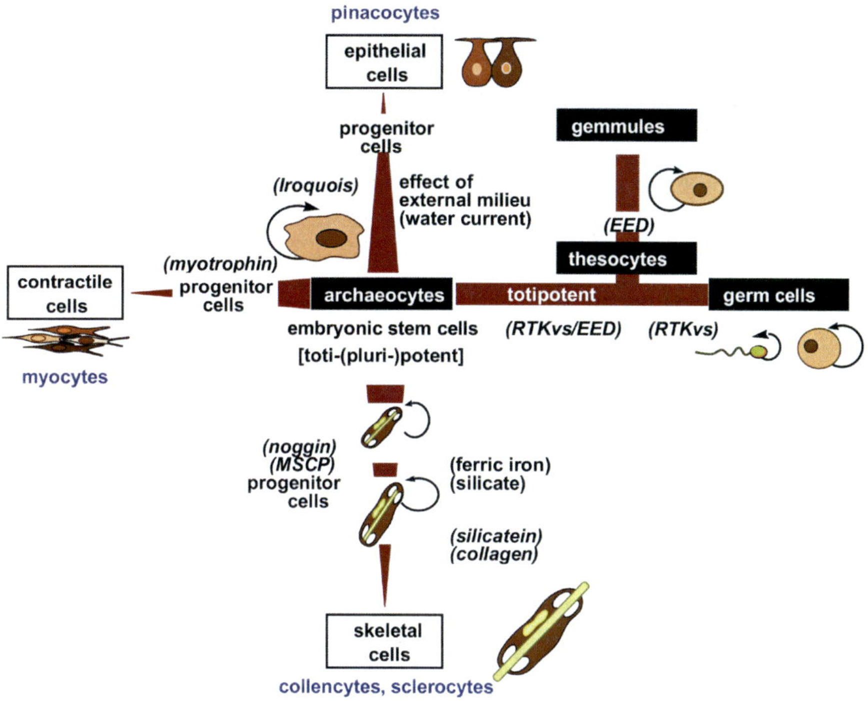

Fig. 13.3 Sponge stem cell system. Schematic outline of the development of the toti-/pluripotent sponge embryonic stem cells, the archaeocytes, to the germ cells on one side and to the three major differentiated cell types, the epithelial-, the contractile- and the skeletal cells on the other side. It is indicated that during these transitions progenitor cells characteristic for these lineages have to be passed. The (potential) factors, e.g. noggin and the mesenchymal stem cell-like protein (MSCP) on the path to the skeletal cells, which trigger the differentiation are shown. In addition it is outlined that committed progenitor cells are formed which respond to the silicate/Fe(+++) stimulus through differentiation to skeletal cells, the sclerocytes (=skeletal cells). Without losing the high level of stem cell propensity the archaeocytes change to germ cells and also to thesocytes, the dominant cells in gemmules (a sexual propagation bodies). The triangles schematically indicate the decrease in stem cell propensity during differentiation

13.2.1 Marker Genes

In order to underline the view that the metazoan stem cell concept can also be applied to Porifera, characteristic marker genes have been cloned from *S. domuncula*. The first cDNA identified whose deduced protein shares sequence similarity to mammalian stem cell markers was the mesenchymal stem cell-like protein (MSCP). *MSCP* is present in mesenchymal human stem cells; experimental evidence exists that *MSCP* is expressed in osteogenic mesenchymal stem cells (Müller et al. 2003a). The functional studies revealed that in sponges the expression of this gene is under positive control of the morphogenetic inorganic elements silicon and ferric iron (Krasko et al. 2002; Müller et al. 2003b). In addition, two

Fig. 13.4 The demosponge *S. domuncula*. It can be cultivated for over two years in aquaria (**A**). *S. domuncula* can propagate sexually (**B**- a free floating egg) and also asexually, via gemmules (**C**- a cross-section though gemmules [gem] which developed on a shell on which the sponge lives). (**D**) Dissociated single cells from this species can form aggregates which differentiate to primmorphs (**E**) after an incubation period of more than 5 days. (**F**) A primmorph which developed on galectin-coated culture dishes; it shows canal-like structures. (**G**) A gemmule which had been treated with LPS, as described under "Materials and Methods". *Size bars* for **A–E** are given; **F,G**: 1 mm

further potential genes, involved in the differentiation of stem cells in sponges, were isolated; noggin and the glia maturation factor (Schröder et al. 2003). Noggin is a glycoprotein that binds bone morphogenetic proteins selectively and antagonizes their effects. It was initially isolated from *Xenopus* and found to be expressed in the Spemann's organizer. In vertebrate development noggin is involved in the formation of dorsal mesoderm derivatives, e.g. the skeletal muscles (Smith and Harland 1992; Smith et al. 1993; Valenzuela et al. 1995; Ogawa et al. 2002; Müller et al. 2003b).

In the initial phase of formation, the *in vitro* 3D-cell primmorph system contains predominantly the toti/pluri-potent archaeocytes which can be stimulated to differentiate into four main tissue-specific directions (Müller 2006). If they are induced by the inorganic factors silicate or ferric iron, the archaeocytes give rise to the skeletal cells through an increased expression of genes encoding the structural proteins silicatein and collagen; a process which is mediated by noggin and *MSCP* (Fig. 13.3); Müller et al. 2004. Second, if archaeocytes are exposed to a morphogen (myotrophin) they are directed towards the contractile cell lineage (Krasko et al. 2002; Müller 2006). A third lineage, which gives rise to the aquiferous canal system, is induced by a physical factor; there *Iroquois* gene expression is induced as a result of an increased water current which is paralleled with the formation of canal-like pores in the primmorphs (Perovic et al. 2003). In higher metazoans the expression of the *Iroquois* genes is thought to confer identity to a particular region, and hence it can be classified to the homeotic selector genes (see: Perovic et al. 2003). The early functions of *Iroquois* in insects include the definition of the eye and notum territories, and in vertebrates the formation of the neural ectoderm. Late functions of *Iroquois* are subdividing the territories; the dorsal-lateral subdivision in

Drosophila notum and the patterning of notum bristles and wing veins, and in vertebrates the subdivision of neural tube and heart patterning. The sponge *S. domuncula* not only contains in its genome an *Iroquois* gene, which comprises not only remarkable sequence similarity to those sequences described from triploblastic animals, but also is expressed during the formation of one major morphogenetic remodeling process, the construction of the aquiferous canal system (Fig. 13.3).

Two further genes have been cloned which are of relevance for the presented work here, the thymidine kinase and the perforin-like protein. In vertebrate cells, salvage pathway phosphorylation of thymidine is catalyzed by two thymidine kinases, the cell-cycle regulated cytoplasmic thymidine kinase-1 and the constitutively expressed mitochondrial thymidine kinase-2 (Segura-Pena et al. 2007). In cellular metabolism, thymidine kinase-1 and thymidine kinase-2 serve to maintain sufficient dTTP for DNA replication and repair. Hence the level of thymidine kinase-1 transcripts reflects the extent of cell proliferation of a given cell population. We have cloned from *S. domuncula* the cytosolic thymidine kinase-1 (Fig. 13.5), since it is a well known cell-cycle-regulated enzyme of importance in nucleotide

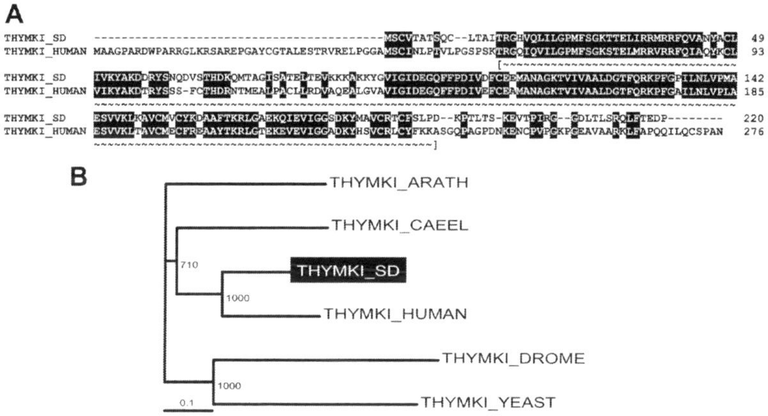

Fig. 13.5 Cytoplasmic thymidine kinase from *S. domuncula*. (**A**) The deduced sponge protein was aligned with the human soluble thymidine kinase 1 (THYMKI_HUMAN; accession number NP_003249.2; Segura-Pena et al. 2007). Amino acids, identical or similar among two sequences, are in inverted type. The borders of the thymidine kinase (EC 2.7.1.21) [tdk] domain (hamap:MF_00124; motif scan database [http://myhits.isb-sib.ch/cgi-bin/motif_scan] are marked [~~~~~~]). (**B**) These two sequences were compared with the *C. elegans* thymidine kinase family member (thk-1) (THYMKI_CAEEL; NP_501886.1), the *D. melanogaster* the putative protein CG40300-PC (THYMKI_DROME; NP_001036627.1), the *S. cerevisiae* suppressor of var1 protein (THYMKI_YEAST; 151942764), and the putative thymidine kinase from *A. thaliana* (THYMKI_ARATH; NP_568426.1). After alignment the phylogenetic tree was built, using the plant sequence as an outgroup; the neighbor-joining method (Saitou and Nei 1987) was used. The degree of support for internal branches was further assessed by bootstrapping (Felsenstein 1993). The numbers at the nodes are an indication of the level of confidence for the branches as determined by bootstrap analysis [1000 bootstrap replicates]

metabolism. The thymidine kinases-1 have a tetrameric structure in which each subunit contains an alpha/beta-domain that is similar to ATPase domains of members of the RecA structural family and a domain containing a structural zinc. The *S. domuncula* cDNA, encoding the thymidine kinase-1, was identified and was cloned as described (Perovic et al. 2003). The deduced protein with a chain length of 220 amino acids, giving a predicted size of 24,251 Da, shares highest sequence similarity to the human soluble thymidine kinase 1 (NP_003249.2); Segura-Pena et al. 2007 (Fig. 13.5A). The two sequences share 45% identical amino acids and 56% similar amino acids. This high score is also reflected by the constructed rooted phylogenetic tree; the human and the sponge gene fall into one branch (Fig. 13.5B). Separated from this branch are the (distantly) related sequences from *Drosophila melanogaster* (3% identical – 7% similar amino acids) and *Saccharomyces cerevisiae* (31–44%). Again separated from these two branches is the thymidine kinase family member from *Caenorhabditis elegans*.

Since sponges are filter feeders that are exposed to large amounts of bacteria present in their surrounding aqueous milieu, they had to develop strategies to resist and defend themselves against attacking microorganisms, among them also the Gram-positive bacteria. In a previous study we could demonstrate that in primmorphs from *S. domuncula*, the expression of a gene, termed perforin-like protein [MARKER GENE FOR DEFENSE], is activated in response to exposure towards a Gram-negative α-proteobacterium (Thakur et al. 2003). The perforin-like protein contains one EGF-like domain cysteine pattern signature and a C2 domain [thought to be involved in Ca^{2+}-dependent phospholipid binding (Davletov and Suedhof 1993).

13.2.2 Expression Pattern of Archaeocytes (Stem Cells): "Reproductive" Cells

Stem cells are self-renewing populations of cells that undergo symmetric and/or asymmetric divisions either to self-renew or to differentiate into different kinds of differentiated progeny. This minimal definition does not allow a clear distinction of stem cells from other dividing and differentiating cells (Cai et al. 2004). Surely, stem cells are provided with a high capacity for cell-cycle, for cellular protective and DNA repair mechanisms and for apoptosis. Recently genetic expression markers have been identified, which can be applied for the identification of "embryonic" cells and tissue in sponges (Müller 2006).

As outlined above, sponges have developed two propagation systems, sexual reproduction (Fig. 13.4B) and asexual reproduction by propagation bodies [gemmules] (Fig. 13.4C). The first study, using molecular markers to determine the restriction of gene expression during embryogenesis in a sponge appeared recently (Perović-Ottstadt et al. 2004b). It was found that in oocytes, morulae and blastulae/larvae from *S. domuncula* distinct genes are expressed, among them a sponge-specific receptor tyrosine kinase (RTKvs). In addition, the sex-determining protein FEM1 and the sperm associated antigen-related protein are highly expressed; in adult animals the levels of expression of these genes are very low (Perović-Ottstadt et al. 2004b).

The asexual reproduction pattern in sponges, i.e. gemmule formation, was already early in focus of developmental biologists (Laurent 1842). Gemmule formation in sponges is induced by environmental factors, e.g. temperature, or dryness (Wagner et al. 1998). The cells in the gemmules are the thesocytes, which form a homogenous population (Fig. 13.3); it has been proposed that these cells also derive from the totipotent archaeocytes (see: Simpson 1984).

The recently identified MARKER GENES FOR TOTIPOTENT CELLS, which are highly expressed either in oocytes or in cells of gemmules, are the receptor tyrosine kinase *RTKvs* (oocytes and early larvae) and the embryonic development protein *EED* (gemmules); Müller 2006. The expression patterns of these genes are applied as tools to distinguish between differentiation levels of the cells. In tissue of adult *S. domuncula* those genes are expressed only in a few cells that are scattered in the pinacoderm. Sequence analyses of the two sponge RTKs show that the extracellular part of the *S. domuncula* kinase RTKvs_SUBDO has no obvious similarity to any other (deduced) protein hitherto submitted to the databases. Therefore, this part of the protein can currently be considered to be truly sponge-specific (Perović-Ottstadt et al. 2004b). The intracellular TK domain, however, possesses the 12 characteristic subdomains. The embryonic development protein EED of *S. domuncula* is a member of the Polycomb-group (PcG) family (Sewalt et al. 1998). *In situ* hybridization demonstrated that RTKvs_SUBDO is highly expressed in eggs and early stages of embryos in *S. domuncula* (Müller 2006). In view of these data, we suggest that the cells expressing these two marker genes represent archaeocytes, which are in the "functional" state in either fertilized eggs or cells constituting early embryos (as is the case for RTKvs_SUBDO), or form the gemmules (EED2_SUBDO); Fig. 13.6.

13.2.3 Expression Pattern of Archaeocytes (Stem Cells): Sclerocyte Lineage [Skeletal Cells]

Sclerocytes are the cells which produce the siliceous spicules, the skeletal elements of sponges. As outlined the mesenchymal stem cell-like protein (MSCP-l) and noggin can be considered as MARKER GENES FOR MULTIPOTENT STEM CELLS (Müller et al. 2003a). During differentiation to sclerocytes the genes *silicatein* and *collagen* undergo strong expression (Krasko et al. 2002).

13.2.4 Expression Pattern of Archaeocytes (Stem Cells): Pinacocyte Lineage [Epithelial Layer]

As reviewed, the pinacocyte surface layer can be looked upon as an epithelium (Müller et al. 2004). One MARKER GENE FOR THE DIFFERENTIATION

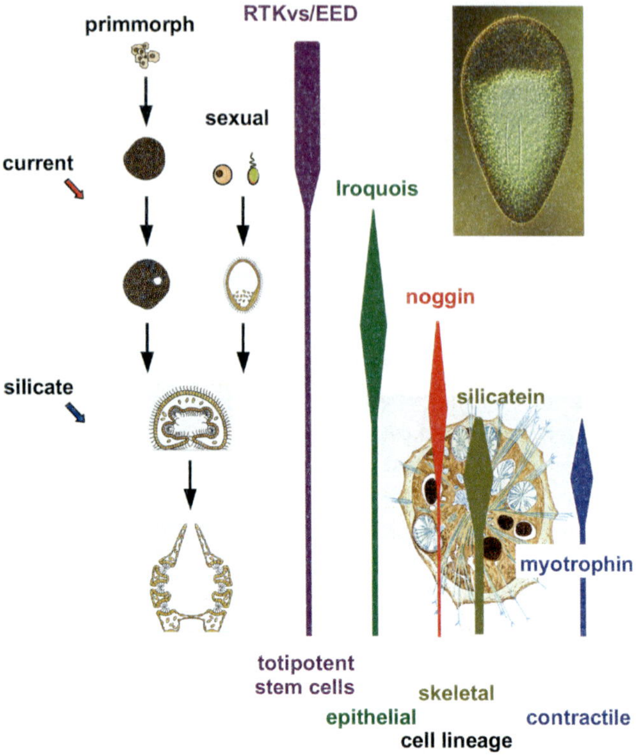

Fig. 13.6 Sequential expression of (putative) stem cell marker genes in *S. domuncula*. In primmorphs as well as in germ cells a high expression of two genes can be identified, the sponge-specific receptor tyrosine kinase (RTKvs) and the embryonic development protein (EED). They might be considered as markers for totipotent stem cells. At exposure of primmorphs to water current, the transcription factor Iroquois is expressed; this process is seen primarily in epithelial cells. Noggin as well as silicatein gene expression are provoked after addition of silicate/Fe(+++) to the culture medium; the expression is prominent in the skeletal (spicule)-forming cell lineage. In contractile cells (myocytes), myotrophin is expressed. Bars are underlaid with a photograph of a larva of the freshwater sponge *L. baicalensis* and a cross section through an entire sponge (*Craniella schmidtii*), showing embryos within the parent (Sollas et al. 1888)

of the archaeocyte stem cells to the pinacocytes has been isolated from *S. domuncula* (Perovic et al. 2003). This gene, *Iroquois* (MARKER GENE FOR THE PINACOCYTE LINEAGE) codes for a putative homeobox gene. The putative *Iroquois* transcription factor was found to be expressed in cells which are adjacent to the canal system; its expression is upregulated in primmorphs which are cultivated in strong water current (Perovic et al. 2003); Fig. 13.3. The finding that also in sponges the expression of the *Iroquois* gene is restricted to a specific tissue region, the epithelial layer of the aquiferous system, adds a further piece to the understanding of the complexity of tissue organization in sponges.

13.2.5 Expression Pattern of Archaeocytes (Stem Cells): Myocyte Lineage

Myocytes in sponges are functionally characterized as cells which synthesize the organic skeletal elements, e.g. collagen. During the progress of archaeocytes to myocytes, myotrophin is expressed in *S. domuncula* (Schröder et al. 2000). Myotrophin was first found in mammalian systems; in cardiac myocytes myotrophin stimulates protein biosynthesis (Sen et al. 1990), suggesting a crucial role in the formation of cardiac hypertrophy (reviewed in: Sil et al. 1998; Schröder et al. 2000). The sponge myotrophin shares the highest sequence similarity with the human molecule. Recombinant sponge myotrophin was found to stimulate protein synthesis by 5-fold (Schröder et al. 2000). Since myotrophin is neither expressed during embryogenesis nor in gemmules, it might be characterized as a MARKER GENE FOR THE MYOCYTE LINEAGE. After incubation of single cells with myotrophin the primmorphs show an unusual elongated, oval appearance. Furthermore, in the presence of myotrophin sponge cells up-regulate collagen gene expression. We assume that the sponge myotrophin causes in homologous cells the same/similar effect as the cardiac myotrophin in mammalian cells, where it is also involved in initiation of cardial ventricular hypertrophy.

13.3 Bacterial Infection

Sponge-bacteria interactions are probably among the oldest host-bacteria interactions known, dating back more than 500 million years in time (Wilkinson et al. 1984). Several studies have revealed that permanent associations exist between certain host sponges and specific microorganisms; however their interactions remained largely unknown (Althoff et al. 1998; Friedrich et al. 1999; Schmidt et al. 2000). Moreover, sponges may also succumb to microbial and fungal infections which result in the disintegration of the sponge fibers/tissue and ultimately lead to sponge death (Vacelet et al. 1994). The fact that sponges are susceptible to microbial infection suggests that they should also be provided with mechanisms to prevent these types of diseases.

It is known that sponges possess molecules resembling those of the mammalian immune system (Müller et al. 1999b). As examples of the innate immune system, scavenger receptor cysteine rich domains and macrophage derived cytokine-like molecules have been identified which are upregulated during auto- and allografting experiments. In addition, the (2′–5′) oligoadenylate synthetase system exists in sponges (Schröder et al. 2008). Precursors of the adaptive immune system have also been identified and were shown to be functional in sponges (Müller et al. 2004). Accordingly, the expression of a lymphocyte-derived cytokine from mammals is up-regulated during non-self grafting experiments in *S. domuncula*. In the sponge *Geodia cydonium*, two immunoglobulin like receptors have been identified which are also up-regulated during grafting experiments. These findings demonstrate that sponges contain elements for innate immune recognition (Müller et al. 1999b).

The lipopolysaccharide (LPS)-mediated pathway is an additional mechanism involved in the mammalian immune response. LPS, an endotoxin derived from the outer cell wall of Gram-negative bacteria binds to the cell surface molecule CD14 (Ulevitch and Tobias 1994). This interaction is mediated by LPS-binding protein(s) (Scott et al. 2000). CD14, a plasma membrane linked molecule presents LPS to a specific transducer resulting in an enhanced production of reactive oxygen metabolites and gene expression (reviewed in: Jiang et al. 2000). Serine-threonine directed MAP (mitogen-activated protein) kinases are essential components of the LPS-mediated pathway. These proteins can be grouped into three main families, the extracellular signal-regulated kinases (ERKs), the p38 kinases and the c-jun N-terminal kinases (JNKs) (Seger and Krebs 1995). In deuterostomes, the latter two kinases are phosphorylated after exposure of cells to LPS (Yang et al. 2000). In view of the recent findings by Müller et al. (1999c) it is hypothesized that a similar LPS-mediated immune response pathway can also be found in sponges.

In a previous study we demonstrated (Böhm et al. 2001), that cells from *S. domuncula*, that had been exposed to LPS, respond with the activation [phosphorylation] of the stress-activated MAP kinases p38 and JNK. Furthermore first insights into the types of microorganisms that are permanently associated with *S. domuncula* were gathered. In addition, we obtained evidence that in sponges LPS interacts with Gram-negative bacteria via a cell-surface bound receptor, the LPS-binding protein (Wiens et al. 2005). The cDNA was isolated and the protein expressed. During binding to LPS the protein dimerizes. Co-immunoprecipitation analysis revealed that this protein interacts with MyD88, after exposure of the animals towards LPS. The sponge MyD88 is composed of two protein interaction domains, a TIR domain (present in Toll-like receptors and in adapter molecules such as MyD88) and a death domain (present in MyD88 and the interleukin-1 receptor-associated kinase 4 IRAK). Northern blot experiments and *in situ* hybridization studies showed that the level of the LPS-binding protein does not change after LPS treatment, while MyD88 expression is strongly upregulated. As an executing molecule of this pathway (LPS-binding protein – MyD88) the macrophage expressed protein, a perforin-like molecule (Mr 74171), was identified. The cDNA was isolated; the gene is highly expressed after LPS treatment, especially at the surfaces of the animals. The recombinant protein comprises biological activity and eliminates Gram-negative bacterium (*E. coli* and BL21 [sponge-associated Gram-negative bacteria] were used); it is inactive against the Gram-positive bacterium *Staphylococcus aureus*. These data indicate that *S. domuncula* is provided with an innate immune system against Gram-negative bacteria; the ligand LPS (a pathogen-associated molecular pattern) is recognized by the pattern-recognition receptor (LPS-binding protein) which interacts with MyD88. A signal transduction is established which results in an elevated expression of both MyD88 and the macrophage expressed protein, as executing proteins.

13.3.1 *Activation of p38 Kinase*

The phosphorylation of the two kinases (p38 and JNK) strongly increases after exposure of the sponge tissue to LPS, a response which is characteristic of the mammalian p38 and JNK activation in response to LPS. The activation occurs rapidly within the first hour after LPS exposure. Moreover, the inhibitory effect of LPS on cells from *S. domuncula* has also been determined by measuring the incorporation of tritium labeled phenylalanine into protein fraction. At a concentration of 3 μg/ml of LPS, the reduction in the incorporation rate is already significant (Böhm et al. 2001). These findings show that the defense pathways are highly conserved between sponges and humans (Böhm et al. 2002).

13.3.2 *ELISA Assay (Method)*

13.3.2.1 Preparation of Sponge Total Protein Extract

Primmorphs are exposed to LPS 10 μg/ml (L2880, Sigma, Deisenhofen, Germany) and then processed as described (Böhm et al. 2000). Tissue is extracted with lysis-buffer (1× TBS [Tris-buffered saline], pH 7.5, 1 mM EDTA [ethylene diamine tetra-acetic acid], 1% Nonidet-P40, 10 mM NaF, protease inhibitor cocktail [1 tablet/10 ml] and 1 mM sodium orthovanadate) after homogenization in a mortar; a clear supernatant is obtained by centrifugation (13,000 × g; 10 min; 4°C) and then subjected to ELISA analysis.

13.3.2.2 Determination of the Phosphorylation Level of p38 by Enzyme-Linked Immunosorbent Assay (ELISA)

An enzyme-linked immunosorbent assay (ELISA) for detection of the phosphorylation level of p38 was developed. This ELISA assay is based on the ability of the anti-phosphorylated 38 (pp38) antibody to bind with high affinity to a peptid containing the TGY recognition motif (Böhm et al. 2000). The total protein extract is covalently bound to a 96-wells polystyrene plate. The complexes formed between proteins bound to the plate and the antibodies against pp38 are detected using a peroxidase-conjugated secondary antibody and the 3,3,5,5-tetramethylbenzidine (TMB) as substrate.

In detail: Flat-bottomed polystyrene 96-well Pro-Bind plates (Becton Dickinson, France) were covered with 50 μl of total protein extract in final concentrations of 5–60 μg/well during incubation for 3 h at room temperature. After washing three times with phosphate buffered saline (PBS), containing 0.05% Tween-20 (PBS/T buffer), the plates were blocked with 3% bovine serum albumin (BSA) in PBS (150 μl/well), overnight at 4°C. The level of pp38 in the protein mixture bound to the plate was detected with a polyclonal antibody against phosphorylated p38 (pp38)

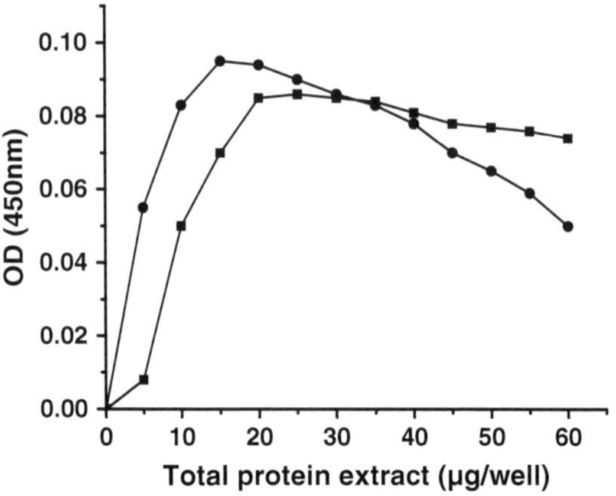

Fig. 13.7 ELISA test for the determination of the level of pp38 in sponge tissue extracts. For this series of experiments extracts from primmorphs had been used. ELISA plates were coated with 5–60 μg/well of sponge protein extract. Subsequently, the level of pp38 in the extract was detected with the commercial polyclonal antibodies against phosphorylated p38 (pp38) from Santa Cruz Biotechnology (■) and the anti-pp38 antibodies directed against the *S. domuncula* peptide (•) (Müller et al. 2002). Immunocomplexes were obtained by addition of a goat anti-rabbit IgG, which had been conjugated with horseradish peroxidase and visualized with the TMB substrate. Photometric determination was performed at 450 nm

from *S. domuncula* (Müller et al. 2002); this antibody was diluted 1:1000. In addition, a second, commercial polyclonal antibody which was raised against pp38 was used (Santa Cruz Biotechnology, California) and applied at a dilution of 1:2000. Again, the antibodies were diluted in PBS/BSA and incubated for 1 h. After blocking with 2% goat serum (Dianova, Hamburg, Germany) and 0.3% BSA in PBS for 30 min, the immunocomplexes were detected using goat anti-rabbit IgG, conjugated with horseradish peroxidase (1:10000; Sigma, Deisenhofen, Germany) under application of the 3,3,5,5-tetramethylbenzidine [TMB] substrate (KPL, Germany). The reaction was stopped after an appropriate time (typically 10 min) by adding 1 M H_2SO_4; the absorbance (OD_{450}) was measured at 450 nm (Titertek MultiScan Plus).

This direct ELISA assay was used to determine the p38 activation [phosphorylation] level in sponge primmorphs exposed to LPS. At first, the optimal concentration of the extract which reveals the highest signal in the photometric test was titrated. There sponge protein extract was added to the wells in concentrations between 5 and 60 μg/well of sponge protein extract (Fig. 13.7). The results revealed that at a concentration between 10 and 20 μg/well the highest OD_{450} can be recorded. Interestingly, the two antibodies used for this series of experiments, the commercial anti-pp38 (Santa Cruz Biotechnology, California) and our anti-pp38 raised against the *S. domuncula* peptide gave the same results.

Subsequently, this ELISA assay was applied to determine in a time-kinetic experiment the time required for the sponges to react to LPS. Primmorphs were exposed to 10 μg/ml of LPS for 5 min to 24 h. Then protein was prepared and subjected to

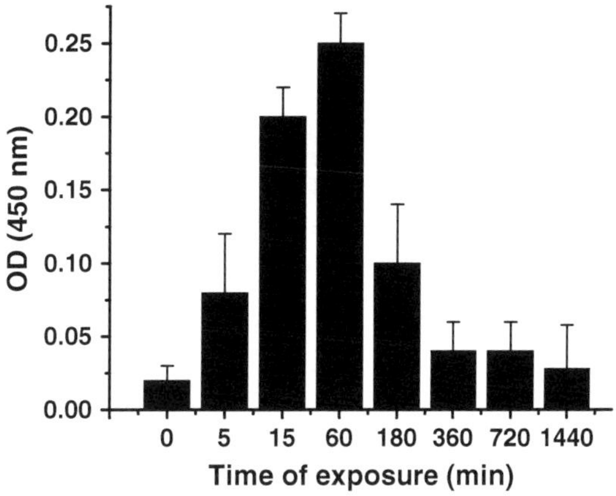

Fig. 13.8 Effect of LPS exposure on activation of pp38 in *S. domuncula* primmorphs. The tissue-like 3D-cell culture had been incubated for 5 min to 1440 min (24 h) with 10 μg/ml of LPS. Then protein was extracted and an aliquot of 20 μg was used for coating per well; then incubation with anti-pp38 (Santa Cruz Biotechnology) followed. The background values, measured with p38 antibodies, were subtracted

the ELISA assay to quantify the level of pp38. The highest level of phosphorylation of p38 (to pp38) was be measured after 60 min (Fig. 13.8).

From these data we conclude (i) firstly that primmorphs from *S. domuncula* respond – also under the experimental conditions used – to the exposure to LPS and (ii) secondly that the highest level of phosphorylation (formation of pp38) is seen after an exposure time of 60 min.

13.3.3 Real-Time Reverse Transcription-PCR (Method)

The technique of real-time reverse transcription-PCR (RT-qPCR) was used to quantify the expression levels of the genes coding for; (i) STEM CELL NUMBER: noggin (noggin-l [*S. domuncula*] NOGG-l_SD; accession number CAD59735; Müller et al. 2003b); (ii) PROLIFERATION: thymidine kinase (THYMKI_SD [*S. domuncula*]; accession number AM905441), (iii) DIFFERENTIATION: integrin-ß (integrin beta subunit [*S. domuncula*] INTb_SD; CAB38100; Wimmer et al. 1999a,b), (iv) LPS-CHALLENGE: perforin (macrophage expressed protein [*S. domuncula*] MPEG_SD; AJ890501; Wiens et al. 2005) and finally (v) THE HOUSEKEEPING GENE: glycerol 3-phosphate dehydrogenase (GAPDH_SD [*S. domuncula*]; accession number AM902265).

13.3.3.1 Total RNA Extraction

Samples were extracted using the TRIzol reagent (Invitrogen Corp., Carlsbad, CA) according to the manufacturer's instructions. Total RNA from primmorphs was isolated by the modified guanidine hydrochloride method (Yoshikawa et al. 2000). The concentration of each RNA sample was measured using a spectrophotometer. Only the RNA samples with a 260/280 ratio between 1.8 and 2.2 and a 260/230 ratio higher than 1.9 were used for the analysis. In addition the integrity of RNA samples was assessed by agarose gel electrophoresis.

One microgram of each mRNA was reversely transcribed using the oligo dT primer and Super Script II (Invitrogen) as described (Yoshikawa et al. 2000).

13.3.3.2 Primer Design

For GAPDH expression studies the following primer pair had been selected: the forward primer was 5-ATCACAGGGGGAGCAAAGAA GGTCAT-3 [26mer] and the reverse primer 5-AAGTGGGGCTAGGC AGTTTGTGGTG-3 [25mer]; for *noggin* expression forward 5-TTCCCTCG ATATTTTTCTGCTGGCTCTT-3 [28mer] and reverse primer 5-CGTCCTTCCTCTTC GTCCTCTTACTATTG-3 [29mer]; for *integrin-ß* expression forward 5-CTTTGGACCTGCTTGCGAGTGTGA-3 [24mer] and reverse primer TCGCAA GCCGTTCCAAAGTAAGGTT-3 [25mer]; for *thymidine kinase* expression forward 5-CTTTCCCGATATTGTAGACTTTTG-3 [24mer] and reverse primer 5-CACACCATACAGACTGCCTTTAG-3 [23mer] and for *perforin* expression forward 5-CAATGAGATGTCCGGGTGGGTTTACT-3 [26mer] and reverse primer 5-GTTCTTTCGCAGGGTGGGCTTAGG-3 [24mer]. The resulting PCR product lengths were *GAPDH* fragment 150 bp [ranging from: 399-424 to 548-524]; *noggin* fragment 183 bp [422-449 to 604-576]; *integrin-ß* fragment 147 bp [1398-1421 to 1544-1520]; *thymidine kinase* fragment 165 bp [327-350 to 491-469] and *perforin* fragment 155 bp [1895-1920 to 2049-2026].

13.3.3.3 qPCR

RT-qPCR was done using the Light Cycler (Roche Diagnostics, Meylan, France), which exploits the ability of SYBR green to fluoresce after hybridization with a double-strand DNA. The analyses were performed in 20 AL glass capillaries using the Light Cycler fast start DNA master SYBR green kit (Roche Diagnostics). Then, 1 mM of each primer and 3 mM of $MgCl_2$ in the total volume of 20 µl were used in each real-time RT-PCR amplification. The real-time RT-PCR cycle started with the initial denaturation at 95°C for 10 min, followed by 45 cycles of denaturation at 95°C for 10 s, annealing at 57°C for 10 s for *GAPDH*, 56°C for 10 s for *noggin*, 57°C for 10 s for *integrin-ß*, 53.0°C for 10 s for *thymidine kinase*, and 57°C for 10 s for *perforin* and finally an elongation at 72°C for 10 s. As an internal quantitative control of the gene expression, the glyceraldehyde-3-phosphate dehydrogenase (GAPDH) gene expression was determined.

13.4 Assessment of the Stem Cell Level in Primmorphs

It is known from earlier studies that primmorphs contain proliferating and differentiating cells (Custodio et al. 1998; Müller et al. 1999a). If these 3D-cell aggregates remain under slight rotation, they do not attach to the substratum (petri dish).

However, if they are placed onto coated plastic dishes (Wiens et al. 2003; Adell et al. 2007) they adhere and start to reconstitute to organ-like assemblies, expressing e.g. genes encoding proteins of the Wnt pathway.

13.4.1 Primmorph Formation

Primmorphs were obtained from single cells as described before (Custodio et al. 1998; Müller et al. 1999a); they reached sizes of 3–5 mm after three days. During this period the cultures were slightly shaken. Then, the primmorphs were transferred into a culture dish and cultivated further in natural seawater, supplemented with 0.2% of RPMI1640 – medium and the optimal silicate concentration (60 μM; Wiens et al. 2004). Incubation of the cultures continued for 1 to 3 days as follows; the primmorphs (i) remained under shaking in this medium in non-coated dishes, or (ii) were transferred into plates coated with recombinant *S. domuncula* galectin (Wiens et al. 2003). Primmorphs kept under such conditions strongly increased the proliferation and differentiation capacity of their cells. Among the genes which underwent high expression were the homeobox genes (Wiens et al. 2003). In the final series of experiments, the primmorphs (iii) were transferred to galectin-coated plates and incubated for 3 to 5 consecutive days with 10 μg/ml of LPS (Sigma L2880; LPS from *E. coli*). Representative images are shown in Fig. 13.4. Those primmorphs which remained under shaking for 3 days remained as ball-like aggregates in the medium, and did not attach to the substrate (Fig. 13.4E). Those primmorphs which could attach to the petri dish on the galectin coat, formed canal-like structures (Fig. 13.4F); and finally the primmorphs which had been incubated for 3 days in the presence of LPS developed apoptotic bodies (Fig. 13.4G).

13.4.2 Expression Pattern in Primmorphs for Noggin, Thymidine Kinase, Integrin and Perforin in Dependence on Adhesion and LPS Treatment

The primmorphs remained either untreated (formation of ball-like, non-adherent 3D-cell aggregates; Fig. 13.4E), or were placed onto galectin-coated culture dishes (attached primmorphs, which show canal-like structures Fig. 13.4F), or were treated with LPS (primmorphs which show "apoptotic" structures; Fig. 13.4G). Since the effect of the different treatments on gene expression should be measured, their duration was set to 1 and 3 days, respectively. After this incubation period the primmorphs were collected, RNA extracted, mRNA purified and finally analyzed by RT-qPCR for the transcript levels. In Table 13.1, the relative expression levels for the different marker genes (*noggin*, *thymidine kinase*, *integrin-ß*, and *perforin*) are summarized.

Table 13.1 Relative quantification of noggin, thymidine kinase, integrin-ß, and perforin in primmorphs under different treatment. The 3D-cell aggregates remained either non-attached (non-treated [culture dishes on a moving platform]), or cultured on galectin-coated dishes (attachment of the primmorphs and formation of canal-like structures), or were treated with LPS (formation of apoptotic bodies). The relative expression of reference genes in primmorphs is given. Assays of transcript levels for each reference gene were normalized by use of the housekeeping gene GAPDH. Each experiment has been performed 5-times; the means and the standard deviations are given

	Duration (days)		qPCR ratio [ratios × 100]			
Primmorphs Treatment:	1	3	Noggin: GAPDH	TK: GAPDH	Integrin-ß: GAPDH	Perforin: GAPDH
None	1		0.025 ± 0.003 [2.5-]	0.051 ± 0.006 [5.1-]	0.018 ± 0.004 [1.8-]	0.001 ± 0.001 [0.1-]
		3	0.031 ± 0.004 [3.1-]	0.047 ± 0.005 [4.7-]	0.024 ± 0.003 [2.4-]	0.003 ± 0.001 [0.3-]
Galectin-coated dishes	1		0.019 ± 0.004 [1.9-]	0.041 ± 0.005 [4.1-]	0.068 ± 0.007 [6.8-]	0.002 ± 0.001 [0.2-]
		3	0.017 ± 0.004 [1.7-]	0.039 ± 0.005 [3.9-]	0.072 ± 0.008 [7.2-]	0.0017 ± 0.002 [0.2-]
LPS	1		0.003 ± 0.001 [0.3-]	0.0009 ± 0.001 [0.1-]	0.0044 ± 0.005 [0.4-]	0.056 ± 0.007 [5.6-]
		3	0.007 ± 0.001 [0.7-]	0.001 ± 0.001 [0.1-]	0.0063 ± 0.007 [0.6-]	0.089 ± 0.010 [8.9-]

13.4.2.1 Expression in Non-treated Primmorphs

The highest expression level of the reference gene in primmorphs which remained in the rotating culture dishes was seen for *noggin*; with a ratio to *GAPDH* of 0.025 [2.5-fold (×100) expression with respect to *GAPDH*; 1 day incubation period] (Table 13.1). This value increased to 3.1-fold after 3 days. Comparably high is – in these primmorphs – also the expression level of the *thymidine kinase*; a 5-1-fold ratio has been measured at day 1, and a 4.7-fold at day 3. A lower level had been determined for *integrin-ß* with 1.8- and 2.4-fold (day 1 and day 3, respectively). Very low is the transcript level of *perforin*; it measures at day 1 0.1-fold and at day 3 0.3-fold.

13.4.2.2 Expression Levels in Attached Primmorphs

As expected, the relative level of *integrin-ß* expression is highest with 6.8-fold (day 1) and 7.2-fold (day 3); Table 13.1. Comparable with the ratios in the non-attached primmorphs are the values for *thymidine kinase* with 4.1-fold (day 1) and 3.9-fold (day 3); lower is the ratio for *noggin* with 1.9-fold (day 1) and 1.7-fold (day 3). The expression ratio for *perforin* 0.2-fold (remained unchanged).

13.4.2.3 Level of Expression in LPS-Treated Primmorphs

A drastic effect on the expression level is seen when primmorphs are incubated with LPS (Table 13.1). The ratios of expression for *noggin*, *thymidine kinase* and *integrin-ß* drop to very low levels, with 0.3-fold (*noggin*), 0.01-fold (*thymidine kinase*) and 0.4 (*integrin-ß*). In contrast, the level for perforin expression strongly increases to 5.6-fold (day 1) and 8.9-fold (day 3) in LPS treated primmorphs.

13.5 Effect of LPS Treatment on the Number of Stem Cells

Exposure of mouse embryo stem cells to LPS results in Fas-mediated apoptosis (Zou et al. 2000). On the other hand, it had been reported that LPS causes an induction of mesenchymal stem cells towards an osteogenic differentiation (Cho et al. 2006). Therefore, it was suggestive to study the effect of the endotoxin LPS on the number of stem cells in the primmorph system.

In order to provide the basis for conclusive results, the number of stem cells in non-differentiated (ball-like non-attached primmorphs) and substrate-caused differentiating primmorphs (attached aggregates) had to be assessed first. To estimate the number of stem cells, the level of expression of noggin (marker for toti-/omni-potent cells) had been quantified by qPCR. In parallel, the transcript level of thymidine kinase (marker for proliferation), integrin-ß (marker for substrate adhesion) and perforin (marker for the anti-bacterial effector protein) had been determined.

In the absence of any additional extracellular effector/mediator the ball-like primmorphs show a high level of expression of *noggin* and *thymidine kinase* genes (Table 13.1) This result confirmed earlier observations that cells in these aggregates show a high proliferation capacity – and no apparent sign of formation of tissue-like assemblies (Custodio et al. 1998; Müller et al. 1999a). A schematic representation is given in Fig. 13.9.

The primmorphs which had been attached to the substrate, the galectin-coated culture dishes, show a high expression of the gene encoding integrin, and considerable levels of *thymidine kinase* and *noggin* transcript (Table 13.1), while the expression of perforin is low (Fig. 13.10).

Based on the recent observation, suggesting that the signal transduction pathway, originating from pattern-recognition receptors, activates also the MAP-kinase system (Lee and Kim 2007; Beisswenger et al. 2007), we investigated whether LPS exposure results in a differential alteration of gene expression in primmorphs. Indeed, it was determined that the expression level of the defense molecule against microorganisms, perforin, strongly increases from 0.1-fold (untreated primmorphs) to 5.6-fold (LPS-exposed primmorphs); Table 13.1. This upregulation is selective, since simultaneously the expression level of *noggin*, *thymidine kinase* and *integrin* undergo a drastic reduction (Table 13.1; Fig. 13.11). It remains to be studied if this effect is reversible. Considering these data – positive impact of anti-bacterial proteins on gene expression under simultaneous down-regulation of expression of those genes which regulate differentiation plasticity and cell differentiation – support the view that the LPS caused effect on cell metabolism results also in an induction of apoptotic signaling pathways.

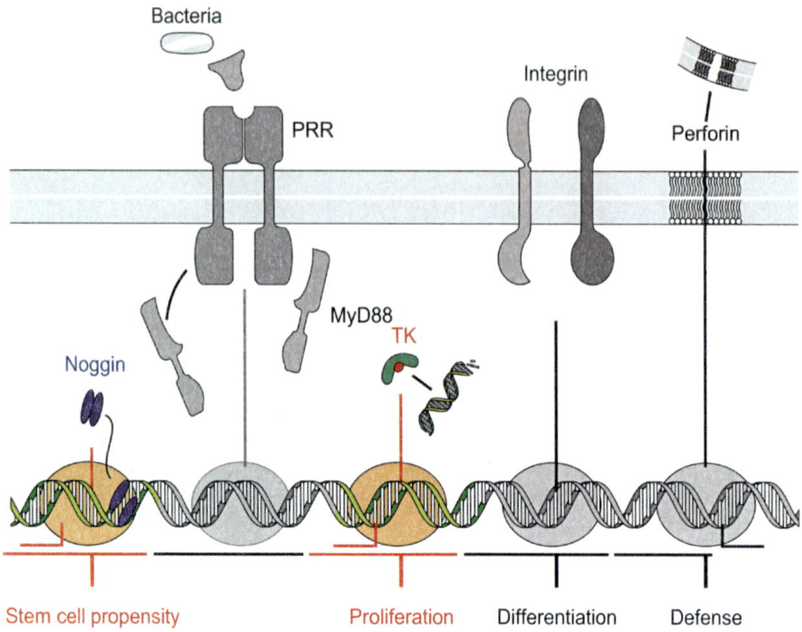

Fig. 13.9 Schematic outline of the signal-transduction pathways and of the gene expression levels in primmorphs which remained in a non-attached state in the culture dish. In these aggregates the cell-surface associated receptors, pattern-recognition receptor (PRR) and the integrins, are (if at all) only insignificantly expressed. Consequently, the underlying genes, *pattern-recognition receptor* and *integrin*, is minimal. Likewise the gene encoding the executing extracellular attacking protein, perforin, is not expressed. In contrast, the transcript level of *noggin* (an indicator for the number of stem cells) and *thymidine kinase* is high, indicating a substantial level of cells that are provided with high differentiation plasticity and proliferation capacity

13.6 Concluding Remarks

There was never a faster progress in the understanding of the differentiation capacity of sponge cells than during the last 10 years when sophisticated molecular biological techniques were developed and applied. It became possible to trace back empirical observations on a causal-analytical basis. Already over 100 years ago the astonishing regenerative power of sponge cells (Schmidt 1862), and of archaeocytes in particular (Wilson 1907), had been recognized. The concept of embryonic unspecialized cells in hexactinellids had been introduced by Schulze (1904) who coined them "sorites". These sorites were described as separate cell types besides the germ cells. The processes of spermatogenesis, oogenesis and embryogenesis of Demospongiae had been described in detail by Noll (1888). Interesting – but at the time not recognized – are the contributions of Diaz (1979) on the transformation of cells in sponges through differentiation and de-differentiation. Especially his view on the plastic differentiation capacity between archaeocytes and

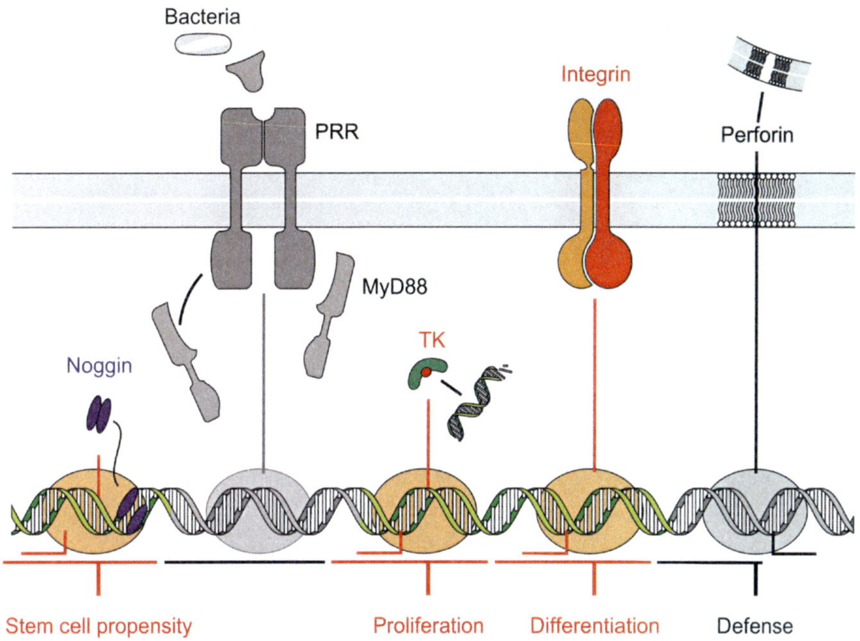

Fig. 13.10 Expression level in primmorphs which had been attached to galectin-coated petri dishes. The *integrin* transcript level is high and also the level for *noggin* and *thymidine kinase* is substantial. Again, the amount of transcripts encoding the toxic protein perforin are low

epithelial cells is here relevant. After the identification of the first protein-coding genes in *G. cydonium* and *S. domuncula* it became suddenly overt that sequences of sponge genes are metazoan-like and share high sequence similarity to those found in humans (Pfeifer et al. 1993). In addition, the *in vitro* primmorph cell culture system became a powerful tool to study gene expression in "embryonic" cells.

Focusing on the stem cell system in sponges the main lessons are; (i) their cells progress from a primordial stage to terminally differentiated stages, (ii) they contain totipotent stem cells, (iii) during the progression from stem cells to differentiated cells genes are expressed among which some share high sequence similarity to those identified in vertebrates. At present the prevailing notion recognizes the high plasticity of stem cells because of the high regeneration/repair capacity of somatic sponge cells. Molecular genetic studies in sponges will continue to assess this view. After all, it is expected that ultimately sponges become the model (Pilcher 2005). Finally it could be demonstrated that an exposure of primmorphs to LPS causes a reduction of the number of stem cells under simultaneous upregulation of the defense molecule perforin.

Note: The following sequences from *Suberites domuncula* have been deposited (EMBL/GenBank): the cDNA for the glycerol 3-phosphate dehydrogenase

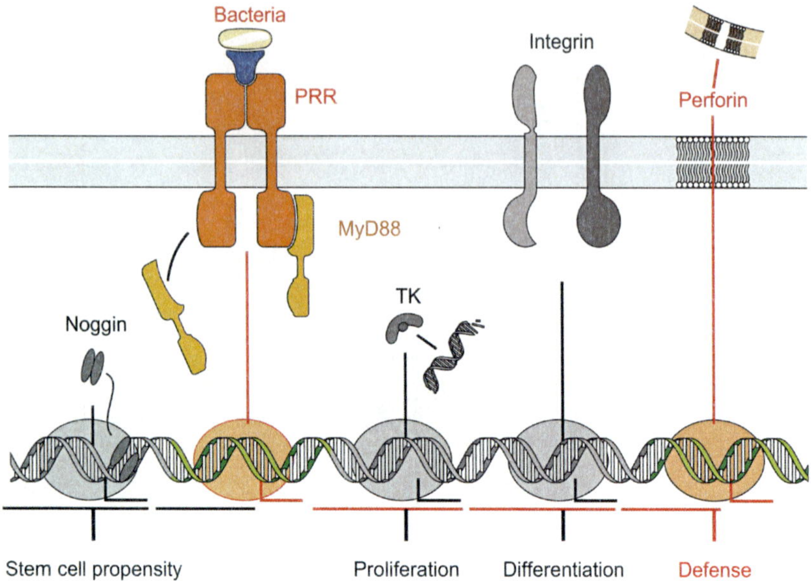

Fig. 13.11 Innate immune response in sponges against Gram-negative bacteria. The simplified model of the LPS-mediated signaling shows that LPS, a pathogen-associated molecular pattern (PAMP), causes (homo)dimerization of the LPS-binding protein, a pattern-recognition receptor (PRR), on the cell surface. After complex formation between PAMP and PRR, PRR initiates a signal transduction cascade, with MyD88 as the first member. Downstream, intermediate members of the signaling pathway, which likely involves IRAKs, TRAFs, NF-κB and also LITAF (to be published), transduce an increased expression of the activation phase's key molecules, PRR and MyD88, and finally of the executing effector/defense molecule, the perforin-like protein

(GAPDH_SD) under AM902265 and for thymidine kinase (THYMKI_SD) under AM905441.

Acknowledgments This work was supported by a grant from the Deutsche Forschungsgemeinschaft (WI 2116/2-2), the Bundesministerium für Bildung und Forschung (project C1 20 20: cooperation project WTZ BRA – Health of marine ecosystems (03F0451A), and Center of Excellence *BIOTEC*marin) and the International Human Frontier Science Program (RG-333/ 96-M).

References

Adell T, Grebenjuk VA, Wiens M, Müller WEG (2003a) Isolation and characterization of two T-box genes from sponges, the phylogenetically oldest metazoan taxon. Dev Genes Evol 213:421–434

Adell T, Nefkens I, Müller WEG (2003b) Polarity factor "Frizzled" in the demosponge *Suberites domuncula*: Identification, expression and localization of the receptor in the epithelium/pinacoderm. FEBS Lett 554:363–368

Adell T, Thakur AN, Müller WEG (2007) Isolation and characterization of Wnt pathway-related genes from Porifera. Cell Biol Int 31:939–949

Althoff K, Schütt C, Steffen R, Batel R, Müller WEG (1998) Evidence for a symbiosis between bacteria of the genus *Rhodobacter* and the marine sponge *Halichondria panicea*: Harbor also for putatively-toxic bacteria? Mar Biol 130:529–536

Beisswenger C, Coyne CB, Shchepetov M, Weiser JN (2007) Role of p38 MAP kinase and transforming growth factor-ß signaling in transepithelial migration of invasive bacterial pathogens. J Biol Chem 282:28700–28708

Bergquist PR (1978) Sponges. University of California Press, Berkeley

Böhm M, Schröder HC, Müller IM, Müller WEG, Gamulin V (2000) The mitogen-activated protein kinase p38 pathway is conserved in metazoans: Cloning and activation of p38 of the SAPK2 subfamily from the sponge *Suberites domuncula*. Biol Cell 92:95–104

Böhm M, Hentschel U, Friedrich A, Fieseler L, Steffen R, Gamulin V, Müller IM, Müller WEG (2001) Molecular response of the sponge *Suberites domuncula* to bacterial infection. Mar Biol 139:1037–1045

Böhm M, Gamulin V, Schröder HC, Müller WEG (2002) Evolution of osmosensing signal transduction in metazoa: Stress activated protein kinases p38 and JNK. Cell Tissue Res 308:431–438

Borojevic R (1966) Étude expérimentale de la différentiation des cellules de l' éponge au cours de son développement. Dev Biol 14:130–153

Borojevic R (1970) Différenciation cellulaire dans l'embryogenèse et la morphogenèse chez les spongiaires. Symp Zool Soc Lond 25:467–490

Borojevic R (1971) Le comportement des cellules d' éponge lors de processus morphogénétique. Ann Biol 10:533–545

Brusca RC, Brusca GJ (1990) Invertebrates. Sunderland, Mass, Sinauer

Cai JC, Weiss ML, Rao MS (2004) In search of "stemness". Exp Hematol 32:585–598

Cho HH, Bae YC, Jung JS (2006) Role of Toll-like receptors on human adipose-derived stromal cells. Stem Cells 24:2744–2752

Cory S, Adams JM (1998) Matters of life and death: Programmed cell death at Cold Spring Harbor. Biochim Biophys Acta 1377:R25-R44

Custodio MR, Prokic I, Steffen R, Koziol C, Borojevic R, Brümmer F, Nickel M, Müller WEG (1998) Primmorphs generated from dissociated cells of the sponge *Suberites domuncula*: A model system for studies of cell proliferation and cell death. Mech Ageing Dev 105:45–59

Davletov BA, Suedhof TC (1993) A single C2 domain from synaptotagmin I is sufficient for high affinity Ca2+/phospholipid binding. J Biol Chem 268:26386–26390

Diaz JP (1977) Transformation histologiques et cytologiques post-traumatiques chez la demosponge *Suberites massa* Nardo. Bull Museum Natural History Nature (Paris) 445:375–396

Diaz JP (1979) Variations, differenciations et functions des categories cellulaires de la demosponge d'eaux saumatres, *Suberites massa*, Nardo, au cours du cycle biologique annuel et dans des conditions experimentales. Thèse, Université des Sciences et techniques du Languedoc, Montpellier

Felsenstein J (1993) PHYLIP, ver. 3.5. University of Washington, Seattle

Friedrich AB, Merkert H, Fendert T, Hacker J, Proksch P, Hentschel U (1999) Microbial diversity in the marine sponge *Aplysina cavernicola* (formerly *Verongia cavernicola*) analyzed by fluorescence in situ hybridisation (FISH). Mar Biol 134:461–470

Galliot B, Miller D (2000) Origin of anterior patterning – how old is our head? Trends Genetics 16:1–5

Garrone R (1978) Phylogenesis of Connective Tissue. Morphological Aspects and Biosynthesis of Sponge Intercellular Matrix. Karger S, Basel

Grunz H (ed) (2004) The Vertebrate Organizer. Springer Verlag, Berlin, Heidelberg, New York

Hoshiyama D, Suga H, Iwabe Y, Koyanagi M, Nikoh N, Kuma K, Matsuda F, Honjo T, Miyata T (1998) Sponge Pax cDNA related to Pax-2/5/8 and ancient gene duplications in the Pax family. J Mol Evol 47:640–648

Jiang Q, Akashi S, Miyake K, Petty HR (2000) Cutting edge: Lipopolysaccharide induces physical proximity between CD14 and Toll-like receptor 4 (TLR4) prior to nuclear translocation of NF-κB. J Immunol 165:3541–3544

Jones L (2001) Stem cells: So what's in a niche? Curr Biol 11:R484–R486

Krasko A, Schröder HC, Batel R, Grebenjuk VA, Steffen R, Müller IM, Müller WEG (2002) Iron induces proliferation and morphogenesis in primmorphs from the marine sponge *Suberites domuncula*. DNA Cell Biol 21:67–80

Laurent L (1842) Recherches sur l'hydre et l'éponge d'eau douce. Arthus Bertrand, Paris

Lee MS, Kim YJ (2007) Pattern-recognition receptor signaling initiated from extracellular, membrane, and cytoplasmic space. Mol Cells 23:1–10

Müller WEG (2001) How was metazoan threshold crossed: The hypothetical Urmetazoa. Comp Biochem Physiol [A] 129:433–460

Müller WEG (2005) Spatial and temporal expression patterns in animals. In: Meyers RA (ed) Encyclopedia of Molecular Cell Biology and Molecular Medicine, vol 13. WILEY-VCH Press, Weinheim. pp 269–309

Müller WEG (2006) The stem cell concept in sponges (Porifera): Metazoan traits. Semin Cell Dev Biol 17:481–491

Müller WEG, Schäcke H (1996) Characterization of the receptor proteintyrosine kinase gene from–the marine sponge *Geodia cydonium*. Progr Molec Subcell Biol 17:183–208

Müller WEG, Blumbach B, Wagner-Hülsmann C, Lessel U (1997) Galectins in the phylogenetically oldest metazoa, the sponges [Porifera]. Trends Glycosci Glycotechnol 9:123–130

Müller WEG, Wiens M, Batel R, Steffen R, Borojevic R, Custodio MR (1999a) Establishment of a primary cell culture from a sponge: Primmorphs from *Suberites domuncula*. Mar Ecol Prog Ser 178:205–219

Müller WEG, Blumbach B, Müller IM (1999b) Evolution of the innate and adaptive immune systems: Relationships between potential immune molecules in the lowest metazoan phylum [Porifera] and those in vertebrates. Transplantation 68:1215–1227

Müller WEG, Perovic S, Wilkesman J, Kruse M, Müller IM, Batel R (1999c) Increased gene expression of a cytokine-related molecule and profilin after activation of *Suberites domuncula* cells with xenogeneic sponge molecule(s). DNA Cell Biol 18:885–893

Müller WEG, Schröder HC, Skorokhod A, Bünz C, Müller IM, Grebenjuk VA (2001) Contribution of sponge genes to unravel the genome of the hypothetical ancestor of Metazoa (Urmetazoa). Gene 276:161–173

Müller WEG, Krasko A, Skorokhod A, Bünz C, Grebenjuk VA, Steffen R, Batel R, Müller IM, Schröder HC (2002) Histocompatibility reaction in the sponge *Suberites domuncula* on tissue and cellular level: Central role of the allograft inflammatory factor 1. Immunogenetics 54:48–58

Müller WEG, Wiens M, Müller IM, Schröder HC (2003a) The chemokine networks in sponges: Potential roles in morphogenesis, immunity and stem cell formation. Prog Mol Subcell Biol 34:103–143

Müller WEG, Korzhev M, Le Pennec G, Müller IM, Schröder HC (2003b) Origin of metazoan stem cell system in sponges: First approach to establish the model (*Suberites domuncula*). Biomol Eng 20:369–379

Müller WEG, Wiens M, Adell T, Gamulin V, Schröder HC, Müller IM (2004) Bauplan of Urmetazoa: Basis for Genetic Complexity of Metazoa. Int Rev Cytol 235:53–92

Noll FC (1888) Beiträge zur Naturgeschichte der Kieselschwämme. Abhandlungen Senckenbergische Naturforscher Gesellschaft Frankfurt 2:1–58

Nüsslein-Volhard C, Wieschaus E (1980) Mutations affecting segment number and polarity in Drosophila. Nature 287:795–801

Ogawa K, Ishihara S, Saito Y, Mineta K, Nakazawa M, Ikeo K, Gojobori T, Watanabe K, Agata K (2002) Induction of a noggin-like gene by ectopic DV interaction during planarian regeneration. Dev Biol 250:59–70

Perovic S, Schröder HC, Sudek S, Grebenjuk VA, Batel R, Štifanic M, Müller IM, Müller WEG (2003) Expression of one sponge *Iroquois* homeobox gene in primmorphs from *Suberites domuncula* during canal formation. Evol Dev 5:240–250

Perović-Ottstadt S, Adell T, Proksch P, Wiens M, Korzhev M, Gamulin V, Müller IM, Müller WEG (2004a) A (1→3)-ß-D-glucan recognition protein from the sponge *Suberites domuncula*:

Mediated activation of fibrinogen-like protein and epidermal growth factor gene expression. Eur J Biochem 271:1924–1937
Perović-Ottstadt S, Ćetković H, Gamulin V, Schröder HC, Kropf K, Moss C, Korzhev M, Diehl-Seifert B, Müller IM, Müller WEG (2004b) Molecular markers for germ cell differentiation in the demosponge *Suberites domuncula*. Int J Dev Biol 48:293–305
Pfeifer K, Haasemann M, Gamulin V, Bretting H, Fahrenholz F, Müller WEG (1993) S-type lectins occur also in invertebrates: High conservation of the carbohydrate recognition domain in the lectin genes from the marine sponge *Geodia cydonium*. Glycobiology 3:179–184
Pilcher H (2005) Back to our roots. Nature 435:1022–1023
Saitou N, Nei M (1987) Neighbor-joining method: A new method for reconstructing phylogenetic trees. Mol Biol Evol 4:406–425
Schmidt EW, Obraztsova AY, Davidson SK, Faulkner DJ, Haygood MG (2000) Identification of the antifungal peptide-containing symbiont of the marine sponge *Theonella swinhoei* as a novel delta-proteobacterium, 'Candidatus Entotheonella palauensis'. Mar Biol 136: 969–977
Schmidt O (1862) Die Spongien des Adriatischen Meeres. Engelmann W, Leipzig
Schröder HC, Krasko A, Batel R, Skorokhod A, Pahler S, Kruse M, Müller IM, Müller WEG (2000) Stimulation of protein (collagen) synthesis in sponge cells by a cardiac myotrophin-related molecule from *Suberites domuncula*. FASEB J 14:2022–2031
Schröder HC, Krasko A, Le Pennec G, Adell T, Hassanein H, Müller IM, Müller WEG (2003) Silicase, an enzyme which degrades biogenous amorphous silica: Contribution to the metabolism of silica deposition in the demosponge *Suberites domuncula*. Prog Mol Subcell Biol 33:249–268
Schröder HC, Natalio F, Wiens M, Tahir MN, Shukoor MI, Tremel W, Belikov SI, Krasko A, Müller WEG (2008) The 2'-5'-oligoadenylate synthetase in the lowest Metazoa: Isolation, cloning, expression and functional activity in the sponge *Lubomirskia baicalensis*. Mol Immunol 45:945–953
Schulze FE (1904) Hexactinellida. Wissenschaftliche Ergebnisse der Deutschen Tiefsee-Expedition, vol 4. Fischer-Verlag, Jena
Scott MG, Vreugdenhil AC, Buurman WA, Hancock RE, Gold MR (2000) Cutting edge: Cationic antimicrobial peptides block the binding of lipopolysaccharide (LPS) to LPS-binding protein. J Immunol 164:549–553
Seger R, Krebs EG (1995) The MAPK signaling cascade. FASEB J 9:726–735
Segura-Pena D, Lutz S, Monnerjahn C, Konrad M, Lavie A (2007) Binding of ATP to TK1-like enzymes is associated with a conformational change in the quaternary structure. J Mol Biol 369:129–141
Sen S, Kundu G, Mekhail N, Castel J, Misono K, Healy B (1990) Myotrophin: Purification of a novel peptide from spontaneously hypertensive rat heart that influences myocardial growth. J Biol Chem 265:16635–16643
Sewalt RGAB, Vlag J, Gunster MJ, Hamer KM, Blaauwen JL, Satijn DPE, Hendrix T, Driel R, Otte AP (1998) Characterization of interactions between mammalian Polycomb-group proteins Enx/EZH2 and EED suggests the existence of different mammalian Polycomb-group protein complexes. Mol Cell Biol 18:3586–3595
Sil P, Kandaswamy V, Sen S (1998) Increased protein kinase C activity in myotrophin-induced myocyte growth. Circ Res 82:1173–1188
Simpson TL (1984) The cell biology of sponges. Springer-Verlag, New York
Smith WC, Harland RM (1992) Expression cloning of noggin, a new dorsalizing factor localized to the Spemann organizer in *Xenopus* embryos. Cell 70:829–840
Smith WC, Knecht AK, Wu M, Harland RM (1993) Secreted noggin protein mimics the Spemann organizer in dorsalizing mesoderm. Nature 361:547–549
Sollas WJ (1888) Tetractinellida. In: Wyville Thomson C (ed) Report on the scientific results of the voyage of HMS Challenger. Eyre & Spottiswoode, London, pp 1–453

Thakur NL, Hentschel U, Krasko A, Pabel CT, Anil AC, Müller WEG (2003) Antibacterial activity of the sponge *Suberites domuncula* and its primmorphs: Potential basis for the defense. Aquat Microb Ecol 31:77–83

Ulevitch RJ, Tobias PS (1994) Recognition of endotoxin by cells leading to transmembrane signaling. Curr Opin Immunol 6:125–130

Vacelet J, Vacelet E, Gaino E, Gallissian MF (1994) Bacterial attack of spongin skeleton during the 1986–1990 Mediterranean sponge disease. In Van Soest RWM, Van Kempen TMG, Braekman JC (eds) Sponges in time and space. Balkema, Rotterdam, pp 355–362

Valenzuela DM, Economides AN, Rojas E, Lamb TM, Nunez L, Jones P, Ip NY, Espinosa R, Brannan CI, Gilbert DJ, Copeland NG, Jenkins NA, LeBeau MM, Harland RM, Yancopoulos GD (1995) Identification of mammalian noggin and its expression in the adult nervous system. J Neurosci 15:6077–6084

Wagner AJ, Weissman IL (2004) Plasticity of adult stem cells. Cell 116:639–648

Wagner C, Steffen R, Koziol C, Batel R, Lacorn M, Steinhart H, Simat T, Müller WEG (1998) Apoptosis in marine sponges: A biomarker for environmental stress (cadmium and bacteria). Mar Biol 131:411–421

Wiens M, Mangoni A, D'Esposito M, Fattorusso E, Korchagina N, Schröder HC, Grebenjuk VA, Krasko A, Batel R, Müller IM, Müller WEG (2003) The molecular basis for the evolution of the metazoan bodyplan: Extracellular matrix-mediated morphogenesis in marine demosponges. J Mol Evol 57:S60-S75

Wiens M, Perović-Ottstadt S, Müller IM, Müller WEG (2004) Allograft rejection in the mixed cell reaction system of the demosponge *Suberites domuncula* is controlled by differential expression of apoptotic genes. Immunogenetics 56:597–610

Wiens M, Korzhev M, Krasko A, Thakur NL, Perović-Ottstadt S, Breter HJ, Ushijima H, Diehl-Seifert B, Müller IM, Müller WEG (2005) Innate immune defense of the sponge *Suberites domuncula* against bacteria involves a MyD88-dependent signaling pathway: Induction of a perforin-like molecule. J Biol Chem 280:27949–27959

Wiens M, Belikov SI, Kaluzhnaya OV, Krasko A, Schröder HC, Perovic-Ottstadt S, Müller WEG (2006) Molecular control of serial module formation along the apical-basal axis in the sponge *Lubomirskia baicalensis*: Silicateins, mannose-binding lectin and Mago Nashi. Dev Genes Evol 216:229–242

Wilkinson CR, Garrone R, Vacelet J (1984) Marine sponges discriminate between food bacteria and bacterial symbionts: Electron microscope radioautography and in situ evidence. Proc R Soc Lond B 220:519–528

Wilson HV (1907) On some phenomena of coalescence and regeneration in sponges. J Exp Zool 5:245–258

Wimmer W, Blumbach B, Diehl-Seifert B, Koziol C, Batel R, Steffen R, Müller IM, Müller WEG (1999a) Increased expression of integrin and receptor tyrosine kinase genes during autograft fusion in the sponge *Geodia cydonium*. Cell Adhes Commun 7:111–124

Wimmer W, Perovic S, Kruse M, Krasko A, Batel R, Müller WEG (1999b) Origin of the integrin-mediated signal transduction: Functional studies with cell cultures from the sponge *Suberites domuncula*. Eur J Biochem 178:156–165

Wolpert L (1998) Principles of development. Oxford University Press, Oxford

Yang H, Young DW, Gusovsky F, Chow JC (2000) Cellular events mediated by bipopolysaccharide-stimulated toll-like receptor 4. MD-2 is required for activation of mitogen-activated protein kinases and Elk-1. J Biol Chem 275:20861–20866.

Yoshikawa T, Nagasugi Y, Azuma T, Kato M, Sugano S, Hashimoto K, Masuho Y, Muramatsu M, Seki N (2000) Isolation of novel mouse genes differentially expressed in brain using cDNA microarray. Biochem Biophys Res Comm 275:532–537

Zou GM, Reznikoff-Etiénat MF, Milliez J, Léon W, Hirsch F (2000) Fas-mediated apoptosis of mouse embryo stem cells: Its role during embryonic development. Am J Reprod Immunol 43:240–248

Chapter 14
Defence Mechanisms and Stem Cells in *Holothuria polii* and *Sipunculus nudus*

Giuseppina D'Ancona Lunetta

Abstract *Holothuria polii* (Phylum Echinodermata) and *Sipunculus nudus* (Phylum Sipuncula) have been used as model systems for studying invertebrate immunity. In the coelomic fluid of *Holothuria polii*, coelomocytes (i.e. amoebocytes, Type I, Type II, Type III spherule cells) and stem cells are present. After an injection of formalin-fixed sheep erythrocytes into the coelomic cavity, structural modifications in the animal's, such as the coelomic cavity, stone canal and periesophageal ring were observed. Cellular events observed within the coelomic cavity include the recruitment of a large number of coelomocytes from the haemopoietic areas, phagocytosis carried out by amoebocytes and the formation of brown masses. The stone canal is not an immunocompetent organ but it is involved in the production and activity of the amoebocytes. The periesophageal ring is an organ which produces spherule cells. Haemerythrocytes, urn cell complexes, brown bodies, Type I and Type II granulocytes and the lamine of connective tissue are all distinguishable in the fluid coelomic of *Sipunculus nudus*. Clusters of Type II granulocytes at various differentiation stages are also present. The consequences of a cutaneous wound in *Holuthuria polii* and *Spinuculus nudus* and the cells involved in healing were studied. Type I granulocytes in *Sipunculus nudus* are capable of extracellular digestion and they are immunoreactive to antibodies directed against IL-4, IL-10 and EGF. The results obtained regarding the healing of the wound, the presence and location of stem cells are comparable to the results obtained from vertebrates.

Keywords Defence mechanisms · *Holothuria polii* · *Sipunculus nudus* · Stem cell · Wound repair

---

G. D'Ancona Lunetta (✉)
Institute of Hystology and Embryology, Department of Biology, Viale delle Scienze, Edificio 16, University of Palermo, Palermo, Italy
e-mail: dancona@unipa.it

B. Rinkevich, V. Matranga (eds.), *Stem Cells in Marine Organisms*,
DOI 10.1007/978-90-481-2767-2_14, © Springer Science+Business Media B.V. 2009

14.1 Invertebrate Immunity

Invertebrates which are challenged by foreign agents respond by invoking innate immunity, either from the humoral or cellular arms of the organism. Regarding the humoral arm, molecules of various types (such as agglutinine, lectin and lysine) possess a defensive function which involves a defensive function and it confers an immune capacity to invertebrates (Nicosia 1979; Yeaton 1981; Rheins et al. 1980; Ey and Jenkin 1982; Amirante and Basso 1984; Olafsen 1986; Parrinello and Arizza 1988, 1989; Canicattì 1989, 1990, 1991; Canicattì and D'Ancona 1990; Parrinello 1991; Arizza et al. 1991, 1993; Cammarata et al. 1993; Vasquez et al. 1996; Leclerc 1996; Fragkiadakis 2000; Glinski and Jarosz 2000; Raftos et al. 2001; Vasta et al. 2001; Tiscar and Mosca 2004).

A simple complement system, which is analogous to the alternative pathways of vertebrates has been highlighted in tunicates by Smith et al. (1999) and in sea urchins (Echinodermata) by Gross et al. (1999) and by Smith et al. (2001). The immune responses to introduced foreign material or injected particulate material are mediated principally by coelomocytes. Phagocytosis, cytotoxic activity, the formation of dark brown masses or brown bodies comprise various reactions of coelomocytes in invertebrates. A large number of studies have demonstrated various manifestations of reactions to coelomocytes in invertebrates. Numerous studies have been performed on tunicates but there have been few studies on echinoderms and, in particular, on the Holothuroidea (Hyman 1955; Parrinello et al. 1976, 1979; Smith 1981; Cooper 1982; Shinn 1985; Dybas and Frankboner 1986; Beck et al. 1993). Research into the immune cell response of *Holothuria polii* (Holothuroidae) and *Sipunculus nudus* (Sipunculoidea) has also been carried out. *Holothuria polii* and *Sipunculus nudus* provide us with two interesting models with which to study and compare the defence mechanisms with those of vertebrates, which are phylogenetically distant. Tissues in adult mammals, in which stem cells divide and differentiate into adult cells, are present. A typical example is haemopoietic tissue which is cabably of periodically substituting older haematic cells, which have been destroyed by haemocateretic organs. *Holuthuria polii* and *Sipunculus nudus* also possess coelomic fluid in which cells, with the functions of respiratory exchange and defence, are located. Since it can be proposed that the vitality of such cells is limited over time, it has also been hypothesized that *Holuthuria polii* and *Sipunculus nudus* possess stem cells which supply new coelomocytes. Confirming the presence of stem cells has been made possible by observing defence mechanisms following an injection of extraneous material into the coelomic cavity during the healing process of an induced cutaneous wound. The results obtained have been compared with those obtained from mammals.

The cellular events involved in the defense mechanisms of *Holothuria polii* have been observed after an injection of sheep formalinized erythrocytes (fSRBC), as antigens, into the coelomic cavity of the animal (Canicattì and D'Ancona 1989), Polian vesicles (D'Ancona et al. 1989), the stone canal (D'Ancona Lunetta 1996) and the periesaphageal ring (D'Ancona and Michelucci 2001).

14.2 *Holothuria polii*

Holothuria polii belongs to the class Holothuroidae in the phylum Echinodermata. This species has a cylindrical, elongated body with radial symmetry. The oral and aboral areas are located at the two extreme ends of the body and the pharynx, oesophagus, stomach and intestine are located in a posterior position to the mouth. The latter is surrounded by a crown of tentacles, that is, blind extrusions of the water system which are used by the animal to capture food. The water system is a complicated system of internal canals and external appendices which allow the animal to move and it is composed of the periesophageal ring, which is positioned around the oesophagus. The stone canal and Polian vesicles (on the opposite side) are connected to the periesophageal ring (Fig. 14.1). Coelomic fluid, containing numerous coelomocytes, runs throughout all the cavities of *Holothuria polii* (the perivisceral coelom, the water haemal and perihaemal systems).

14.2.1 Coelomocytes in Holothuria polii

In many invertebrates, cells, defined as leucocytes, amoebocytes and haemocytes, circulate in body fluid and they play an important function in the innate immune processes (Cooper 1976). In echinoderms, cells, called coelomocytes, circulate in the coelomic fluid of animals, in the haemal system (Hyman 1955) and they have the function of transporting, accumulating and clotting (Endean 1966). Coelomocytes also have the function of defending against invaders (Karp and Coffaro 1982). With the aim of furthering our understanding of the cells involved in the immune reactions of *Holothuria polii*, cells of the coelomic fluid were studied (Canicattì et al. 1989b; D'Ancona and Canicattì 1990). They were observed under the microscope, after they had been subjected to cytohemical and ctyoenzyme techniques. The results are summarized in Table 14.1.

These cells were also microstructurally studied using electron microscopy with the aim of observing organelles and possible intracellular inclusions, and to improve

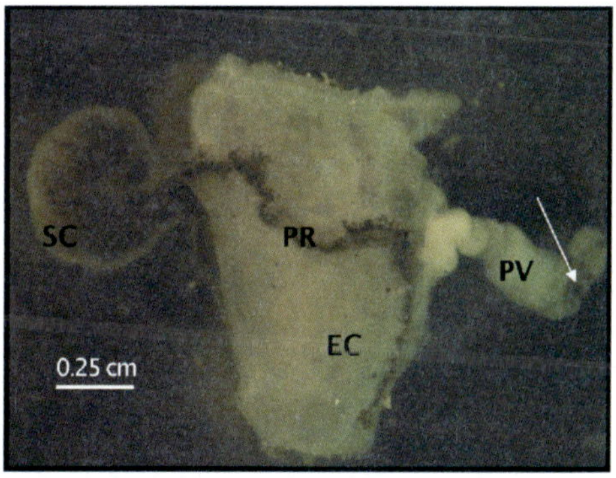

Fig. 14.1 A piece of *Holothuria polii* esophageal canal (EC), showing the stone canal (SC), periesophageal ring (PR) and the Polian vesicle (PV)

Table 14.1 Cytochemical reactions of spherule cells and amoebocytes

Reactions*	Type I	Type II	Type III	Amoeb.
0.1% Toluidine ethanol	+++	+	–	–
Alcian blue pH1	+++	+	–	–
Alcian blue	++	+	+++	++
Stempien	+++	–	n.d.	–
PAS	+	+++	+++	–
Ninidrina-Schiff	–	n.d.	n.d.	++
Chèvremont	+++	n.d.	n.d.	n.d
Biebrich scarlet	+	+	+++	++
Sudan black	+	–	+	–
Lillie	–	–	++	–
Schmorl	–	–	+++	–
Hueck	–	–	+	–
Acid phophatase	+	+–	+++	+
Alcaline phosphatase	+	+–	+	+
Chloracetate-Esterase	+++	+–	++	+++
Peroxidase	+++	+–	+	–

*Strength of reactions was rated on a + scale:
(–) negative
(+) weakly positive
(++) moderately positive
(+++) strongly positive
(+–) uncertain reaction
(n.d.) not determined

our understanding of the function of the coelomocytes; amoebocytes, spherule cells and stem cells were also highlighted. The amoebocytes varied markedly in size (12–25 μm) and form, revealing two major types: bladder and filipodial amoebocytes. The former are round in shape with a vacuolated cytoplasm and small protruding extensions; the latter display longer cytoplasmic extensions. The amoebocytes reveal a spherical nucleus with a small nucleolus, endoplasmatic reticulum and well-developed Golgi apparatus. The vacuoles in the cytoplasm are rich in engulfed material.

Amoebocyte morphology is similar to that described in other holothuroidae (Hetzel 1963; Fontane and Lambert 1977) and in other echinoderm species (Boolootian and Giese 1958; Endean 1966; Smith 1981; Kanungo 1984). The presence of phagocytic vacuoles indicates that the cells participate in the elimination of foreign material (Canicattì et al. 1989b) (Fig. 14.2A). The presence of RER in the amoebocytes of *Holothuria polii* confirms their capacity to produce haemolysins (Canicattì and Ciulla 1988), which are involved in internal defence mechanisms (Canicattì and Parrinello 1985). The spherule cells are spherical cells, 5–20 μ in diameter, which can be classified into three types: Type I basophilic spherule cells (Fig. 14.2B); Type II methachromatic spherule cells (Fig. 14.2C); and Type III cells with void vesicles (Fig. 14.2D).

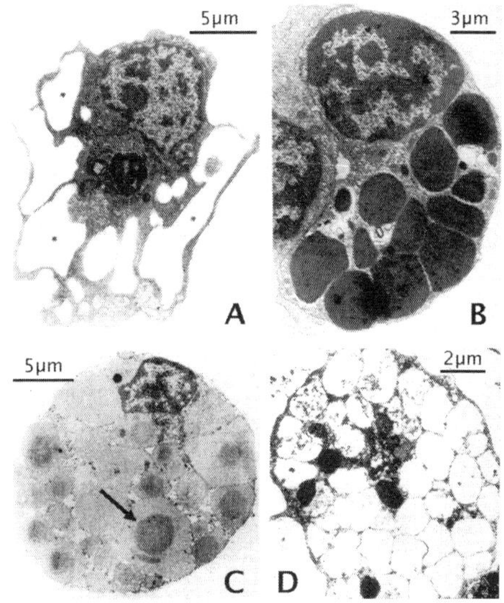

Fig. 14.2 Coelomocytes of *Holothuria polii*. (**A**) ultrastructural appearance of a phagocytic amoebocyte. Note the bladder-vesicles (∗). (**B**) electron micrograph of a Type I spherula cell. Note the electron-dense material constituting the granules. (**C**) electron micrograph of a Type II spherula cell. Note the granules with an inner electron-dense core. (**D**): electron micrograph of a Type III spherule cell.

Type I and Type II spherule cells act together with amoebocytes to produce an antigen-encapsulating structure, the brown body. These spherule cells contain lysosomial hydrolytic enzymes, such as arylsulphatase (Canicattì 1988). These enzymes could contribute to the destruction of material encapsulated during the formation of brown bodies. Type III spherule cells are involved in the synthesizing processes of melanin-like pigments, which are responsible for the characteristic brown colour of brown bodies. Stem cells, spherical in shape with a diameter of 6–8 μ, possess a large round nucleus, which is surrounded by a fine cytoplasmatic ring (Fig. 14.3A).

14.2.2 Cellular Aspects of the Immune Response of Holothuria polii

In 1989 Canicattì and D'Ancona observed cellular events occurring during the immune response of *Holothuria polii*. After an injection of formalinized sheep erythrocytes (fSRBC) into the coelomic cavity, brown masses, brown bodies, large quantities of free coelomocytes and "rosettes" were identified. Each rosette was made up of amoebocytes with 4–25 adhering erythrocytes, and this phase was the first part of the process of phagocytosis and it was followed by the internalisation and digestion of erythrocytes.

The recognition mechanisms are probably mediated by haemolysins in the coleomic fluid which could act in a similar manner to opsonins. Bertheussen (1983) has also attributed the function of opsonin to haemolysin in the coelomic fluid of sea urchins. The first phase accumulation of coelomocytes could be an indication

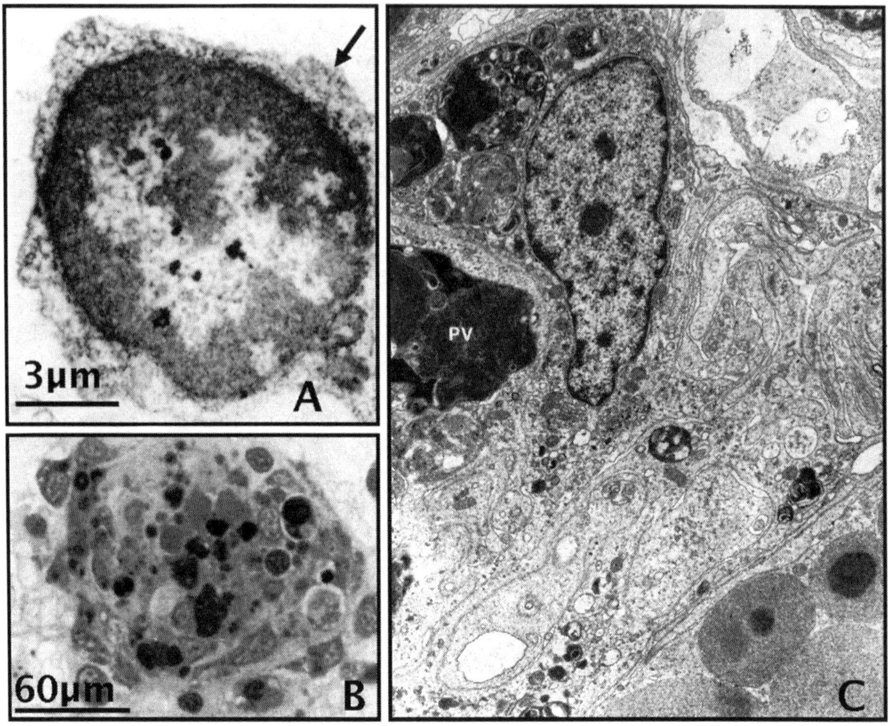

Fig. 14.3 (**A**) Electron micrograph of a stem cell. Note the large nucleus surrounded by a fine, cytoplasmic layer; (**B**) brown mass composed of phagocyting amoebocytes; (**C**) electron micrograph of fSRBC induced modular structure. The different cells maintain their individuality and no fusion occurs among them. (PV) phagocytic vacuoles with engulfed fSRBC. ×5430

of an immediate inflammatory response in *Holothuria polii*. The cells conglomerated and merged into syncitial structures, which can be defined as brown masses, rich in melanin (Fig. 14.3B). The latter are made up of ameobocytes containing endocytated erythrocytes and acidophilus spherule cells. The active transformations of the engulfed materials occur in the brown masses. Similar structures, defined as brown bodies, have been highlighted in other echinoderms by Hyman 1955; Hetzel 1965; and Dybas and Frankboner 1986. Canicattì et al. (1989a) studied natural and induced brown bodies by injecting fSRBC into the coelomic cavity of *Holothuria polii*. Natural brown bodies are constituded by variable numbers of nodules containing encapsulated invaders (gamontocysts of gregarine protozoans, or turbellarian egg capsules), immersed in a extracellular granular material in which spherule cells Type I, Type II and III were present. Induced brown bodies present the same characteristic as the natural brown bodies. However, the nodules are constituted by aggregated amoebocytes with engulfed material in their cytoplasm. Also described in the inflammatory processes of molluscs (Cheng and Galloway 1970), brown bodies have been subsequently held to be polikaryons arising from cell fusion (Anderson 1987). In both cases (natural and induced), brown bodies are made up of amoebocytes, which are surrounded by an extra-cellular matrix containing spherule cells. Commonly discovered by Mariano and Spector (1974) in vertebrates as a response to inflammatory processes, it seems that brown bodies may originate from cell fusion (Murch et al. 1982). In contrast no syncitial or junctional complexes have been identified by the electron microscope in the brown bodies of *Holothuria polii*. The amoebocytes appear to be aggregated and compacted due to the presence of numerous superimposed laminar protrusions between the cells (Fig. 14.3C). Moreover, the cohesion of the cells is probably made possible by the presence of extra-cellular granular material, which has a cementing function.

14.2.3 The Stone Canal

The stone canal is a short, non-calcified canal, approximately 1.28 cm in length, which is not connected to external body surfaces. It displays a proximal extremity, connected to a water ring and a distal part (the madreporic swelling), hanging into the coelom. A central tube descends to the underlying stone canal and the canal is surrounded by a very thick layer of connective tissue (Fig. 14.4A). The connective tissue at the proximal extremity of the stone canal is compact, displaying a smooth surface. The connective tissue of the madreporic swelling is loose and covered by a multi-layered epithelium with numerous pores (Fig. 14.4B). Amoebocytes containing fragments of Type II spherule cells can be observed between the undulating shape of the fibres of the connective tissue. It is important that the amoebocytes, which perform an phagocytic activity, are continuously substituted by new coelomocytes originating from the multi-layered epithelium. The epithelium is made up of cube-shaped cells which assume a cylindrical shape before undergoing mitosis and they migrate to the underlying connective tissue (Fig. 14.4C); here these cells differentiate into amoebocytes. Following an injection of fSRBC, the external epithelium of the stone canal thins out and the underlying connective tissue becomes slack and devoid of cells. An injection attracts the amoebocytes in the connective tissue and induces the production of new cells by the external epithelium of the stone canal. Numerous cells migrate into the cavity of the stone canal (Fig. 14.4D). 96 h after an injection of fSRBC into the ceolomic cavity, the quantity of red cells to be eliminated diminishes. No other amoebocytes migrate and subsequently the connective tissue and external epithelium are found to be enriched with cells (Fig. 14.4E). The ameobocytes in *Holothuria polii* are immunocompetent cells, reacting against foreign material. They possess the characteristic of engulfing particulate material from the same animal forming brown masses which are composed of ameobocytes and Type II spherule cells (Canicattì and D'Ancona 1989). The stone canal is an organ which, due to its structure and cellular composition, is capable of performing limited phagocytosis. The main importance of the stone canal is its capacity to supply and release new amoeboycytes into the coelomic cavity. Therefore,

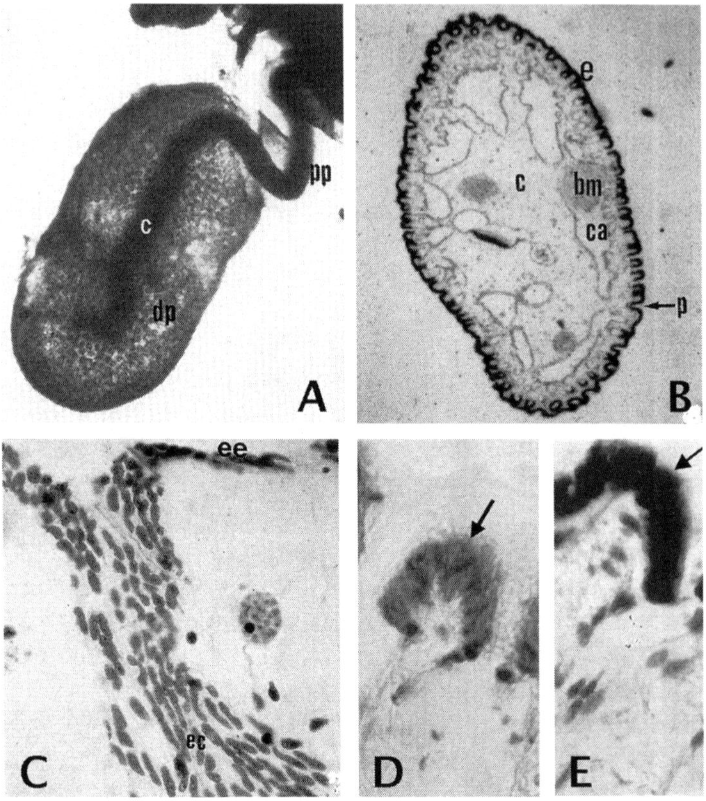

Fig. 14.4 (**A**) General view of the stone canal, pp= proximal part, dp= distal part of madreporic swelling, c= central canal. ×17.3; (**B**) cross section of the madreporic swelling showing the external epithelium (e) with numerous pores (p) communicating with the central cavities (ca) present in the loose connective tissue; brown masses (bm) are also visible. ×40; (**C**) epithelial cells (ec) migrating into the lower connective tissue (c); ee, external epithelium. ×400; (**D**) cross section of the madreporic swelling in a control and in a sample injected with fSRBC after 24 h (**E**)×160

the stone canal is a haemopoietic organ with the multi-layered epithelium of the madreporic swelling representing a continuous source of stem cells (D'Ancona Lunetta 1996).

14.2.4 Polian Vesicles

The Polian vesicles in *Holothuria polii* were examined as part of a study on the tissues and organs which are active in the inflammatory response after an injection of erythrocytes (fSRBC). Passing from the external to internal walls, the structure of the vesicle walls is composed of a layer of connective tissue containing coelomocytes, a layer of circular muscular fibres and, finally, a multi-layered epithelium

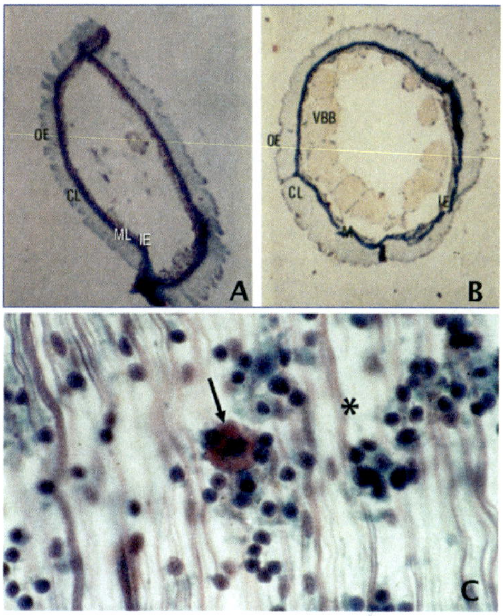

Fig. 14.5 Gomori triple staining of Polian vesicle from (**A**) undisturbed and (**B**) fSRBC injected animals; OE, outer epithelium; CL, connective layer; ML, muscular layer; IE, inner epithelium layer; VBB, vesicular brown body. ×10 (**C**) selected image of the reacting muscular layer. Note the large acidophil spherule in the center and numerous coelomocytes migrating in muscular layer ×800

lining a wide lumen (Fig. 14.5A). It has been demonstrated that circulating coelomocytes are immunocompetent cells, capable of reacting against foreign agents (Karp and Coffaro 1982; Canicattì and D'Ancona 1989). Polian vesicles also react against foreign agents.

After an injection of fSRBC into the coelomic cavity, a series of reactions occurred which could be traced back to an inflammatory process: the connective layer thickened but it was of a looser consistency (Fig. 14.5B). Numerous coelomocytes, originating from the external layer, crossed the muscular layer towards the internal epithelium (Fig. 14.5C). The latter is modified, becoming involved in the formation of numerous brown bodies, which are similar to those present in the coelomic cavity (Canicattì and D'Ancona 1989) (Fig. 14.6). The cause of the inflammatory process is not clear. Since the Polian vesicles are in communication via the periesophageal ring with the coelomic cavity, the hypothesis that the red blood cells (fSRBC) in the coelomic cavity easily pass into the Polian vesicles seems probable; there they would activate a defence reaction. No cells in the process of mitosis were observed in the Polian vesicles. There are, therefore, no stem cells in the Polian vesicles.

14.2.5 Periesophageal Ring

The inflammatory reaction in *Holothuria polii* was also studied in the periesophageal ring after injecting fSRBC into the coelomic cavity (D'Ancona and Michelucci 2001). The periesophageal ring in *Holothuria polii* is a wide canal, approximately 30 μm thick, and it is made up of connective tissue. It is the structure

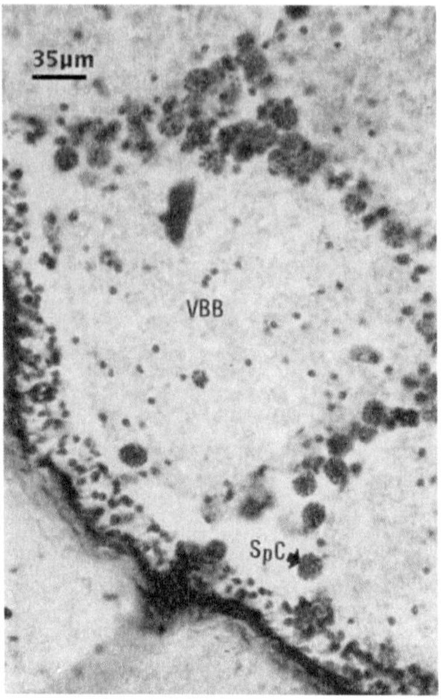

Fig. 14.6 Selected image showing Type I spherule cells (SpC) in vesicular brown bodies

which carries the water system and is connected to the Polian vesicles, the stone canal and the coelomic cavity (Fig. 14.7A). The periesophageal ring is made up of loose connective tissue in which Type I, Type II and Type III spherule cells are present and distributed in a homogeneous way (Fig. 14.7B). There is no significant numerical difference in the spherule cells of the proximal and distal parts in control animals. The term "proximal" refers to the extreme part of the ring towards the mouth; "distal" refers to the opposite part.

Holothuria polii is a benthic animal (Tortonese 1985), in which numerous commensal and parasitic animals are found, including protozoa and metazoa (Jangoux 1987a, b). The animal reacts by forming brown bodies with spherule cells and amoebocytes. After stimulation with fSRBC, Type I and Type II spherule cells are reduced in number over a period of 48 h. The connective tissue of the periesophageal ring becomes looser and small spherical cells, some in the process of mitosis, are highlighted (Fig. 14.7D). Type I and Type II spherule cells appear to be more transparent, small and disorganised than Type II spherule cells. Some Type III spherule cells are devoid of granules (Fig. 14.7C).

Ohuye (1938) has proposed that in various invertebrates, including two Holothurian species, lymphocytes represent primitive pluripotent cells, from which all other coelomocyte types are derived. Endean (1958) has hypothesized that amoebocytes in *Holothuria leucospilata* originate from the lining epithelium of respiratory trees, after which they migrate into the coelomic fluid where they are differentiated into spherule cells. According to Hetzel (1965), some lymphocyte-similar cells, which have differentiated from the haematic vessel walls of the mesenchymal connective tissue, become spherule cells. No mitotic spherule cells in circulation have been recorded in either the coelomic cavity, the stone canal or

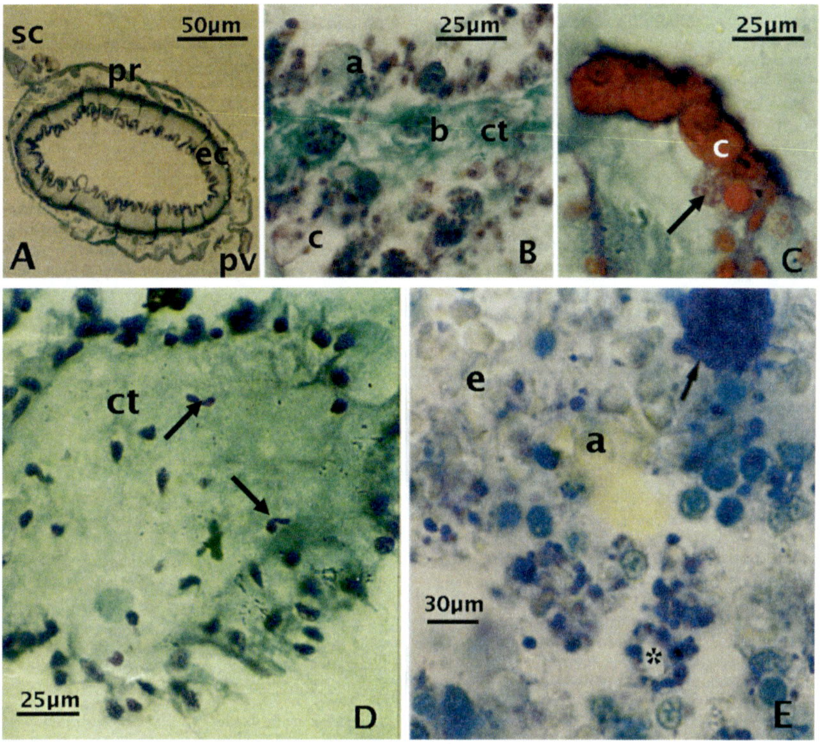

Fig. 14.7 (A/C) Non injected specimens. **A**: cross section of the esophageal canal (ec) with the periesophageal ring (pr); (sc), stone canal; (pv), Polian vesicle. **B**: selected image of the periesophageal ring. Note Type I spherule cells (b), Type II spherule cells (a) and Type III spherule cells (c) in the connective tissue (ct). **C**: selected image of Type III spherule cells (c) some of wich are degranulated *arrow*. **D**: injected specimen. Cells undergoing mitosis *arrow* in the connective tissue (ct) of the periesophageal ring. (A/D) Coloured with Gomori triple staining. **E**: cellular aspect of the immune response as visible in the space between the wall of the alimentary canal and the periesophageal ring. Type I spherule cells are almost immature (X) and do not show any metachromatic granules. Coloured with toluidine blue; e, erythrocytes; a, amebocytes

the Polian vesicles in *Holothuria polii*. Such organs are composed of tissue which is not suitable for differentiation into spherule cells. However, small transparent cells in the periesophageal ring, displaying mitotic activity, can be defined as stem cells (D'Ancona et al. 1989); these are responsible for the production of spherule cells. The periesophageal ring is, therefore, a haemopoietic organ.

14.3 The Healing of Wounds in *Holothuria polii*

Confirming the presence of stem cells in *Holothuria polii* has been made possible by observing the cytological and histological features by integument healing assays where incisions of approximately 3 cm were made into the integument (D'Ancona unpublished). The first sign of healing in the control animals was marked 24 h after the incision as the migration of Type I and Type II spherule cells from the ceolomic cavity towards the apical zone of the cut occurred. Whilst the majority of the cells was degranulated, they formed a fine superficial film which ensured the continuity of the damaged tissue. Moreover, the cut began to be filled with new collagen fibres which were not distributed in a parallel fashion and they were always in association with type II spherule cells. Cells similar to fibroblasts and epithelial cells came together to heal the wound. Fine collagen filaments, delineated by a few Type II spherule cells, were present in animals which had been subject to antigen stimulation by fRSBC and the healing took place at a slower rate compared to that of the control animals. The number of coelomocytes present in the wounds of treated animals was reduced after a period of 24 h, resembling the cellular reaction of *Holothuria polii* after challenged by injection. Type II and Type III spherule cells, together with amoebocytes, isolated the host and formed brown bodies (Canicattì and D'Ancona 1989). Type III spherule cells located in the wound also protect the animal from possible foreign agents and/or the hydrolysis of components in the connective tissue. Indeed, such spherule cells along the wound are often associated to fragments of partially-digested collagen fibre. This is enhanced by the presence of acidic phosphatases and peroxidases, which are highlighted in Type III spherule cells (D'Ancona and Canicattì 1990).

Previous studies on Holothuroidea have been performed on *Stichopus tremulus* (Rollefsen 1965), *Stichopus badionotus* (Cowden 1968) and *Tyone briareus* (Menton and Eisen 1974). In all these cases the healing process was compared to the type of incision made and the time elapsed between the incision and observation. According to Rollefsen (1965), the healing of the wound is due to the production of new collagen fibres by morula cells. Having increased in number, the morula cells can be considered homologous to vertebrate mastocytes. Cowden (1968) has also attributed great importance to collagen fibres but not to morula cells. Menton and Eisen (1974) have described the presence of coelomocytes which, together with epithelial cells, are responsible for the healing of the wound. Morula cells and cells similar to fibroblasts flocked to the wound but no mitosis was observed. The absence of mitotic spherule cells in *Holothuria polii* and the column formation of the same spherule cells in the coelomic cavity towards the wound confirm that the site of spherule cell production is the periesophageal ring.

14.4 *Sipunculus Nudus*

Sipunculus nudus belongs to the phylum Sipuncula (known as peanut worms or star worms) in which approximately 300 ubiquitous species exist (Ghirardelli 1995; Brusca and Brusca 1996). Most sipunculids are infaunal organisms which lodge themselves into crevices, burrow into soft sediments, or bore into calcareous substrata. The Sipunculi are akin to molluscs and echinoderms, having in common with the former the typical cross-shaped rosette, which becomes visible during

Fig. 14.8 *Sipunculus nudus* (approximately 10 cm in length) (*arrow*) with retracted introvert. The specimens were imported from Vietnam in plastic boxes containing synthetic sponges and seawater

embryonic development. The relationship of Sipunculi to echinoderms is suggested by the similarity between the tentacles of Sipunculi and the Holothurioidee. The size of sipunculi tentacles varies but they all possess a thick trunk and a more slender introvert with a mouth at its end, which is surrounded by tentacles and hooks. In addition to the digestive system, one or two internal compensation sacs extend into the perivisceral coelom. Passing from the external to the internal, the walls of *Sipunculus nudus* are made up of a cube-shaped epithelium which secretes an outer cuticle, a derma, a layer of circular musculature; the latter follows longitudinal musculature. The derma contains fine fibres, connective cells, coelomate cells and longitudinal canals. These canals communicate with each other and the general coelom (Hyman 1959). Containing coelomic fluid, the coelomic cavity is delineated by a peritoneum, which is composed of flat cells (Fig. 14.8).

14.4.1 The Coelomatic Fluid of Sipunculus nudus

The coelomocytes of the Sipuncula are classified into 5 categories (Dybas 1981a, b) but not all the cellular categories are present throughout the species. In *Sipunculus nudus* young haemerythrocytes, old amoebocytes, enigmatic vesicles, and motile and fixed urn cell complexes (Herubel 1908) are present. Free-swimming ciliated urn cell complexes have been observed by Selensky (1908) and Cantacuzène (1928). Matozzo et al. (2001) have observed ciliated urn cell complexes secreting a sticky mucus, by which foreign cells can be recognised and trapped. Such cells are subsequently absorbed and lysed by other coelomocytes (Cantacuzène 1922; Bang and Bang 1962, 1965, 1974, 1975, 1976, 1980). Granulocytes have been observed by Cantacuzène (1922), Volkonsky (1933), Valembois and Boiledieu (1980) and

Matozzo et al. (2001). D'Ancona et al. (2004) have studied the cellular components of the coelomic fluid and corpuscular components contained therein. The aim was to look for immunocompentent cells and possible sites of stem cells. Haemerythrocytes, "signet ring cells" and urn cell complexes comprise the cellular components of the ceolomic fluid. Finely granular acidophilous material, vesicle fragments, the laminae of connective tissue, stem cell aggregates and brown bodies were also observed in the coelomic fluid. Haemerythrocytes, containing respiratory pigments, for example, haemerithrin, formed into spheres with round nuclei which measured 10–13 μm in diameter (Fig. 14.9A). When the haemerythrocytes matured, the nucleus became picnotic, the cytoplasm diminished and the external membranes assumed an undulating shape. In these conditions, the haemerythrocytes migrated in rows towards the brown bodies (Fig. 14.9I).

There are two types of granulocytes: Type I and Type II. The former possess cytoplasmatic extensions with adhering granules (Fig. 14.9E). Many small acidophilous granules in this Type I cells are present (Fig. 14.9D,E). Type II granulocytes reveal large compartments (Fig. 14.9G), with a central basophile core (Fig. 14.9F). Type I and Type II granulocytes contain chloracetate-esterase, alkaline phosphatase and acid phosphatase. Type I granulocytes perform their function outside the cell by releasing their granules near the material which is to be digested or acted upon in some way, thereby producing different effects. However, they do not possess a phagocytory function. Type II granulocytes often contain corpuscular components/particulate material, iron and lipids in their cytoplasm. They also possess hydrolases and are, therefore, involved in phagocytic activity and intra-cellular digestion.

Signet-ring cells are rare cells (12 μm in diameter), consisting of a large central vacuole, surrounded by a thin cytoplasm layer, which is thicker around the nucleus. The central part of the cell contains sulphated acid MPS (Fig. 14.9B). The entire cell reacts positively to alkaline and acid phosphatase and chloroacetate esterase. Type II granulocytes could be responsible for engulfing and digesting large amounts of organic material after which they become signet-ring cells. The urn cell complexes are 50–200 μm in diameter and they possess adhering granulocytes and particulate material, and reacted positively with chloroacetate esterase (Fig. 14.9C,H).

Sipunculus nudus displays brown bodies which are not associated esclusively with defence mechanisms. The external part of the brown bodies are composed of Type I and Type II layers of granulocytes, invading pathogens and parasites, unicellular organisms or crystal fragments, all of which penetrate the coelomic cavity. The brown bodies accumulate aged components (such as haemerythrocytes), which are typical of coelomic fluid. They, contain large quantities of iron (Fig. 14.9L)

The material released by Type I spherule cells performs a different function on the base of the peritoneal membrane, thereby causing the separation of laminae from the underlying musculature. Laminae (Fig. 14.10A) composed of fragments of connective tissue, are present in the coelomic fluid, in which small, colourless cells (7 μm in diameter) are found. On one side, there are smaller cells, which increase in size from the bottom to the top of the laminae. There exists a slope of cell maturity which commences at from the peritoneal membrane towards the coelomic cavity.

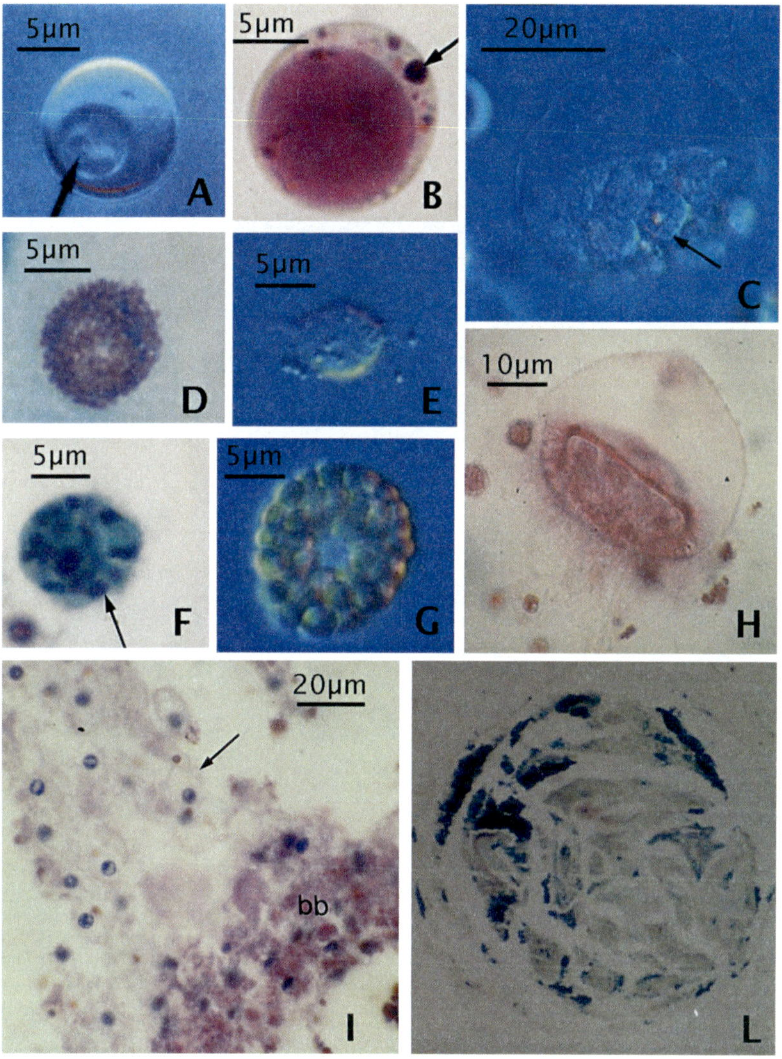

Fig. 14.9 Particulate components of the *Sipunculus nudus* coelomic fluid. (**A**) haemerithrocyte in interference microscope; the *arrow* indicates the nucleus; (**B**) toluidine blue-positive signet-ring cell; small methachromatic spots are evident in the cytoplasm (*arrow*); (**C**) interference microscope observation of urna cell complexed; the *arrow* indicates the adhering granulocytes; (**D**) Gomori triple staining-positive Type I granulocyte; (**E**) interference microscope image of Type I granulocyte; thin cytoplasmatic protrusions are also present near the granules; (**F**) Gomori triple staining type II granulocyte; the *arrow* indicates a compartment with a basophil granule; (**G**) interference microscope image of Type II granulocyte; (**H**) chloroacetate reaction-positive urna cell complexes on the concave side; (**I**) Mann-Dominici staining; numerous aged haemerythrocytes (*arrow*) are directed toward the cortical side of the brown body; (**L**) positive Perl's reaction of a brown body section. ×81

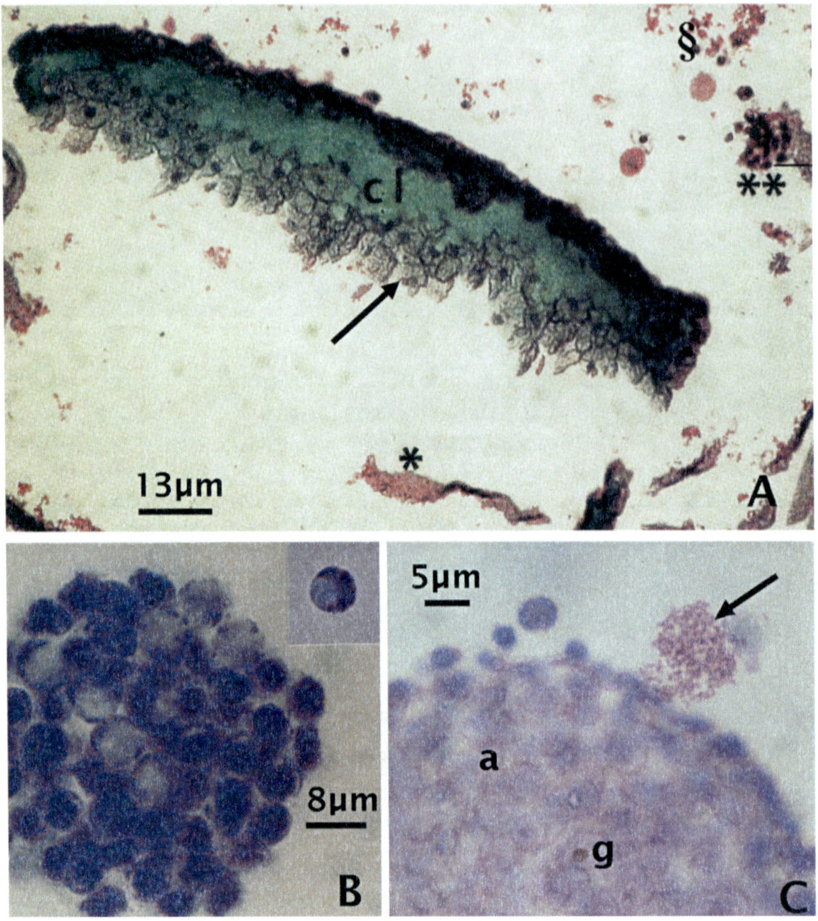

Fig. 14.10 (**A**) Some particulate components present in the coelomatic cavity: connective lamina (cl), Type I granulocyte undergoing degranulation (**), material degranulated by Type I granulocytes (***). In the lower side of connective lamina it is possible to see stem cells colourless. Gomori triple staining. The *arrow* indicates the muscular connective; (**B**) aggregate stem cells in differentiation stained with toluidine blue. In insert: an almost mature granulocyte of about 6 μm, found near cell aggregate; (**C**) detail of germinal aggregate stained with Mann-Dominici. The *arrow* indicates a Type I granulocyte undergoing degranulationh; its acidophil material (a) moves around each aggregate cell (g)

These cells could be the stem cells of hyalinocytes, observed in Phascolosoma vulgare by Volkonsky (1933) and in *Sipunculus nudus* by Valembois and Boiledieu (1980) and Matozzo et al. (2001). Transparent aggregates, making up the centre and peripheral parts of the cell, can be observed in the coelomic fluid of *Sipunculus nudus* with variable quantities of methachromatic granules (Fig. 14.10B). It has been hypothesized that these could be Type II granulocytes at various stages of differentiation. This is validated by the various sizes of cells and the varying quantity of granules. Colourless cells, which are localized in the cluster centre, could be hyalinocytes undergoing differentiation. Aggregates of germ cells are also present in the coelomic cavity and, on the surface of some of these aggregates, Type I spherule cells, which are degranulating the acidophilous material, have been observed. The same type of material is distributed around a single germ cell (Fig. 14.10C).

14.5 The Healing of Wounds in *Sipunculus nudus*

An incision performed longitudinally at a site in opposition to the anus of *Sipunculus nudus* involved the integument and the underlying circular musculature (D'Ancona Lunetta 2005), probably following the production *in situ* of various chemotactic substances (Abercrombie 1972; Postlethwaite et al. 1976). Large quantities of Type I granulocytes migrated from the coelomic cavity towards the wound where they degranulated, flattened out and adhered along the lower margins of the wound to form laminae. Similar in function to a plug, the role of the laminae is probably to create a structure which can impede the external leaking of coelomic material and block the entrance of possible pathogenic agents. Type I granulocytes fulfil a haemostatic function (Fig. 14.11A). In contrast to the haemocytes present in *Limulus maximus* during the healing of a wound (Franchini and Ottaviani 2000), Type I granulocytes are not capable of engulfing and digesting material but they secrete hydrolases enzymes into the surrounding environment (Matozzo et al. 2001; D'Ancona et al. 2004). The latter are capable of digesting cellular detritus and damaged tissue. Other cells, such as haemerythrocytes and Type II granulocytes, are trapped between Type I granulocytes. The haemerythrocytes probably flock around the wound to supply oxygen, which is necessary for metabolic activity (Steins et al. 2001). Morule cells in *Sticopus tremulus* are responsible for the production of the intercellular substances of connective tissue (Rollefsen 1965). They are also responsible for the production of fibres (Endean 1966) and the healing of wounds in connective tissue (Smith 1981). In other echinoderms the same cytological characteristics relating to connective tissue are found in the granular inclusions of the spherule cells (Endean 1958; Hetzel 1963; Johnson 1969; D'Ancona and Canicattì 1990). Moreover, Type II granulocytes in *Sipunculus nudus* possess the same cytological characteristics as the morule cells of *Sticopus tremulus* and the Type I spherule cells observed in *Holothuria polii* (D'Ancona and Canicattì 1990). Type II granulocytes display the same histochemical affinity as connective tissue and it can, therefore, be hypothesized that they are the main producers of connective tissue (Fig. 14.11B).

In mammals and some invertebrates, growth factors, such as PDGF and TGF-β, play an important role in the healing of wounds, being chemotactic and proliferating in nature (Postlethwaite et al. 1976; Deuel et al. 1982; Seppa et al. 1982; Senior et al. 1983; Wahl et al. 1987; Clark et al.1997; Ottaviani et al. 1997; Franchini and Ottaviani 2000). Of all the cells present in the coelomic fluid of *Sipunculus nudus*,

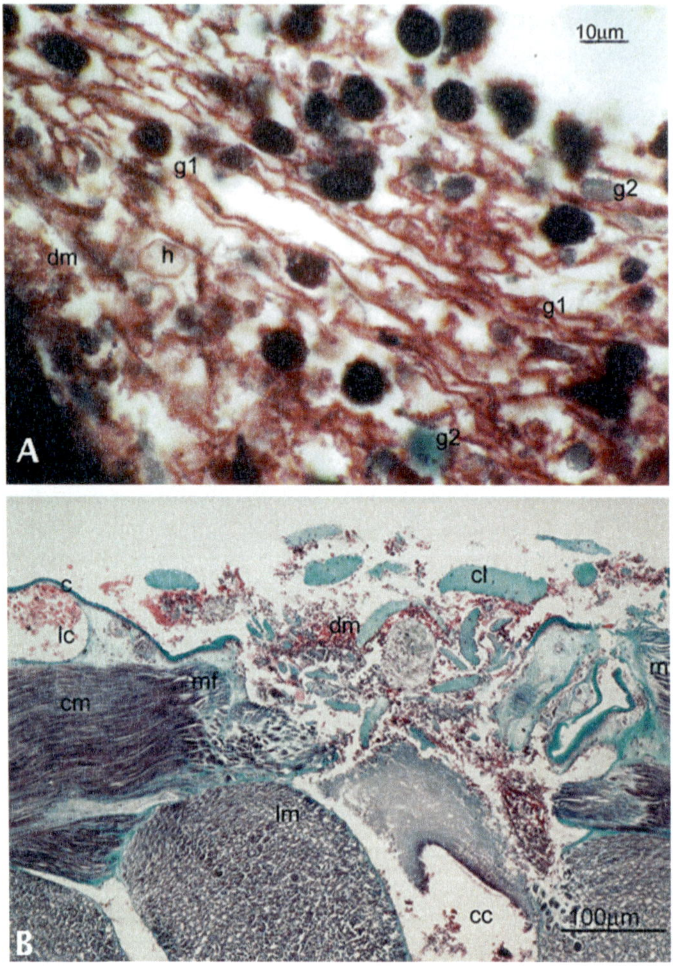

Fig. 14.11 A: Histological section stained with Gomori triple staining 24 h after wound. Note a barrier made up of flattened Type I granulocytes (1), wich have trapped haemerithrocytes (h), Type II granulocytes (g2) and degranulated material (dm); B: histological section 30 h after a deep incision, stained with Gomori triple staining. Note longitudinal canal (lc), connective lamina distributed over 2-3 layers (cl), degranulated acidophilic material (*), coelomocytes (c), circular muscle fibres with rounded extremities (cm), coelomatic cavity (cc), longitudinal muscle (lm), outer part of the wound (op)

only Type I granulocytes are intensely immunoreactive to anti EGF (Fig. 14.12A), IL-4 (Fig. 14.12B) and IL-10 (Fig. 14.12C); the degranulated material is also immunoreactive (D'Ancona Lunetta 2005). EGF has been considered an important factor due to its effect on differentiation and the division of epithelial and mesenchymal cells in culture (Cohen 1962; Honegger et al. 1990); in invertebrates EGF displays effects on epithelial organs (Carpenter and Cohen 1979) and on new-born

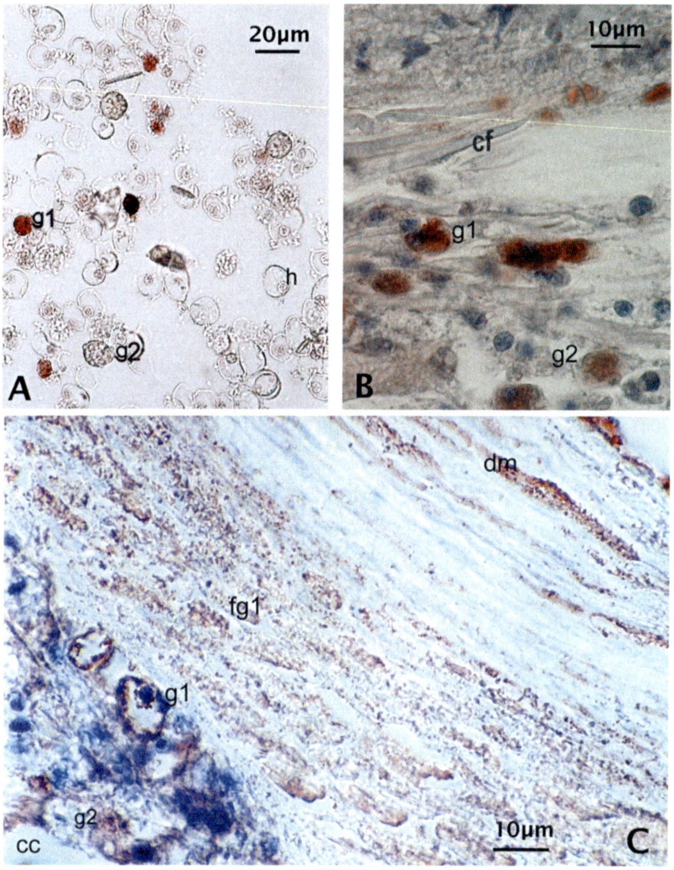

Fig. 14.12 (**A**) Immunocytochemical reaction with anti-EGF Ab 3 h after incision. Positive Type I granulocytes (g1); negative Type II granulocytes (g2) and haemerythrocytes (h) in the coelomatic cavity; (**B**) immunocytochemical reaction with anti-IL-4 Ab 24 h after incision.Type I granmulocytes (g1) positive and Type II granulocytes (g2) negative to anti IL-4- Ab; (cf)= collagen fibres. (**C**) Immunocytochemical reaction with anti-IL-10 Ab 24 h after incision. Partially degranulated Type I granulocytes (g1) were moderately reactive whilst the degranulated material (dm) was highly reactive. Negative Type II ganulocyte (g2); fg1, flattened granulocytes; ct, connective tissue; cc, coelomatic cavity; op, outer part of the wound

animals (Carpenter and Cohen 1990). According to Tsutsumi et al. (1986), EGF in mice induces an increase in spermatozoa; it is also found in platelets. However, EGF is currently an example of an important molecule as a factor involved in cell growth (Adamo et al. 2002).

The mitotic effect of EGF on the granuloctyes present in the wound in *Sipunculus nudus* can be excluded and no mitotic granulocytes are highlighted. It is probable that EGF acts on stem cells which are present in the peritoneal membrane, thereby stimulating the multiplication and differentiation of granulocytes (D'Ancona et al.

2004). Regarding its other functions, IL-4 produces inflammatory reactions in vertebrates and it stimulates the expression of various molecules which adhere to endothelial cells (in particular, VCAM-1) and between lymphocytes, monocytes and eosinophyles. Even if not supported by the experimental method, it could be hypothesized that molecules in *Sipunculus nudus* which are capable of creating connections between the cells, are released. Haemerythrocytes, for example, in the wound are flattened and tightly bound one to the other, thereby forming a laminar structure. In the coelomic cavity, each of the germ cells is associated with the degranulated material of Type I spherule cells, which give rise to spherical aggregates. This is not to exclude that such degranulated material could also stimulate the maturation of germ cells. IL-10 is a cytochine which is produced in man, particularly by monocytes and, in lesser quantities, by lymphocytes. It transforms monocytes into macrophages, thereby stimulating their endocytic and phagocitory activity. The presence of IL-10 in *Sipunculus nudus* could be related to the enzymatic activity of Type I granulocytes. Since this activity is extra-cellular, it can be hypothesized that only in the presence of material to be digested does IL-10 intervene to render the enzymes of Type I granulocytes active. Considering the enhanced inflammatory processes, which result from an induced wound, IL-10 could have an anti-inflammatory action. This would be to suppose, also in *Sipunculus nudus*, a pleiotropic effect of IL-10, as in man. This has been previously hypothesized in other invertebrates by Terkeltaub (1999) and Pinderki Oslund et al. (1999), and of other cytochines by Ottaviani et al. (2004).

14.6 The Defence Mechanisms of Vertebrates and Invertebrates

Each vertebrate organism has its own chemical equilibrium which can be easily modified by foreign agents such as viruses, bacteria, fungi, parasites and tissue transplants. The external (skin) and internal (mucous membranes) continuous epithelial lining of the body form the first barrier of defence against a potentially hostile environment. When this barrier is damaged, for example, having been cut or lacerated, foreign molecules can pass into the body and the immune system is activated. Two types of immunity exist: the first is present from birth, that is, an innate or a specific immunity. This immunity is not modified following subsequent contact with foreign agents, functioning by means of neutrophils, macrophages (both possess a phagocytic function) and various lymphocytes, the so-called "natural killers" (NK). Soluble factors, such as lysozymes, the reactive C protein and the complement are also involved in this type of innate defence. The complement is a complex system of proteins and it often binds itself to foreign particles by means of opsonin molecules. The II immunity which is specific or acquired is endowed with an immunological memory. Specific immunity can be humoral or cell-mediated; the former includes an antigen capable of stimulating an immune response at the moment when it comes into contact with immuno-competent cells, that is, B lymphocytes are transformed into plasmacells and they produce antibodies. The latter neutralise the antigen effect. There also exist T lymphocytes wich are responsible for cell-mediated immunity. In this case, antigen presenting cells, (APCs) for example macrophages, have the task of internally processing foreign protein. Being bound to MHC, fragments of such proteins form complexes which, once exposed on the surface of the APC, are recognised by the lymphocytes, which in turn are endowed with specific receptors. The subsequent production of cytochine activates the macrophages and destroys the complexes. It is clear that the defence mechanisms of vertebrates require the intervention of molecules and numerous cells (Abbas et al. 2000).

Invertebrates possess an innate immunity, producing soluble molecules with cells whose defensive functions are correlated to cytotoxic processes. Unlike vertebrates, invertebrates possess few cell types and those that they do possess, however, constitute a true defence system since these cells protect the organism from external invaders; they react to tissue transplants of extraneous tissue and the encourage the healing of the wound. Smith et al. (2001) have studied the evolutionary history of the complement and they identified two components of this system in sea urchins. They observed that the number of components increased with increasing complexity of the animal. The opsonisation of foreign material provoked the subsequent phagocitary intervention in sea urchin coelomocytes. This activity was similar to that observed in homologous proteins in the vertebrate complement system. In the same year therefore, Beschin et al. (2001) excluded a possible phylogenetic relationship between the cytokine-like molecules of invertebrates and their vertebrate counterparts. They confirmed a functional affinity between the cytokine activity of vertebrates and the activity of the cytokine-like molecules of invertebrates. Subsequently, Beschin et al. (2004) have discovered functional analogies between the TNF of mammals and the defence molecule CCF of the invertebrate *Eisenia fetida*. This analogy was based on a similar lectin-like activity/domain and it could be considered a point of convergence from an evolutionary point of view. Following a comparative and functional study on growth and cytokine factors, it was seen that growth factors similar to cytokine exist, for example TGF, CTGF and VEGF, and the interleukine IL-8 and the CC chemokine ligand 2/monocyte chemoattractant protein -1(CCL2). These molecules mediate and regulate various regenerative processes in vertebrates and invertebrates (Tettamanti et al. 2006).

In adult mammals, the healing of a large wound will occur if its edges are brought together and then sown, after which a scar forms. Generally, tissue repair in vertebrate embryos is rapid, efficient, perfect and leaves no scar, characteristics which are lost as the embryo develops. Small, cutaneous wounds in chicken embryos close up without forming a scar due to the contraction of an actin purse-string, made up of fibroblasts, which cause a rapid assembling of front row epithelial cells (Martin and Lewis 1992). The contraction of connective tissue in adult chickens takes place via contracting myofibroblasts (Nodder and Martin 1997). According to Redd et al. (2004), a key difference between healing in adults and embryos, which could explain the appearance or otherwise of a scar, can be related to the inflammatory response of the adult, which is absent in the embryo. Since apoptosis is responsible for reduced cell production, it has been hypothesized that the absence of a scar in vertebrate embryos can be attributed to apoptosis. It has been demonstrated that no correlation exists between the reduced infiltration of inflammatory cells and the absence of a scar in young embryos (Bae et al. 2004). According to Gillitzer et al. (2000) and Gillitzer and Goebeler (2001), the process of healing a cutaneous wound is regulated by growth factors which stimulate the proliferation of resident cells, (that is, keratinocytes and endolethial cells) and their synthesis of components of the extracellular matrix. Neutrophils, macrophages, lymphocytes and mastocytes participate in wound healing not only as immunological-effector cells, but also as sources of inflammatory and growing cytokines. The presence of receptors on the resident cells indicates that cytokines are important in controlling the growth of new epithelium, the reconstruction of tissue and angiogenesis. Cytokines are, therefore, important in integrating inflammatory events in the process of healing. Fibroblasts in *Hirudo medicinalis* (Anellidae, Hirudinea) take part in wound healing and angiogenesis, increasing to form a collagen scaffold, particularly under the effect of EGF, a factor which is also involved in the healing of wounds in vertebrates (Tettamanti et al. 2004). The processes of assembling the structural pattern of collagen fibrils are similar to those observed in vertebrates, in particular, regarding fish scales, the skin of amphibians and human cornea. The collagen structures in the leech have, therefore, been highly conservative features throughout the evolutionary process (Tettamanti et al. 2005).

Grimaldi et al. (2006) have found similarities in the healing processes of the hirudinea and vertebrates, in particular, between the antigen expressions of haemopoietic cells, endothelial cells and cytochine secretions. Previously, Kindred (1924) observed that the removal of a fragment of the body wall in echinoderms led to the healing of the wound by the proliferation and infiltration of cells from the surrounding connective tissue. Anderson (1962, 1965) observed in *Asterias*, the participation of coelomocytes and ameobocytes in the healing of the wound, both of which came from tissue adjacent to the wound. Cowden (1968), in *Sticopus badionotus* instead observed rapid healing with complete fibrogenesis without the intervention of morula cells. Regarding the healing of wounds in the sea cucumbers *Sticopus chloronatus*, an active role has been attributed to fatty acids (Fredalina et al. 1999).

An excision made in the mantle of the ostrich *Pinctada fucata* (the Akoya pearl oyster) produces a large wound which, within 1 h of the excision, reduces in size following muscular contraction. The number of haemocytes increases, the synthesis of connective tissue begins and the epithelial cells proliferate to cover the area of the wound (Acosta-Salmòn and Southgate 2006). In *Holothuria polii*, the healing of the wound is carried out by spherule cells which are present in the coelomic fluid of the animal (D'Ancona unpublished). In *Sipunculus nudus* the same process is performed by Type I and Type II granulocytes, the latter which also circulate in the coelomic fluid. The healing time for *Sipunculus nudus* is quicker than that for *Holothuria polii*, probably due to the reduced thickness of the surface of the cut (D'Ancona Lunetta 2005). An incision in vertebrate skin produces a cascade of events which can be outlined as an initial inflammatory phase, in which the platelets excrete cytochines, growth factors and molecules which promote the amassing of the aforementioned platelets in order to prevent blood loss.

Thereafter, neutrophils, macrophages and mast cells are absorbed by the blood circulating the tissue surrounding the wound. Neutrophils cleanse the wound by secreting proteases which break down damaged tissue. Neutrophils usually undergo apoptosis once they have completed their task and they are then engulfed and degraded by macrophages (Martin and Leibovich 2005).

There follows a prolific phase in which fibroblasts begin to enter the wound site, they multiply and excrete matrices and collagen fibres. The granular tissue thus formed consists of new blood vessels, fibroblast, inflammatory cells, endothelial cells, myofibroblasts and the components of a new provisional extracellular matrix. This phase of epithelium-building commences with the base cells. Approximately 10 days after the wound begins to heal, fibroblasts have differentiated into myofibroblast and the wound begins to contract (Eichler and Carlson 2005).

An inflammatory response also occurs in *Sipunculus nudus* following an induced wound. A comparison between healing in these invertebrates and vertebrates has highlighted the different number of cells involved in the healing of the wound. Type I and Type II granulocytes are capable of healing a wound without the intervention of other cells. This depends of the capacity of the Type I granulocytes in performing various functions during the healing process. Type I granulocytes in the wound flatten out and join up, thereby creating a structure which is comparable to the blood white clot, which is formed by platelets in mammals. Such a structure represents a defensive barrier for the animal as it impedes the loss, for example, of cells in the coelomic cavity towards the external environment and it also avoids extraneous material from entering the ceolomic cavity. By means of molecules present in secreted granules in their surroundings, Type I granulocytes also perform a defensive activity by destroying possible extraneous agents. Type I granulocytes, by means of litic enzymes which they possess, eliminate tissue detritus, which is formed when the incision is made. The second phase of healing involves the synthesis of connective and muscular tissue by Type II granulocytes. There could be various molecules in Type I granulocytes which, possessing a paracrine effect, induce Type II granulocytes to synthesize tissue necessary for healing.

Numerous factors and different types of cytochine are present in the vertebrates. Type I granulocytes have also highlighted the presence of EGF, cytochine-similar molecules which are reactive to IL-4 and IL-10; it should not be excluded that other molecules, as yet undiscovered, could also be present. As with other invertebrates, we can find cells and molecules in *Sipunculus nudus* and *Holothuria polii* which perform functions which are comparable to those in more evolved organisms. It appears evident that the first studies performed on invertebrates attributed, after incision, the exclusive intervention of coelomocytes, epithelial cells and collagen fibres regarding the healing of a wound. In subsequent years, the use or research of cytochines and factors of varying types has attributes a fundamental or complementary role to these molecules regarding the healing of wounds, as occurs with the vertebrates.

14.7 Stem Cells

Rossi et al. (2006) have hypothesized that the aging of the body could be caused by a reduced capacity of stem cells to produce new cells. It is necessary for mechanisms which permit not only the self-maintenance of the reserve of stem cells to be active but also the differentiation of mature cells in the tissue. If the cells in the tissue are not renewed, immune cells, for example, will be absent and the defense of the organism will be compromised.

Stem cells are unspecialised cells which, on division, produce another stem cell in addition to another cell, the latter which is capable of dividing and forming differentiated cells. They are multipotent if they give rise to numerous types of cells; unipotent if they give rise to only one type of cell. In mammals pluripotent stem cells are present in the mass of cells in the embryonic knot of blastocysts, which are being implanted in the embryo and the foetus during development. They are also found in adult individuals but they are limited to various parts of the organism: the spinal cord, seminifer epithelium of the male gonads, retina, epithelium and brain. Mesenchymal stem cells are pluripotent cells which are present in adult bone marrow and which are replicated as undifferentiated cells. Mesenchymal stem cells are differentiated into mesenchymal tissue, such as bone, cartilage, fat, tendons, muscle and stroma marrow. These adult stem cells could be produced in culture to differentiate into adypocytes, condrocytes and osteocytes (Pittenger et al. 1999). According to Krause et al. (2001) adult bone marrow cells possess a differentiative capacity as they can also differentiate into the epithelial cells of the liver, lung and skin. Stem cells can also be obtained from human fat following liposuction, and these can produce the cells of cartilaginous tissue, bone, muscle tissue and neuron-like cells (De Ugarte et al. 2003).

Cossu and Bianco (2003) have observed in mice that a part of the cells which contribute to the development of skeleton muscle during the foetal period and after birth are not derived from somites. Totipotent stem cells, defined as mesoangioblasts, are present and distributed in the blood vessels of mice bone marrow. The mesoangioblasts migrate via the blood circulation and play a role throughout the entire mesoderm. Cossu and Bianco (2003) have hypothesised that these stem cells are differentiated in a different way according to their environment. The capacity to give rise to one or more types of specialised cells is controlled by secreted proteins and transcriptional factors (Ying et al. 2003; Lee et al. 2004; Maye et al. 2004). Similar research has been performed on Drosophila (Parisi and Lin 1999; Cerutti et al. 2000) and mammals (Sharma et al. 2001; Spassov and Jurecic 2003). The hypothesis of environmental influence on the destination of stem cells has been reviewed more recently by Sieburg et al. (2006). They hypothesized that adult mouse stem cells in blood are not all the same since haemopoietic stems cell (HSCs) consist of a limited number of distinct HSC (haematopoietic stem compartment) subsets. The various types of behaviour of stem cells are not stimulated by their environment but by their internal structure. Intestinal cells of the fruit fly have the same genetic programme as that found in man. Indeed, stem cells in enterocytes are multipotent, like those of vertebrates, capable of transforming themselves into different cells types (Ohlstein 2005).

No genetic-based research has yet been carried out on *Holothuria polii* and *Sipunculus nudus*, nor have the molecular mechanisms which control the differentiation of their stem cells been studied. It is not, therefore, possible to seek structural internal similarities between the stem cells of *Holothuria polii* and *Sipunculus nudus* and other more evolved animals. One difference between these invertebrates and vertebrates regards the location of stem cells. In *Sipunculus nudus*, a sole site of stem cells (the coelomic epithelium) is present. Two sites (the periesophageal ring and stone canal) are present in *Holothuria polii* and it is only from these sites that stem cells can originate: they differentiate into coelomocytes in the coelomic cavity, where they complete their differentiation. Finally, they will reach sites where they will perform their functions.

Vertebrate bone marrow and the lymphatic organs are they greatest producers of stem cells, which migrate far from their place of origin with different destinations. Stem cells in adult vertebrates are also found in differentiated tissue such as spinal cord, seminifer epithelium, the retina, epithelium and brain. These differentiated tissue stem cells can differentiate in locations which are advantageous to the tissue in which they are found, having the function of substituting the used cells. Moreover, cultured stem cells originating from differentiated tissue can also trans-differentiate into cells of other tissues. It can, therefore, be affirmed that, whilst simple in *Sipunculus nudus* and *Holothuria polii*, the organisation of stem cells becomes increasingly complex not only in relation to the different phylogenetic position of the taxa but, above all, on the basis of the gene expression of the stem cells.

References

Abbas AK, Lichtman AH, Pober JS (2000) Cellular and molecular immunology. WB Saunders Company, Philadelphia

Abercrombie M (1972) Behavior of cells toward one another. In: Montagna W, Billingham RE (eds) Advances in biology of skin V: Wound healing. Academic Press Inc, London, pp 95–112

Acosta-Salmòn H, Southgate PC (2006) Wound healing after excision of mantle tissue from the Akoya pearl oyster, *Pinctata fucata*. Comp Biochem Physiol, Part A: Mol Integrative Physiol 143(2):264–268

Adamo S, Carinci P, Molinaro M, Siracusa G, Stefanini M, Riparo F (2002) Istologia. Piccin, Padova

Amirante GA, Basso V (1984) Syntesis of lectins in *Squilla mantis* L. (Crustacea Stomatopoda) using monoclonal antibodies. Dev Comp Immunol 8:721–726

Anderson JM (1962) Studies on visceral regeneration in sea-stars. I. Regeneration of pyloric caeca in *Henricia leviscula* (Stimpson). Biol Bull 122:321–342

Anderson JM (1965) Studies on visceral regeneration in sea-stars. II. Regeneration of pyloric caeca in Asteriidae, with notes on the source of cells in regenerating organs. Biol Bull 122:1–123

Anderson RS (1987) Polycarion formation by *Mercenaria mercenaria* hemocytes. Biol Bull 172:236–245

Arizza V, Parrinello V, Cammarata M (1993) Immunocytochemical localization of cellular lectins in *Phallusia mamillata* hemocytes. Animal Biol 2:15–22

Arizza V, Parrinello N, Schimmenti S (1991) In vitro release of lectins from hemocytes of *Phallusia mamillata* . Dev Comp Immunol 15:219–226
Bae TH, Kim HK, Park DS, Kim WS, Kim SH, Lee TJ (2004) Expression of Fas, Fas ligand, bcl-2 in fetal wound healing of rats. J Korean Soc Plastic Reconstruction Surg 31(2):243–248
Bang BG, Bang FB (1965) Mucus hypersecretion in a normally isolated non-innervated cell. Cahiers de Biologie Marine 6:257–264
Bang BG, Bang FB (1974) Invertebrate model for study of macromolecular regulators of mucus secretion. Lancet 2:1292–1294
Bang BG, Bang FB (1975) Cell recognition by mucus secreted by urn cell of *Sipunculus nudus* . Nature 253:634–635
Bang BG, Bang FB (1976) The mucous secretory apparatus of the free urn cell of *Sipunculus nudus* . Cahiers de Biologie Marine 17:423–432
Bang BG, Bang FB (1980) The urn cell complex of *Sipunculus nudus* a model for study of mucus-stimulating substances. Biol Bull 159:571–581
Bang FB, Bang G (1962) Studies on sipunculid blood: Immunologic properties of coelomic fluid and morphology of "urn cell". Cahiers de Biologie Marine 3:363–374
Beck G, O'Brien R, Habicht GS, Stilman DL, Cooper EL, Raftos DA (1993) Invertebrate cytokines III: Invertebrate interleukin-1- like molecules stimulate phagocytosis by tunicate and echinoderm cells. Cell Immunol 146:284–299
Bertheussen K (1983) Complement-like activity in sea urchin coelomic fluid. Dev Comp Immunol 7:21–33
Beschin A, Bilej M, Magez S, Lucas R, De Baetselier P (2004) Functional convergence of invertebrate and vertebrate cytokine-like molecules based on a similar lectin-like activity. Progr Mol Subcell Biol 34:145–163
Beschin A, Bilej M, Torreale E, De Baetselier P (2001) On the existence of cytokines in invertebrates. Cell Mol Life Sci 58:801–814
Boolootian RA, Giese AA (1958) Coelomic corpuscles of echinoderms. Biol Bull 115:53–63
Brusca RC, Brusca GJ (1996) Invertebrati. Zanichelli, Bologna
Cammarata M, Parrinello N, Arizza V (1993) In vitro release of lectins from *Phallusia mamillata* hemocytes after their fractionation on a density gradient. J Exp Zool 266:319–327
Canicattì C (1988) Arylsulphatase in coelomocytes of *Holothuria polii* (Echinodermata). Experientia 44:1011–1013
Canicattì C (1989) Evolution of lytic system in echinoderms. II Naturally occurring haemolytic activity in *Marthasterias glacialis* (Asteroidea). Comp Biochem Physiol 93A:587–591
Canicattì C (1990) Hemolysins pore-forming proteins in invertebrates. Experientia 46:239–244
Canicattì C (1991) Binding properties of *Paracentrotus lividus* (Echinoidea) hemolysin. Comp Biochem Physiol 98A:463–468
Canicattì C, Ciulla D (1988) Studies on *Holothuria polii* (Echinodermata) coelomocyte lysate. II. Isolation and characterization of coelomocyte hemolysins. Dev Comp Immunol 12:55–64
Canicattì C, D'Ancona G (1990) Biological protective substances in *Marthasterias glacialis* (Asteroidea) epidermal secretion. J Zool London 222:445–454
Canicattì C, D'Ancona G (1989) Cellular aspects of *Holothuria polii* immune response. J Invert Pathol 53:152–158
Canicattì C, Parrinello N (1985) Hemagglutinin and hemolysin levels in the coelomic fluid from *Holothuria polii* (Echinodermata) following sheep erythrocyte injection. Biol Bull 168: 175–182
Canicattì C, D'Ancona G, Farina-Lipari E (1989a) The *Holothuria polii* brown bodies. Boll Zool 56:275–283
Canicattì C, D'Ancona G, Farina-Lipari E (1989b) The coelomocytes of *Holothuria polii* (Echinodermata). Light and electron microscopy. Boll Zool 56 :29–36
Cantacuzène J (1922) Sur le rôle agglutinant des urnes chez *Sipunculus nudus*. Comptes Rendus des Seances de la Societé de Biologie Paris 87:259–263
Cantacuzène J (1928) Recherches sur les reactions d'immunité chez les Invertebrates. Arch Roumaines Pathol Experimentale Microbiologie 1:7–80

Carpenter G, Cohen S (1979) Epidermal grow factor. Ann Rev Biochem 48(1):193–216

Carpenter G, Cohen S (1990) Epidermal grow factor. J Biol Chem 265(14):7709–7712

Cerutti L, Mian N, Bateman A (2000) Domains in gene silencing and cell differentiation proteins: The novel PAZ domain and redefinition of the Piwi domain. Trends Biochem Sci 25:481–482

Cheng TC, Galloway PC (1970) Transplantation immunity in molluscs: The histoincompatibility of *Helisoma duryi* normale with allografts and xenografts. J Invertebr Pathol 15:177–192

Clark RA, McCoy GA, Folvord JM, McPherson JM (1997) TGF-â 1 stimulates cultured human fibroblasts to proliferate and produce tissue-like fibroplasia: A fibronectin matrix-dependent event. J Cell Physiol 170:69–80

Cohen S (1962) Isolation of a mouse submaxillary gland protein accelerating incisor eruption and eyelid opening in the new-born animal. J Biol Chem 237:1555–1562

Cooper EL (1976) The earthworm coelomocyte: A mediator of cellular immunity. In: Wright RK, Cooper EL (eds) Phylogeny of thymus and bone marrow Bursa cells. Elsevier, Amsterdam, New York, Oxford, pp 9–18

Cooper EL (1982) The reticuloendothelial system. In: Cohen M, Sigel MM (eds) Phylogeny and Ontogeny, vol 3. Plenum, New York, pp 1–35

Cossu G, Bianco P (2003) Mesoangioblasts, vascular progenitors for extra-vascular mesoderm. Curr Opin Gen Dev 13:1–6

Cowden R (1968) Cytological and histochemical observation on connective tissue cells and cutaneous wound healing in the sea cucumber *Sticopus badionatus*. J Invertebr Pathol 10:151–153

D'Ancona G, Canicattì C (1990) The coelomocytes of *Holothuria polii* (Echinodermata) II. Cytochemical staining properties. Basic Appl Histochem 34:209–218

D'Ancona G, Rizzuto L, Canicattì C (1989) Engagement of Polian vesicles during *Holothuria polii* response to erythrocyte injection. J Invertebr Pathol 54:300–305

D'Ancona Lunetta G (1996) A haemopoietic organ in *Holothuria polii* (Holothuroidea). Animal Biol 5:47–53

D'Ancona Lunetta G (2005) Wound repair in the marine worm *Sipunculus nudus* (Sipunculidae). Invertebr Surv J 2:124–131

D'Ancona Lunetta G, Farina E, Manione R (2004) *Sipunculus nudus*: Particulate components of the coelomic fluid and its relationship with brown bodies. Ital J Zool 71:191–199

D'Ancona Lunetta G, Michelucci ML (2001) Engagement of the periesophageal ring during *Holothuria polii* response to erythrocyte injection. Europ J Histochem 46:185–194

De Ugarte DA, Ashjian PH, Elbarbary A, Hedrick MH (2003) Future of fat as raw material for tissue regeneration. Ann Plast Surg 50(2):215–219

Deuel TYF, Senior RM, Huang JS, Griffin GL (1982) Chemotaxis of monocytes and neutrophils to platelet-derived growth factor. J Clin Invest 69:1046–1049

Dybas L (1981a) Sipunculans and echiuroids. In: Ratcliffe NA, Rowley AF (eds) Invertebrate blood cells, vol I. Academic Press, London, pp 161–168

Dybas L (1981b) Cellular defense reactions of *Phascolosoma agassizii* a Sipunculan worm: Phagocytosis by granulocytes. Biol Bull 161:104–114

Dybas L, Frankboner PN (1986) Holothurian survival strategies: Mechanisms for the maintenance of bacteriostatic environment in the coelomic cavity of the sea cucumber *Parasticopus californicus*. Dev Comp Immunol 10:311–330

Eichler MJ, Carlson MA (2005) Modelling dermal granulation tissue with the linear fibroblast populated collagen matrix: A comparison with the round matrix model. J Dermatol Sci 41(2):97–108

Endean R (1958) The coelomocytes of *Holothuria leucospilata*. Q J Microsc Sci 99:47–60

Endean R (1966) The coelomocytes and coelomic fluid. In: Boolootian RA (ed) Physiology of Echinodermata. Wiley, New York, pp 301–328

Ey PL, Jenkin CR (1982) Molecular basis of self-non self discrimination in the invertebrate. In: Cohen N, Sigel MM (eds) The reticuloendothelial system, vol 3. Plenum press, New York/London, pp 360–365

Fontane AR, Lambert P (1977) The fine structure of the leukocytes of the holothurian *Cucumaria miniata*. Canad J Zool 55:1530–1544
Fragkiadakis GA (2000) Isolation of lectins from hemolymph of decapod crustaceans by adsorbition on formalinized erythrocytes. J Biochem Biophys Methods 44:109–114
Franchini A, Ottaviani E (2000) Repair of molluscan tissue injury: Role of PDGF and TGF-ß. Tissue Cell 32(4):312–321
Fredalina BD, Ridzwan BNH, Abidin AA, Kaswandi MA, Zaiton H, Zali I, Kittakoop P, Jais AM (1999) Fatty acid compositions in local sea cucumber, *Stichopus chloronotus*, for wound healing. Gen Pharmacol 33(4):337–40
Ghirardelli E (1995) Phylum sipunculi (Sipuncula). In: Baccetti B et al (eds) Zoologia vol 2. Grasso, Bologna, pp 477–486
Gillitzer R, Toksoy A, Voss A (2000) Role of chemokines in human skin wound healing. Zentralbl chir 125(1):56–9
Gillitzer R, Goebeler M (2001) Chemokines in cutaneous wound healing. J Leukocyte Biol 69(4):513–521
Glinski Z, Jarosz J (2000) Immune phenomena in echinoderms. Archivum Immunologiae et Therapiae Experimentalis (Warsz) 48(3):189–193
Grimaldi A, Tettamanti G, Perletti G, Valvassori R, de Enguileor M (2006) Hematopoietic cell formation in leech wound healing. Curr Pharmac Design 12(24):3033–3041
Gross PS, Al-Sharif WZ, Clow LA, Smith LC (1999) Echinoderm immunity and the evolution of the complement system. Dev Comp Immunol 23:429–442
Herubel MA (1908) Recherches sur les sipunculides. Memoirs Societè Zoologique de France 20:107–418
Hetzel HR (1963) Studies on holothurian coelomocytes. I. A survey of coelomocyte types. Biol Bull 125:289–301
Hetzel HR (1965) Studies on holoturian coelomocytes. II. The origin of coelomocytes and the formation of brown bodies. Biol Bull 128:102–111
Honegger AM, Schmidt A, Ullrich A, Schlessinger J (1990) Evidence for epidermal growth factor (EGF)-induced intermolecular autophosphorylation of the EGF receptors in living cells. Mol Cell Biol 10(8):4035–4044
Hyman LH (1955) The invertebrates: Echinoderms, vol. IV. McGraw-Hill, New York, pp 121–691
Hyman LH (1959) The invertebrates: Smaller coelomocytes groups. McGraw-Hill, New York
Jangoux M (1987a) Disease of Echinodermata. I. Agents microorganism and protistants. Dis Acquat Organ 2:147–162
Jangoux M (1987b) Disease of Echinodermata. II. Agents metazoan (Mesozoa to Bryozoa). Dis Acquat Organ 2:205–234
Johnson PT (1969) The coelomic elements of sea urchin. Histochemistry 17:213–231
Kanungo K (1984) The coelomocytes of asteroid echinoderms. In: Cheng TC (ed) Comparative pathobiology, vo. 6. Plenum Press, New York/London, pp 8–33
Karp RD, Coffaro KA (1982) Cellular defence system of the Echinodermata. In: Cohen N, Sigel MM (eds) The reticuloendothelial system, vol 3. Plenum Press, New York/London, pp 257–281
Kindred JE (1924) The cellular elements in the perivisceral fluid of echinoderms. Biol Bull 50: 147–154
Krause DS, Theise ND, Collector MI, Henergariu O, Hwang S, Gardner R, Neutzel S, Sharkis SJ (2001) Multi-organ, multi lineage engraftment by a single bone marrow-derived stem cell. Cell 105:369–377
Leclerc M (1996) Humoral factors in marine invertebrates. In: Rinkevich B, Muller WEG (eds) Progress in Molecular and Subcellular Biology: Invertebrate Immunology. Springer-Verlag, Berlin/Heidelberg, pp 1–9
Lee HY, Kleber M, Hari L, Brault V, Suter U, Taketo MM, Kemler R, Sommer L (2004) Instructive role of wnt/beta-catenin in sensory fate specification in neural crest stem cells. Science 303:1020–1023
Mariano M, Spector WG (1974) The formation and properties of macrophage polykarions (inflammatory giant cells). J Pathol 113:1–19

Martin P, Lewis J (1992) Actin cables and epidermal movement in embryonic wound healing. Nature 360:179–183

Martin P, Leibovich SJ (2005) Inflammatory cells during wound repair: The good, the bad and the hugly. Trend Cell Biol 15(11):599–607

Matozzo V, Perin L, Cima F, Ballarin L (2001) Phagocytic and enzymatic activities of cells and urn cell complexes in the coelomic fluid of the marine worm *Sipunculus nudus* (Sipunculida). Ital J Zool 68:273–280

Maye P, Becker S, Siemen H, Thorne J, Byrd N, Carpentino J, Grabel L (2004) Hedgehog signalling is required for the differentiation of ES cells into neurectoderm. Dev Biol 265:276–290

Menton DN, Eisen AZ (1974) Cutaneous wound healing in the sea cucumber *Thyone briareus*. J Morphol 141:185–204

Murch AR, Grounds MD, Marshall CA, Papadimitrion JM (1982) Direct evidence that inflammatory multinucleated giant cells form by fusion. J Pathol 137:177–180

Nicosia SV (1979) Lectin-induced mucus release in the urn cell complex of the marine invertebrate *Sipunculus nudus* (Limnaeus). Scienze 206(4419):698–700

Nodder S, Martin P (1997) Wound healing in embryos. Anatomy and Embryology 195(3):215–228. Springer, Berlin

Ohlstein B (2005) Le staminali di un insetto. http:www.lescienze.it/index.html

Ohuye T (1938) On the coelomic corpuscles in the body fluid of some invertebrates. X. Science Reports of the Tohoku University Ser 4 Biology 12:593–622

Olafsen JA (1986) Invertebrate l: Biochemical heterogeneity as possible key to their function. In: Brehelin M (ed) Immunity in Invertebrates. Springer-Verlag, Berlin, pp 94–111

Ottaviani E, Franchini A, Kletsas D, Bernardi M, Genedani S (1997) Involvement of PDGF and TGF-ß in cell migration and phagocytosis in invertebrate and human immunocytes. Anim Biol 6:91–95

Ottaviani E, Malagodi D, Franchini E (2004) Invertebrate humoral factors: Cytokines as mediators of cell survival. Progr Mol Subcell Biol 34:1–23

Parisi M, Lin H (1999) The Drosophila pumilio gene encodes two functional protein isoforms that play multiple roles in germline development, gonadogenesis, oogenesis and embryogenesis. Genetics 153:235–250

Parrinello N (1991) Animal lectins and immune recognition: A phylogenetic approach. In: Lanzavecchia G, Valvassori R (eds) Form and function in Zoology. Selected Symposia and Monographs, vol 5. Mucchi Editore, Modena, pp 43–68

Parrinello N, Arizza V (1988) Sugar-specific lectins from the tunicate *Ascidia Malaca*: Subunit characterization and hemocyte surface distribution. Dev Comp Immunol 12:495–507

Parrinello N, Arizza V (1989) Sugar-specific cellular lectins of *Phallusia mamillata* hemocytes: Purification, characterization and evidence for cell-surface localization. Dev Comp Immunol 13:113–121

Parrinello N, Canicattì C, Rindone D (1976) Naturally occurring hemagglutinins in the coelomic fluid of the Echinoderms *Holothuria polii* Delle Chiaje and *Holothuria tubolosa* Gmelin. Boll Zool 43:259–271

Parrinello N, Rindone D, Canicattì C (1979) Naturally occurring hemolysins in the coelomic fluid of *Holothuria polii* Delle Chiaje (Echinodermata). Dev Comp Immunol 3:45–54

Pinderki Oslund LJ, Hedrick CC, Olvera T, Hagenbaugh A, Territo MM, Berliner JA, Fyfe AI (1999) Interleukin-10 blocks atherosclerotic events in vitro and in vivo. Arteriosclerosis, Thrombosis Vascular Biol 19:2847–2853

Pittenger MF, McKay AM, Beck SC, Jaiswal RK, Douglas R, Mosca JD, Moorman MA, Simonetti DW, Craig S, Marshak DR (1999) Multilineage potential of adult human mesenchymal stem cells. Science 284:143–147

Postlethwaite AE, Snyderman R, Kang AH (1976) The chemotactic attraction of human fibroblasts to a lymphocyte-derived factor. J Exp Med 144:1188–1203

Raftos DP, Green D, Mahajan R, Newton S, Pearce R, Teters R, Robbins J, Nair S (2001) Diverse lectin in tunicates and the proteolytic activation of complement. Adv Exp Med Biol 484: 229–236

Redd MJ, Cooper L, Wood W, Stramer B, Martin P (2004) Wound healing and inflammation: Embryos reveal the way to perfect repair. Philos Trans R Soc Lond B Biol Sci 29:777–784

Rheins LA, Karp RD, Butz A (1980) Induction of specific humoral immunity to soluble proteins in the American cockroach (*Periplaneta Americana*). I. Nature of the primary response. Dev Comp Immunol 4:447–458

Rollefsen I (1965) Studies on the mastcell-like morula cells of holothurian *Sticopus tremulus*. Nature 8:3–13

Rossi DJ, Bryden D, Weissman IL (2006) Hematopoietic stem cell aging: Mechanism and consequence. Exp Gerontol 42:385–390

Selensky W (1908) Untersuchungen uber die Urnen der Sipunculiden. Zoologischer Anzeiger 32:329–336

Senior RM, Griffin GL, Huang JS, Walz DA, Deuel TF (1983) Chemotactic activity of platelet alpha granule proteins for fibroblast. J Cell Biol 96:382–385

Seppa H, Grotendorst G, Seppa S, Schiffmann E, Martin GR (1982) Platelet-derived growth factor is chemotactic for fibroblast. J Cell Biol 92:584–588

Sharma AK, Nelson MC, Brandt JE, Wessman M, Mahmud N, Weller KP, Hoffman R (2001) Human CD34(+) stem cells express the hiwi gene, a human homologue of the Drosophila gene piwi. Blood 97:426–434

Shinn GL (1985) Reproduction of *Anoplodium hymanae*, a turbellarian flatworm (Neorhabdocoela, Umagillidae) inhabiting the coelom of sea cucumbers; production of egg capsules, and escape of infective stages without evisceration of the host. Biol Bull 169:182–198

Sieburg HB, Cho RH, Dystra B, Uchida N, Eaves CJ, Muller-Sieburg CE (2006) The ematopoietic stem compartment consists of a limited number of discrete stem cell subsets. Blood 107(6):2311–2316

Smith LC, Azumi K, Nonaka M (1999) Complement systems in invertebrates. The ancient alternative and lectin pathways. Immunopharmacology 42(1–3):107–120

Smith LC, Clow LA, Terwilliger DP (2001) The ancestral complement system in sea urchins. Immunol Rev 180:16–34

Smith UZ (1981) The echinoderm. In: Ratcliffe NA, Rowley AF (eds) Invertebrate blood cells, vol 2. Accademic Press, London, pp 513–562

Spassov DS, Jurecic R (2003) Mouse Pum1 and Pum2 genes, members of the Pumilio family of RNA-binding proteins, show differential expression in fetal and adult hematopoietic stem cells and progenitors. Blood Cells Mol Diseases 30:55–69

Steins A, Hahn M, Junger M (2001) Venous leg ulcers and microcirculation. Clin Hemorheol Microcirc 24:147–153

Terkeltaub RA (1999) IL-10: An immunologic scalpel for atherosclerosis? Arterioscler Thromb Vasc Biol 19:2823–2825

Tettamanti G, Grimaldi A, Congiu T, Perletti G, Raspanti M, Valvassori R, de Enguileor M (2005) Collagen reorganization in leech wound healing. Biol Cell 97(7):557–568

Tettamanti G, Grimaldi A, Rinaldi L, Arnaboldi F, Congiu T, Valvassori R, de Eguileor M (2004) The multifunctional role of fibroblasts during wound healing in *Hirudo medicinalis* (Annelida, Hirudinea). Biol Cell 96(6):443–455

Tettamanti G, Malagoli D, Benelli R, Albini A, Grimaldi A, Perletti G, Noonan DM, de Eguileor M, Ottaviani E (2006) Growth factors and chemokines: A comparative functional approach between invertebrates and vertebrates. Curr Med Chem 13(23):2737–2750

Tiscar PG, Mosca F (2004) Patologie emergenti in Molluschicoltura. http://www.pg.izs.it/webzine.html

Tortonese E (1985) Echinodermata. Calderoni, Bologna

Tsutsumi O, Kurachi H, Oka T (1986) A physiological role of epidermal growth factor in male reproductive function. Science 233:975–977

Valembois P, Boiledieu D (1980) Fine structure and function of haemerythrocytes and leucocytes of *Sipunculus nudus* . J Morphol 163:69–77

Vasquez L, Jaramillo L, Lascurain R, Cooper EL, Rosas P, Zenteno E (1996) Bacterial agglutination by the sialic acid specific serum lectin from *Macrobrachium rosembergii*. Comp Biochem Physiol 113B:355–359

Vasta GR, Quesenberry MS, Ahmed H, O'Leary N (2001) Lectins from tunicates: Structure-function relationship in innate immunity. Adv Exp Med Biol 484:275–287

Volkonsky M (1933) Digestion intracellulaire et accumulation des colorants acids. Etude cytologique des cellules sanguines des sipunculides. Biol Bull 67:135–199

Wahl SM, Hunt DA, Wakefield LM, McCartney-Francis N, Wahl LM, Roberts AB, Sporn MB (1987) Transforming growth factor type â induces monocyte chemotaxis and growth factor production. Proc Natl Acad Sci USA 84:5788–5792

Yeaton RW (1981) Invertebrate lectins: I. Occurrence. Dev Comp Immunol 5:391–402

Ying QL, Nichols J, Chambers I, Smith A (2003) BMP induction of Id proteins suppresses differentiation and sustains embryonic stem cell self-renewal in collaboration with STAT3. Cell 115:281–292

Index

B. Rinkevich, V. Matranga (eds.), *Stem Cells in Marine Organisms*,
DOI 10.1007/978-90-481-2767-2_BM2, © Springer Science+Business Media B.V. 2009